T0192551

# VISCOELASTIC MATERIALS

Understanding viscoelasticity is pertinent to design applications as diverse as earplugs, gaskets, computer disks, satellite stability, medical diagnosis, injury prevention, vibration abatement, tire performance, sports, spacecraft, and music. This book fits a one-semester graduate course on the properties, analysis, and uses of viscoelastic materials. Those familiar with the author's precursor book, *Viscoelastic Solids*, will see that this book contains many updates and expanded coverage of the materials science, causes of viscoelastic behavior, properties of materials of biological origin, and applications of viscoelastic materials. The theoretical presentation includes both transient and dynamic aspects, with an emphasis on linear viscoelasticity to develop physical insight. Methods for the solution of stress analysis problems are developed and illustrated. Experimental methods for characterization of viscoelastic materials are explored in detail. Viscoelastic phenomena are described for a wide variety of materials, including viscoelastic composite materials. Applications of viscoelasticity and viscoelastic materials are illustrated with case studies.

Roderic Lakes is a Distinguished Professor in the Department of Engineering Physics at the University of Wisconsin–Madison. He is a Fellow in the American Association for the Advancement of Science (AAAS) and has been honored by becoming a Fellow in the American Society of Mechanical Engineers (ASME). He has won numerous teaching awards and is the author and coauthor of more than 194 archival publications; three books, including *Viscoelastic Solids* and *Biomaterials* (2nd and 3rd editions, with J. B. Park); and fourteen book chapters. The author's articles in *Science* and *Nature* are of particular note because they have led to numerous synergistic publications in a variety of disciplines.

# Viscoelastic Materials

Roderic Lakes

*University of Wisconsin–Madison*

# CAMBRIDGE
## UNIVERSITY PRESS

32 Avenue of the Americas, New York NY 10013-2473, USA

Cambridge University Press is part of the University of Cambridge.

It furthers the University's mission by disseminating knowledge in the pursuit of education, learning and research at the highest international levels of excellence.

www.cambridge.org
Information on this title: www.cambridge.org/9781107459786

First published 2009
Reprinted 2012
First paperback edition 2014

*A catalogue record for this publication is available from the British Library*

*Library of Congress Cataloguing in Publication data*

Lakes, Roderic S.
Viscoelastic materials / Roderic Lakes.
    p.    cm.
Includes bibliographical references and index.
ISBN 978-0-521-88568-3 (hardback)
1. Viscoelastic materials. 2. Viscoelasticity. I. Title.
TA418.2.L349 2009
620.11'232–dc22 2009000949

ISBN 978-0-521-88568-3 Hardback
ISBN 978-1-107-45978-6 Paperback

In ancient rivalry with fellow spheres the sun still sings its glorious song and completes with tread of thunder the journey it has been assigned.

*Goethe*

# Contents

# Preface

This book is intended to be of use in a one-semester graduate course on the properties, analysis, and uses of viscoelastic materials. A precursor book, *Viscoelastic Solids*, has been used as a text in such a course. This book contains many updates, expanded coverage of the materials science of the causes of viscoelastic behavior, and of the properties of materials of biological origin, and applications of viscoelastic materials. The objective is to make the subject accessible and useful to students in a variety of disciplines in engineering and physical science. To that end, the coverage is intentionally broad. For research scientists and engineers or graduate students who pursue the subject via self-study, many references have been included to provide links to the literature. The subject may be profitably studied by undergraduate students, particularly those who are interested in vibration abatement, biomechanics, and the study of materials. Most of the book should be accessible to people who have completed an intermediate or an elementary course on the mechanics of deformable bodies. Exposure to elasticity theory, materials science, and vibration theory is helpful but not necessary.

A development of the theory is presented, including both transient and dynamic aspects, with emphasis on linear viscoelasticity. The structure of the theory is presented with the aim of developing physical insight. Methods for the solution of stress analysis problems in viscoelastic objects are developed and illustrated. Experimental methods for characterization of viscoelastic materials are explored in detail. Viscoelastic phenomena are described for a wide variety of materials, including polymers, metals, ceramics, geological materials, biological materials, synthetic composites, and cellular solids. High-damping alloys and composites are considered as well as materials that resist creep. To illustrate the sources of viscoelastic phenomena, we describe and analyze causal mechanisms, with cases of materials of extremely high damping and extremely low damping. The theory of viscoelastic composite materials is presented, with examples of various types of structures and the relationships between structure and mechanical properties. Many applications of viscoelasticity and viscoelastic materials are illustrated, with case studies and analysis of particular cases. Viscoelasticity is pertinent to applications as diverse as earplugs, gaskets, computer disks, satellite stability, medical diagnosis, injury prevention, vibration abatement, tire performance, sports, spacecraft, and music. Examples of the use of viscoelastic materials in the prevention and alleviation of human suffering are explored. Supplementary material, including audio and video demonstrations of viscoelastic behavior, class resources, and research links, are provided on the following Website: `http://silver.neep.wisc.edu/~lakes`.

I thank the University of Wisconsin–Madison and the Vezzetti family for encouragement and support during a portion of the time that this book was in preparation. I am also grateful to my wife, Diana, for her patience and support. Finally, I thank Professor J. Giacomin and many students for their helpful suggestions.

# Introduction: Phenomena

## 1.1 Viscoelastic Phenomena

Most solid materials are described, for small strains, by Hooke's law of linear elasticity: stress $\sigma$ is proportional to strain $\epsilon$. In one dimension, Hooke's law is as follows:

$$\sigma = E\epsilon, \tag{1.1}$$

with $E$ as Young's modulus. Hooke's law for elastic materials can also be written in terms of a compliance $J$:

$$\epsilon = J\sigma. \tag{1.2}$$

Consequently, the elastic compliance $J$ is the inverse of the modulus $E$:

$$J = \frac{1}{E}. \tag{1.3}$$

In contrast to elastic materials, a viscous fluid under shear stress obeys $\sigma = \eta d\epsilon/dt$, with $\eta$ as the viscosity. In reality, all materials deviate from Hooke's law in various ways, for example, by exhibiting, both viscous-like and elastic characteristics. *Viscoelastic* materials are those for which the relationship between stress and strain depends on time (Figure 1.1). *Anelastic* solids represent a subset of viscoelastic materials: anelastic solids have a unique equilibrium configuration and ultimately recover fully after the removal of a transient load.

The stiffness and strength of materials is often illustrated by a stress–strain curve, which is obtained by applying a constant rate of strain to a bar of the material. If the material is linearly elastic, the curve is a straight line with a slope proportional to the elastic modulus (the heavy line in the right diagram in Figure 1.2). For a sufficiently large stress (the yield stress $\sigma_y$), the material exhibits yield as shown in Figure 1.2. This is a threshold phenomenon. A linearly viscoelastic material, by contrast, gives rise to a curved stress–strain plot (the left diagram in Figure 1.2) as demonstrated in §2.5. The reason for this rise is that during constant strain rate deformation, both time and strain increase together. The viscoelastic material is sensitive to time. Consequently, the curve on the left becomes steeper if the strain rate is increased. The residual strain eventually recovers to zero in a viscoelastic solid

Figure 1.1. Droop with time (clock on left) of a viscoelastic pyramid.

(see below); a viscoelastic fluid undergoes a permanent residual strain. The elastic–plastic material is not sensitive to time or rate, but it has a threshold stress: the yield stress. There is always a residual strain $\epsilon_r$ after load removal if the yield stress has been exceeded. When testing and describing viscoelastic materials, it is preferable to apply a step strain or step stress in time rather than a ramp (constant rate of strain) because the effect of time is then isolated from any nonlinearity. The response to step strain is stress relaxation, and the response to step stress is creep.

Some phenomena in viscoelastic materials are:

(1) if the stress is held constant, the strain increases with time (creep);
(2) if the strain is held constant, the stress decreases with time (relaxation);
(3) the effective stiffness depends on the rate of application of the load;
(4) if cyclic loading is applied, hysteresis (a phase lag) occurs, leading to a dissipation of mechanical energy;
(5) acoustic waves experience attenuation;
(6) rebound of an object following an impact is less than 100 precent; and
(7) during rolling, frictional resistance occurs.

All materials exhibit some viscoelastic response. In common metals, such as steel or aluminum, as well as in quartz, at room temperature and at small strain, the behavior does not deviate much from the behavior of linearly elastic materials. Synthetic polymers, wood, and human tissue, as well as metals, at high temperature display large viscoelastic effects. In some applications, even a small viscoelastic response can be significant. To be complete, an analysis or design involving such materials must incorporate their viscoelastic behavior.

Knowledge of the viscoelastic response of a material is based on measurement. The mathematical formulation of viscoelasticity theory is presented in the following chapters with the aim of enabling prediction of the material response to arbitrary load histories. Even at present, it is not possible to calculate viscoelastic response

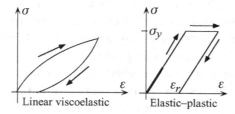

Figure 1.2. Stress–strain plots for deformation at constant strain rate followed by unloading. The plot on the left shows the behavior of a linearly viscoelastic material. The plot on the right shows the behavior of an ideal elastic–plastic material.

from models based on atomic and molecular models alone [1], although considerable understanding has evolved regarding causal mechanisms for viscoelastic behavior (see Chapter 8).

## 1.2 Motivations for Studying Viscoelasticity

The study of viscoelastic behavior is of interest in several contexts. First, materials used for structural applications of practical interest may exhibit viscoelastic behavior that has a profound influence on the performance of that material. Materials used in engineering applications may exhibit viscoelastic behavior as an unintentional side effect. In applications, one may deliberately make use of the viscoelasticity of certain materials in the design process, to achieve a particular goal. Second, the mathematics underlying viscoelasticity theory is of interest within the applied mathematics community. Third, viscoelasticity is of interest in some branches of materials science, metallurgy, and solid state physics because it is causally linked to a variety of microphysical processes and can be used as an experimental probe of those processes (Chapter 8). Fourth, the causal links between viscoelasticity and microstructure are exploited in the use of viscoelastic tests as an inspection tool as well as in the design of materials. Many applications of viscoelastic behavior are discussed in Chapter 10.

## 1.3 Transient Properties: Creep and Relaxation

### 1.3.1 Viscoelastic Functions $J(t)$, $E(t)$

#### Creep

Creep is a progressive deformation of a material under constant stress. In one dimension, suppose the history of stress $\sigma$ as it depends on time $t$ to be a step function with the magnitude $\sigma_0$, beginning at time zero:

$$\sigma(t) = \sigma_0 \mathcal{H}(t). \tag{1.4}$$

$\mathcal{H}(t)$ is the unit Heaviside step function defined as zero for $t$ less than zero, one for $t$ greater than zero, and $1/2$ for $t = 0$ (see Appendix §A.1.3). The strain $\epsilon(t)$ in a viscoelastic material will increase with time. The ratio,

$$J(t) = \frac{\epsilon(t)}{\sigma_0}, \tag{1.5}$$

is called the *creep compliance*. In linearly viscoelastic materials, the creep compliance is independent of stress level. The intercept of the creep curve on the strain axis is ascribed by some authors to instantaneous elasticity. However, no load can be physically applied instantaneously. If the loading curve is viewed as a mathematical step function, we remark that the region around zero time contains an infinite

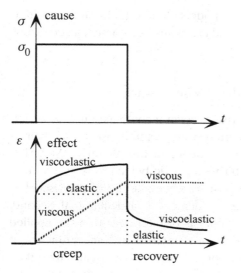

Figure 1.3. Creep and recovery. Stress $\sigma$ and strain $\epsilon$ versus time $t$.

domain on a logarithmic scale, a topic we will return to later. If the load is released at a later time, the strain will exhibit recovery or a progressive decrease of deformation. Strain in recovery may or may not approach zero, depending on the material. The recovery phase is not included in Equations 1.4 and 1.5 but will be treated in §2.2.

The creep response in Figure 1.3 is shown beginning at the same time as the stress history, which is the *cause*. The corresponding functional form is $J(t) = j(t)\mathcal{H}(t)$, with $j(t)$ as a function defined over the entire time scale. This functional

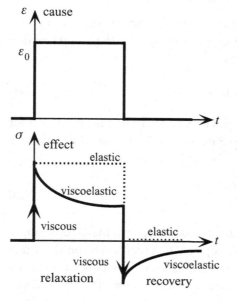

Figure 1.4. Relaxation and recovery.

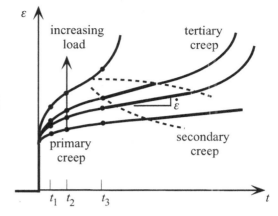

Figure 1.5. Regions of creep behavior. Strain $\epsilon$ versus time $t$, for different load levels.

form for $J(t)$ follows from the physical concept of causality, that is, the effect does not precede the cause.

Creep curves may exhibit three regions (Figure 1.5), *primary creep* in which the curve is concave down, *secondary creep* in which deformation is proportional to time, and *tertiary creep* in which deformation accelerates until creep rupture occurs. Tertiary creep is always a manifestation of nonlinear viscoelasticity, and secondary creep is usually nonlinear as well. Although secondary creep is represented by a straight line in a plot of strain versus time (constant strain rate), that straight line has nothing whatever to do with linear viscoelasticity. Linear response involves a linear relationship between cause and effect: stress and strain at a given time in the case of creep. Specifically, data taken at different load levels may be compared by considering isochronals or data at the same time. Data points at times $t_1$, $t_2$, and $t_3$ are illustrated in Figure 1.5. If the plot of stress versus strain at constant time is a straight line, the material may be linear. Secondary creep almost always entails nonlinear viscoelasticity. The nature of linear viscoelasticity and the distinction between linear and nonlinear behavior are presented in detail in §2.12 and §6.2.

### Plotting Creep Results

Has the creep leveled off? From the top graph in Figure 1.6 (0 to 10 seconds), one might surmise that the creep strain is leveling off and is approaching an asymptotic value. The data used here extend over a wider range of time than shown in the top graph; the plot of the same data in the bottom graph on a scale 0 to 1,000 seconds shows that the creep has not leveled off but continues to progressively longer times. One might also surmise that there is an instantaneous elasticity corresponding to the intercept at time zero. However, use of a linear time scale fails to show processes that occur at short times. Comparing the two plots, it is evident that there is much creep occurring near time zero. The data used in the plots are from a power law: $\epsilon(t) = 10^{-3} t^{1/6}$. Such creep is representative of some experimental results gathered

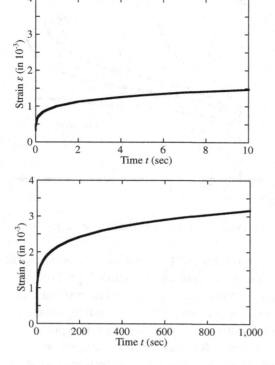

Figure 1.6. Creep on different time scales for the same creep strain $\epsilon(t) = 10^{-3}t^{1/6}$.

over a broad range of time. Therefore, it is customary to plot creep and relaxation results versus log time.

### Relaxation

Stress relaxation is the gradual decrease of stress when the material is held at constant strain. If we suppose the strain history to be a step function of magnitude $\epsilon_0$ beginning at time zero: $\epsilon(t) = \epsilon_0 \mathcal{H}(t)$, the stress $\sigma(t)$ in a viscoelastic material will decrease as shown in Figure 1.4. The ratio,

$$E(t) = \frac{\sigma(t)}{\epsilon_0}, \tag{1.6}$$

is called the *relaxation modulus*. In linear materials, it is independent of strain level, so $E(t)$ is a function of time alone. The symbol $E$ for Young's modulus as stiffness in uniaxial tension and compression is used in subsequent sections because the introductory presentations are restricted to one dimension.

Creep and relaxation can occur in shear or in volumetric deformation as well. The relaxation function for shear stress is called $G(t)$. For volumetric deformation, the elastic bulk modulus is called $B$ (also called $K$). A corresponding relaxation function $B(t)$ may be defined as above, but with the stress as a hydrostatic stress. A similar distinction is made in the creep compliances, $J_G(t)$ for creep in shear, $J_E(t)$ for creep in extension, and $J_B(t)$ for creep in volumetric deformation.

The relaxation curve is drawn as decreasing with time, and the creep curve is drawn as increasing with time; it is natural to ask whether this must be so. To address this question, let us consider only passive materials and perform a thought experiment. A *passive* material is one without any external sources of energy; for a passive mechanical system, the only energy stored in the material is strain energy, and in a dynamical system, kinetic energy as well. An analytical definition of a passive material is given in §2.3. Let an initially unstrained specimen be deformed in creep under dead-weight loading. A material that raises the weight can do so only by performing positive work on the weight; this is impossible in a passive material. So for passive materials $J(t)$ is an increasing function.

A distinction may be made between aging and nonaging materials: in aging materials, properties change with time, typically time as measured following the formation or transformation of the material. Concrete, for example, is an aging material. The discussion here is, for the most part, restricted to nonaging materials.

### 1.3.2 Solids and Liquids

*Elastic solids* constitute a special case for which the creep compliance is $J(t) = J_0 \mathcal{H}(t)$, with $J_0$ as a constant, which is the elastic compliance. Elastic materials exhibit immediate recovery to zero strain following release of the load. Viscoelastic materials that exhibit complete recovery after sufficient time following creep or relaxation are called *anelastic materials*. Viscous fluids constitute another special case in which the creep compliance is $J(t) = (1/\eta)t\mathcal{H}(t)$, with $\eta$ as the viscosity. Creep deformation in viscous materials is unbounded.

In the modulus formulation, a *viscoelastic solid* is a material for which $E(t)$ tends to a finite, nonzero limit as time $t$ increases to infinity; in a viscoelastic liquid, $E(t)$ tends to zero. In the compliance formulation, a viscoelastic solid is a material for which $J(t)$ tends to a finite limit as time $t$ increases to infinity; in a viscoelastic fluid, $J(t)$ increases without bound as $t$ increases.

The time scale extends from zero to infinity. In practice, creep or relaxation procedures in certain regions of the time scale are difficult to accomplish. For example, the region $10^{-10}$ sec to 0.01 sec is effectively inaccessible to most kinds of transient experiment because the load can be applied only so suddenly. Observation of the behavior of materials at longer times is limited by the patience and ultimately by the lifetime of the experimenter. In this vein, one may define [2] the dimensionless Deborah number $D$:

$$D \equiv \frac{\text{time of creep or relaxation}}{\text{time of observation}}. \qquad (1.7)$$

If $D$ is large, we perceive a material to be a solid even if it ultimately relaxes to zero stress. The difficulty in discriminating solids from liquids is a result of the finite lifetime of the human experimenter. Longer-term observations are also possible, as described by the Biblical prophetess Deborah: The mountains flowed before the Lord [3]. The original language may be translated as "flowed" [2], but some

Table 1.1. *Material properties*

| |
| --- |
| Empty space: Quantum vacuum |
| **Gases** |
| **Viscous liquids** |
| *Viscoelastic materials* |
| **Elastic solids** |

translations give "quaked" [RSV] or "melted" [KJV]. The flow of mountains is extremely slow, so that they appear solid to human observers, but are observed to flow before God. There is also an intermediate time scale of interest to engineers: we may wish that the materials used in structures behave as solids over the lifespan of human civilizations, which is longer than that of an individual. Geologists infer behavior of flowing rock over a time scale longer than that of human civilizations. We will return to these topics in discussions of experimental methods and of applications.

Viscoelastic materials are considered in the broader context of physical properties as follows (Table 1.1). Elastic solids support both shear stress and hydrostatic stress and their properties are independent of time or frequency. Viscoelastic materials exhibit time and frequency dependence. Viscous liquids support static hydrostatic stress; they generate shear stress only if the strain is changing with time. Gases are also viscous but they are orders of magnitude more compressible than liquids. Gases and liquids become indistinguishable at the critical point which corresponds to a particular temperature and pressure. In empty space, attractive force between conducting or dielectric surfaces is known as the *Casimir effect*. Isotropic elastic solids are describable by two elastic constants, for example, the shear and bulk modulus. Liquids and gases are also describable by two constants, the viscosity and the compressibility (inverse bulk modulus). By contrast, viscoelastic materials require a *function* of time or frequency to describe the behavior. Therefore, a rich set of physical phenomena can occur.

## 1.4 Dynamic Response to Sinusoidal Load: $E^*$, tan$\delta$

Suppose the stress $\sigma(t)$ is varying (Figure 1.7) sinusoidally in time $t$, as follows:

$$\sigma(t) = \sigma_0 \sin(2\pi \nu t). \tag{1.8}$$

The frequency (in cycles per second or Hertz, abbreviated Hz) is called $\nu$ or $f$. The strain response of a linearly viscoelastic material is also sinusoidal in time, but the response will lag the stress by a phase angle $\delta$

$$\epsilon(t) = \epsilon_0 \sin(2\pi \nu t - \delta). \tag{1.9}$$

The period $T$ of the waveform is the time required for one cycle: $T = 1/\nu$. The phase angle is related to the time lag $\Delta t$ between the sinusoids by $\delta = 2\pi(\Delta t)/T$.

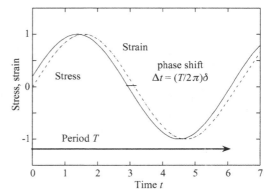

Figure 1.7. Stress and strain versus time *t* (in arbitrary units) in dynamic loading of a viscoelastic material.

To see this, the argument in Equation 1.9 may be written as

$$2\pi\nu t - \delta = 2\pi\nu t - \frac{2\pi\nu\delta}{2\pi\nu} = 2\pi\nu(t - \frac{\delta}{2\pi\nu}) = 2\pi\nu(t - \Delta t). \qquad (1.10)$$

Therefore,

$$\Delta t = \frac{\delta}{2\pi\nu}, \qquad (1.11)$$

with frequency as the inverse of period,

$$\nu = \frac{1}{T}$$

$$\delta = \frac{2\pi\Delta t}{T}. \qquad (1.12)$$

As a result of the phase lag between stress and strain, the dynamic stiffness can be treated as a complex number $E^*$. *"Dynamic," in this context, refers to oscillatory input, not to any inertial effects:*

$$\frac{\sigma}{\epsilon_0} = E^* = E' + iE''. \qquad (1.13)$$

The single and double primes designate the real and imaginary parts; they do not represent derivatives; $i = \sqrt{-1}$. The loss angle $\delta$ is a dimensionless measure of the viscoelastic damping of the material. The dynamic functions $E'$, $E''$, and $\delta$ depend on frequency. The tangent of the loss angle is called the *loss tangent*: $\tan\delta$. In an elastic solid, $\tan\delta = 0$. The relationship between the transient properties $E(t)$ and $J(t)$ and the dynamic properties $E'$, $E''$, and $\tan\delta$ is developed in §3.2.2. Dynamic viscoelastic behavior, in particular $\tan\delta$ and its consequences, is at times referred to as *internal friction* or as *mechanical damping*.

## 1.5 Demonstration of Viscoelastic Behavior

Several commonly available materials may be used to demonstrate viscoelastic behavior. For example, Silly Putty®, sold as a toy, may be formed into a long rod and hung from a support so that it is loaded under its own weight. It will creep without limit, behaving as a liquid. It will also bounce like a rubber ball, behaving nearly elastically at a sufficiently high strain rate. Another example is a foam used for earplugs [4]. This foam can be compressed substantially and will recover most of the deformation in a period of a minute or so; simple creep experiments and demonstrations can also be performed with this material. As for the rate of decay of vibration, an aluminum tuning fork can be used to demonstrate the free decay of vibration. Following an impact, the fork is audible for many seconds, hence for thousands of cycles, demonstrating the low loss tangent of aluminum. A similarly shaped fork made of a material, such as a stiff plastic or wood (or a plastic ruler mounted as a cantilever and set into vibration) with a higher $\tan\delta$, will damp out its vibrations much more quickly. Sounds and waveforms for tuning forks of various materials can be found on the Internet at `http://silver.neep.wisc.edu/~lakes/Demo.html`.

## 1.6 Historical Aspects

A scientific awareness of viscoelastic behavior dates at least to the late eighteenth century. Coulomb [5] (1736–1806) reported studies of the torsional stiffness of wires by a torsional vibration method [6]. He also discussed damping of vibration and demonstrated experimentally that its principal cause was not air resistance but was a characteristic of the wire. As for creep and relaxation [7], Vicat in 1834 [8] surveyed the sagging of wires and of suspension bridges. Weber [9] and Kohlrausch [10] found deviations from perfect elasticity in galvanometer suspensions. Upon the release of torque on the galvanometer suspension, the instrument did not return to zero immediately; instead it converged gradually. This creep recovery was referred to as the *'elastic after effect'*; see also Zener [11]. Creep behavior was observed in silk threads under load in 1841 [12].

Viscoelastic behavior has been studied by eminent figures, such as Boltzmann, Coriolis, Gauss, and Maxwell [7]. Early mathematical modeling of relaxation processes included a stretched exponential formalism (discussed in Chapter 2), which was used to model creep in silk, glass fibers, and rubber. Maxwell [13] developed a relaxation analysis of gas viscosity that is also applicable to viscoelasticity. The integral representation of Boltzmann [14] for the stress–strain relationship forms the basis of the linear theory of viscoelasticity as it is currently understood [15]. The theory of integral equations and of functional analysis as developed by Volterra [16] provides much of the mathematical underpinning of viscoelastic behavior.

As polymeric materials assumed technological importance, intensive study of viscoelasticity began in the 1930s. Leaderman first suggested that an increase in temperature has the effect of contracting the time scale of creep in polymers [17]. Data at different temperatures were used by Tobolsky and Andrews in 1945 to prepare master curves to infer behavior over extended time scales [18]. Ferry [19] soon after presented a theoretical interpretation of the temperature effect. Plazek [20] reviewed many experimental efforts in connection with temperature dependence of polymer viscoelasticity, particularly a polyisobutylene provided by the U.S. National Bureau of Standards. As for terminology, reference to a history of strain is due to Maxwell [21], while reference to hereditary phenomena is due to Volterra [22].

## 1.7 Summary

Viscoelastic behavior manifests itself in creep, or a continued deformation of a material under constant load; and in stress relaxation, or a progressive reduction in stress while a material is under constant deformation. These observed phenomena are the basis for the constitutive equations developed in Chapter 2, for experimental methods for characterizing materials, and are of concern in applications of materials.

## 1.8 Examples

**Example 1.1**
A creep curve is fitted well by a straight line. Does that mean the material is linearly viscoelastic?
**Answer**
To have linear viscoelasticity, it is necessary that the strain response at a given time is proportional to the load. To make a plot of stress versus strain at constant time, a set of creep curves, each one obtained under a different stress, is required. *If creep strain increases linearly with time, the response is viscous.*

**Example 1.2**
The creep compliance is defined as the time-dependent ratio of strain to stress in response to a stress which is a step function in time. Why not define the viscoelastic response in terms of a stress–strain curve? After all, the stress–strain curve is a standard way of expressing the elastic, plastic, and fracture characteristics of a material.
**Answer**
In a viscoelastic material, stress, strain, and time are coupled while in an elastic or an elastic–plastic material, there is no time dependence. One can interpret the initial part of a stress–strain curve in terms of linear viscoelasticity as presented in §2.5. However experiments that provide a stress–strain curve are ordinarily conducted at a constant strain rate. Therefore, as time increases, so does strain. At sufficiently large strains, material nonlinearity, damage, and fracture occur. The creep and

relaxation procedures decouple the dependence on time from any dependence on load level, simplifying interpretation.

## 1.9 Problems

1.1. How will the sound from tuning forks made of brass, aluminum, and plastic differ?

1.2. Discuss five materials that you have observed to be viscoelastic.

1.3. Under what conditions may an aging material be regarded as nonaging?

1.4. Can you think of a material that is not passive?

1.5. How does a linearly viscoelastic material differ from an elastoplastic material?

### BIBLIOGRAPHY

[1] Plunkett, R., Damping Analysis: An Historical Perspective, M$^3$D: Mechanics and Mechanisms of Material Damping, ASTM STP 1169, V. K. Kinra and A. Wolfenden, eds., Philadelphia, PA: ASTM, 1992, pp. 562–569.

[2] Reiner, M., The Deborah Number, *Physics Today*, 17, 62–64, 1969.

[3] Judges 5:5

[4] EAR, division of Cabot Corp., 7911 Zionsville Rd., Indianapolis, IN.

[5] Coulomb, C. A., Recherches théoretiques et experimentales sur la force de torsion et sur l'élasticité des fils de métal, Mém. Paris: Acad. Sci., 1784.

[6] Timoshenko, S. P., *History of Strength of Materials*, NY: Dover, 1983.

[7] Scher, H., Shlesinger, M. F., and Bendler, J. T., Time-Scale Invariance in Transport and Relaxation, *Physics Today*, 44, 26–34, 1991.

[8] Vicat, L. T., Note sur l'allongement progressif du fil de fer soumis à diverses tensions, *Annales, Ponts et Chausées, Mémoires et Docum*, Vol. 7, 1834.

[9] Weber, W., Über die Elastizität der Seidenfäden, *Ann Phys Chem* (Poggendorf's), 34, 247–257, 1835.

[10] Kohlrausch, R., Ueber das Dellmann'sche Elektrometer, *Ann Phys* (Leipzig), 72, 353–405, 1847.

[11] Zener, C., *Elasticity and Anelasticity of Metals*, Chicago: University of Chicago Press, 1948.

[12] Weber, W., Über die Elastizität fester Körper, *Ann Phys Chem* (Poggendorf's), 54, 247–257, 1841.

[13] Maxwell, J. C., On the Dynamical Theory of Gases, *Phil Trans Royal Soc Lond*, 157, 49–88, 1867.

[14] Boltzmann, L., Zur Theorie der Elastischen Nachwirkungen, Sitzungsber. Kaiserlich Akad Wissen Math. *Naturwissen*, 70, 275–306, 1874.

[15] Markovitz, H., Boltzmann and the Beginnings of Linear Viscoelasticity, *Trans Soc Rheology*, 21, 381–398, 1977.

[16] Volterra, V., *Theory of Functionals and of Integral and Integro-Differential Equations*, NY: Dover, 1959.

[17] Leaderman, H., *Elastic and Creep Properties of Filamentous Materials and Other High Polymers*, Washington, DC: The Textile Foundation, 1943.

[18] Tobolsky, A. V., and Andrews, R. D., Systems Manifesting Superposed Elastic and Viscous Behavior, *J Chem Phys*, 13, 3–27, 1945.

[19] Ferry, J. D., Mechanical Properties of Substances of High Molecular Weight. VI. Dispersion in Concentrated Polymer Solutions and Its Dependence on Temperature and Concentration, *J Am Chem Soc*, 72, 3746–3752, 1950.

[20] Plazek, D. J., Oh, Thermorheological Simplicity, Wherefore Art Thou? *J Rheology*, 40, 987–1014, 1996.

[21] Maxwell, J. C., Constitution of Bodies, *Scientific Papers*, 1877, pp. 616–624.

[22] Volterra, V., Sur la Théorie Mathématique des Phénomènes Héréditaires, *J Math Pures et Appl*, 7, 249–298, 1928.

# 2

## Constitutive Relations

### 2.1 Introduction

For some applications of viscoelastic materials, it is sufficient to understand creep and relaxation properties. For example sometimes structural elements are maintained under steady load or constant extension. In other applications the response to an arbitrary load or strain history is required. To predict this response, *constitutive equations* that incorporate all possible responses are of use. Various mathematical tools are used in the development of these equations. The viscoelastic functions in the equations are obtained by *experimentation*.

### 2.2 Prediction of the Response of Linearly Viscoelastic Materials

#### 2.2.1 Prediction of Recovery from Relaxation $E(t)$

The creep and relaxation properties described above in §1.3 permit one to predict the response of the material to a step function stress or strain. To predict the response of the material to *any* history of stress or strain (i.e., stress or strain as a function of time), constitutive equations are developed.

The following is restricted to isothermal deformation in one dimension. The symbol $E$ is used to represent an elastic modulus, however, the analysis applies equally to shear deformation corresponding to a shear modulus $G$ or volumetric deformation corresponding to a bulk modulus $B$, which is sometimes called $K$.

To develop the constitutive equation for linear materials, we use the *Boltzmann superposition principle*, which states that the effect of a compound cause is the sum of the effects of the individual causes. This principle is a statement of linearity. First, we consider the strain associated with a relaxation and recovery experiment, with the intention to use the idea of linearity as embodied in the Boltzmann superposition principle to predict the resulting stress history. Recall that stress relaxation refers to the time variation of stress, $\sigma(t) = \epsilon_0 E(t)$, in response to a strain history, which is a step function $\mathcal{H}(t)$ in time (Appendix §A.1.3): $\epsilon(t) = \epsilon_0 \mathcal{H}(t)$; $E(t)$ is the relaxation modulus. *Recovery* refers to the response following the removal of the strain. The

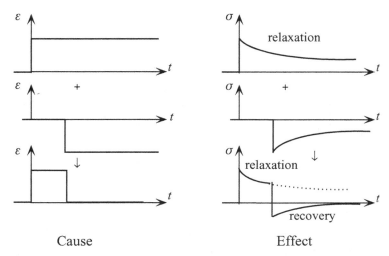

Figure 2.1. Left: Summation of step functions of strain $\epsilon$ versus time $t$, to synthesize a box function strain history for relaxation and recovery. Right: The recovery curve of stress $\sigma$ is generated from a superposition of individual relaxation curves if the material is linear.

assumed strain may be written as a superposition of a step up followed by a step down (Figure 2.1):

$$\epsilon(t) = \epsilon_0[\mathcal{H}(t) - \mathcal{H}(t - t_1)] \tag{2.1}$$

so the stress is (Figure 2.1), from the Boltzmann superposition principle,

$$\sigma(t) = \epsilon_0[E(t) - E(t - t_1)]. \tag{2.2}$$

Here, the stress due to a delayed step in strain $\epsilon_0\mathcal{H}(t - t_1)$ follows the same form of time history $\epsilon_0 E(t - t_1)$ because the stress $\epsilon_0 E(t)$ due to an earlier step in strain, $\epsilon_0\mathcal{H}(t)$ only delayed in time. This is only true if the material in question is nonaging, that is, its mechanical properties do not change with time. Aging materials are considered later. The stress may or may not recover to zero as time $t$ becomes large, depending on the material. Recovery is also called the *elastic after-effect* [1].

### 2.2.2 Prediction of Response to Arbitrary Strain History

Now consider an *arbitrary* history of strain $\epsilon(t)$ as a function of time $t$, assumed to be zero prior to time zero, as shown in Figure 2.2. Consider a segment of this history from time $t - \tau$ to time $t - \tau + \Delta\tau$ (with $\tau$ as a time variable; it plays the same role as the constant $t_1$ in the above recovery example), also shown in Figure 2.2. The segment of strain history may be written as

$$\epsilon(t) = \epsilon(\tau)[\mathcal{H}(t - \tau) - \mathcal{H}(t - \tau + \Delta\tau)]. \tag{2.3}$$

The strain at time $t$ "now" from this pulse is zero.

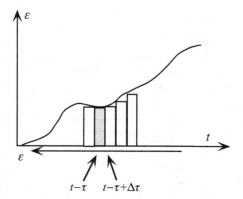

Figure 2.2. Analysis of arbitrary strain history: decomposition into pulse functions. Strain $\epsilon(t)$ is a function of time $t$.

As in Equation 2.2, the increment of stress at time $t$ due to the strain pulse in the past is, from the Boltzmann superposition principle,

$$d\sigma(t) = \epsilon(\tau)[E(t - \tau) - E(t - \tau + \Delta\tau)]. \tag{2.4}$$

Since $\frac{dE(t-\tau)}{d\tau} = \lim_{\Delta\tau \to 0} \frac{E(t-\tau+\Delta\tau)-E(t-\tau)}{\Delta\tau}$, this stress increment may be written as

$$d\sigma(t) = -\epsilon(\tau)\frac{dE(t - \tau)}{d\tau}d\tau. \tag{2.5}$$

The entire strain history may be decomposed into such pulses. The stress at time $t$ is the summation of the stress effects of each of these pulses. We invoke the principle of causality and include only pulses prior to or up to the present time. Pulses in the future are considered to have no effect in the present. In the limit, as the pulse width $\Delta\tau$ becomes small, the summation converges to an integral:

$$\sigma(t) = -\int_0^t \epsilon(\tau)\frac{dE(t - \tau)}{d\tau}d\tau + E(0)\epsilon(t). \tag{2.6}$$

The second term arises as follows: after one strain pulse or an arbitrary number of strain pulses, the strain is zero. However, in an arbitrary strain history, the final strain $\epsilon(t)$ at time $t$ (now) is not, in general, zero. The final increment of stress from zero to $\epsilon(t)$ is $E(0)\epsilon(t)$, because the stress has no time to relax.

Integrate by parts to obtain the final form of the *Boltzmann superposition integral*, which expresses a convolution. Here, $t$ is time, and $\tau$ is a time variable of integration.

$$\boxed{\sigma(t) = \int_0^t E(t - \tau)\frac{d\epsilon(\tau)}{d\tau}d\tau.} \tag{2.7}$$

If the role of stress and strain are interchanged and the above arguments are repeated, a complementary relation may be obtained:

$$\boxed{\epsilon(t) = \int_0^t J(t - \tau)\frac{d\sigma(\tau)}{d\tau}d\tau.} \tag{2.8}$$

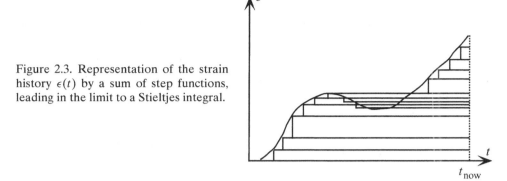

Figure 2.3. Representation of the strain history $\epsilon(t)$ by a sum of step functions, leading in the limit to a Stieltjes integral.

Consequently, if the response of a material to step stress or strain has been determined experimentally, the response of a linearly viscoelastic material to *any* load history can be found for the purpose of analysis or design. Observe, however, that if we input the same step strain in Equation 2.7 that was used in the definition of $E(t)$, the derivative does not exist at the step discontinuity, so the stress cannot be obtained. The difficulty can be overcome by writing the strain history and its derivatives as *distributions*, such as the *delta* function (Appendix §A.1.4; Example 2.1). Distributions are more general than functions. One can also express the Boltzmann integral as a Stieltjes integral (see below) and avoid the use of derivatives.

An alternative way of obtaining the Boltzmann superposition integral makes use of a Stieltjes integral rather than a Riemann integral. The strain history is decomposed into step functions (Figure 2.3) rather than into pulse functions. Each step component in strain gives rise to a relaxing component in stress in view of the definition of the relaxation function. So, $\sigma(t) = \int_0^t E(t - \tau)d\epsilon$.

Here, the strain history need not be a differentiable function of time. If it is a differentiable function, the differential $d\epsilon$ can be written $d\epsilon = (d\epsilon/d\tau)d\tau$ to obtain the Boltzmann superposition integral in Equation 2.7. The Boltzmann superposition integral can also be derived using the concepts of functional analysis [2] or from general principles of continuum mechanics [3].

## 2.3 Restrictions on the Viscoelastic Functions

### 2.3.1 Roles of Energy and Passivity

The relaxation and creep functions are not arbitrary functions. Restrictions on the form of these functions have been derived based on physical principles. For example, the stored energy density $W$ for elastic materials of modulus $E$ at strain $\epsilon$ in one dimension has the form:

$$W(t) = \frac{1}{2}E\epsilon^2,$$
(2.9)

and for linearly viscoelastic materials, it has the form [4]:

$$W(t) = \frac{1}{2} \int_{-\infty}^{t} \int_{-\infty}^{t} E(2t - \tau_1 - \tau_2) \frac{d\epsilon}{d\tau_1} \frac{d\epsilon}{d\tau_2} d\tau_1 d\tau_2, \tag{2.10}$$

in which $\tau_1$ and $\tau_2$ are time variables. In three dimensions, for dilatation, $\epsilon$ is interpreted as $\epsilon_{kk}$, and $E$ as the bulk relaxation modulus; for shear, $\epsilon$ is interpreted as the deviatoric strain $e_{ij}$, and $E$ as the shear relaxation modulus $2\mu(t)$.

The corresponding dissipation inequality is [4]

$$-\frac{1}{2} \int_{-\infty}^{t} \int_{-\infty}^{t} \frac{\partial}{\partial t} E(2t - \tau_1 - \tau_2) \frac{d\epsilon}{d\tau_1} \frac{d\epsilon}{d\tau_2} d\tau_1 d\tau_2 \geq 0. \tag{2.11}$$

This embodies the thermodynamic principle that the dissipation rate of energy is nonnegative. If one requires the energy density to be nonnegative during relaxation (step–strain history), then

$$E(t) \geq 0. \tag{2.12}$$

Similarly, to satisfy the dissipation inequality during relaxation,

$$\frac{dE(t)}{dt} \leq 0. \tag{2.13}$$

The relaxation function, therefore, decreases monotonically with time as was informally demonstrated in §1.3.

Alternatively, one may define [5, 6] an energy function $w$, such that

$$w(t) = \int_{-\infty}^{t} \sigma(\tau) \frac{d\epsilon}{d\tau} d\tau. \tag{2.14}$$

A *passive* system is one for which

$$w(t) \geq 0 \tag{2.15}$$

for all time $t$. The physical interpretation is that of a physical system that can absorb energy, but which is unable to generate energy within itself. A *dissipative* material is defined [8, 9] as one for which the energy function $w(t) \geq 0$, for any path starting from the virgin state (of no prior deformation). Dissipativity has been demonstrated to be a consequence of the second law of thermodynamics [8]. Moreover, for a dissipative material [6], $w(t) \geq 0$ for every closed strain path. Closed strain paths are of particular interest in understanding the dynamic response, are discussed further in §3.4.

### 2.3.2 Fading Memory

In a material that exhibits *fading memory* the magnitude of the effect of a more recent cause (such as a strain) is greater than the magnitude of the effect of a cause of identical magnitude in the distant past. For fading memory it is sufficient [4] that

$$\left|\frac{dE(t)}{dt}\right|_{t=t_1} \leq \left|\frac{dE(t)}{dt}\right|_{t=t_2}, \quad \text{for} \quad t_1 > t_2 > 0. \tag{2.16}$$

To demonstrate this, observe in the alternate form of the Boltzmann integral, Equation 2.6, that the strain history in the integral is weighted by the time derivative of the relaxation modulus. Consequently, if that derivative decreases with time, the material exhibits fading memory. Therefore,

$$\frac{d^2 E(t)}{dt^2} \geq 0,  \tag{2.17}$$

that is, the relaxation curve is concave up.

To have a *unique* solution of boundary-value problems for viscoelastic materials, it is a necessary condition that the initial value of the relaxation functions be positive [7, 8].

## 2.4 Relation between Creep and Relaxation

### 2.4.1 Analysis by Laplace Transforms: $J(t) \leftrightarrow E(t)$

A relationship between the creep function $J(t)$ and the relaxation function $E(t)$ is desired. These functions appear in the Boltzmann superposition integrals above. Such integral equations may be manipulated with the aid of integral transforms (Appendix §A.2), such as the Laplace transform (Appendix §A.2.1). Specifically, Laplace transformation of a linear integral equation or a linear differential equation converts it into an algebraic equation.

We use the derivative theorem and convolution theorem for the Laplace transform to convert constitutive equations 2.7 and 2.8 to $\sigma(s) = sE(s)\epsilon(s)$ and $\epsilon(s) = sJ(s)\sigma(s)$, respectively. Here, $s$ is the transform variable. So,

$$\frac{\sigma(s)}{\epsilon(s)} = sE(s), \quad \frac{\sigma(s)}{\epsilon(s)} = \frac{1}{sJ(s)}.  \tag{2.18}$$

Setting these stress–strain ratios as equal,

$$E(s)J(s) = \frac{1}{s^2}.  \tag{2.19}$$

By taking the inverse transform, using the convolution theorem and the relation $\mathcal{L}[t] = 1/s^2$, are obtain

$$\int_0^t J(t - \tau)E(\tau)d\tau = \int_0^t E(t - \tau)J(\tau)d\tau = t.  \tag{2.20}$$

A different form can be obtained from a rearrangement of the relation in the Laplace plane, and from $s\mathcal{L}[t] = 1/s$,

$$1 = J(0)E(t) + \int_0^t E(t - \tau)\frac{dJ(\tau)}{d\tau}d\tau.  \tag{2.21}$$

The relationships are implicit. Explicit relationships can be developed via Laplace transformations provided a specific analytical form is given for $E(t)$ or $J(t)$; examples are given in subsequent sections. If a viscoelastic function is known in numerical form, the integral can be approximated as a sum, and the calculation

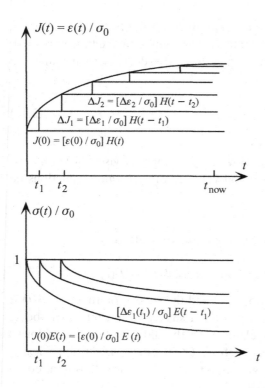

Figure 2.4. Top: decomposition of a creep function $J(t)$ as a sum of immediate and delayed Heaviside step functions in time $t$. Bottom: constant stress as a corresponding sum of relaxing components.

performed numerically (Appendix §A.4). Approximate interrelations can also be developed (see §4.3).

## 2.4.2 Analysis by Direct Construction: $J(t) \leftrightarrow E(t)$

The result obtained by Laplace transforms can be obtained by direct construction, and visualized graphically, as follows. Write the strain $\epsilon(t)$ due to a constant stress $\sigma_c$ as a sum of immediate $\mathcal{H}(t)$ and delayed Heaviside step functions in time. $\epsilon(t) = \epsilon(0)\mathcal{H}(t) + \sum_{i=0}^{N} \Delta\epsilon_i \mathcal{H}(t - t_i)$. Each step strain in the summation (Figure 2.4, top) gives rise to a relaxing component of stress (Figure 2.4, bottom). For a linear material, the modulus is independent of strain, and there is no interaction between the effect of steps in the summation. The stress due to this strain history is, by superposition of the corresponding relaxing components, $\sigma_c = \epsilon(0)E(t) + \sum_{i=0}^{N} \Delta\epsilon_i E(t - t_i)$. Dividing by the stress $\sigma_c$ and using the definition of creep compliance, $1 = J(0)E(t) + \sum_{i=0}^{N} \Delta J_i E(t - t_i)$. Passing to the limit of infinitely many fine step components gives a Stieltjes integral with $\tau$ as a time variable of integration, $1 = J(0)E(t) + \int_0^t E(t - \tau)\frac{dJ(\tau)}{d\tau}d\tau$. This is the same result obtained by Laplace transforms in Equations 2.21, §2.4.1.

## 2.5 Stress versus Strain for Constant Strain Rate

Suppose the material is subjected to a constant strain-rate history beginning at time zero:
$\epsilon(t) = 0$ for $t < 0$, $\epsilon(t) = \dot{\epsilon}t$ for $t \geq 0$,

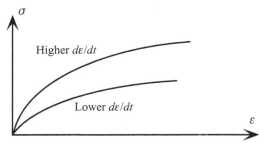

Figure 2.5. Stress $\sigma$ versus strain $\epsilon$ of a *linearly* viscoelastic material in response to constant strain rate, for several rates.

with $\dot{\epsilon}$ as the strain rate. Substituting in the Boltzmann integral, and changing variables, with **T** as a variable of integration,

$$\sigma(t) = \int_0^t E(t - \tau)\frac{d\epsilon}{d\tau}d\tau = \dot{\epsilon}\int_0^t E(t - \tau)d\tau = -\dot{\epsilon}\int_0^t E(\mathbf{T})d\mathbf{T}, \qquad (2.22)$$

so by the Liebnitz rule (see Appendix §A.1.7),

$$\frac{d\sigma(t)}{dt} = \dot{\epsilon}\,E(t), \qquad (2.23)$$

so the *slope* of the stress–strain curve of a linearly viscoelastic material decreases with time and hence with strain (Figure 2.5). The stress–strain curve is nonlinear even though the material is assumed to be linearly viscoelastic. The reason is that although the stress and strain are linearly related at a particular time $t$, the constant strain-rate test involves strain and the time changing simultaneously. Viscoelastic materials depend on time in their stiffness.

One cannot distinguish from a single stress–strain curve (assuming it is concave down) whether the material is linearly viscoelastic, nonlinearly viscoelastic, nonlinearly elastic, or nonlinearly inelastic. These possibilities could be distinguished by performing a series of creep experiments at different stress levels or a series of relaxation experiments at different strain levels as is further discussed in §6.2.

The slope of the stress–strain curve can also be evaluated via Laplace transforms. The Laplace transform of the Boltzmann integral (Equation 2.7) is $\sigma(s) = sE(s)\epsilon(s)$, but $\epsilon(t) = \dot{\epsilon}t$, so $\epsilon(s) = \dot{\epsilon}/s^2$, so $\sigma(s) = sE(s)\dot{\epsilon}/s^2$. Transform back to the time domain to obtain $\frac{d\sigma(t)}{dt} = \dot{\epsilon}E(t)$.

## 2.6 Particular Creep and Relaxation Functions

In this section, exponentials, series of exponentials, power laws, stretched exponentials, and logarithmic functions are considered.

### 2.6.1 Exponentials and Mechanical Models

Among the simplest transient response functions are those that involve exponentials. In relaxation, we may consider

$$E(t) = E_0 e^{-t/\tau_r}, \qquad (2.24)$$

with $\tau_r$ called the *relaxation time*. In creep, the corresponding relation is

$$J(t) = J_0(1 - e^{-t/\tau_c}), \tag{2.25}$$

with $\tau_c$ called the *creep* or *retardation time*. Exponential response functions arise in simple discrete mechanical models composed of springs, which are perfectly elastic ($\sigma_s = E\epsilon_s$), and dashpots, which are perfectly viscous ($\sigma_d = \eta d\epsilon_d/dt$, with $\eta$ as viscosity). The dashpot may be envisaged as a piston–cylinder assembly in which motion of the piston causes a viscous fluid to move through an aperture. Some people find spring-dashpot models to be useful in visualizing how viscoelastic behavior can arise, however they are not necessary in understanding or using viscoelasticity theory. *Warning.* Spring–dashpot models have a pedagogic role, however real materials in general are not describable by models containing a small number of springs and dashpots.

The *Maxwell model* consists of a spring and dashpot in series. Because deformation is assumed to be quasistatic, inertia is neglected and the force or stress is the same in both elements; the total deformation or strain is the sum of the strains. So, for the Maxwell model,

$$\frac{d\epsilon}{dt} = \frac{d\epsilon_s}{dt} + \frac{d\epsilon_d}{dt} = \frac{1}{E}\frac{d\sigma}{dt} + \frac{\sigma}{\eta}. \tag{2.26}$$

This differential equation can be rewritten in terms of the relaxation time $\tau \equiv \eta/E$ as

$$E\frac{d\epsilon}{dt} = \frac{d\sigma}{dt} + \frac{\sigma}{\tau}. \tag{2.27}$$

If we input step strain, the relaxation response is found to be, with $\tau = \tau_r$,

$$E(t) = E_0 e^{-t/\tau_r}, \tag{2.28}$$

however, if we input step stress, the creep response is

$$J(t) = \frac{1}{E} + \frac{t}{\eta}. \tag{2.29}$$

This form is unrealistic for primary creep because the predicted creep response versus time is a straight line, in contrast to curves that are observed experimentally.

Now the *Voigt model*, also called the Kelvin model, consists of a spring and dashpot in parallel so that they both experience the same deformation or strain and the total stress is the sum of the stresses in each element. So, $\sigma = E\epsilon + \eta\frac{d\epsilon}{dt}$, or

$$\frac{\sigma}{E} = \epsilon + \tau_c\frac{d\epsilon}{dt}, \tag{2.30}$$

in which $\tau_c = \eta/E$ is referred to as the retardation time. The creep response of the Voigt model is

$$J(t) = \frac{1}{E}(1 - e^{-t/\tau_c}), \tag{2.31}$$

but the relaxation response is a constant plus a delta function.

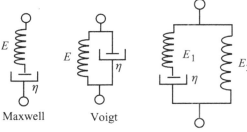

Figure 2.6. Spring–dashpot models. $E$ represents spring stiffness, and $\eta$ represents dashpot viscosity.

Maxwell          Voigt

Standard linear solid

More realistic behavior involving a single exponential in both creep and relaxation can be modeled by the *standard linear solid*, [10] which contains three elements (e.g., Figure 2.6). The left side of the model is a Maxwell model for which

$$E_1 \frac{d\epsilon}{dt} = \frac{d\sigma_1}{dt} + \frac{\sigma_1}{\tau}, \tag{2.32}$$

as above with $\tau \equiv \eta/E_1$. For the right side of the model, $\sigma_2 = E_2\epsilon$, so,

$$E_2 \frac{d\epsilon}{dt} = \frac{d\sigma_2}{dt}. \tag{2.33}$$

The total load (written as a stress) is the sum of the loads on the left side and right side of the diagram:

$$\sigma = \sigma_1 + \sigma_2. \tag{2.34}$$

Construct from the above

$$\frac{d\epsilon}{dt}(E_1 + E_2) = \frac{d\sigma_1}{dt} + \frac{\sigma_1}{\tau} + \frac{d\sigma_2}{dt} \tag{2.35}$$

and add $\tau^{-1}E_2\epsilon = \tau^{-1}\sigma_2$ to both sides to eliminate the individual stresses and to obtain the differential equation for the standard linear solid:

$$\boxed{\frac{d\epsilon}{dt}(E_1 + E_2) + \frac{\epsilon E_2}{\tau} = \frac{\sigma}{\tau} + \frac{d\sigma}{dt}}. \tag{2.36}$$

To obtain the *relaxation function* of the standard linear solid, take the Laplace transform of the differential equation to obtain the following:

$$(E_1 + E_2)(s\epsilon(s) - \epsilon(0)) + \frac{\epsilon(s)E_2}{\tau} = \frac{\sigma(s)}{\tau} + s\sigma(s) - \sigma(0). \tag{2.37}$$

The strain and stress histories are assumed to begin after time zero, which may be considered to occur in the remote past. Then, the surface terms vanish, simplifying the analysis. So,

$$\sigma(s) = \frac{[s(E_1 + E_2) + \frac{E_2}{\tau}]\epsilon(s)}{(s + \frac{1}{\tau})}. \tag{2.38}$$

But, from above (§2.2), $\sigma(s) = s\,E(s)\epsilon(s)$ is the Boltzmann integral in the Laplace plane, so,

$$E(s) = \frac{E_2}{s} + \frac{E_1}{(s + \frac{1}{\tau})}.\tag{2.39}$$

Transforming back to obtain the relaxation modulus, we identify $\tau = \tau_r$, the relaxation time.

$$\boxed{E(t) = E_2 + E_1 e^{-t/\tau_r}}.\tag{2.40}$$

So the relaxation function for the standard linear solid is a decreasing exponential. To obtain the *creep function* corresponding to $E(t) = E_2 + E_1 e^{-t/\tau_r}$, consider the Laplace transform

$$E(s) = \frac{E_2}{s} + \frac{E_1}{(s + \frac{1}{\tau_r})},\tag{2.41}$$

and recall that, from Equation 2.19, $E(s)J(s) = \frac{1}{s^2}$, so,

$$J(s) = \frac{1}{s^2 E(s)},\tag{2.42}$$

so,

$$J(s) = \frac{1}{s\,E_2 + E_1 \frac{s^2}{s + \frac{1}{\tau_r}}}$$

$$= \frac{s + \frac{1}{\tau_r}}{s\,E_2(s + \frac{1}{\tau_r}) + E_1 s^2}$$

$$= \frac{s + \frac{1}{\tau_r}}{s^2(E_2 + E_1) + E_2 \frac{s}{\tau_r}}$$

$$= \frac{1}{E_2 + E_1} \frac{s + \frac{1}{\tau_r}}{s^2 + \frac{E_2}{E_2 + E_1} \frac{s}{\tau_r}}.\tag{2.43}$$

Write the last expression in the form:

$$J(s) = \frac{P_a}{s} + \frac{P_b}{(s + \frac{1}{P_c \tau_r})} = \frac{P_a(s + \frac{1}{P_c \tau_r}) + P_b s}{s(s + \frac{1}{P_c \tau_r})},\tag{2.44}$$

and compare terms to find $P_a, P_b, P_c$.

But $P_a = 1/E_2$ by consideration of the asymptotic behavior for large time, so,

$$J(s) = \frac{\frac{1}{E_2 P_c \tau_r} + (\frac{1}{E_2} + P_b)s}{s(s + \frac{1}{P_c \tau_r})}.\tag{2.45}$$

Compare with the above,

$$P_c = \frac{E_1 + E_2}{E_2}, \quad P_b = \frac{1}{E_1 + E_2} - \frac{1}{E_2} = -\frac{E_1}{E_2(E_1 + E_2)}.\tag{2.46}$$

Transforming back to obtain the creep function corresponding to the exponential relaxation function in Equation 2.40,

$$\boxed{J(t) = \frac{1}{E_2} - \frac{E_1}{E_2(E_1 + E_2)} e^{-t/\tau_c},} \qquad (2.47)$$

with

$$\tau_c = \tau_r \frac{(E_1 + E_2)}{E_2}, \qquad (2.48)$$

called the *creep* or *retardation time*.

Observe that the retardation time is *not* equal to the relaxation time; it is larger. Physical insight is developed in Chapter 4 in connection with the distinction between creep and relaxation. A comparative graph is shown in Figure 4.4. The ratio of retardation to relaxation times depends on the *relaxation strength*, which is defined as the change in stiffness during relaxation divided by the stiffness at long time. The relaxation strength is illustrated here for a single exponential but is defined below for any relaxation function:

$$\Delta = \frac{E_1}{E_2}. \qquad (2.49)$$

So,

$$\tau_c = \tau_r(1 + \Delta). \qquad (2.50)$$

If the relaxation strength is small ($E_1 \ll E_2$) then the retardation and relaxation times are approximately equal. We may also write the relaxation strength in terms of the asymptotic behavior of any relaxation function as

$$\boxed{\Delta = \frac{E(0) - E(\infty)}{E(\infty)}.} \qquad (2.51)$$

By analysis of Equation 2.47 for $J(t)$, the relaxation strength in the compliance formulation is obtained for any creep function as

$$\Delta = \frac{J(\infty) - J(0)}{J(0)}. \qquad (2.52)$$

The single exponential response functions do not well approximate the behavior of most real materials. Specifically, a single exponential relaxation function undergoes most of its relaxation over about one decade (a factor of ten) in time scale. Real materials relax or creep over many decades of time scale (see Chapter 7). It is of course possible to generalize the differential equations to include derivatives of arbitrarily high-order $n$. For example, differential operators may be defined as follows [7] in terms of the time derivative operator: $D = \frac{d}{dt}$.

The constitutive equation can then be written as a polynomial of the differential operator with $p_k$ and $q_k$ as coefficients,

$$P(D) = \sum_{k=0}^{N} p_k \frac{d^k}{dt^k}, \quad \text{and} \quad Q(D) = \sum_{k=0}^{N} q_k \frac{d^k}{dt^k}. \tag{2.53}$$

Suppose [7]

$$q_r = \sum_{n=r}^{N} p_n \frac{d^{n-r} E}{dt^{n-r}} |_0, \tag{2.54}$$

with these derivatives to be evaluated for an argument of zero and that the stress and strain meet initial conditions

$$\sum_{r=k}^{N} p_r \frac{d^{r-k}\sigma}{dt^{r-k}} |_0 = \sum_{r=k}^{N} q_r \frac{d^{r-k}\epsilon}{dt^{r-k}} |_0, \tag{2.55}$$

in which these derivatives are to be evaluated for an argument of zero. Suppose further that the material obeys the linear differential equation

$$P(D)\sigma(t) = Q(D)\epsilon(t). \tag{2.56}$$

Apply a Laplace transform to the differential equation:

$$P(s)\sigma(s) = Q(s)\epsilon(s). \tag{2.57}$$

The stress–strain ratio in the Laplace plane, $\frac{\sigma(s)}{\epsilon(s)} = \frac{Q(s)}{P(s)}$, when compared with the Laplace–transformed Boltzmann integral, Equation 2.18, gives

$$E(s) = \frac{Q(s)}{s P(s)}. \tag{2.58}$$

The relaxation function $E(t)$ is obtained by inversion of the transform by partial fractions to achieve a sum of terms of the type $\frac{1}{s+a}$. Each such term, following an inverse transform, gives rise to an exponential. The transient response function associated with a linear differential equation of order $n$ is a sum of $n$ exponentials. It may be written as the following [2]:

$$E(t) = \sum_{n=0}^{N} \lambda_n e^{-t/\tau_n}. \tag{2.59}$$

such a sum is called a *Prony series*.

Differential equations, such as Equation 2.56, can be considered to arise from complicated assemblages of springs and dashpots, however, such a visualization is hardly useful because the springs and dashpots are not identified with physical features within the material.

### 2.6.2 Exponentials and Internal Causal Variables

The purpose of this segment is to show how single exponentials in the transient response can arise via the coupling between strain and a physical quantity, known as an *internal variable*, which can be shown to relax, by a flow process, when perturbed. Examples of internal variables are temperature, electric field, and pore pressure of fluid in a porous material (see Chapter 8).

Suppose that the stress $\sigma$ or strain $\epsilon$ is coupled to a field variable $\xi$, called an *internal variable*; $\Xi$ is another field variable. $J$ is a compliance and $\kappa$, $K$ and $\Psi$ are material constants. Here, $K$ represents coupling between stress and an internal variable; it is not a bulk modulus,

$$\epsilon = J\sigma + \kappa\xi \tag{2.60}$$

$$\Xi = K\sigma + \Psi\xi. \tag{2.61}$$

Assume further that the internal variable, when perturbed, approaches its equilibrium value $\xi_e$ (assumed to be zero for zero stress; $\xi_e = \sigma b$ with $b$ characteristic of the material) according to a first-order equation,

$$\frac{d\xi}{dt} = -\frac{(\xi - \xi_e)}{\tau}, \tag{2.62}$$

with $\tau$ as the relaxation time. Combining these relations to eliminate $\xi$ yields a standard linear solid, for which the relaxation and creep functions contain single exponentials. Analyses for specific kinds of internal variables are given in the discussion of relaxation mechanisms in Chapter 8.

If there are multiple internal variables, then a relaxation function as a discrete sum of exponentials is expected

$$E(t) = \sum_{n=0}^{N} \lambda_n e^{-t/\tau_n}. \tag{2.63}$$

Here, the $\lambda_n$ are weighting coefficients. One may also envisage a distribution or *spectrum* $H(\tau)$ of relaxation times $\tau$ and associated exponential terms, as follows (see §4.2)

$$E(t) - E_e = \int_{-\infty}^{\infty} H(\tau)e^{-t/\tau} d\ln\tau = \int_0^{\infty} \frac{H(\tau)}{\tau} e^{-t/\tau} d\tau. \tag{2.64}$$

### 2.6.3 Fractional Derivatives

Simple constitutive equations covering many decades of time scale have been constructed using fractional derivatives [12], which are operators that generalize the order of differentiation to fractional orders $\beta$, with $0 < \beta < 1$. The concept of fractional derivative can be visualized via the Laplace transform relation (in which $\mathcal{L}(f(t)) = F(s)$) with vanishing initial values and initial slopes of $f(t)$:

$$\mathcal{L}\{\frac{d^\beta f(t)}{dt^\beta}\} = s^\beta F(s). \tag{2.65}$$

This relation can be demonstrated for integer order via the definition of the Laplace transform. For fractional order, it may be regarded as a definition of the fractional derivative. Fractional derivatives can also be defined by the following integral representation [12]:

$$\frac{d^\beta x(t)}{dt^\beta} \equiv \frac{1}{\Gamma(1-\beta)}\frac{d}{dt}\int_0^t \frac{x(\tau)}{(t-\tau)^\beta}d\tau. \tag{2.66}$$

An element obeying a constitutive equation of the form:

$$\sigma(t) = E\tau_r^\beta \frac{d^\beta \epsilon(t)}{dt^\beta}, \tag{2.67}$$

with $E$ as a stiffness constant, $\tau_r$ as a time constant, and $\beta$ as a material parameter, can be used to obtain the relaxation function [13, 15]. For relaxation, the strain begins at time zero following a Heaviside step function $\mathcal{H}(t)$. The relaxation function is

$$E(t) = [E_1 - E_2(\frac{t}{\tau_r})^\beta]\mathcal{H}(t). \tag{2.68}$$

So, the relaxation function contains a power law in time. Terms of this type can be superposed to obtain sufficient generality to model much material behavior. It has been found [12] that only one or two fractional derivative terms in the constitutive relation suffice to describe behavior of actual materials over many decades of time scale.

### 2.6.4 Power-Law Behavior

For some materials [11, 12] the relaxation function is approximated by $E(t) = At^{-n}$. The plot of this relaxation function on a log–log scale is a straight line. The relaxation can be expressed in terms of percentage per decade. Use the above Laplace transform relations for the power law to determine $J(t)$. Taking the Laplace transform of $E(t)$,

$$E(s) = \frac{A\Gamma(-n+1)}{s^{-n+1}}, \tag{2.69}$$

but $E(s)J(s) = \frac{1}{s^2}$ (Equation 2.19) is the relation between creep and relaxation properties, from above, so,

$$J(s) = \frac{1}{s^2 E(s)} = \frac{1}{A\Gamma(-n+1)}s^{-n+1-2} = \frac{1}{A\Gamma(1-n)}\frac{1}{s^{n+1}}. \tag{2.70}$$

But

$$\frac{1}{s^{n+1}} = \frac{\mathcal{L}[t^n]}{\Gamma(n+1)}, \tag{2.71}$$

so,

$$J(t) = \frac{1}{A\Gamma(1-n)\Gamma(n+1)}t^n. \tag{2.72}$$

So, the creep function is also a power law.

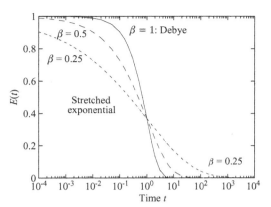

Figure 2.7. Stretched exponential relaxation functions. Arbitrary units of stiffness and time.

Andrade creep is a particular form of power law for which $J(t) = J_0 + At^{1/3}$. It was originally developed for metals at large strain but is also used to model creep in glassy polymers [14].

### 2.6.5 Stretched Exponential

Relaxation in complex materials with strongly interacting constituents often follows [16–20] the stretched exponential or KWW (after Kohlrausch, Williams, Watts) form:

$$E(t) = (E_0 - E_\infty)e^{-(t/\tau_r)^\beta} + E_\infty, \qquad (2.73)$$

with $0 < \beta \le 1$, $E_0$ and $E_\infty$ as constants and $\tau_r$ as a characteristic relaxation time. Behavior of this type is commonplace [19]. The relaxation of the stretched exponential occurs over a wider range of time on a logarithmic scale than does the exponential form ($\beta = 1$), which occurs mostly within one decade, a factor of ten in the time scale, as shown in Figure 2.7. Dynamic behavior corresponding to this relaxation function can be obtained analytically for $\beta = 0.5$ [21], but, for general values of $\beta$, numerical methods are appropriate. In viscoelastic solids, $E_\infty > 0$ and $\tan\delta$ forms a broad peak with $\tan\delta \approx \nu^{-n}$ for frequencies $\nu$ well above the peak, which can be demonstrated numerically using the transformation relations developed in §3.4. As for interrelations of the transient functions, if relaxation follows a stretched exponential, then creep will follow a different time dependence [22] unless the exponent is near 0.5.

### 2.6.6 Logarithmic Creep; Kuhn Model

Creep of many solids including rubber, glass, and concrete, has been modeled with a logarithmic time function [23]:

$$J(t) = J_A + J_B ln(t). \qquad (2.74)$$

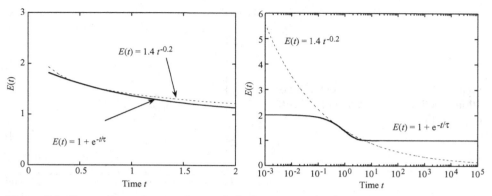

Figure 2.8. Comparison of power law and single exponential (Debye) relaxation functions (left) on a linear time scale over one decade and (right) over a logarithmic time scale over eight decades. Arbitrary units of stiffness and time.

To avoid a singularity at $t = 0$, Kuhn [24] suggested

$$\frac{dJ(t)}{dt} = \frac{J_B}{t}(1 - e^{-tC_\omega}),\tag{2.75}$$

with $J_A$, $J_B$, and $C_\omega$ as constants. Kuhn obtained the creep function as an integral with a lower limit of zero for rubbery materials in which the compliance at zero time is orders of magnitude smaller than at a later time. A modified version [23], suitable for a wider range of materials is as follows, expressed as a distribution of retardation times (see Chapter 4):

$$J(t) = J_A + J_B \int_{1/C_\omega}^{\infty} \frac{1 - e^{-t/\tau}}{\tau} d\tau.\tag{2.76}$$

The relaxation function has been evaluated for the modified Kuhn model, as an integral requiring numerical evaluation.

### 2.6.7 Distinguishing among Viscoelastic Functions

To make meaningful comparisons, it is considered helpful to use logarithmic time scales in the plotting of relaxation results. A comparison of single exponential (Debye) and power law relaxations is shown on a linear scale over one decade in the left graph in Figure 2.8 and on a logarithmic scale over eight decades in the right graph in Figure 2.8. The curves appear to be similar when plotted on a linear scale because only about one decade (a factor of 10) of time is readily viewed. They differ considerably when viewed on a logarithmic scale.

### 2.7 Effect of Temperature

The development thus far has been under the assumption of isothermal conditions. The viscoelastic functions of materials can be expected to depend on temperature $T$, as well as time $t$: the relaxation function may be written $E = E(t, T)$. Envisage a material in which the viscoelasticity arises from a molecular rearrangement process, which occurs under stress, or from a diffusion process under stress. The speed of such processes depends on the speed of molecular motion of which temperature is a measure. If all the processes contributing to the viscoelasticity of the material are accelerated to the same extent by a temperature rise, then, we have a relaxation function of the following form:

$$E(t, T) = E(\zeta, T_0), \qquad (2.77)$$

with $\zeta = \frac{t}{a_T(T)}$, in which $\zeta$ is called the *reduced time*, $T_0$ is the reference temperature, and $a_T(T)$ is called the *shift factor*. For such materials, a change in temperature stretches or shrinks the effective time scale. Since viscoelastic properties are usually plotted versus log time or log frequency, a temperature change for such materials corresponds to a horizontal *shift* of the material property curves on the log time or of log frequency axis. Materials that behave according to Equation 2.77 obey the *time–temperature superposition* and are called *thermorheologically simple* [25].

The shift factor $a_T(T)$ depends on temperature and on the material as discussed in §7.6.1. Many materials exhibit temperature dependent creep behavior that follows the *Arrhenius* relation [1],

$$\tau^{-1} = \nu_0 \exp\{-\frac{U}{kT}\}, \qquad (2.78)$$

with $\tau$ as a time constant, $\nu_0$ as a characteristic frequency, $T$ as absolute temperature, $U$ as the *activation energy*, and $k$ as the Boltzmann constant. This form gives rise to thermorheologically simple behavior with the following shift factor $a_T$ as is shown in Example 6.9.

$$\ln a_T = \frac{U}{k}\{\frac{1}{T_2} - \frac{1}{T_1}\}. \qquad (2.79)$$

The time–temperature shift for polymers tends to follow the empirical WLF equation [26] (after Williams, Landel, and Ferry [27]),

$$\log a_T = -\frac{C_1(T - T_{\text{ref}})}{C_2 + (T - T_{\text{ref}})}. \qquad (2.80)$$

in which $T_{\text{ref}}$ is the reference temperature and the logarithm is of base 10. In polymers near the glass transition temperature, $a_T(T)$ varies sufficiently rapidly with temperature that an isothermal approximation may not be warrantable in the presence of small temperature variations. The constants $C_1$ and $C_2$ depend on the particular polymer. Even so, the behavior of different amorphous polymers is sufficiently similar in normalized coordinates that Ferry has proposed the following "universal" constants, for $T_{\text{ref}}$ taken as $T_g$ the glass transition temperature: $C_1 = 17.44$, and $C_2 = 51.6$.

Figure 2.9 shows representative relaxation data over three decades of time, and for several temperatures. The curve constructed by time–temperature shifts is called a *master curve*. The following procedure is used. A temperature called the *reference*

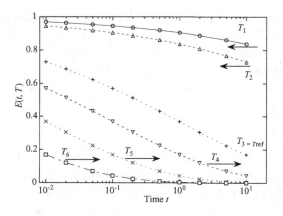

Figure 2.9. Relaxation curves over three decades of time, and for several temperatures, with $T_3 > T_2 > T_1$. Arbitrary units of stiffness and time.

*temperature*, is chosen, for which there is no shift. The master curve is generated by horizontally shifting the experimental curve for temperatures just above or below the reference temperature until they coincide. The directions of the shifts along the log-time axis in Figure 2.9 are shown by arrows. Then, the curve for the next higher or lower temperature is shifted, and so on until all the data are analyzed. If overlap cannot be achieved by this procedure, then the material is not thermorheologically simple. The master curve obtained from shifting the three decade curves in Figure 2.9 is shown in Figure 2.10. The 10-decade range of this master curve comes about as a result of the shift of the individual curves along the abscissa. For some materials, particularly polymers, a modest change in temperature can give rise to a very large change in the relaxation times, as shown in Figure 2.11. The numbers used to generate this WLF plot are representative of a Hevea rubber [26], with the simplification of assuming a single relaxation time process.

Many amorphous polymers appear to be thermorheologically simple, however crystalline polymers and most composite materials are not thermorheologically simple; the different phases each have different temperature dependencies of their viscoelastic behavior. In any material that has a multiplicity of relaxation mechanisms, thermorheological simplicity is not to be expected. Moreover if a phase

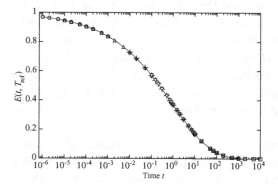

Figure 2.10. Master curve generated by shifting the above relaxation curves. Arbitrary units of stiffness and time.

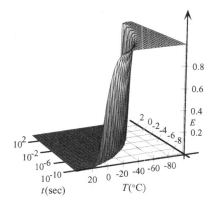

Figure 2.11. Plot of the WLF equation for shift of relaxation times, assuming $T_{\text{ref}} = -73°C$, $C_1 = 16.8$, $C_2 = 53.6$, and assuming a single exponential relaxation process.

transformation (such as melting or freezing) occurs in any part of the temperature range of interest and in any portion of the material, the material will not be thermorheologically simple. If the material burns or decomposes in part of a temperature range, the material will clearly not be thermorheologically simple.

To experimentally determine if a material is thermorheologically simple, one may perform a set of creep or relaxation tests at different temperatures, and plot the results. If the various curves can be made to overlap by horizontal shifts on the log-time axis, the material is considered to be thermorheologically simple. If the curves do not overlap, the material is not thermorheologically simple. The concept of time–temperature superposition has been of greatest use in polymers. Plazek [28] observes that given data within a fairly narrow experimental window (three decades or less) the test for thermorheological simplicity can only be definitive in its failure. The experiment is capable of demonstrating thermorheological complexity, but it cannot demonstrate simplicity. The reason is that deviations can and do appear if the material is examined over a greater number of decades of time or frequency.

## 2.8 Three-Dimensional Linear Constitutive Equation

In three dimensions, Hooke's law of linear elasticity is given by

$$\sigma_{ij} = C_{ijkl}\epsilon_{kl}, \qquad (2.81)$$

with $C_{ijkl}$ as the elastic modulus tensor, and the usual Einstein summation convention assumed in which repeated indices are summed over [29–31] as follows:

$$\sigma_{ij} = \sum_{k=1}^{3}\sum_{l=1}^{3} C_{ijkl}\epsilon_{kl}. \qquad (2.82)$$

There are 81 components of $C_{ijkl}$, but taking into account the symmetry of the stress and strain tensors, only 36 of them are independent. If the elastic solid is describable by a strain energy function, the number of independent elastic constants is reduced to 21. An elastic modulus tensor with 21 independent constants describes an anisotropic material with the most general type of anisotropy, triclinic symmetry. Anisotropy refers to dependence of properties upon direction. Materials with orthotropic symmetry are invariant to reflections in two orthogonal planes and are describable by nine elastic constants. Materials with axisymmetry, also called *transverse isotropy* or *hexagonal symmetry*, are invariant to 60 degree rotations about an axis and are describable by five independent elastic constants. Materials with cubic symmetry are describable by three elastic constants. Isotropic materials, with properties that are independent of direction, are describable by two independent elastic constants.

Examples of materials with these symmetry classes are as follows: a unidirectional fibrous material may have orthotropic symmetry if the fibers are arranged in a rectangular packing, or hexagonal symmetry if the fibers are packed hexagonally. Wood has orthotropic symmetry, while human bone has hexagonal symmetry. A composite with an equal number of fibers in three orthogonal directions is cubic. A fabric with fibers in a square array is two-dimensionally cubic. Table salt crystals are cubic. Glass, cast glassy polymers, and cast polycrystalline metals exhibit approximately isotropic behavior.

For linearly viscoelastic materials the following is obtained by conducting for each component, arguments identical to those given for one dimension:

$$\sigma_{ij}(t) = \int_0^t C_{ijkl}(t - \tau)\frac{d\epsilon_{kl}}{d\tau}d\tau. \tag{2.83}$$

As with the elastic case, this equation is sufficiently general to accommodate any degree of anisotropy. Each independent component of the modulus tensor can have a different time dependence. Many practical materials are approximately isotropic (their properties are independent of direction). For isotropic elastic materials, the constitutive equation is

$$\sigma_{ij} = \lambda\epsilon_{kk}\delta_{ij} + 2\mu\epsilon_{ij}, \tag{2.84}$$

in which $\lambda$ and $\mu$ are the two independent Lamé elastic constants, $\delta_{ij}$ is the Kronecker delta (1 if $i = j$, 0 if $i \neq j$), and $\epsilon_{kk} = \epsilon_{11} + \epsilon_{22} + \epsilon_{33}$. Engineering constants, such as Young's modulus $E$, shear modulus $G$, Poisson's ratio $v$ can be extracted from these tensorial constants. Specifically (see also Appendix §A.5) [31],

$$G = \mu, \tag{2.85}$$

$$E = \frac{G(3\lambda + 2G)}{(\lambda + G)}, \tag{2.86}$$

$$v = \frac{\lambda}{2(\lambda + G)}. \tag{2.87}$$

As for the interpretation of $C_{1111}$ for isotropic elastic materials, we note that [29, 30],

$$C_{1111} = E\frac{1 - v}{(1 + v)(1 - 2v)}. \tag{2.88}$$

See Example 5.9. Observe that $C_{1111} \neq E$, unless Poisson's ratio is zero. For comparison, the bulk modulus is

$$B = 2G \frac{1 + \nu}{3(1 - 2\nu)}. \tag{2.89}$$

In isotropic viscoelastic solids,

$$\sigma_{ij}(t) = \int_0^t \lambda(t - \tau) \delta_{ij} \frac{d\epsilon_{kk}}{d\tau} d\tau + \int_0^t 2\mu(t - \tau) \frac{d\epsilon_{ij}}{d\tau} d\tau. \tag{2.90}$$

There are two independent viscoelastic functions that can have different time dependence. As in the case of elasticity, we may consider engineering functions rather than tensorial ones, for example, relaxation functions, such as $E(t)$, $G(t) = \mu(t)$, $\nu(t)$ and $B(t)$.

## 2.9 Aging Materials

Aging materials are those in which the material properties themselves change with time. The origin of the time scale for aging is the time of creation of the material. This time may be the time of mixing in the case of concrete, the time of initiation of polymerization or the time of last solidification in the case of polymers, or the birth of an organism in the case of biological materials. In a linearly viscoelastic material that ages, the constitutive equation for one spatial dimension is

$$\epsilon(t) = \int_0^t J(t, \tau) \frac{d\sigma}{d\tau} d\tau. \tag{2.91}$$

The creep function in this case is a function of two variables, and it can be represented as a surface or as a family of curves. In some classes of material [32, 33], aging time $t_a$ slows the characteristic retardation times $\tau_c$ by a factor:

$$\tau_c = \tau_0 t_a^{\mu_{str}}, \tag{2.92}$$

in which $\mu_{str}$ is the Struik shift factor (nothing to do with the shear modulus) and $\tau_0$ is a retardation time. In such cases, the effect of aging is represented by a shift of the creep curves along the log-time axis.

We may consider a general creep function of the form $J(t, \tau)$, which allows aging, and imposes the restriction that, over a time scale of interest, the material properties do not depend on time. So, the creep function is unchanged by a shift on the time axis. This corresponds to, $J(t + t_0, \tau + t_0) = J(t, \tau)$, or $J(t, \tau) = J(t - \tau)$ [34]. This is the form for nonaging materials developed in §2.2.

## 2.10 Dielectric and Other Forms of Relaxation

Materials have other physical properties, not only mechanical properties. For example [35], the strain $\epsilon_{ij}$ can depend on the stress $\sigma_{kl}$ via the elastic compliance $J_{ijkl}$, on the electric field $\mathcal{E}_k$ (in piezoelectric materials with modulus tensor $d_{kij}$ at constant temperature), and on the temperature change $\Delta T$ (in most materials, via the thermal expansion $\alpha_{ij}$). Moreover, the electric displacement vector $\mathcal{D}_i$ depends on electric field via $K_{ij}$, which is the dielectric tensor at constant stress and temperature, and in some materials it can depend on temperature change $\Delta T$ via the pyroelectric effect; $p_i$ is the pyroelectric coefficient at constant stress. These phenomena are incorporated in the following linear constitutive equations (without relaxation):

$$\epsilon_{ij} = J_{ijkl}\sigma_{kl} + d_{kij}\mathcal{E}_k + \alpha_{ij}\Delta T, \tag{2.93}$$

$$\mathcal{D}_i = d_{ijk}\sigma_{jk} + K_{ij}\mathcal{E}_j + p_i\Delta T. \tag{2.94}$$

Here, the usual Einstein summation convention over repeated subscripts is used. The above equations are linear, and they incorporate no relaxation.

Relaxation can occur in these physical properties in real materials. The linear constitutive equations with relaxation can be constructed as follows by using the same arguments given above, for each tensor element in the above relations:

$$\epsilon_{ij}(t) = \int_0^t J_{ijkl}(t-\tau)\frac{d\sigma_{kl}}{d\tau}d\tau + \int_0^t d_{kij}(t-\tau)\frac{d\mathcal{E}_k}{d\tau}d\tau + \int_0^t \alpha_{ij}(t-\tau)\frac{d\Delta T}{d\tau}d\tau, \tag{2.95}$$

$$\mathcal{D}_i(t) = \int_0^t d_{ijk}(t-\tau)\frac{d\sigma_{jk}}{d\tau}d\tau + \int_0^t K_{ij}(t-\tau)\frac{d\mathcal{E}_j}{d\tau}d\tau + \int_0^t p_i(t-\tau)\frac{d\Delta T}{d\tau}d\tau. \tag{2.96}$$

Time dependence of $K_{ij}$ is referred to as dielectric relaxation. Relaxation can also occur in the magnetic properties of materials, in optical properties, and in strain-optical retardation [36]. Electrical and magnetic properties, as well as viscoelastic properties (Chapter 3), can also be described in the frequency domain.

## 2.11 Adaptive and "Smart" Materials

Materials can be time dependent in ways other than viscoelastic response. For example, a structural member can gradually become stiffer and stronger if additional substance is added in response to heavy loading. It can become less dense and weaker if material is removed in response to minimal loading. Biological materials, such as bone [37–39], muscle and tendon behave in this way. Such behavior is called *adaptive behavior*. Constitutive equations have been developed for the adaptive elasticity of bone [40] [41]. The stress strain relation is a modification of Hooke's law in that the proportionality between the stress $\sigma$ and the strain $\epsilon$ is dependent on the volume fraction of material present. The change in stiffness of the bone is assumed to depend on its porosity:

$$\sigma_{ij} = (\xi_0 + e)C_{ijkl}(e)\epsilon_{kl}, \tag{2.97}$$

Specifically, a change $e$ in the solid volume fraction of a porous material with respect to a reference volume fraction $\xi_0$ drives the adaptation process:

$$\frac{de}{dt} = a(e) + A_{ij}(\epsilon_{ij}). \tag{2.98}$$

Here, $a$ and $A$ are coefficients that depend on the type of bone. This formulation does not account for the viscoelastic response of bone. Conventional engineering materials, such as steel, aluminum, concrete and polymers, are *not* adaptive.

*Smart* materials are those that can respond in an active way to mechanical loads. A smart material may contain embedded sensors that provide information concerning the strain, temperature, or degree of damage in the material [42]. Sensor information can inform the user that an overload condition is present or that repair is necessary. To achieve true smart response, the sensor data are input to a control system, which may contain a microprocessor that controls actuators within the material to reduce the deformation or repair the damage. Sensors and actuators are discussed in the context of experimental methods in §6.4. For example, piezoelectric materials generate an electrical signal when strained and also deform in response to electrical excitation. Piezoelectric materials have been used both as sensors and as actuators. Another class of actuator material is electrorheological fluids, which have a viscosity that can be controlled by an electric field [43]. They are fluid-solid composites. A representative electrorheological fluid contains particles of cornstarch dispersed in silicone oil.

Ideally, a smart material would emulate such biological characteristics as self-repair, adaptability to conditions, self-assembly, homeostasis, and the capacity for regeneration. Thus far, no synthetic material has such characteristics. Currently, there are few examples of smart materials; the subject is largely a research topic. Some examples are given in §10.6.

## 2.12 Effect of Nonlinearity

In this section, constitutive equations are considered other than that of linear viscoelasticity. The aim is to gain the ability to distinguish between several different types of mechanical response; experimental aspects are discussed in §6.2. We remark that a linear material obeys linear integral equations or linear differential equations. For a material to be linear, it is necessary, but not sufficient, for the measured creep or relaxation function to be independent of stress or strain.

### 2.12.1 Constitutive Equations

*Simple Nonlinear Elasticity and Flow*

Consider first elastic behavior. Linear elasticity is expressed by the elementary Hooke's law in one dimension and Equation 2.81 in three dimensions. In one dimension, nonlinear elasticity may be expressed as

$$\sigma = f(\epsilon)\epsilon, \qquad (2.99)$$

with $f(\epsilon)$ allowed to be a nonlinear function of strain. Time is not involved in either linear or nonlinear elasticity, so there is no creep or relaxation and the material recovers fully and instantaneously, and the path for unloading is identical to the

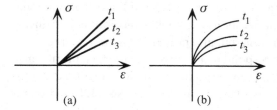

Figure 2.12. Stress $\sigma$ versus strain $\epsilon$ at constant time (isochronals) for linear and nonlinear materials. In linear materials (a), the isochronals are straight lines, and in nonlinear materials (b), the isochronals are curved.

path for loading. Isochronals, or data at constant time, taken from creep curves, all coincide in an elastic material, whether it be linear or nonlinear.

Consider an elastoplastic material. At sufficiently small strain the material behaves elastically, so there is no time-dependent creep or relaxation behavior, and the material recovers fully and instantaneously. If the yield point is exceeded, there is still no time dependent creep or relaxation behavior because time is not included in elastoplasticity. However, recovery is incomplete: there is some residual strain and the path for unloading differs from the path for loading. The residual strain is constant in time.

Nonlinear viscoelasticity gives rise to curved isochronals in the stress–strain diagram in Figure 2.12. Several constitutive equations are available for the modeling of nonlinear viscoelasticity. The simplest of these are restricted to describing creep. A simple equation that is commonly used is the Bailey–Norton relation intended to model primary and secondary creep:

$$\epsilon(t, \sigma) = A\sigma^m t^n. \tag{2.100}$$

Creep formulations of this type do not account for recovery or history effects.

### Nonlinear Superposition and QLV

The following simple, nonlinear relation allows for prediction of history dependence. This single-integral form is called *nonlinear superposition*, which allows the relaxation function to depend on strain level:

$$\sigma(t) = \int_0^t E(t - \tau, \epsilon(\tau)) \frac{d\epsilon}{d\tau} d\tau. \tag{2.101}$$

A similar equation may be written for stress-dependent creep in the compliance formulation.

$$\epsilon(t) = \int_0^t J(t - \tau, \sigma(\tau)) \frac{d\sigma}{d\tau} d\tau. \tag{2.102}$$

If a series of relaxation tests is done at different strain levels, relaxation will be observed, but the functional form of the relaxation curves will depend on the strain level. It is also possible to consider a separable kernel, such as $E(t, \epsilon) = E(t) f(\epsilon)$, in which the kernel is a product of a time-dependent part $E(t)$ and a strain-dependent part $f(\epsilon)$. This is called *quasilinear viscoelasticity* (QLV), originally proposed by

Fung [44]. QLV has been used to model the behavior of soft tissues [44]. Nonlinear superposition, Equation 2.101 describes a specific kind of nonlinearity. There is no sensitivity to multiple steps. QLV is even more specific in that the shape of the relaxation curves is independent of strain level. Real materials may not, in fact, be so simple.

*Other Single Integral Forms: BKZ and Schapery*

A single integral nonlinear relation, which has been used for rubbery materials, is the BKZ relation of Bernstein, Kearsley, and Zapas [48]:

$$\sigma(t) = \int_0^t A(t - \tau) f(\epsilon_g(\tau)) d\tau, \tag{2.103}$$

in which $A(t)$ is a function and $\epsilon_g$ is a generalized strain measure appropriate for large deformation. For elastomers the relaxation behavior takes the form:

$$\frac{\sigma(t)}{\lambda^2 - \lambda^{-1}} = (\lambda^2 - 1)\{\frac{1}{2} A_1(t) + A_2(t)\} + \frac{1}{\lambda}\{A_1(t) + A_2(t)\} + A_3 - A_1(t), \tag{2.104}$$

in which $\lambda$ is the extension ratio defined as the ratio of deformed length to initial length; $A_1(t)$ and $A_2(t)$ are functions of time, and $A_3$ is a constant. A strain measure appropriate for finite (large) deformations, in terms of the extension ratio, is [14]:

$$\epsilon_g = \frac{1}{2}(\lambda^2 - 1). \tag{2.105}$$

Again, Equation 2.104 describes a specific kind of nonlinearity. A creep formulation used by Schapery [50] was derived using principles of irreversible thermodynamics. There is a stress-induced shift in the time scale, as follows:

$$\epsilon(t) = J_0 g_0 \sigma(t) + g_1 \int_0^t J(\zeta(t) - \zeta_\tau(\tau)) \frac{dg_2 \sigma(\tau)}{d\tau} d\tau. \tag{2.106}$$

Here, $J_0$ is a time-independent compliance, $J$ is a creep compliance, which depends on stress indirectly through the reduced time variables $\zeta(t)$ and $\zeta_\tau(\tau)$ defined below, and $g_0, g_1, g_2$, are material properties that are functions of stress

$$\zeta(t) = \int_0^t \frac{d\xi}{a_\sigma[\sigma(\xi)]}, \tag{2.107}$$

$$\zeta_\tau(\tau) = \int_0^\tau \frac{d\xi}{a_\sigma[\sigma(\xi)]}. \tag{2.108}$$

The quantity $a_\sigma[\sigma(\xi)]$ is a shift factor that depends on stress, which in turn depends on time. This formulation, because it contains five functions and a constant, is quite adaptable [49].

*Multiple Integral Forms*

More general equations for nonlinear viscoelasticity have been proposed [51–54]. In one dimension for small strain, in the modulus formulation,

$$\sigma(t) = \int_0^t E_1(t-\tau)\frac{d\epsilon}{d\tau}d\tau + \int_0^t \int_0^t E_2(t-\tau_1, t-\tau_2)(\frac{d\epsilon}{d\tau_1})(\frac{d\epsilon}{d\tau_2})d\tau_1 d\tau_2 + \cdots.$$

(2.109)

In the compliance formulation, one may write:

$$\epsilon(t) = \int_0^t J_1(t-\tau)\frac{d\epsilon}{d\tau}d\tau + \int_0^t \int_0^t J_2(t-\tau_1, t-\tau_2)(\frac{d\sigma}{d\tau_1})(\frac{d\sigma}{d\tau_2})d\tau_1 d\tau_2 + \cdots. \quad (2.110)$$

Here, $\tau_1$ and $\tau_2$ are time variables of integration. To experimentally distinguish materials that obey nonlinear constitutive equations, single creep or relaxation tests are usually inadequate. Such equations predict relaxation behavior, which can depend on strain level. They may however be distinguished by the use of strain histories containing more than one step. For example, a relaxation and recovery test contains two steps, and similarly for a creep and recovery test. In linear materials and in materials obeying nonlinear superposition, initial recovery can be predicted from the time and strain dependence of the preceding relaxation or form of the time and stress dependence of creep. In materials obeying multiple integral formulations, such as Equation 2.109, recovery can differ significantly from relaxation since the multiple integrals generate interaction terms between time steps. The kernels $E_1$, $E_2$, $E_3 \cdots$ in the higher order terms in Equation 2.109 can be extracted by performing a series of experiments with multiple steps at different strain levels and with different time lapses between the steps. Such experiments (§6.2) are laborious. The multiple integral formulation [53, 54] of nonlinear viscoelasticity is applicable to a broad range of nonlinear systems in addition to viscoelastic materials.

### 2.12.2 Creep–Relaxation Interrelation: Nonlinear

Nonlinearly viscoelastic materials cannot be analyzed via Laplace transforms because the Laplace transform is a linear operator. The interrelation between creep and relaxation is developed here by direct construction. Write the time-dependent strain due to a constant stress $\sigma_c$ as a sum of immediate and delayed Heaviside step functions in time $\mathcal{H}(t)$,

$$\epsilon(t) = \epsilon(0)\mathcal{H}(t) + \sum_{i=0}^N \Delta\epsilon_i \mathcal{H}(t - t_i).$$

(2.111)

Based on the definition of the relaxation function each step strain in the summation (top) gives rise to a relaxing component of stress (Figure 2.13, bottom). Consider single-integral type nonlinear response and exclude response which must be

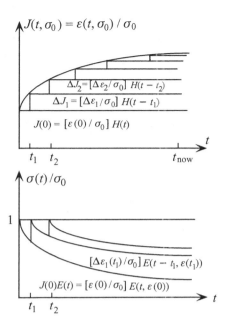

Figure 2.13. Top: decomposition of a creep function as a sum of immediate and delayed Heaviside step functions in time $t$. Bottom: constant stress as a corresponding sum of relaxing components for a solid obeying nonlinear superposition.

describable by a multiple integral formulation. Hence, assume there is no effect from the interaction between the step components:

$$\sigma_c = \epsilon(0)E(t, \epsilon) + \sum_{i=0}^{N} \Delta\epsilon_i E(t - t_i, \epsilon). \tag{2.112}$$

Divide by $\sigma_c$, and use the definition of the creep compliance,

$$1 = J(0)E(t, \epsilon) + \sum_{i=0}^{N} \Delta J_i E(t - t_i, \epsilon(t_i)). \tag{2.113}$$

Pass to the limit of infinitely many fine step components to obtain a Stieltjes integral, with $\tau$ as a time variable of integration:

$$1 = J(0)E(t, \epsilon) + \int_0^t E(t - \tau, \epsilon(\tau))\frac{dJ(\tau, \sigma_c)}{d\tau}d\tau. \tag{2.114}$$

The relationship is implicit and is analogous to the one for the linear case. To develop an explicit form, assume a specific form for the creep behavior.

First, assume the creep function to be separable into a stress-dependent portion and a power law in time. Such a form, called *quasi-linear viscoelasticity* (QLV), has been used widely in biomechanics; the interrelation has been analyzed in the context of interpretation of experiments with ligaments [45]:

$$J(t, \sigma) = g(\sigma)t^n = [g_1 + g_2\sigma + g_3\sigma^2 + \cdots]t^n. \tag{2.115}$$

Here, $g_1, g_2, g_3$ are constants. Consider the following relaxation function, with $f_1, f_2, f_3$ are constants, as a trial solution. A separable form does not give rise to a solution;

$$E(t, \epsilon) = f_1 t^{-n} + f_2 \epsilon(t) t^{-2n} + f_3 \epsilon(t)^2 t^{-3n} + \cdots. \tag{2.116}$$

The creep strain is $\epsilon(t) = J(t)\sigma_c = \epsilon_0 t^n$, with $\epsilon_0$ as the strain amplitude. Substitute Equations 2.115 and 2.116 into Equation 2.114, and recognize that $J(0)$ for the power law vanishes,

$$1 = \int_0^t [f_1(t - \tau)^{-n} + f_2 \epsilon_0 (t - \tau)^n (t - \tau)^{-2n}$$

$$+ f_3 \epsilon_0^2 (t - \tau)^{2n} (t - \tau)^{-3n} + \cdots]$$

$$[g_1 + g_2 \sigma + g_3 \sigma^2 + \cdots] n \tau^{n-1} d\tau. \tag{2.117}$$

Factor the stress-dependent and time-dependent portions,

$$1 = [g_1 + g_2 \sigma + g_3 \sigma^2 + \cdots]$$

$$[f_1 + f_2 \epsilon_0 + f_3 \epsilon_0^2 + \cdots]$$

$$\int_0^t n(t - \tau)^{-n} \tau^{n-1} d\tau. \tag{2.118}$$

The integral portion gives results ($\frac{1}{n\pi} \sin n\pi$) identical to the linear case by Laplace transformation of the integral and an identity involving the gamma function. The stress-dependent portion is related to the strain dependent portion by inversion of a power series [46]:

$$f_1 = 1/g_1$$

$$f_2 = -g_2/g_1^3$$

$$f_3 = (2g_2^2 - g_1 g_3)/g_1^5. \tag{2.119}$$

Although the creep behavior was assumed separable, the corresponding relaxation behavior is not separable. Therefore, if creep can be modeled with QLV then relaxation will not obey QLV but will require a more general nonlinear model for its description.

Interrelations between creep and relaxation for materials that obey nonlinear superposition with a nonseparable kernel are derived using similar methods [47]. For example, if

$$J(t, \sigma) = g_1 t^n + g_2 t^m, \tag{2.120}$$

then,

$$E(t, \epsilon) \approx f_1 t^{-n} + f_2 \epsilon(t) t^{-q}, \tag{2.121}$$

with (in which $\Gamma$ is the gamma function)

$$\epsilon(t) = g_1 \sigma t^n + g_2 \sigma^2 t^m$$

$$f_1 = [1/g_1][\sin n\pi / n\pi]$$

$$q = 3n - m$$

$$f_2 = \frac{-f_1 g_2 m \Gamma(-n+1)\Gamma(m)}{n g_1^2 \Gamma(-2n+m+1)\Gamma(n)}. \tag{2.122}$$

So, simple nonlinear creep functions containing power-law terms can be related to a relaxation function containing power law terms, provided the material obeys a single-integral, nonlinear superposition, constitutive equation.

## 2.13 Summary

Constitutive equations for viscoelastic materials have been developed. The concept of linearity, as embodied in the Boltzmann superposition principle, was used to obtain an integral equation, the Boltzmann superposition integral, as the constitutive equation for linear viscoelasticity. Manipulations of such equations are facilitated by the Laplace transform. We find that in linearly viscoelastic materials, the stress–strain curve for constant strain rate is not a straight line. Other types of experiment, such as creep or relaxation, are usually more informative. A brief consideration of simple causes for relaxation led to single exponential creep and relaxation functions and to the concept of spectra.

## 2.14 Examples

**Example 2.1**

Suppose the stress follows a step function in time, $\sigma_0 \mathcal{H}(t)$. Obtain the strain history using the Boltzmann integral and delta functions.

**Solution**

The step occurs at zero, therefore, to include the response to the full stress history, the time scale for integration must begin prior to zero:

$\epsilon(t) = \int_{-\infty}^{t} J(t-\tau)\frac{d\sigma(\tau)}{d\tau}d\tau = \sigma_0 \int_{-\infty}^{t} J(t-\tau)\delta(\tau)d\tau = \sigma_0 J(t).$

The last expression follows via the sifting property of the delta function. Use of the Dirac delta function permits one to handle the singularity within the integral to obtain the creep response.

**Example 2.2**

Suppose that the stress follows an impulse function in time, $A\,\delta(t)$. Obtain the strain history using the Boltzmann integral.

**Solution**

Substituting the delta, $\epsilon(t) = \int_{-\infty}^{t} J(t-\tau)\frac{d\sigma(\tau)}{d\tau}d\tau = A\int_{-\infty}^{t} J(t-\tau)\frac{d\delta(\tau)}{d\tau}d\tau = A\frac{dJ(t)}{dt}$. To understand the sifting, recognize that the derivative of the delta is called the *doublet* (Appendix §A.1.4),

$$\frac{d\delta(\tau)}{d\tau} = \psi.$$

The doublet sifts the negative of the derivative, as follows:

$$\int_{-\infty}^{\infty} f(x)\frac{d\delta(x-a)}{dx}dx = -\frac{df(x)}{dx}\big|_{x=a}.$$

To demonstrate the sign, perform the substitution $\mathbf{T} = t - \tau$, $d\mathbf{T} = -d\tau$, so

$$\epsilon(t) = A\int_{-\infty}^{t} J(t-\tau)\frac{d\delta(\tau)}{d\tau}d\tau = A\int_{t-\mathbf{T}=-\infty}^{t-\mathbf{T}=t} J(\mathbf{T})\frac{d\delta(t-\mathbf{T})}{(-d\mathbf{T})}(-1)d\mathbf{T}. \qquad (2.123)$$

Moreover, the doublet is an odd function so $\psi(-x) = -\psi(x)$, so,

$$\epsilon(t) = A\frac{dJ(t)}{dt}. \qquad (2.124)$$

Therefore, the doublet sifts the derivative of the creep function from the integral.

**Example 2.3**

Suppose a material is subjected to zero stress and strain for time $t < $ zero, followed by a constant strain $\epsilon_0$ for $0 < t < t_1$, followed by zero stress for $t > t_1$. Obtain an expression for the stress and strain for all times and consider the particular case of the strain at infinite time. The stress and strain for $t < 0$ are given as zero. For $0 < t < t_1$, the strain is constant, so that stress relaxation occurs, and the stress is $\sigma(t) = \epsilon_0 E(t)$.

**Solution**

To evaluate strain for $t > t_1$, take a Laplace transform of the Boltzmann integral:

$$\sigma(t) = \int_0^t E(t-\tau)\frac{d\epsilon(\tau)}{d\tau}d\tau. \qquad (2.125)$$

$$\frac{\sigma(s)}{s} = E(s)\epsilon(s), \qquad (2.126)$$

so,

$$\int_0^t \sigma(\tau)d\tau = \int_0^t \epsilon(\tau)E(t-\tau)d\tau. \qquad (2.127)$$

Substitute $\sigma(t) = \epsilon_0 E(t)$ to obtain for $t > t_1$,

$$\epsilon_0 \int_0^{t_1} E(t)d\tau = \int_0^t \epsilon(\tau)E(t-\tau)d\tau. \qquad (2.128)$$

This is an implicit form for the strain. It may be deconvoluted numerically if needed. To obtain the limit for long time, recognize that

$$\int_0^t \epsilon(\tau)E(t-\tau)d\tau = \int_0^t \epsilon(t-\tau)E(\tau)d\tau \qquad (2.129)$$

$$\frac{\epsilon(\infty)}{\epsilon_0} = \frac{\int_0^{t_1} E(\tau)d\tau}{\int_0^\infty E(\tau)d\tau}. \qquad (2.130)$$

The denominator on the right may be thought of as an asymptotic viscosity $\eta_{\text{asymp}}$ for the following reason. This viscosity is infinite for an elastic solid or a viscoelastic solid in contrast to a viscoelastic liquid. Because viscosity is defined as the ratio of stress to strain rate, consider a constant strain rate history with a strain rate $\dot\epsilon$.

$$\sigma(t) = \int_0^t E(t-\tau)\frac{d\epsilon(\tau)}{d\tau}d\tau = \dot\epsilon \int_0^t E(t-\tau)d\tau. \qquad (2.131)$$

For long time, the ratio of stress to strain rate, hence, the asymptotic viscosity, is

$$\eta_{\text{asymp}} = \frac{\sigma(t)}{\dot\epsilon} = \int_0^\infty E(t-\tau)d\tau. \qquad (2.132)$$

We remark that Hopkins and Kurkjian [55] discussed this problem in the context of asymptotic viscosity in glass fibers; experiments were adduced in which glass fibers were rolled about a cylinder for several months, then unwound. Their curvature persisted for years.

**Example 2.4**
Suppose a material is subjected to zero stress and strain for time $t <$ zero, followed by a constant strain $\epsilon_0$ for $0 < t < t_1$, followed by zero stress for $t > t_1$. Find an explicit expression for the strain for $t \gg t_1$.
**Solution**
The stress history, for $t \gg t_1$, may be regarded as a pulse $\sigma(t) = k\delta(t)$, with $k = \sigma_{\text{avg}}t_1$ in which

$$\sigma_{\text{avg}} = \frac{1}{t_1}\int_0^{t_1}\sigma(\tau)d\tau = \frac{1}{t_1}\epsilon_0\int_0^{t_1}E(\tau)d\tau, \qquad (2.133)$$

since during $0 < t < t_1$, we have stress relaxation. The response to a pulse stress is $\epsilon(t) = k\frac{dJ(t)}{dt}$ from Example 2.2, so,

$$\epsilon(t) = \epsilon_0\int_0^{t_1}E(\tau)d\tau\frac{dJ(t)}{dt}, \qquad (2.134)$$

for $t \gg t_1$.

To evaluate asymptotic behavior as time becomes large, we can decompose the creep function into the sum of a transient, a constant, and a linear function of time:

$$J(t) = J_{\text{trans}}(t) + J(\infty) + t/\eta_{\text{asymp}}, \qquad (2.135)$$

so, for a sufficiently long time after the transient has effectively vanished, $dJ(t)/dt$ approaches $1/\eta_{asymp}$, so

$$\epsilon(t \to \infty) \approx \frac{k}{\eta_{asymp}} = \frac{\epsilon_0}{\eta_{asymp}} \int_0^{t_1} E(\tau) d\tau. \qquad (2.136)$$

This result is identical with the long time limit in Example 2.3. For a viscoelastic solid, $1/\eta_{asymp} = 0$, so only the transient term contributes to the time dependence.

### Example 2.5
Show that $J(0)E(t) + \int_0^t E(t - \tau) \frac{dJ(\tau)}{d\tau} d\tau = 1$.
### Solution
Take Laplace transformations of the Boltzmann integrals for creep and relaxation Equations 2.7 and 2.8, to obtain:

$$\epsilon(s) = sJ(s)\sigma(s), \sigma(s) = sE(s)\epsilon(s). \qquad (2.137)$$

Substituting, $s^2 J(s)E(s) = 1$, or

$$sJ(s)E(s) = \frac{1}{s}. \qquad (2.138)$$

Taking the inverse Laplace transform,

$$J(0)E(t) + \int_0^t E(t - \tau)\frac{dJ(\tau)}{d\tau} d\tau = 1, \qquad (2.139)$$

as desired.

### Example 2.6
Show that $J(t)E(t) \leq 1$.
### Solution
Begin with the exact result obtained above,

$$1 = J(0)E(t) + \int_0^t E(t - \tau)\frac{dJ(\tau)}{d\tau} d\tau, \qquad (2.140)$$

so,

$$1 \geq J(0)E(t) + \int_0^t E(t)\frac{dJ(\tau)}{d\tau} d\tau, \qquad (2.141)$$

since for $\tau > 0$,

$$E(t - \tau) > E(t), \qquad (2.142)$$

since $E$ is a decreasing function.

$$1 \geq J(0)E(t) + E(t)[J(t) - J(0)] = E(t)J(t), \qquad (2.143)$$

since $E$ is a decreasing function. So,

$$J(t)E(t) \leq 1, \tag{2.144}$$

as desired.

**Example 2.7**
Consider the following constitutive equations with $A, B, m$, and $n$ as constants. Discuss the physical meaning of the equations:

$$\sigma(t) = \int_0^\infty E(t - \tau) \frac{d\epsilon(\tau)}{d\tau} d\tau, \tag{2.145}$$

$$\sigma(t) = \int_0^t [(t - \tau)^{-n} A + (t - \tau)^{-m} B \epsilon(t - \tau)] \frac{d\epsilon(\tau)}{d\tau} d\tau. \tag{2.146}$$

**Solution**
For Equation 2.145 the integral includes times greater than $t$, so that the material is responding to strains at times in the future. Such behavior is called *noncausal*, and has not been physically observed. By contrast the Boltzmann integral represents causal behavior.

For 2.146 the modulus function in the [ ] brackets depends on $(t - \tau)$ only, not an arbitrary function of $t$ and $\tau$, so the material is nonaging. The second term confers strain dependence, so that the material is a particular type of nonlinearly viscoelastic solid.

**Example 2.8**
Determine the strain response to a stress history that is triangular in time:
$\sigma(t) = 0$ for $t < 0$,
$\sigma(t) = (\sigma_0/t_1)t$, for $0 < t < t_1$,
$\sigma(t) = 2\sigma_0 - (\sigma_0/t_1)t$, for $t_1 < t < 2t_1$, and
$\sigma(t) = 0$ for $2t_1 < t < \infty$.
   Plot $\epsilon(t)$ and $\sigma(\epsilon)$ for the particular creep function $J(t) = J_0(1 - e^{-t/\tau_c})$.

**Solution**
Substitute in the Boltzmann integral. The slopes are piecewise constant.
$\epsilon(t) = 0$, for $t < 0$,
$\epsilon(t) = \frac{\sigma_0}{t_1} \int_0^t J(t - \tau) d\tau$, for $0 < t < t_1$,
$\epsilon(t) = \frac{\sigma_0}{t_1}[\int_0^{t_1} J(t - \tau) d\tau - \int_{t_1}^t J(t - \tau) d\tau]$, for $t_1 < t < 2t_1$, and
$\epsilon(t) = \frac{\sigma_0}{t_1}[\int_0^{t_1} J(t - \tau) d\tau - \int_{t_1}^{2t_1} J(t - \tau) d\tau]$, for $2t_1 < t < \infty$.
   To generate a plot for the given creep function, substitute the given creep function and decompose the exponential as follows:
$\int_a^b [1 - \exp(-\frac{t-\tau}{\tau_c})] d\tau = b - a - \tau_c[\exp[-\frac{t-b}{\tau_c}] - \exp[-\frac{t-a}{\tau_c}]]$.
   The strain history for $\tau_c = 10$ time units is shown on the left in Figure 2.14 and the stress–strain relation on the right.

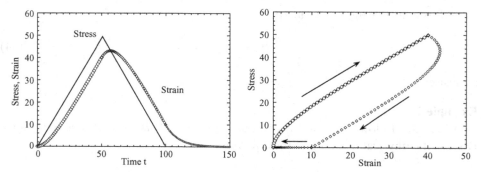

Figure 2.14. Left: Strain history $\epsilon(t)$ due to a triangular pulse of stress, assuming $J(t) = J_0(1 - e^{-t/10})$. Right: Stress versus strain due to a triangular pulse of stress.

## Example 2.9

Obtain the creep compliance for the Voigt model from the differential equation via Laplace transforms.

## Solution

For the Voigt model, the differential equation is

$\sigma = E\epsilon + \eta\frac{d\epsilon}{dt}$.

Take a Laplace transform,

$\sigma(s) = E\epsilon(s) + \eta s\epsilon(s)$. But the time constant is $\tau = \eta/E$, so

$\sigma(s) = E\epsilon(s)(1 + \tau s)$. But $J(s)E(s) = 1/s^2$ from Equation 2.19, so,

$$J(s) = \frac{1}{sE(1 + \tau s)},\tag{2.147}$$

$J(s) = \frac{1}{E}\frac{\frac{1}{\tau}}{s(s+\frac{1}{\tau})} = \frac{1}{E}\frac{s-s+\frac{1}{\tau}}{s(s+\frac{1}{\tau})} = \frac{1}{E}[\frac{s+\frac{1}{\tau}}{s(s+\frac{1}{\tau})} - \frac{s}{s(s+\frac{1}{\tau})}] = \frac{1}{E}[\frac{1}{s} - \frac{1}{(s+\frac{1}{\tau})}]$.

Take the inverse transform,

$J(t) = \frac{1}{E}(1 - e^{-t/\tau})$.

An alternate solution may be developed by constructing $sJ(s)$ from Equation 2.147 for $J(s)$.

$sJ(s) = \frac{1}{E\tau}\frac{1}{(s+\frac{1}{\tau})}$,

so, following an inverse transform,

$\frac{dJ(t)}{dt} = \frac{1}{E\tau}e^{-t/\tau}$.

Integrating, and recognizing, from the Voigt model that the limiting compliance at infinite time is $1/E$,

$J(t) = \frac{1}{E}(1 - e^{-t/\tau})$, as above.

## Example 2.10

Why bother with integral equations when differential equations are more familiar to most people?

## Answer

It is indeed possible to describe viscoelastic behavior with differential equations. The drawback is that a differential description of material behavior over a wide

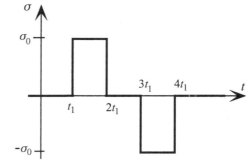

Figure 2.15. Given stress history.

range of time or frequency usually requires differential equations of high order and containing many terms. Integral equations are simpler for such a situation.

**Example 2.11**
Find the stress rate corresponding to a constant strain-rate history beginning at time zero. Use Laplace transforms.
**Solution**
The strain history is

$$\epsilon(t) = \dot{\epsilon} t \mathcal{H}(t). \tag{2.148}$$

Take a Laplace transform,

$$\epsilon(s) = \dot{\epsilon}/s^2, \tag{2.149}$$

but the Laplace transform of the Boltzmann integral is $\sigma(s) = sE(s)\epsilon(s)$, so $\sigma(s) = \dot{\epsilon} E(s)/s$, so,

$$s\sigma(s) = \dot{\epsilon} E(s). \tag{2.150}$$

Transforming back,

$$\frac{d\sigma}{dt} = \dot{\epsilon} E(t) \tag{2.151}$$

as was found by evaluation of the integral.

**Example 2.12**
Suppose a linearly viscoelastic material is subjected to the following stress history
$\sigma(t) = 0$, for $t < t_1$,
$\sigma(t) = \sigma_0$, for $t_1 < t < 2t_1$,
$\sigma(t) = 0$, for $2t_1 < t < 3t_1$,
$\sigma(t) = -\sigma_0$, for $3t_1 < t < 4t_1$, and
$\sigma(t) = 0$, for $4t_1 < t < \infty$.

Determine the strain response in general. For $t \gg 4t_1$, develop an approximate simple solution for the strain response.

## Solution

Write the stress history as a sum of Heaviside step functions: $\sigma(t) = \sigma_0[\mathcal{H}(t - t_1) - \mathcal{H}(t - 2t_1) - \mathcal{H}(t - 3t_1) + \mathcal{H}(t - 4t_1)]$.

Apply the Boltzmann superposition principle. If the stress follows a step function in time, the strain is proportional to the creep compliance. Because the stress is written as a sum of step functions, the strain follows a sum of creep curves,

$$\epsilon(t) = \sigma_0[J(t - t_1) - J(t - 2t_1) - J(t - 3t_1) + J(t - 4t_1)],$$

with the creep function understood to have a value of zero for a negative argument.

To obtain an approximate solution for $t > 4t_1$, rewrite this, with shifted time variables $t' = t - 2t_1$, $t'' = t - 4t_1$, using the definition of the derivative, as

$$\epsilon(t) = \sigma_0 t_1 \left[ \frac{J(t'+t_1)-J(t')}{t_1} - \frac{J(t''+t_1)-J(t'')}{t_1} \right].$$

The primes here represent shifted variables, not derivatives. In the limit as $t_1 \to 0$ (compared with $t$),

$$\epsilon(t) \approx \sigma_0 t_1 \left[ \frac{dJ(t')}{dt'} - \frac{dJ(t'')}{dt''} \right] = \sigma_0 t_1 \left[ \frac{dJ(t''+2t_1)}{dt'} - \frac{dJ(t'')}{dt''} \right]$$

$$= \sigma_0 t_1^2 \left[ \frac{1}{2t_1} \left[ \frac{dJ(t''+2t_1)}{dt''} - \frac{dJ(t'')}{dt''} \right] \right].$$

$$\epsilon(t) \approx \sigma_0 2 t_1^2 \frac{d^2 J(t-4t_1)}{dt^2}.$$

As $t_1 \to 0$ in comparison with $t$, $J(t - 4t_1) \to J(t)$. Observe that the sifting property of the doublet functional can also be used to obtain a second derivative approximation.

## Example 2.13

Why not just substitute $E(t)$ for the elastic modulus $E$ to obtain the linear viscoelastic response?

## Answer

When we assume linearity, superposition can be applied to segments of the strain history, giving rise to a convolution equation, so the suggested substitution contradicts linearity.

## Example 2.14

What is the distinction between relaxation and recovery?

## Answer

These words are often used interchangeably in informal language, but they have very different meaning in viscoelasticity. Relaxation is the stress response in time to an applied step function strain in time. Recovery is the response following the removal of an applied stress or strain.

## Example 2.15

Does a residual strain after deformation always involve plasticity?

## Answer

Residual strain can also occur in a linearly viscoelastic fluid. Such a material can appear solid if the dominant relaxation times are longer than the longest time of observation. In contrast to plasticity, there is no threshold stress in linear viscoelasticity.

## 2.15 Problems

2.1. Predict the strain response of a linearly viscoelastic material to a stress, which is approximated as a doublet (derivative of a delta function) in time, assuming any needed material properties are known.

2.2. A linearly viscoelastic material is subjected to the following history: for $t < 0$, $\sigma = 0, \epsilon = 0$; for $0 < t < t_1, \epsilon = \epsilon_0$ ; for $t > t_1, \sigma = 0$. Find the stress and strain at all times. Discuss the physical significance of this stress and strain history. How could it be applied?

2.3. Determine analytically the stress strain curve for a viscoelastic solid of known properties under constant stress rate.

2.4. Perform an informal experiment to determine the creep properties of a foam earplug (or other available viscoelastic material). Sophisticated electronic apparatus is not required to do a meaningful creep test. Use available materials to build a simple apparatus. Although the experiment is informal, careful diagrams should be made of your apparatus, and the procedure should be described systematically. Plot creep compliance $J(t)$ versus log time. Is a simple spring-dashpot model appropriate? Estimate error bars for your creep compliance.

2.5. Predict the strain response of a linearly viscoelastic material of known but arbitrary properties to a stress history proportional to $A\delta(t) - B\delta(t + t_1)$ in which $t$ is time and $A$, $B$, and $t_1$ are positive constants.

2.6. A durometer is a device used to measure the stiffness of rubbery materials. The durometer has a flat surface that is pressed to the rubber specimen. A protruding probe causes an indentation in the rubber. The observer reads the value of the indentation from a dial gage, which reads from zero to 100, linked to the probe. A spring within the durometer provides the indenting force exerted by the probe. If the rubber is viscoelastic, the dial reading changes with time after the durometer is pressed to the rubber specimen. Example 6.1 shows that if the specimen material is so stiff that most of the deformation occurs in the spring, the spring displacement as a function of time follows the creep function of the material.

Suppose that the specimen material is very compliant, so that most of the deformation occurs in the specimen and little in the spring, so the durometer reading is near zero. For that case, show that the normalized spring displacement $x_1(t)$ follows the time dependence of the relaxation function.

2.7. Must recovery at zero stress following some arbitrary stress history always be monotonic?

2.8. Show that $J(t) \geq \frac{t}{\int_0^t E(\tau)d\tau}$. *Hint:* use $\int_0^t E(t - \tau)J(\tau)d\tau = t$.

2.9. Does any physical law require a material to exhibit fading memory? If not, give an example of a material that does not exhibit fading memory.

2.10. Plot the strain response for time $t > 0$ of a linearly viscoelastic material subjected to the stress history given in Problem 2.5 for the cases of a standard linear solid, a power-law material and a material that exhibits stretched-exponential transient properties. Discuss.

2.11. Would it not be easier to describe viscoelasticity by $\sigma = E\epsilon + \eta d\epsilon/dt$ rather than a Boltzmann integral?

2.12. Why is it that $E(t) \neq 1/J(t)$?

## BIBLIOGRAPHY

[1] Nowick, A. S., and Berry, B. S., *Anelastic Relaxation in Crystalline Solids*, New York: Academic, 1972.

[2] Christensen, R. M., *Theory of Viscoelasticity*, New York: Academic, 1982.

[3] Coleman, B. D., and Noll, W., Foundations of Linear Viscoelasticity, *Rev Mod Phys*, 33, 239–249, 1961.

[4] Christensen, R. M., Restrictions upon Viscoelastic Relaxation Functions and Complex Moduli, *Trans Soc Rheology*, 16, 603–614, 1972.

[5] Zemanian, A. H., *Distribution Theory and Transform Analysis*, New York: Dover, 1965.

[6] Day, W. A., Restrictions on Relaxation Functions in Linear Viscoelasticity, *Quart J Mech Appl Math*, 24, 487–497, 1971.

[7] Gurtin, M. E., and Sternberg, E., On the Linear Theory of Viscoelasticity, *Arch Rational Mech Analy*, 11, 291–356, 1962.

[8] Fabrizio, M., and Morro, A., *Problems in Linear Viscoelasticity*, Philadelphia, PA: SIAM, 1992.

[9] Gurtin, M. E., and Herrera, I., On Dissipation Inequalities and Linear Viscoelasticity, *Quart Appl Math*, 23, 235–245, 1965.

[10] Zener, C., *Elasticity and Anelasticity of Metals*, Chicago: University of Chicago Press, 1948.

[11] Pipkin, A. C., *Lectures on Viscoelasticity Theory*, Heidelberg Springer Verlag; London: George Allen and Unwin, 1972.

[12] Torvik, P. J., and Bagley, R. L., On the Appearance of Fractional Derivatives in the Behavior of Real Materials, *J Appl Mech*, 51, 294–298, 1984.

[13] Koeller, R. C., Applications of Fractional Calculus to the Theory of Viscoelasticity, *J Appl Mech*, 51, 299–307, 1984.

[14] Ward, I. M., and Hadley, D. W., *Mechanical Properties of Solid Polymers*, New York: J. Wiley, 1993.

[15] Rabotnov, Yu. N. *Elements of Hereditary Solid Mechanics*, Moscow: Mir Publishers, 1980.

[16] Nutting, P. G., A New General Law of Deformation, *J Franklin Inst*, 191, 679–685, 1921.

[17] Kohlrausch, R., Ueber das Dellmann'sche Elektrometer, *Ann Phys* (Leipzig) 72, 353–405, 1847.

[18] Jonscher, A. K., The "Universal" Dielectric Response, *Nature*, 267, 693–679, 1977.

[19] Palmer, R. G., Stein, D. L., Abrahams, E., and Anderson, P. W., Models of Hierarchically Constrained Dynamics For Glassy Relaxation, *Phy Rev Lett* 53, 958–961, 1984.

[20] Ngai, K. L., Universality of Low-Frequency Fluctuation, Dissipation, and Relaxation Properties of Condensed Matter. I, Comments. *Solid State Physics*, 9, 127–140, 1979.

[21] Williams, G., and Watts, D. C., Non-symmetrical Dielectric Relaxation Behaviour Arising from a Simple Empirical Decay Function, *Trans Faraday Soc*, 66, 80–85, 1970.

[22] Berry, G. C., and Plazek, D. J., On the Use of Stretched Exponential Functions for Both Linear Viscoelastic Creep and Stress Relaxation, *Rheol Acta*, 36, 320–329, 1997.

[23] Lubliner, J., and Panoskaltis, V. P., The Modified Kuhn Model of Linear Viscoelasticity, *Int J Solids Structures*, 24, 3099–3112, 1992.

[24] Kuhn, W., Kunzle, O., Preissmann, A., Relaxationszeitspektrum Elastizität und Viskosität von Kautschuk, *Helv Chim Acta*, 30, 307–328; 464–486, 1947.

[25] Schwarzl, F., and Staverman, A. J., Time–Temperature Dependence of Linear Viscoelastic Behavior, *J Appl Phy*, 23, 838–843, 1952.

[26] Ferry, J. D., *Viscoelastic Properties of Polymers*, 2nd ed., New York: J. Wiley, 1970.

[27] William, M. L., Landel, R. F., and Ferry, J. D., The Temperature Dependence of Relaxation Mechanisms in Amorphous Polymers and Other Glass Forming Liquids, *J Am Chem Soc* 77, 3701–3707, 1955.

[28] Plazek, D. J., Oh, Thermorheological Simplicity, Wherefore Art Thou? *J Rheology*, 40, 987–1014, 1996.

[29] Sokolnikoff, I. S., *Theory of Elasticity*, Malabar, FL: Krieger, 1983.

[30] Timoshenko, S. P., and Goodier, J. N., *Theory of Elasticity*, New York: McGraw Hill, 1982.

[31] Fung, Y. C., *Principles of Solid Mechanics*, Englewood Cliffs, NJ: Prentice Hall, 1968.

[32] Hodge, I. M., Physical Aging in Polymeric Glasses, *Science*, 267, 1945–1947, 1995.

[33] Struik, L. C. E., *Aging in Amorphous Polymers and Other Materials*, Amsterdam, New York: Elsevier Scientific, 1978.

[34] Flügge, W., *Viscoelasticity*, Waltham, MA: Blaisdell, 1967.

[35] Nye, J. F., *Physical Properties of Crystals*, Oxford: Clarendon, 1976.

[36] Stein, R. S., Rheo-Optical Studies of Rubbers, *Rubber Chemistry and Technology*, 49, 458–535, 1976.

[37] Wolff, J., Gesetz der Transformation der Knochen, Hirschwald, Berlin, 1892.

[38] Frost, H. M., *Bone Remodelling and Its Relation to Metabolic Bone Diseases*, Springfield, IL: C. Thomas, 1973.

[39] Currey, J., *Mechanical Adaptations of Bones*, Princeton, NJ: Princeton University Press, 1984.

[40] Cowin, S. C., and Hegedus, D. H., Bone Remodeling I: Theory of Adaptive Elasticity, *J Elasticity*, 6, 313–326, 1976.

[41] Hegedus, D. H., and Cowin, S. C., Bone Remodeling II: Small Strain Adaptive Elasticity, *J Elasticity*, 6, 337–352, 1976.

[42] Culshaw, B., *Structures and Materials*, Boston: Artech, 1996.

[43] Gandhi, M. V., and Thompson, B. S., *Materials and Structures*, Chapman and Hall, London, 1992.

[44] Fung, Y. C., Stress-Strain History Relations of Soft Tissues in Simple Elongation, in Biomechanics, *Its Foundations and Objectives*, Y. C., Fung, N. Perrone, and M. Anliker, eds., Englewood Cliffs, NJ: Prentice Hall, 1972.

[45] Lakes, R. S., and Vanderby, R., Interrelation of Creep and Relaxation: A Modeling Approach for Ligaments, *J Biomech Eng*, 121, 612–615, 1999.

[46] Abramowitz, M., and Stegun, I., *Handbook of Mathematical Functions*, NY: Dover, 1965.

[47] Oza, A., Lakes, R. S., and Vanderby, R., Interrelation of Creep and Relaxation for Nonlinearly Viscoelastic Materials: Application to Ligament and Metal, *Rheologica Acta*, 42, 557–568, 2003.

[48] Bernstein, B., Kearsley, E. A., and Zapas, L. J, A Study of Stress Relaxation with Finite Strain, *Trans Soc Rheol* 7, 391–410, 1963.

[49] Findley, W. N., Lai, J. S., and Onaran, K., *Creep and Relaxation of Nonlinear Viscoelastic Materials*, Amsterdam: North Holland, 1976.

[50] Schapery, R. A., On the Characterization of Nonlinear Viscoelastic Materials, *Polymer Engineering and Science*, 9, 295–340, 1969.

[51] Green, E., and Rivlin, R. S., The Mechanics of Non-Linear Materials with Memory, *Arch Rational Mech Anal*, 1, 1–21, 1957.

[52] Ward, I. M., and Onat, E. T., Non-Linear Mechanical Behaviour of Oriented Polypropylene, *J Mech Phys Solids*, 11, 217–229, 1963.

[53] Schetzen, M., Volterra, and Wiener, *Theories of Nonlinear Systems*, New York: John Wiley, 1980.

[54] Lockett, F. J., *Nonlinear Viscoelastic Solids*, New York: Academic, 1972.

[55] Hopkins, I. L., and Kurkjian, C. R., Relaxation Spectra and Relaxation Processes in Solid Polymers and Glasses, in *Acoustics*, W. P. Mason, ed., New York: Academic Press, 1965.

# 3

---

# Dynamic Behavior

## 3.1 Introduction and Rationale

The response of viscoelastic materials to sinusoidal load is developed in this section, and this response is referred to as *dynamic*. *Dynamic in this context has no connection with inertial terms or resonance*. The dynamic behavior is of interest because viscoelastic materials are used in situations in which the damping of vibration or the absorption of sound is important. The results developed in this section have a bearing on applications of viscoelastic materials, on experimental methods for their characterization, and on the development of physical insight regarding viscoelasticity.

The frequency of the sinusoidal load on an object or structure may be so slow that inertial terms do not appear (the subresonant regime), or high enough that resonance of structures made of the material occurs. At a sufficiently high frequency, dynamic behavior is manifested as wave motion. This distinction between ranges of frequency does not appear in the classical continuum description of a homogeneous material, because the continuum view deals with differential elements of material.

Oscillatory stress and strain histories are represented by sinusoid functions. Suppose we have $\sigma(\omega t) = \sigma_0 \sin\omega t$, in which $t$ is time, $\sigma_o$ is the amplitude, and $\omega$ is the angular frequency. The sine function repeats every $2\pi$ radians. So $\sigma(\omega t + 2\pi) = \sigma(\omega t)$. The time $T$ required for the sine function to complete one cycle is obtained from $\omega T = 2\pi$, or $T = 2\pi/\omega$. $T$ is called the *period*, the number of seconds required for one cycle. The inverse of $T$, the number of cycles per second (also called Hertz, Hz), is called the *frequency*, $\nu = 1/T$. The relationship between frequency and angular frequency is $\omega = 2\pi\nu$. Frequency is sometimes denoted as $f$.

In a viscoelastic material, stress and strain sinusoids are out of phase. To represent the phase shift consider two sinusoids, $\sin\omega t$ and $\sin(\omega t + \delta)$. The quantity $\delta$ is called the *phase angle*. In a plot of the two waveforms, the sinusoids are shifted with respect to each other on the time axis (Figure 3.1). Recall that the cosine function is $\pi/2$ radians out of phase with the sine function. Sinusoidal functions that represent oscillatory quantities in which phase is important are commonly written in *complex exponential notation* [1]. A complex number may be written in rectangular representation as $Z = a + bi$, with $i = \sqrt{-1}$ (called $j$ in the electrical engineering

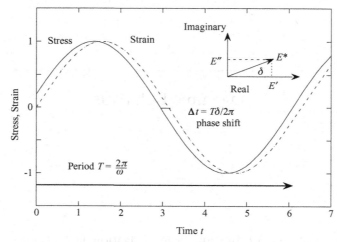

Figure 3.1. Stress and strain versus time, with a phase shift. Arbitrary units of stress and time.

community). The magnitude is $|Z| = \sqrt{a^2 + b^2}$. The complex number can be written in polar form $Z = |Z|e^{i\theta}$. The connection is given by Euler's equation, $e^{i\theta} = \cos\theta + i \sin\theta$, with $i = \sqrt{-1}$. $\theta$ is called the *phase* and represents an angle in the complex plane. Euler's equation may be rewritten $e^{i\omega t} = \cos\omega t + i \sin\omega t$. The quantity $e^{i\omega t}$ is a complex number. The physical meaning is that the sine and cosine functions are $\pi/2$ radians out of phase, so that the imaginary term represents a quantity $\pi/2$ radians out of phase.

The stress strain plot for a linearly viscoelastic material under sinusoidal load is elliptical, as demonstrated in §3.2. The shape of the ellipse is independent of stress as shown in Figure 3.2. By contrast, an elastic–plastic material exhibits a threshold. Below the threshold yield stress, the material is elastic, and its stress–strain plot is a straight line. Above the yield stress, irreversible deformation occurs in the elastic–plastic material.

## 3.2 The Linear Dynamic Response Functions $E^*$, $\tan\delta$

In this section, we obtain the dynamic material behavior (for a strain that is sinusoidal in time) from the transient behavior, assumed to be known. Let the history

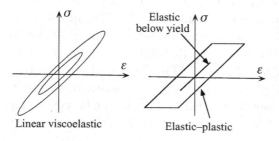

Figure 3.2. Stress strain curves for cyclic deformation at two different amplitudes. Left: linearly viscoelastic material. Right: ideal elastic–plastic material.

of strain $\epsilon$ to be sinusoidal (Figure 3.1). In complex exponential form, $\epsilon(t) = \epsilon_0 e^{i\omega t}$. $\omega$ is the angular frequency in radians per second and is related to the frequency $\nu$ in cycles per second (Hz) by $\omega = 2\pi\nu$.

### 3.2.1 Response to Sinusoidal Input

We make use of the Boltzmann superposition integral, with the lower limit taken as $-\infty$ since a true sinusoid has no starting point:

$$\sigma(t) = \int_{-\infty}^{t} E(t-\tau)\frac{d\epsilon}{d\tau}d\tau. \tag{3.1}$$

To achieve explicit convergence of the integral, decompose the relaxation function into the sum $E(t) \equiv \hat{E}(t) + E_e$ with $E_e = \lim_{t\to\infty} E(t) = E(\infty)$ called the *equilibrium modulus* (in the context of polymers) and substitute the strain history in Equation 3.1. Recall (§1.3) that $E_e > 0$ for solids and $E_e = 0$ for liquids in

$$\sigma(t) = E_e\epsilon_0 e^{i\omega t} + i\omega\epsilon_0 \int_{-\infty}^{t} \hat{E}(t-\tau)e^{i\omega\tau}d\tau. \tag{3.2}$$

Make the substitution of a new time variable $t' = t - \tau$,

$$\sigma(t) = \epsilon_0 e^{i\omega t}[E_e + \omega\int_0^{\infty}\hat{E}(t')\sin\omega t'dt' + i\omega\int_0^{\infty}\hat{E}(t')\cos\omega t'dt']. \tag{3.3}$$

If the strain is sinusoidal in time, so is the stress. The stress and strain are no longer in phase (Figure 3.1) as anticipated in Figure 1.7. The stress–strain relation becomes

$$\sigma(t) = E^*(\omega)\epsilon(t) = (E' + iE'')\epsilon(t), \tag{3.4}$$

in which, with $\hat{E}(t) = E(t) - E(\infty)$,

$$\boxed{E'(\omega) \equiv E_e + \omega\int_0^{\infty}\hat{E}(t')\sin\omega t'dt'} \tag{3.5}$$

is called the storage modulus,

$$\boxed{E''(\omega) \equiv \omega\int_0^{\infty}\hat{E}(t')\cos\omega t'dt'} \tag{3.6}$$

is called the *loss modulus*, and the loss tangent (dimensionless) is

$$\boxed{\tan\delta(\omega) \equiv \frac{E''(\omega)}{E'(\omega)}}. \tag{3.7}$$

Equations 3.5 and 3.6 give the dynamic, frequency dependent mechanical properties in terms of the relaxation modulus. $E'$ is the component of the stress–strain ratio in phase with the applied strain; $E''$ is the component 90 degrees out of phase. The single and double primes do not represent derivatives; they are a conventional notation for the real and imaginary parts, respectively. Equations 3.5 and 3.6 are one-sided Fourier sine and cosine transforms. The relations may be inverted [2] to obtain

$$E(t) = E_e + \frac{2}{\pi} \int_0^\infty \frac{(E' - E_e)}{\omega} \sin\omega t d\omega, \tag{3.8}$$

$$E(t) = E_e + \frac{2}{\pi} \int_0^\infty \frac{E''}{\omega} \cos\omega t d\omega. \tag{3.9}$$

Physically the quantity $\delta$ represents the phase angle between the stress and strain sinusoids. The dynamic stress–strain relation can be expressed as

$$\sigma(t) = |E^*(\omega)|\epsilon_0 e^{i(\omega t + \delta)}, \tag{3.10}$$

with $E^* \equiv E' + iE''$.

One can also consider the dynamic behavior in the compliance formulation. In the modulus formulation $\sigma(t) = E^*(\omega)\epsilon(t)$, determined above, is an algebraic equation, for sinusoidal loading, the strain is $\epsilon(t) = \frac{1}{E^*(\omega)}\sigma(t)$. But the complex compliance $J^*$ is defined by the equation:

$$\epsilon(t) = J^*(\omega)\sigma(t), \tag{3.11}$$

with

$$J^* = J' - iJ''. \tag{3.12}$$

So, the relationship between the dynamic compliance and the dynamic modulus is

$$\boxed{J^*(\omega) = \frac{1}{E^*(\omega)}.} \tag{3.13}$$

This is considerably simpler than the corresponding relation for the transient creep and relaxation properties. Note that $\tan\delta(\omega) = J''(\omega)/J'(\omega)$.

In the compliance formulation, as with the modulus formulation, exact interrelations are available [2] between the creep and dynamic functions, as follows: here, $J_e = \lim_{t\to\infty} J(t) = J(\infty)$ is the *equilibrium compliance* at infinite time. Recall (§1.3) that for solids, $J_e < \infty$ and for liquids, $J_e$ diverges.

$$J'(\omega) = J_e - \omega \int_0^\infty [J_e - J(t')] \sin\omega t' dt', \tag{3.14}$$

$$J''(\omega) = \omega \int_0^\infty [J_e - J(t')] \cos\omega t' dt'. \tag{3.15}$$

These are for solids; for liquids, terms involving the asymptotic viscosity appear.

The frequency range extends from zero to infinity. Frequencies below 0.1 Hz are associated with seismic waves and with Fourier spectra of quasistatic events. Vibration of structures and solid objects occur from about 0.1 Hz to 10 kHz depending on the size of the structure. Stress waves from 20 Hz to 20 kHz are perceived as sound. Waves above 20 kHz are referred to as ultrasonic; ultrasonic frequencies between 1 MHz and 10 MHz are commonly used in the nondestructive evaluation of engineering materials and for diagnostic ultrasound in medicine. Frequencies above about $10^{12}$ Hz correspond to molecular vibration and represent an upper limit for stress waves in real solids. Due to the wide range of frequencies often considered to be of interest, plots of properties versus log frequency are commonly used. A factor of 10 on such a scale is referred to as a decade.

Plots of dynamic viscoelastic functions may assume a variety of forms. First, one may plot the dynamic properties versus frequency, with the frequency scale ordinarily shown logarithmically. Second, one may plot the imaginary part versus the real part: $E''$ versus $E'$ or $J''$ versus $J'$. Such a plot is often used in dielectric relaxation studies and is referred to as a *Cole–Cole plot*. A single relaxation time process gives a semicircle in a Cole–Cole plot. Third, one may plot stiffness versus loss: $|E^*|$ versus tan$\delta$, a stiffness-loss map.

### 3.2.2 Dynamic Stress–Strain Relation

#### *The Loss Tangent,* tan$\delta$

Consider now the relation between stress and strain in dynamic loading of a linearly viscoelastic material. First, the shape of the stress–strain curve is determined, and a relation for the loss tangent is developed. Above, we have considered stress and strain as they depend on time; see also §1.4. Suppose we have

$$\epsilon = B \sin\omega t, \tag{3.16}$$

and

$$\sigma = D \sin(\omega t + \delta). \tag{3.17}$$

These are parametric equations for an elliptic Lissajous figure (Figure 3.3). The elliptic shape is a consequence of linearly viscoelastic behavior. For some materials, the curve of stress versus strain in dynamic loading is not elliptical, but has pointed ends; this behavior is a manifestation of material nonlinearity.

The loop is called a *hysteresis loop*. *Hysteresis*, in general, refers to a lag between cause and effect. In some contexts, such as the study of metals, a specific view of hysteresis is taken, that it represents a frequency-independent damping, as discussed in Chapter 7. To interpret the dimensions and intercepts of the figure, first let the intercept with the $\epsilon$ axis be called $A$, assumed to occur at time $t_1$. Then, $A = B \sin\omega t_1$ and

$$0 = D \sin(\omega t_1 + \delta) = D \sin\omega t_1 \cos\delta + D \cos\omega t_1 \sin\delta. \tag{3.18}$$

Substituting $\sin\omega t_1 = \frac{A}{B}$ and $\cos\omega t_1 = \frac{1}{B}\sqrt{B^2 - A^2}$ into the last expression,

$$0 = D\frac{A}{B} \cos\delta + \frac{D}{B}(\sqrt{B^2 - A^2}) \sin\delta. \tag{3.19}$$

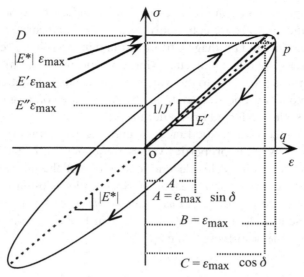

Figure 3.3. Stress $\sigma$ versus strain $\epsilon$ for a linearly viscoelastic material under oscillatory load-ing. Illustration of slopes and intercepts. A material with a rather large value of $\tan\delta \approx 0.4$ is shown for illustration. The material could be a viscoelastic rubber.

Squaring the last expression,

$$A^2 \cos^2\delta = B^2 \sin^2\delta - A^2 \sin^2\delta, \qquad (3.20)$$

so $A^2(\sin^2\delta + \cos^2\delta) = B^2 \sin^2\delta$, so that

$$\boxed{\sin\delta = \frac{A}{B}}. \qquad (3.21)$$

One may also in Equations 3.16 and 3.17, let $\omega t = -\delta$, so $\sigma = 0$. Then, $\epsilon_A = B\sin(-\delta)$ at the intercept. However, the ellipse is symmetric, so $A = B\sin\delta$, giving Equation 3.21.

Now, with the aim of obtaining a different relation between $\delta$ and the parame-ters of the ellipse, consider the point of maximum stress, which occurs when

$$\sin(\omega t + \delta) = 1. \qquad (3.22)$$

Then,

$$\omega t + \delta = \frac{\pi}{2}, \qquad (3.23)$$

so, referring to Figure 3.3, $C = \epsilon(\sigma_{\max})$,

$$C = \epsilon(\sigma_{\max}) = B\sin(\frac{\pi}{2} - \delta) = B[\sin\frac{\pi}{2}\cos(-\delta) + \cos\frac{\pi}{2}\sin(-\delta)], \qquad (3.24)$$

so,

$$\epsilon(\sigma_{\max}) = B\cos\delta, \qquad (3.25)$$

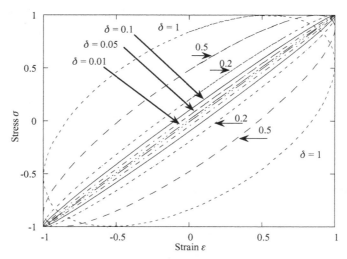

Figure 3.4. Stress $\sigma$ versus strain $\epsilon$ for a linearly viscoelastic material under oscillatory loading. Elliptic Lissajous figures for various $\delta$. For most structural metals, $\delta < 0.01$; see Chapter 7.

Since $A = B \sin\delta$,

$$\boxed{\tan\delta = \frac{A}{C}}. \tag{3.26}$$

The width of the elliptic Lissajous figure, then, is a measure of the loss angle $\delta$ of a linearly viscoelastic material. Interpretation of intercepts of the stress–strain plot are shown in Figure 3.3, and the Lissajous figures for various $\delta$ are shown in Figure 3.4. For a linearly elastic solid $\delta = 0$, and for a viscous fluid, $\delta = \pi/2$ radians.

### The Storage Modulus $E'$

To obtain a relation for the storage modulus $E'$ in connection with the elliptic stress strain curve, suppose $\epsilon = \epsilon_{\max} \sin\omega t$. Then, the stress $\sigma$ at the maximum strain $\epsilon_{\max}$ is $\sigma = E'\epsilon_{\max} \sin\omega t + E''\epsilon_{\max} \cos\omega t$. For $\epsilon_{\max}$, $\omega t = \pi/2$, then $\sigma = E'\epsilon_{\max}$ so the slope of the line from the origin to the point of maximum strain is $E'$, as shown in Figure 3.3. The stress value proportional to the storage modulus is also marked on the ordinate in Figure 3.3.

### The Storage Compliance $J'$

To obtain a relation for the storage compliance $J'$ in connection with the elliptic stress–strain curve, suppose $\sigma = \sigma_{\max} \sin\omega t$. Then, the strain at the maximum stress is, by the definition of $|J^*|$, $\epsilon = (J' - iJ'')\sigma_{\max}$. $\epsilon = J'\sigma_{\max} \sin\omega t - J''\sigma_{\max} \cos\omega t$, because $i$ refers to a 90-degree phase shift. For $\sigma_{\max}$, $\omega t = \pi/2$, then $\epsilon = J'\sigma_{\max}$ or

$\sigma_{max} = \epsilon/J'$. So, the slope of the line from the origin to the point of maximum stress is $1/J'$, as shown in Figure 3.3. Observe that $1/J' \geq E'$; equality occurs if the material is elastic ($\delta = 0$).

### *The Loss Modulus E''*

Consider time $t = 0$, so, $\epsilon = 0$, so, $\sigma(t) = D \sin\delta$, but $\sigma(t) = \epsilon_{max}[E' \sin\omega t + E'' \cos\omega t]$, so, for $t = 0$,

$$\boxed{\sigma(0) = \epsilon_{max} E''}, \tag{3.27}$$

which is marked as an intercept on the ordinate in Figure 3.3.

### *The Dynamic Modulus $|E^*|$*

Also observe that $\sigma_{max} = |E^*|\epsilon_{max}$, so, again referring to Figure 3.3, this modulus corresponds to a line of intermediate slope between that for $E'$ and that for $1/J'$.

### 3.2.3 Standard Linear Solid

The *dynamic functions* of the standard linear solid, that exhibits a relaxation that is exponential in time,

$$E(t) = E_2 + E_1 e^{-t/\tau_r}, \tag{3.28}$$

with $\tau_r$ called the relaxation time, are obtained from Equations 3.5 and 3.6 developed above and from the following standard integrals $\int_0^\infty e^{-ax} \sin bx\, dx = \frac{b}{a^2+b^2}$, $\int_0^\infty e^{-ax} \cos bx\, dx = \frac{a}{a^2+b^2}$.

So, for the standard linear solid,

$$\boxed{E'(\omega) = E_2 + E_1 \frac{\omega^2 \tau_r^2}{1 + \omega^2 \tau_r^2}} \tag{3.29}$$

$$\boxed{E''(\omega) = E_1 \frac{\omega \tau_r}{1 + \omega^2 \tau_r^2}}. \tag{3.30}$$

These forms may also be obtained by substitution of a sinusoidal stress history in the differential equations shown in §2.6. The shape of $E''(\omega)$ is referred to as a *Debye peak. This peak has nothing to do with resonance because there are no inertial terms.* In terms of the relaxation strength $\Delta = \frac{E_1}{E_2}$, we have

$$\boxed{\tan\delta = \frac{\Delta}{\sqrt{1 + \Delta}} \frac{\omega \tau_m}{1 + \omega^2 \tau_m^2}}, \tag{3.31}$$

$$\tau_m = \tau_r \sqrt{1 + \Delta}, \tag{3.32}$$

as is demonstrated in Example 3.10. A plot of several viscoelastic functions corresponding to the standard linear solid is shown in Figure 4.2. We remark that if the loss is small, the peak value of tan$\delta$ is $\frac{1}{2}\Delta$.

## 3.3 Kramers–Kronig Relations

Because the dynamic properties $E'$ and $E''$ (or $J'$ and $J''$) of a linear material are derived from the same transient function, they cannot be independent. The relationships between them are known as the Kramers–Kronig relations, in honor of the scientists who developed them in connection with electromagnetic phenomena [3, 4]; see also [5].

To derive the Kramers–Kronig relations, we recognize that the dynamic response is a Fourier transform of a causal response in the time domain [3, 6]. Causality entails an effect that cannot precede the cause; causality was implicitly included in the definitions of the creep and relaxation functions. Under these conditions, the dynamic compliance function is analytic in the complex $\omega$ plane. Apply Cauchy's integral theorem:

$$\oint_c \frac{f(z)}{z-a} dz = 2\pi i f(a) \cdot \frac{1}{2},\tag{3.33}$$

if $a$ is on contour $c$ (principal part)

· 1 if $a$ inside $c$;

· 0 if $a$ outside $c$

Here, $f(z)$ is a complex analytic function of a complex variable $z$. Now define $\zeta = \xi + i\eta$; $f = u + iv$. Cauchy's integral theorem is used to evaluate the integral using a contour along the real line and avoiding the singularity at $z = a$, via a semicircular arc. The integral is considered as a principal part, denoted by $\wp$. To calculate the principal part, one evaluates the integral by integrating almost up to the singularity, then one continues integrating from just after the singularity. Finally, one takes the limit as the distance $q$ from the singularity tends to zero:

$$\wp \int_0^\infty f(x)dx = \lim_{q\to 0}\int_0^{x-q} f(x)dx + \lim_{q\to 0}\int_{x+q}^\infty f(x)dx,\tag{3.34}$$

$$f(\zeta) = \frac{1}{i\pi}\wp\oint_c \frac{f(z)}{z-\zeta}dz\tag{3.35}$$

Decompose this into its real and imaginary parts:

$$u(\xi,0) = \frac{1}{\pi}\wp\int_{-\infty}^\infty \frac{v(x,0)}{x-\xi}dx, \quad \text{and} \quad v(\xi,0) = \frac{-1}{\pi}\wp\int_{-\infty}^\infty \frac{u(x,0)}{x-\xi}dx,\tag{3.36}$$

$$u(\xi,0) = \frac{1}{\pi}[\wp\int_0^\infty \frac{v(x,0)}{x-\xi}dx + \wp\int_{-\infty}^0 \frac{v(x,0)}{x-\xi}dx]\tag{3.37}$$

$$= \frac{1}{\pi}[\wp\int_0^\infty \frac{v(x,0)}{x-\xi}dx - \wp\int_0^{-\infty} \frac{v(x,0)}{x-\xi}dx].\tag{3.38}$$

But $f(\zeta)$ is the Fourier transform of a real function, so, $v(x) = -v(-x)$, so

$$u(\xi, 0) = \frac{1}{\pi} [\wp \int_0^\infty \frac{v(x, 0)}{x - \xi} dx + \wp \int_0^\infty \frac{v(x, 0)}{x + \xi} dx] = \qquad (3.39)$$

$$u(\xi, 0) = \frac{1}{\pi} [\wp \int_0^\infty \frac{v(x)(x + \xi) + v(x)(x - \xi)}{x^2 - \xi^2} dx] = \qquad (3.40)$$

$$\frac{2}{\pi} [\wp \int_0^\infty \frac{x v(x)}{x^2 - \xi^2} dx]. \qquad (3.41)$$

By similar arguments, and with $u(-x) = u(x)$,

$$v(\xi) = \frac{2\xi}{\pi} [\wp \int_0^\infty \frac{u(x)}{x^2 - \xi^2} dx]. \qquad (3.42)$$

Let the analytic function $f(z)$ be the complex dynamic compliance $J^*(\omega)$, with proper consideration of the limiting behavior at infinity for an analytic function. Then, the Kramers–Kronig relations for the dynamic compliance are [2, 7]:

$$\boxed{J'(\omega) - J'(\infty) = \frac{2}{\pi} [\wp \int_0^\infty \frac{\varpi J''(\varpi)}{\varpi^2 - \omega^2} d\varpi]}, \qquad (3.43)$$

and

$$\boxed{J''(\omega) = \frac{2\omega}{\pi} [\wp \int_0^\infty \frac{J'(\varpi) - J'(\infty)}{\omega^2 - \varpi^2} d\varpi]}. \qquad (3.44)$$

These expressions indicate that the storage and loss components of the dynamic compliance are intimately related: the presence of loss entails *dispersion*, which is a variation of compliance with frequency. Moreover, the relationship is nonlocal in that the presence of loss at one frequency is linked to dispersion over the entire frequency range. Illustrations of the relation between loss and dispersion may be found in Examples 3.4, 3.5, 3.10, Figures 4.1 and 4.2, and Chapter 7.

Alternate approaches to deriving the Kramers–Kronig relations are available [7, 8]. For example, one may use the fact that the Fourier transform of the step function $\mathcal{H}(t)$ is $\mathcal{F}[\mathcal{H}(t)] = \lim_{\epsilon \to 0^+} \frac{i}{(\omega + i\epsilon)} = \wp \frac{i}{\omega} + \pi \delta(\omega)$. Again, assuming causality, the transient response function must have the form: $J(t) = j(t)\mathcal{H}(t)$, in which $j(t)$ is defined for the entire time scale. Take a Fourier transform of the above product, using the convolution theorem. Because the value of $j(t)$ for $t < 0$ is immaterial, one may choose $j(-|t|) = j(|t|)$ so that $J^*(\omega)$ is purely real, to get Equation 3.44. Similarly, one may choose $j(-|t|) = -j(|t|)$ so that $J^*(\omega)$ is purely imaginary, to get Equation 3.43. In a related vein, Golden and Graham [9] remark that causality means the relaxation function $E(t)$ is defined for $t > 0$, so the real part $\Re[E^*]$ is even in frequency and the imaginary part $\Im[E^*]$ is odd in frequency.

## 3.4 Energy Storage and Dissipation

The elliptical plot of stress versus strain in dynamic loading of a linear material encloses some area. The area within the ellipse represents an energy per volume dissipated in the material, per cycle. Let us examine the mechanical damping as it relates to the storage and dissipation of energy. To that end, consider [2, 10]:

$$\sigma = \sigma_0 \sin\omega t, \epsilon = \epsilon_0 \sin(\omega t - \delta). \tag{3.45}$$

The period $T = 1/\nu = 2\pi/\omega$ corresponds to one full cycle. The energy stored over one full cycle is zero because the material returns to its starting configuration. To find the stored energy, integrate over a quarter cycle, or $t = \frac{\pi}{2\omega}$:

$$\int_0^{\epsilon_0} \sigma \, d\epsilon = \int_0^{\pi/2\omega} \sigma \frac{d\epsilon}{dt} dt$$

$$= \omega\epsilon_0\sigma_0 \int_0^{\pi/2\omega} [\cos\omega t \, \sin\omega t \, \cos\delta + \sin^2\omega t \, \sin\delta] dt$$

$$= \epsilon_0\sigma_0 [\frac{\cos\delta}{2} + \frac{\pi \, \sin\delta}{4}]. \tag{3.46}$$

The first term represents stored energy; the second term (which vanishes for $\delta = 0$) represents dissipated energy. Since $E' = |E^*| \cos\delta$, the energy stored elastically in the material is

$$W_s = \int_0^{\epsilon_0} E'\epsilon \, d\epsilon = \frac{1}{2} E'\epsilon_0^2. \tag{3.47}$$

This stored energy corresponds to the area in triangle *opq* in Figure 3.3. The stored energy is calculated for a quarter cycle rather than a full cycle since the material returns to its original condition after a full cycle. The dissipated energy for a full cycle is proportional to the area within the ellipse in Figure 3.3. The dissipated energy over a quarter cycle is determined by taking one-quarter of the integral over a full cycle. Recall $E'' = |E^*| \sin\delta$,

$$W_d = \frac{1}{4} \int_0^{2\pi/\omega} \sigma \frac{d\epsilon}{dt} dt = \frac{\pi}{4} E''\epsilon_0^2. \tag{3.48}$$

So,

$$\frac{W_d}{W_s} = \frac{\pi}{2} \tan\delta. \tag{3.49}$$

The physical meaning of the loss tangent is associated with the ratio of energy dissipated to the energy stored in dynamic loading. This is why it is considered a measure of "internal friction" or mechanical dissipation. Friction, a surface effect that always has a threshold and is always nonlinear [12], is different from internal friction that represents energy dissipation per volume and usually refers to a linear or perhaps weakly nonlinear effect. Hysteresis loops for ideal surface friction resemble those for an elastic–plastic material, on the right in Figure 3.2.

Some authors use a damping measure $\Psi$ known as the *specific damping capacity*:

$$\Psi = 2\pi \tan\delta. \tag{3.50}$$

$\Psi$ refers to the energy ratio for a full cycle. The quantity $\Psi$ is meaningful for nonlinear materials as well as linear ones because the energies can be calculated from the stress–strain loop even if its shape is not elliptical. Because there are several possible ways of defining the stored energy, there are also several expressions for $\Psi$ [11]. In the above, the stored elastic component of energy was used, in harmony with the results of Ferry [2] and Nashif, Jones, and Henderson [13]. One can also consider as stored energy the total strain energy at maximum deformation [14] or the total kinetic plus potential energies in vibrating systems with inertia [15].

Restrictions on the dynamic viscoelastic functions can be derived [16] from various energy principles. For example, the requirement of a nonnegative rate of dissipation of energy gives the conclusion $E'' \geq 0$. Demonstration of $E' \geq 0$ requires the assumption of both nonnegative stored energy and nonnegative rate of dissipation of energy. Although experiments show $dE'(\omega)/d\omega \geq 0$, analytical demonstration of such a requirement has proven elusive [16]. Energy relations may also be considered in connection with the elliptic stress–strain curve. Because the dissipated energy per cycle is $\pi E'' \epsilon_0^2$, a negative value of $E''$ would correspond to a gain of mechanical energy per cycle, and a traversal of the elliptic curve in the opposite direction. In passive materials (see §2.3), there is no external or internal source of energy that could supply this gain. Under these assumptions, we have $E'' \geq 0$.

Energy concepts [17] can be introduced as follows. One may assume that work must be done to deform a viscoelastic solid from its virgin state (prior to the application of any stress). Constitutive laws that have the following property are called *dissipative*. $\int_0^t \sigma_{ij}(\tau)\epsilon_{ij}(\tau)d\tau \geq 0$, for all sufficiently smooth strain histories that satisfy $\epsilon_{ij}(0) = 0$. This criterion differs from limits on the dynamic functions that are defined for steady state sinusoidal loading in that it is an energy integral for an initially undeformed material. The concept of dissipative material does not imply a relaxation modulus function that is monotonically decreasing. An example is given of a (nonphysical) relaxation modulus in dimensionless variables, $E(t) = \cos t$ for $0 \leq t \leq \infty$. This can be shown to be dissipative [17], but it is nonmonotonic.

Nonlinearly viscoelastic materials also dissipate energy. Tan$\delta$ as defined above only applies to linear materials. In nonlinear materials, an energy ratio $\Psi$ continues to have a clear physical interpretation even though the stress–strain curve is no longer elliptical.

If energy is stored within a material or structure, the real part of the modulus or spring constant can be negative, giving rise to a negative slope in the stress–strain plot as shown in the center diagram in Figure 3.5. If such a material is passive, energy is dissipated in cyclic deformation, so that the plot is traversed in a clockwise direction as indicated by the arrows. Materials provided with a source of power can convert that power into mechanical output, as indicated by the arrows in the

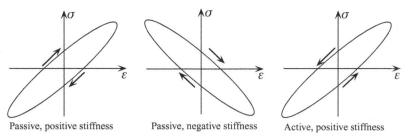

Figure 3.5. Stress–strain curves for cyclic deformation of linearly viscoelastic materials. Left: Lissajous figure for passive material, positive stiffness. Center: passive material, negative stiffness. Right: active material.

diagram on the right. A structure with a negative spring constant is unstable by itself but can be stabilized by a stiff structure. Similarly, an object made of isotropic material with a negative bulk modulus can be stabilized by embedding it in a stiff material, provided that the modulus $C_{1111}$ is positive.

## 3.5 Resonance of Structural Members

Resonance is particularly relevant in the applications of linearly viscoelastic materials, and in experimental methods for their characterization, and it is referred to extensively in later sections. Torsion is considered here since the distributed system for torsion is comparatively simple. In contrast, tension-compression vibration of a rod is dispersive even in the elastic case as a result of the Poisson effect.

### 3.5.1 Resonance, Lumped System

Let us consider a linearly viscoelastic rod, fixed at one end and attached to a rigid disk of rotary inertia $I$ at the other end, with the aim of determining the dynamic compliance as a function of frequency. This system is lumped in the sense that all the inertia is considered to reside in the disk (because it is large compared with the rod) and all the compliance is considered to reside in the rod (because it is slender and made of a more flexible material). An external torque,

$$M_{\text{ext}} = M_0 e^{i\omega t}, \tag{3.51}$$

sinusoidal in time with angular frequency $\omega$, is applied to the disk. From Newton's second law,

$$M_{\text{ext}} - M_r = I \frac{d^2\theta}{dt^2}, \tag{3.52}$$

with $M_r$ as the torque of the rod upon the disk, and $\theta$ as the angular displacement. Because the rod is viscoelastic, with complex shear modulus $G^*$,

$$M_r = G^* k_g \theta = G'(1 + i\tan\delta)k_g\theta, \tag{3.53}$$

with $k$ as a geometrical factor which depends on the cross-sectional size and shape of the rod. For a round, straight rod, $k_g = \pi d^4/32L$, with $d$ as the rod diameter and $L$ as the rod length. So,

$$I\frac{d^2\theta}{dt^2} + G'(1 + i\tan\delta)k_g\theta - M_{\text{ext}} = 0. \tag{3.54}$$

Because the applied torque is sinusoidal in time, so is angular displacement: $\theta = \theta_0 e^{i\omega t}$, with $\theta_0$ as a complex amplitude to represent the possibility of a phase shift. Substituting,

$$[M_{\text{ext}} - G'(1 + i\tan\delta)k_g\theta] = I\frac{d^2\theta}{dt^2}, \tag{3.55}$$

$[M_0 - G'(1 + i\tan\delta)k_g\theta_0] = -I\omega^2\theta_0$.
   So,

$$\frac{\theta_0}{M_0} = \frac{1}{G'(1 + i\tan\delta)k_g - I\omega^2}. \tag{3.56}$$

The $i\tan\delta$ term indicates the angular displacement has a phase shift with respect to the driving torque even at very low frequency for which the inertial term $I\omega^2$ is negligible. The ratio of angular displacement amplitude to driving torque amplitude $\theta_0/M_0$ represents a structural compliance; conversely, $M_0/\theta_0$ is a structural rigidity. By contrast, the shear modulus $G$ is a material rigidity, and $J = 1/G$ is a material compliance. Define the normalized dynamic torsional compliance as follows:

$$\Gamma \equiv \frac{G'k_g\theta_0}{M_0}. \tag{3.57}$$

Calculate $\Gamma$ from Equation 3.56, keeping in mind the fact that $G'$ and $\tan\delta$ are functions of frequency,

$$\Gamma = \frac{1}{1 + i\tan\delta - (\omega/\omega_1)^2} \tag{3.58}$$

in which

$$\omega_1 \equiv \sqrt{\frac{G'(\omega_1)k_g}{I}} \tag{3.59}$$

is the natural angular frequency. When $\omega = \omega_1$, $\Gamma$ attains its maximum value. The dynamic compliance is a complex quantity that has a magnitude and a phase associated with it. The peak in dynamic structural compliance becomes larger as the damping $\tan\delta$ becomes smaller, as shown in Figure 3.6. The *structural* phase angle $\phi$ between torque and angular displacement may be determined from the ratio of imaginary to real parts,

$$\tan\phi = \Im(\Gamma^{-1})/\Re(\Gamma^{-1}), \tag{3.60}$$

$$\boxed{\tan\phi = \frac{\tan\delta}{1 - (\omega/\omega_1)^2}.} \tag{3.61}$$

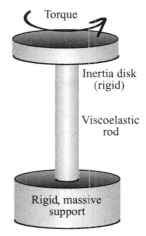

Torque

Inertia disk (rigid)

Figure 3.6. Diagram of a lumped torsional oscillator.

Viscoelastic rod

Rigid, massive support

This equation shows that the structural phase angle $\phi$ is approximately the same as the material phase $\delta$ for sufficiently low-frequency $\omega \ll \omega_1$; $\phi$ is $\pi/2$ radians at $\omega_1$, and $\phi$ is $\pi$ radians for high-frequency $\omega \gg \omega_1$. The structural phase angle $\phi$ undergoes large changes near resonance though the material phase angle $\delta$ was assumed to be constant, as shown in Figure 3.7.

The magnitude of $\Gamma$ is

$$|\Gamma| = \sqrt{\Gamma\Gamma^*}, \tag{3.62}$$

with $\Gamma^*$ as the complex conjugate of $\Gamma$.

$$\Gamma\Gamma^* = \frac{1}{[1 - (\omega/\omega_1)^2]^2 + \tan^2\delta}. \tag{3.63}$$

The dependence of $|\Gamma|$ upon frequency is referred to as a Lorentzian function, sometimes called a *Lorentzian line shape* in view of the form of the spectral lines

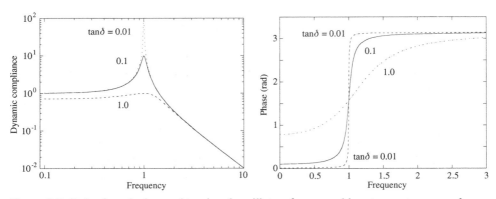

Figure 3.7. Behavior of a lumped torsional oscillator, for several loss tangents assumed constant in frequency. Frequency is normalized to the natural frequency. Left: normalized dynamic structural compliance versus normalized frequency. Right: phase $\phi$ versus normalized frequency.

associated with electromagnetic radiation (light) from atomic resonances. At the natural frequency $\omega = \omega_1$, the maximum value of $|\Gamma|$ is $|\Gamma| = \frac{1}{\tan\delta}$, so the compliance at resonance diverges in the elastic case ($\delta \to 0$). Observe that the structural compliance of the resonance of the lumped torsional system (for small loss) is a factor $1/\tan\delta$ larger than the compliance under static conditions. This may be expressed as the *magnification factor* associated with resonance to be $1/\tan\delta$.

Consider now the width of the resonance curve $|\Gamma|$ versus frequency. When $|\Gamma|$ is at half its maximum value, $|\Gamma\Gamma^*|$ is at one-quarter of the maximum. So,

$$[1 - (\omega/\omega_1)^2]^2 = 3\tan^2\delta. \tag{3.64}$$

Now an approximation is used to obtain an expression for the width,

$$[1 - (\omega/\omega_1)^2] = \frac{(\omega_1^2 - \omega^2)}{\omega_1^2} = \frac{(\omega_1 - \omega)(\omega_1 + \omega)}{\omega_1^2} \approx \frac{2(\omega_1 - \omega)}{\omega_1} \equiv \frac{\Delta\omega}{\omega_1}, \tag{3.65}$$

with the approximation $\omega_1 + \omega \approx 2\omega_1$ valid if the resonance curve is narrow. So, the relative width of the resonance curve is

$$\boxed{Q^{-1} = \frac{\Delta\omega}{\omega_1} \approx \sqrt{3}\tan\delta}. \tag{3.66}$$

This $\Delta\omega$ represents the *full* width of the resonance curve at half maximum amplitude, or twice the half excursion represented by $(\omega_1 - \omega)$. The relative width of the resonance curve is referred to as $Q^{-1}$ with $Q$ as the *quality factor*, a term also used to describe electronic resonant circuits. Some writers consider the frequency at which the response is $1/\sqrt{2}$ of maximum, rather than half of maximum for which $Q^{-1} = \frac{\Delta\omega}{\omega_1} \approx \tan\delta$. Because $20\log_{10}\sqrt{2} = 3.01$, the corresponding frequency is referred to as a *3 dB point* in connection with the logarithmic decibel (dB) scale. Derivation of exact relations between $Q$ and $\tan\delta$ generally requires a model of the viscoelastic response of the material. For example, the exact expression [11] for a material that can be described by a complex modulus and with damping independent of frequency is

$$Q^{-1} = \sqrt{1 + \tan\delta} - \sqrt{1 - \tan\delta}. \tag{3.67}$$

Observe that the width of the resonance curve for small loss is independent of the details of the inertia disk and rod stiffness and dimensions; it depends only on the loss tangent. Since an approximation was made that the resonance curve is narrow, this expression is valid only for materials with a small loss tangent. For that reason, resonant vibration methods are commonly used in experimental procedures to determine the loss tangent of low-loss materials. One can develop an exact relationship, valid for large loss, between $\tan\delta$ and the relative width $Q^{-1}$ of the resonance curve. To do so requires knowledge of the functional form of the viscoelastic properties as they depend on frequency [11].

Observe also that the response near resonance depends sensitively on $\tan\delta$. For low loss materials (e.g., $\tan\delta < 10^{-2}$), a subresonant test gives a Lissajous figure, which is difficult to distinguish from a straight line. In such a test, materials with $\tan\delta$

between about $10^{-3}$ and $10^{-9}$ cannot be easily distinguished (unless the signals are rather free of noise). Yet, the resonant behavior of such materials differs dramatically. Applications are discussed in §10.6 and §10.9.

The torque–angle relations found here are structural properties in that they depend on the geometry of the object and the attached mass or inertia. The resonance peak is not to be confused with a Debye peak. The Debye peak frequency depends on a time constant within the material; the peak occurs in the same place regardless of the inertia added to the specimen.

### 3.5.2 Resonance, Distributed System

Consider torsional oscillation of a circular cylindrical rod of a linearly viscoelastic material, fixed at one end and free of constraint, with an applied moment, sinusoidal in time, at the other end. This is referred to as a *fixed-free configuration* in the experimental literature. The stiffness and inertia are distributed throughout the rod. There are an infinite number of degrees of freedom. The dynamic structural rigidity is the ratio of applied external moment $M$ to the angular displacement $\theta$; the ratio is complex since $M$ and $\theta$ need not be in phase. The rod length is $L$ and its radius is $R$. End inertia is not considered here. The dynamic structural rigidity is [18, 19],

$$\frac{M}{\theta} = [\frac{\pi}{2} R^4][\rho\omega^2 L]\frac{\text{ctn}\omega^*}{\omega^*} \tag{3.68}$$

in which the material has a complex shear modulus $G^*$,

$$\Omega^*(\omega) = \sqrt{\frac{\rho\omega^2 L^2}{G^*(\omega)}}. \tag{3.69}$$

Methods for the solution of boundary value problems, such as this, are treated in Chapter 5; the derivation of the above expression is presented in §5.8. The normalized compliance based on Equation 3.68 is plotted in Figure 3.8 as a function of normalized frequency for several values of $\tan\delta$. Observe that the resonance peaks are in the ratios 1, 3, 5, 7, 9.

We develop in the following, simple expressions for the rigidity at resonance and determine the width of the resonance curve under the assumption of small loss. The aim is to cultivate physical insight and to provide simple expressions for the interpretation of experiments.

Case 1. Let us find the rigidity at the first resonance, which has the lowest frequency $\omega_1$. Assume the loss is small, $\tan\delta \ll 1$. Expand the cotangent as a Taylor series about $\pi/2$ that corresponds to the lowest resonance frequency:
$f(x) = f(a) + f'(a)(x - a) + f''(a)(x - a)^2/2! + \cdots$
Observe that $d[\text{ctn}x]/dx = -csc^2 x$
So,
$\text{ctn}\Omega^* \approx \text{ctn}[\frac{\pi}{2}] + [-csc^2(\frac{\pi}{2})][\Omega^* - \frac{\pi}{2}] + \cdots \approx [\frac{\pi}{2} - \Omega^*]$

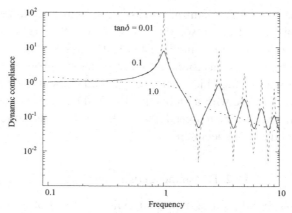

Figure 3.8. Normalized dynamic structural compliance for a distributed torsional oscillator (a cylindrical rod fixed at one end and driven at the free end) for several loss tangents (assumed constant in frequency) versus normalized frequency. The plot is based on Equation 3.68, without appeal to small loss approximation.

so $\operatorname{ctn}\Omega^* \approx [\frac{\pi}{2}] - \sqrt{\frac{\rho\omega^2 L^2}{G'(1+i\tan\delta)}}$ , but at the first resonance, the real part of $\Omega^*$ is $\frac{\pi}{2}$ to minimize the structural rigidity:

$$\sqrt{\frac{\rho\omega_1^2 L^2}{G'}} = \frac{\pi}{2},$$

so, writing a Taylor expansion of the square root and of the inverse, and retaining the lowest-order terms to obtain a simple form for low loss,

$$\operatorname{ctn}\Omega^* \approx [\frac{\pi}{2}][1 - (1 - \frac{1}{2}i\tan\delta)] = [\frac{\pi}{4}]i\tan\delta, \tag{3.70}$$

so the dynamic rigidity at the first resonance is

$$\frac{M}{\theta} = [\frac{\pi}{2}R^4][\rho\omega_1^2 L][\frac{1}{2}]i\tan\delta \tag{3.71}$$

in which $i$ indicates a 90-degree phase shift at resonance. Here, $\omega_1$ is the angular frequency at the first resonance. At resonance, the dynamic rigidity is proportional to the loss tangent so the dynamic compliance becomes large as $\tan\delta$ becomes small.

Case 2. Let us find the rigidity at the next highest resonance of the viscoelastic rod. For an elastic rod fixed at one end and free to rotate at the other end (fixed-free condition), the next resonance occurs at an angular frequency $\omega_3$, about three times that of the lowest resonance. Assume the loss is small, $\tan\delta \ll 1$. Expand the cotangent about $3\pi/2$, which corresponds to the next highest resonance frequency:

$$\operatorname{ctn}\Omega^* \approx \operatorname{ctn}[\frac{3\pi}{2}] + [-csc^2(\frac{3\pi}{2})][\Omega^* - \frac{3\pi}{2}] + \cdots = \tag{3.72}$$

$$0 - 1\{\sqrt{\frac{\rho\omega^2 L^2}{G'(1 + i\tan\delta)}} - \frac{3\pi}{2}\}$$

$$= \frac{3\pi}{2}[1 - (1 + i\tan\delta)^{-1/2}],$$

Figure 3.9. Forced vibration of an elastic string fixed to a viscoelastic support (after S. Harris, with permission; ScienceCartoonsPlus.com).

since

$$\sqrt{\rho \omega_3^2 L^2 / G'} = \frac{3\pi}{2}, \tag{3.73}$$

so,

$$\operatorname{ctn}\Omega^* \approx \frac{3\pi}{4} i \tan\delta, \tag{3.74}$$

so,

$$\frac{M}{\theta} = [\frac{\pi}{2} R^4][\rho \omega_3^2 L][\frac{1}{2}] i \tan\delta, \tag{3.75}$$

which has the same form as in the case of the lowest resonance. For small loss, $G'$ is nearly independent of frequency, so the angular frequency $\omega$ of the next highest resonance is about three times that of the first, so that the dynamic rigidity $M/\theta$ at this resonance is nine times greater than that of the first resonance. The higher resonant modes, therefore, will appear at much smaller amplitudes than the first mode, assuming a constant drive torque amplitude $M$. These higher modes occur at frequencies in the ratios 1, 3, 5, 7, 9 ... provided that the loss tangent is small. This resonance structure in the fixed-free is also observed in musical instruments, such as the clarinet, in which one end is effectively closed and the other end is open. By contrast, a rod in a free–free configuration has both ends unconstrained. The resonant frequencies are then in the ratios 1, 2, 3, 4, 5.... This resonance structure is observed in the flute, which is effectively open at both ends and in other instruments (Figure 3.9).

Case 3. Consider the region near the first resonance with the aim of determining the width of the resonance curve. Let $\omega_1$ be the angular frequency of the first resonance,

$$\omega_1 = [\frac{\pi}{2}][G'/\rho L^2]^{1/2}, \tag{3.76}$$

so,

$$\text{ctn}\Omega^* \approx [\frac{\pi}{2}][1 - \{\omega/\omega_1\}(1 - \frac{1}{2}i\tan\delta)]. \tag{3.77}$$

Evaluate the magnitude of the rigidity,

$$|\text{ctn}\Omega|^2 \approx [\frac{\pi}{2}]^2[(1 - \omega/\omega_1)^2 + \{\omega/2\omega_1\}^2 \tan^2\delta]. \tag{3.78}$$

Find $\omega$ such that the dynamic compliance is half of its value at resonance. For $\omega = \omega_1$, $|\text{ctn}\Omega^*|^2 \approx [\frac{\pi}{2}]^2[\frac{1}{4}] \tan^2\delta$, so

$$4[(\omega_1 - \omega)/\omega_1]^2 \approx 3\tan^2\delta. \tag{3.79}$$

So, the full width of the resonance curve at half maximum is

$$\boxed{\frac{\Delta\omega}{\omega_1} \approx \sqrt{3}\tan\delta}. \tag{3.80}$$

This is the same result obtained in the case of lumped inertia, even though the exact expression for the shape of the resonance curve is different. Neither the rod diameter nor the density appear in Equation 3.80; the loss is obtained from the *width* of the resonance curve (dynamic rigidity or compliance versus frequency). We remark that Equation 3.80 applies to the case of small damping, $\tan\delta \ll 1$.

## 3.6 Decay of Resonant Vibration

Suppose that in the lumped torsional system considered in §3.5, the external sinusoidal torque is suddenly removed, so $M_{\text{ext}} = 0$ in Equation 3.54. Suppose, moreover, that the viscoelastic material is passive. So,

$$I\frac{d^2\theta}{dt^2} + G'(1 + i\tan\delta)k_g\theta = 0. \tag{3.81}$$

Consider a solution for the angular displacement of the following form: it is conventional to use $\lambda$ in this context as a variable that is to be determined. It has nothing to do with a wavelength,

$$\theta(t) = \theta_0 e^{i\lambda t}. \tag{3.82}$$

Then,

$$\frac{d^2\theta}{dt^2} = -\theta_0\lambda^2 e^{i\lambda t} = -\theta(t)\lambda^2, \tag{3.83}$$

so,

$$G'(1 + i\tan\delta)k_g = I\lambda^2, \tag{3.84}$$

$$\lambda = \omega_1\sqrt{1 + i\tan\delta}, \tag{3.85}$$

with $\omega_1 \equiv \sqrt{\frac{G'k_g}{I}}$. So,

$$\theta(t) = \theta_0 \exp\{i\omega_1 t\sqrt{1 + i\tan\delta}\}. \tag{3.86}$$

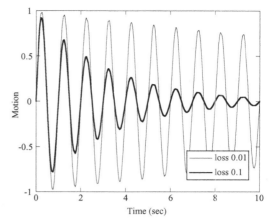

Figure 3.10. Free decay of resonant vibration motion with time for materials with $\tan\delta = 0.1$ (thick line) and 0.01 (thin line).

If we again assume the loss to be small, $\tan\delta \ll 1$, then expanding the square root as a power series about one, and retaining the lowest-order term, $\lambda \approx \omega_1(1 + \frac{1}{2}i\tan\delta)$, so,

$$\theta(t) \approx \theta_0 e^{i\omega_1 t} e^{-(\omega_1 t/2)\tan\delta}. \tag{3.87}$$

This form shows the effect of the natural frequency and the mechanical damping separately; it represents a damped sinusoid, which decays in amplitude with time (Figure 3.10). The sign in the square root solution is chosen to correspond to a passive material, which damps the vibrations rather than amplifying them. When $(\omega_1 t/2)\tan\delta = 1$, the amplitude has decayed to $1/e$ of its value at time zero; the corresponding time is $t_{1/e} = 2/\omega_1 \tan\delta$. The number $n_{1/e}$ of cycles required for the amplitude to decay by a factor of $1/e$ is

$$\boxed{n_{1/e} = t_{1/e}\frac{\omega_1}{2\pi} = t_{1/e}\nu_1 = \frac{1}{\pi\tan\delta}}. \tag{3.88}$$

The loss tangent may be expressed in terms of the time $t_{1/e}$ for decay of vibration and the period $T$ of oscillation:

$$\boxed{\tan\delta = \frac{1}{\pi}\frac{T}{t_{1/e}}}. \tag{3.89}$$

To evaluate the rate of decay, consider adjacent cycles, at time $t_1 = 0$ and $t_2 = 1/\nu = 2\pi/\omega_1$. The corresponding angular displacement amplitudes based on Equation 3.87 are $A_1 = \theta_0$ and $A_2 = \theta_0 \exp\{-(\omega_1 2\pi \tan\delta)/2\omega_1\}$.

Define the *log decrement* as

$$\Lambda \equiv ln\frac{A_1}{A_2} \tag{3.90}$$

$$\boxed{\Lambda = \pi\tan\delta}. \tag{3.91}$$

Experimentally (see §6.9), free decay of vibration is often used to determine the loss tangent of low-loss materials. As discussed in §1.5, viscoelasticity can be demonstrated by observing the free decay of vibration of tuning forks or cantilevers of various materials.

In the study of differential equations, the equation,

$$m\frac{d^2u}{dt^2} + b\frac{du}{dt} + Cu = 0, \tag{3.92}$$

is often solved [20]. Here $u$ is a variable and $m$, $b$, and $C$ are positive constants. This equation is seen as equivalent to Equation 3.81 (if $\tan\delta$ is constant), in which $i = \sqrt{-1}$ shifts the phase of a sinusoid by 90 degrees and so, for sinusoids, the $i$ represents a first derivative. The differential equation may be rewritten as

$$\frac{d^2u}{dt^2} + \frac{b}{m}\frac{du}{dt} + \omega_1^2 u = 0, \tag{3.93}$$

with $\omega_1 = \sqrt{C/m}$.

Substitution of the following trial solution (with $U_0$ as a constant)

$$u(t) = U_0 e^{Pt}, \tag{3.94}$$

into the above differential equation leads to the following algebraic equation:

$$P^2 + \frac{b}{m}P + \omega_1^2 = 0. \tag{3.95}$$

So,

$$P = -\frac{1}{2}\frac{b}{m} \pm \sqrt{\frac{1}{4}(b/m)^2 - \omega_1^2}. \tag{3.96}$$

The case $(b/m) < 2\omega_1$ gives rise to a complex exponential that represents a damped sinusoid. This is the case considered in Equation 3.96 and is referred to as underdamped. If $(b/m) = 2\omega_1$, then $u(t) = U_0 e^{-\omega_1 t}$, and there are no vibrations, just a gradual return to equilibrium. This case is called critical damping. If $(b/m) > 2\omega_1$, the two roots for $P$ are both real and negative, and $u(t)$ becomes a sum of two decaying real exponentials. Again, there are no vibrations. This case is called over-damping or heavy damping. For $(b/m) \gg \omega_1$, the response approximates a single exponential. This condition is considered to be very heavy damping.

If the damping coefficient $b$ has its origin in viscoelastic behavior rather than simple viscous drag, we must recognize that the loss tangent, in general, depends on frequency. Also, in a general viscoelastic material, the stiffness $C$ must depend on frequency by virtue of the Kramers–Kronig relations. In study of a system near resonance, it may be expedient to ignore the frequency dependence of $\tan\delta$ provided that its magnitude is small. To understand this, let us examine the Fourier transform [21] of a damped sinusoid:

$$\mathcal{F}\{\mathcal{H}(t)e^{-at}\sin\omega_1 t\} = \frac{\omega_1}{(a + i\omega_1)^2 + \omega_1^2}. \tag{3.97}$$

Since $a = (\frac{1}{2}\omega_1)\tan\delta$, a small loss corresponds to a narrow peak in the frequency content. So even if $\tan\delta$ varies with frequency, only a small range of frequencies occur in the signal, and $\tan\delta$ is effectively constant over that range. Free decay for small $\tan\delta$ approximates a sinusoid at one frequency. The case of heavy damping with nonconstant loss is more complicated and can be handled with transform methods [22].

## 3.7 Wave Propagation and Attenuation

Consider the following simple one-dimensional problem of waves in a thin rod of linearly viscoelastic material. Waves [23, 24] are transmitted down a long uniform rod of length $L$, cross-sectional area $A$, dynamic modulus $E^*$, and density $\rho$. In the following, a one-dimensional analysis is performed on the rod; neglect of Poisson effects is warrantable if the wavelength of stress waves in the rod is much greater than the rod diameter. The stress–strain relation is taken as

$$\sigma = E^*\epsilon = E^*\frac{\partial u}{\partial z}, \tag{3.98}$$

in which $z$ is a coordinate along the rod axis and $u$ is a displacement in that direction. Consider now a differential element of length $dz$ along the rod. Applying Newton's second law,

$$\rho A dz \frac{\partial^2 u}{\partial t^2} = A\sigma(z + dz, t) - A\sigma(z, t), \tag{3.99}$$

dividing by $dz$,

$$\rho A \frac{\partial^2 u}{\partial t^2} = \frac{\partial A\sigma}{\partial z}, \quad \text{and} \tag{3.100}$$

incorporating the stress–strain relation,

$$\frac{\partial^2 u}{\partial t^2} = \frac{E^*}{\rho}\frac{\partial^2 u}{\partial z^2}. \tag{3.101}$$

This is a *wave equation*. Consider a trial solution of the form (with $k$ as wave number and $\omega$ as angular frequency):

$$u(z, t) = u_0 e^{i(kz-\omega t)}, \tag{3.102}$$

Then,

$$\omega^2 u_0 = k^2 u_0 \frac{E^*}{\rho}, \tag{3.103}$$

$$k\sqrt{\frac{E'}{\rho}(1 + i\tan\delta)} = \omega, \tag{3.104}$$

$$k\sqrt{\frac{E'}{\rho}}\sqrt{1 + i\tan\delta} = \omega. \tag{3.105}$$

Suppose that the loss is small, then expand the square root containing the loss tangent and retain the lowest-order terms. Moreover, define

$$c = \sqrt{\frac{E'}{\rho}}, \tag{3.106}$$

$$k \approx \frac{\omega}{c}(1 - \frac{1}{2}i\tan\delta). \tag{3.107}$$

Then, Equation 3.100 has a solution,

$$u(z, t) = u_0 \exp\{i\omega(\frac{z}{c} - t)\} \exp\{-z\frac{\omega}{2c}\tan\delta\}. \tag{3.108}$$

In the complex exponential (left), the argument $z/c - t$ is constant if $dz/c = dt$, or $dz/dt = c$. So, the wave speed $c$ is interpreted as the *phase velocity* of the wave. As time $t$ increases, $z$ must increase to keep the phase argument constant, so the wave moves to the right (direction of increasing $z$). A left moving wave would be written as $u(z, t) = u_0 \exp i\omega\{z/c + t\}$. As for the interpretation of $k$ (as a real number, for zero loss), observe that a sinusoid goes through a complete cycle when its argument goes through $2\pi$. So, $kz = 2\pi$ corresponds to one spatial cycle in $z$, or one *wavelength*, given the name $\lambda$. So, $k = 2\pi/\lambda$. Similarly, at a given point in space, one cycle occurs when $\omega t = 2\pi$. The time for one cycle is called the *period* $T$ of the wave, and the inverse of the period is the number of cycles per unit time, the *frequency* $\nu = 1/T$. So, $\omega = 2\pi\nu$. The second exponential in Equation 3.108 is a real exponential, and signifies a decrease of wave amplitude with distance $z$. This decrease is known as *attenuation*. The attenuation coefficient (per unit length) is defined for small loss as

$$\boxed{\alpha \approx \frac{\omega}{2c}\tan\delta}. \tag{3.109}$$

$\alpha$, the attenuation per unit length, is commonly given in units of nepers per length, in which *neper* is dimensionless. One neper is a decrease in amplitude of a factor of $1/e$. This corresponds to $\alpha z = 1$ in Equation 3.108 with the second term taken as $e^{-\alpha z}$. The word neper is a degradation of the name of the inventor, Napier, of the natural logarithm. The attenuation can also be expressed [23] in decibels (dB) per unit length by the following [7]: $20 \log_{10}(e^{\alpha}) = 8.686\alpha$, so $\alpha(\text{dB/cm}) = 8.68\alpha$ (neper/cm).

If the rod is elastic ($\delta = 0$), the stress waves propagate without dispersion, that is, the wave speed is independent of frequency, provided that the one-dimensional assumption is justified by the wavelength being much larger than the rod diameter. Moreover, the waves propagate without attenuation. If the rod is viscoelastic, wave speed depends on frequency, because $E'$ depends on frequency, and there is attenuation. Exact relations for wave speed $c$ and attenuation $\alpha$ can be extracted with further effort [18],

$$c = \sqrt{\frac{|E^*|}{\rho}}\sec\frac{\delta}{2} \tag{3.110}$$

$$\boxed{\alpha = \frac{\omega}{c}\tan\frac{\delta}{2}}. \tag{3.111}$$

Table 3.1. *Measures of damping*

| Measure | Relation to $\delta$ | Phenomenon |
|---|---|---|
| Loss angle | $\delta$ | Phase, stress, strain |
| Loss tangent | $\tan\delta$ | Tangent of the phase angle |
| $\dfrac{E''}{E'}$ | $\tan\delta$ | Ratio of imaginary part to real part of modulus |
| Quality factor | $Q \approx (\tan\delta)^{-1}$ | Resonance peak width |
| Log decrement | $\Lambda \approx \pi\tan\delta$ | Free decay of vibration |
| Decay time $t_{1/e}$ | $\tan\delta \approx \dfrac{1}{\pi}\dfrac{T}{t_{1/e}}$ | Free decay of vibration |
| Specific Damping Capacity | $\Psi = 2\pi\tan\delta$ | Ratio of energy dissipated to energy stored |
| Attenuation | $\alpha\lambda = 2\pi\tan\dfrac{\delta}{2}$ | Wave attenuation, (neper/$\lambda$) |
| Attenuation | $\alpha = \dfrac{\omega}{c}\tan\dfrac{\delta}{2}$ | Wave attenuation, (neper/m) |

Some authors use units of nepers/wavelength; since $\lambda v = c$, such attenuations give the loss without any frequency term, as follows:

$$\alpha\lambda = 2\pi\tan\frac{\delta}{2}. \tag{3.112}$$

The attenuation per unit wavelength $\lambda$ is also expressed as $\alpha\lambda = \pi/Q$ nepers $\approx 27.3/Q$ dB. Here, $Q$ can be considered in terms of energy dissipated in a wave, or as the number of wavelengths for an attenuation of $\pi$ nepers, or as the number of cycles required for the amplitude to drop to $e^{-\pi}$ of its original value [23].

## 3.8 Measures of Damping

The loss angle $\delta$, or the loss tangent $\tan\delta$, may be considered as the fundamental measure of damping in a linear material. Other measures, such as those developed above, are often cited in analyses of viscoelastic behavior. For the purpose of comparison, these measures of damping and their relationship to the loss angle are presented in Table 3.1.

The approximations are valid for the case of small damping, $\tan\delta \ll 1$. They are within 1 percent of the exact value up to a loss tangent of about 0.2 [11]. Derivation of exact relations (e.g., Equation 3.67) generally requires a model of the viscoelastic response of the material.

## 3.9 Nonlinear Materials

Nonlinearly viscoelastic materials (see §2.12 for constitutive equations) can also be excited by sinusoidal loads. The simplest dynamic example of the effect of nonlinearity is nonlinear elasticity in one dimension. Suppose that

$$\sigma = f(\epsilon). \tag{3.113}$$

Write $f(\epsilon)$ as a power series, so,

$$\sigma = a_1\epsilon^1 + a_2\epsilon^2 + a_3\epsilon^3 + \cdots. \tag{3.114}$$

If the strain is sinusoidal,

$$\epsilon = \cos\omega t, \tag{3.115}$$

trigonometric identities, for example,

$$\cos^2\omega t = \frac{1}{2}(1 + \cos 2\omega t), \tag{3.116}$$

$$\cos^3\omega t = \frac{1}{4}\cos 3\omega t + \frac{3}{4}\cos\omega t, \tag{3.117}$$

may be used [25] to express the stress as a sum of sinusoids:

$$\sigma(t) = \sum_{n=1}^{N} a_n \sin n\omega_n t. \tag{3.118}$$

Consequently the effect of the nonlinearity is to generate higher harmonics (integer multiples) of the driving frequency. If the driving signal contains several frequencies, the material responds at new frequencies corresponding to sums and differences of the drive frequencies [25].

As for a viscoelastic material obeying nonlinear superposition, analysis shows that higher harmonics are generated as well [26]. In the electrical engineering community, this sort of response is called *harmonic distortion*. In a Green–Rivlin solid, which obeys the Green–Rivlin multiple integral expansion (§2.12), the response contains higher harmonics as in the material obeying nonlinear superposition. These two types of material can be distinguished by applying a strain history containing a superposition of several frequencies [27, 28]. In a Green–Rivlin solid, oscillations occur at new frequencies not originally present in the original excitation. In the electrical engineering community, this sort of response is called *intermodulation distortion*.

Response of a nonlinear material to cyclic load may also be visualized via a plot of stress versus strain, as shown in Figure 3.11. In a nonlinear material, the plot is no longer elliptical in shape, as it is in a linear material. Harmonic distortion due to the material gives rise to a nonelliptic plot. Many types of curves are observed in real materials. The area within the closed curve, as in the case of linear materials, represents energy per volume dissipated per cycle.

In the dynamic response of nonlinearly viscoelastic materials, the loss angle $\delta$ can still be defined as the phase angle between the drive signal and the fundamental component of the response at that frequency. However, the interpretation of the loss angle in terms of energy dissipation is no longer the same as it is in linear materials. In nonlinear solids, it is more sensible to speak directly of energy dissipation per cycle.

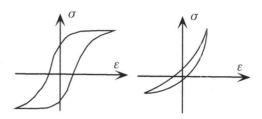

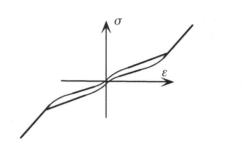

Figure 3.11. Response (stress $\sigma$ vs. strain $\epsilon$) of several nonlinear materials to cyclic load. The loop is no longer elliptical as it is in linear materials.

## 3.10 Summary

The dynamic properties of linearly viscoelastic materials have been developed in terms of the transient functions. In dynamic (sinusoidal) loading the stress–strain ratio or dynamic modulus has a magnitude and phase and can be represented as a complex number. The real and imaginary parts are associated with storage and dissipation of mechanical energy, respectively. The real and imaginary parts of the dynamic modulus and compliance are not independent but are related by the Kramers–Kronig relations. In structural resonance, the material loss tangent manifests itself in a broadening and lowering of the resonance peak, and in the damping of free vibration of structural members. As for low-loss materials, we remark that in a subresonant test using conventional mechanical testing equipment, materials with $\tan\delta$ between about $10^{-3}$ and $10^{-9}$ cannot be easily distinguished. Yet, the resonant behavior of such materials differs dramatically. In wave propagation, the loss tangent is associated with attenuation of waves. The examples considered are directly relevant to experimental methods for determining material properties and to applications of viscoelastic materials.

## 3.11 Examples

**Example 3.1**
Express the dynamic modulus $E^* = E' + iE''$ in terms of the loss tangent.
**Solution**
$$E^* = E' + iE'' = E'(1 + i(\tfrac{E''}{E'})) = E'(1 + i\tan\delta).$$

**Example 3.2**
Express the dynamic modulus $E^* = E' + iE''$ in terms of $|E^*|$ and the phase $\delta$. Also, express $|E^*|$ in terms of $E'$ and $\delta$.

**Solution**

$E^* = \frac{|E^*|(E'+iE'')}{|E^*|} = |E^*|(\cos\delta + i\,\sin\delta)$. This form may be visualized by expressing the complex modulus as a point in the complex plane:

$$|E^*| = \sqrt{(E'+iE'')(E'-iE'')} = \sqrt{E'^2 + E''^2} = E'\sqrt{1 + \tan^2\delta} = \frac{E'}{\cos\delta}.$$

**Example 3.3**

A viscoelastic ball is dropped from a height $h_{\mathrm{drop}}$ upon a rigid floor. To what fraction **f** of that height will it rebound? An approximate value is satisfactory.

**Solution**

As an initial approximation, represent the impact event as half of a cycle of harmonic loading. Proceed by equating the energy dissipated in the material by viscoelastic damping with the reduction in potential energy associated with the difference in height. This procedure neglects any energy loss due to air resistance, radiation of sound energy during the impact, and friction. For a linearly viscoelastic material, following Figure 3.12, the stress–strain curve for harmonic loading is an ellipse. The area of the ellipse represents energy dissipated per unit volume per cycle, and the impact is considered as half of a cycle. This is an approximation because the actual impact begins with zero stress and zero strain; a transient term is neglected. So the energy density $W_d$ dissipated in half of a cycle is proportional to the area within the ellipse,

$$W_d = \frac{1}{2}\pi ab, \tag{3.119}$$

with $a$ as the semiminor axis and $b$ as the semimajor axis of the ellipse. Consider the stress to be normalized to Young's modulus so that the ellipse appears on dimensionless axes for simpler interpretation. We will consider a ratio of energies so that the modulus will not appear in the final result in any case. Referring to Figure 3.3, $b = B\sqrt{2}$ and $a = A/\sqrt{2}$, so from Equation 3.21,

$$W_d = \frac{1}{2}\pi AB. \tag{3.120}$$

But the normalized stored energy is $W_s = \frac{1}{2}B^2\cos\delta$, so

$$\frac{W_d}{W_s} = \pi\frac{A}{B\cos\delta} = \pi\tan\delta. \tag{3.121}$$

Consider now the stored and dissipated energies in terms of ratio **f** of the drop height $h_{\mathrm{drop}}$ to the rebound height $h_{\mathrm{reb}}$, the ball mass $m$, and the acceleration due to gravity $g$,

$$\mathbf{f} = \frac{h_{\mathrm{reb}}}{h_{\mathrm{drop}}}, \tag{3.122}$$

$$W_s = mgh_{\mathrm{drop}}, \quad W_d = mg(h_{\mathrm{drop}} - \mathbf{f}\cdot h_{\mathrm{drop}}) = mgh_{\mathrm{drop}}(1-\mathbf{f}), \tag{3.123}$$

so,

$$\frac{W_d}{W_s} = 1 - \mathbf{f}. \tag{3.124}$$

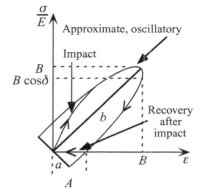

Figure 3.12. Stress $\sigma$ versus strain $\epsilon$ for a half cycle of harmonic loading compared with an impact.

Equating the expressions for energy ratio, $1 - \mathbf{f} \approx \pi \tan\delta$, so,

$$\mathbf{f} \approx 1 - \pi \tan\delta. \tag{3.125}$$

This approximation may be improved somewhat by considering the two small "triangles" at the base of the half ellipse in Figure 3.12. Each of these has area $\frac{1}{2}a^2$. But $a = A/\sqrt{2} = (1/\sqrt{2})B\sin\delta$ so the total area to be subtracted from the half of the ellipse area is $(\frac{1}{2})B^2\sin^2\delta$. The dissipated energy is $W_d = \frac{1}{2}\pi AB - \frac{1}{2}B^2\sin^2\delta = \frac{1}{2}B^2[\pi\sin\delta - \frac{1}{2}\sin^2\delta]$. Again, the normalized stored energy is $W_s = \frac{1}{2}B^2\cos\delta$. $\frac{W_d}{W_s} = 1 - \mathbf{f}$. So,

$$\mathbf{f} = \frac{h_{\text{reb}}}{h_{\text{drop}}} \approx 1 - (\pi\tan\delta - \frac{1}{2}\frac{\sin^2\delta}{\cos\delta}). \tag{3.126}$$

This result still does not take into account the fact that the load history starts at zero and, therefore, neglects an additional area in the diagram. For large $\delta$, the approximation breaks down since large $\delta$ gives rise to a negative height ratio $\mathbf{f}$, which cannot occur.

A better approximation to rebound resilience may be had by considering free decay of vibration following Equation 3.87, for small $\tan\delta$ with $\omega$ as the natural angular frequency. $x(t) \approx x_0 e^{-(\omega t/2)\tan\delta}\sin\omega t$. The moving mass is considered to come in contact with a one-dimensional massless viscoelastic spring at time zero, execute one half of a cycle of damped vibration, then rebound, losing contact at time $t = \pi/\omega$. The velocity is $v = \frac{dx}{dt} = x_0\omega e^{-(\omega t/2)\tan\delta}[\cos\omega t - \frac{1}{2}\tan\delta\sin\omega t]$. The impact velocity $v_0$ is calculated at time $t = 0$, $v_0 = x_0\omega$. The rebound velocity is calculated at time $t = \pi/\omega$, when the displacement is again zero, $v_1 = -x_0\omega e^{-(\pi/2)\tan\delta}$. The height ratio is related to the velocity ratio via the relation between kinetic and potential energy. $\mathbf{f} = \frac{h_{\text{reb}}}{h_{\text{drop}}} = \frac{v_1^2}{v_0^2} = [e^{-(\pi/2)\tan\delta}]^2 = e^{-\pi\tan\delta}$.

So,

$$\boxed{\tan\delta \approx \frac{1}{\pi}ln[\frac{h_{\text{drop}}}{h_{\text{reb}}}]}. \tag{3.127}$$

This approximation makes physical sense at the extremes of damping: the height ratio **f** is unity for zero damping and tends to zero for very high damping. The result is approximate because a simplified form of the vibration equation is used, and because the treatment is one-dimensional. Moreover, energy loss due to radiated sound and stress waves in the substrate is not considered. The actual impact force is a pulsed function of time, which may be Fourier analyzed as a distribution of frequencies, so $\tan\delta$ over a range of frequency contributes to the observed rebound resilience. A more detailed treatment is in §10.11.

Physically, the fraction **f** may be interpreted as a resilience: 100 percent for a perfectly elastic material and zero for the case of zero rebound. This may be contrasted with the *coefficient of restitution*, which is defined as the ratio of *speed v* after an impact to the speed before the impact. Because gravity converts potential energy corresponding to the height $h$ to kinetic energy as the ball is dropped according to $mgh = \frac{1}{2}mv^2$, it is evident that there is a quadratic relationship between the height ratio and the coefficient of restitution. For further analysis and examples, see Chapter 10.

### Example 3.4
Show that the change in storage compliance over the full frequency range is proportional to the area under the loss compliance curve on a log frequency scale.
### Solution
Use the Kramers–Kronig relation, Equation 3.43, with $\omega = 0$ to obtain the full frequency range,

$$J'(0) - J'(\infty) = \int_0^\infty \frac{J''(\varpi)}{\varpi} d\varpi = \int_0^\infty J''(\varpi) d\ln \varpi. \tag{3.128}$$

### Example 3.5
Determine the dynamic behavior associated with an exponential relaxation function,

$$E(t) = (\Delta E)e^{-t/\tau}. \tag{3.129}$$

### Solution
Consider the standard integrals,

$$\int_0^\infty e^{-ax} \cos bx\, dx = \frac{a}{a^2 + b^2}, \tag{3.130}$$

$$\int_0^\infty e^{-ax} \sin bx\, dx = \frac{b}{a^2 + b^2}, \tag{3.131}$$

Then, with Equations 3.5 and 3.6,

$$E''(\omega) = (\Delta E)\omega \int_0^\infty e^{-t/\tau} \cos \omega t\, dt = (\Delta E)\frac{\omega\tau}{1 + \omega^2\tau^2}, \tag{3.132}$$

$$E'(\omega) = E'(0) + (\Delta E)\omega \int_0^\infty e^{-t/\tau} \sin \omega t\, dt = E'(0) + (\Delta E)\frac{\omega^2\tau^2}{1 + \omega^2\tau^2}. \tag{3.133}$$

**Example 3.6**

Show that $J' = \frac{\frac{1}{E'}}{1+\tan^2\delta}$. Observe that $J^* = J' - iJ''$. Discuss the physical interpretation of this in terms of the stress–strain diagram.

**Solution**

$$J^* = \frac{1}{E^*} = \frac{1}{E' + iE''} \tag{3.134}$$

$$= \frac{1}{E' + iE''}\frac{E' - iE''}{E' - iE''}$$

$$= \frac{E'}{E'^2 + E''^2} - i\frac{E''}{E'^2 + E''^2}.$$

Since $\tan\delta = \frac{E''}{E'}$,

$$\frac{E'}{E'^2 + E''^2} = \frac{1}{E'}\frac{1}{1 + \tan^2\delta}, \tag{3.135}$$

so,

$$J' = \frac{1}{E'}\frac{1}{1 + \tan^2\delta}. \tag{3.136}$$

The physical interpretation of this in terms of the stress–strain diagram is that in the elliptic Lissajous figure, the slope $1/J'$ corresponding to the stress–strain ratio at maximum stress is greater than the slope $E'$ corresponding to the stress–strain ratio at maximum strain.

**Example 3.7**

Observe in §3.5.1 that the structural compliance of the resonance of the lumped torsional system (for small loss) is a factor $1/\tan\delta$ larger than the compliance under static conditions. Therefore, the *magnification factor* is $1/\tan\delta$. What is the corresponding magnification factor for the first resonance of the distributed rod in torsion?

**Solution**

At the first resonance of the distributed torsion case for a circular cylinder,

$$\omega_1 = \frac{\pi}{2}\sqrt{\frac{G'}{\rho L^2}}, \tag{3.137}$$

substitute in the relation for dynamic rigidity at the first resonance,

$$\frac{M}{\theta} = [\tfrac{1}{2}\pi R^4][\rho\omega_1^2 L]\tfrac{1}{2}i\tan\delta, \tag{3.138}$$

$$\frac{M}{\theta} = \frac{\pi R^4}{2L}\frac{\pi^2}{8}i\tan\delta, \tag{3.139}$$

but the static rigidity of the rod in torsion is

$$\frac{M}{\theta} = \frac{\pi R^4}{2L}. \tag{3.140}$$

But $8/\pi^2 \approx 0.81057$, so the magnification factor for the first resonance of the distributed system is *not* $1/\tan\delta$, but rather $0.81/\tan\delta$.

## Example 3.8

In §3.4, the principles of nonnegative stored energy and nonnegative rate of dissipation of energy were discussed in connection with inequalities for the dynamic functions. Are these real physical laws? Do they follow from physical laws? Would it be possible to create a material for which $\tan\delta < 0$? If so, what would be required? Is it possible to have negative stiffness?

### Solution

Some of the mathematical principles actually entail assumptions about the material or the nature of the system. Negative stiffness in dynamical systems is certainly possible. As shown in §3.5.1, a phase difference of 180 degrees occurs between load and deformation above resonance in an elastic material, or far above resonance in a viscoelastic material. This phase shift corresponds to a negative *structural* stiffness because motion occurs in the opposite direction as force. In static deformation, negative stiffness entails a condition of unstable equilibrium. Examples include a bead rolling on a smooth bowl, which is concave down; as well as a buckled column in response to a lateral load. When the buckled column is straight, it is in unstable equilibrium and it exhibits negative stiffness for lateral perturbations. Systems with negative stiffness must contain stored energy or they must receive energy from some external source. Such systems are certainly possible, but once prepared, they will diverge from the condition of unstable equilibrium unless restrained.

Negative damping (or negative attenuation) corresponds to a *gain* of energy in vibration or wave motion. Such gain is possible in light of the principle of energy conservation, if there is an external source of energy. In passive materials, there is no such energy source by definition, so $E'' \geq 0$. Mechanical examples are given in §7.4.5. Some more familiar examples in electrical engineering and optics are as follows: an electrical amplifier exhibits gain, and the energy is supplied by the electrical power supply. A laser (the word means light amplification by stimulated emission of radiation) amplifies light, and as with other amplifiers it can serve as an oscillator. The laser is provided with an external source of power.

## Example 3.9

What is the shape of the load-deformation diagram for a lumped system under harmonic load as the natural frequency is approached?

### Solution

Under quasistatic conditions, well below the lowest resonance, the load-deformation diagram follows the stress–strain behavior and is elliptical; the width of the ellipse is related to $\tan\delta$ by Equation 3.26. At resonance, the phase $\phi$ between load and deformation is 90 degrees, following Equation 3.61. This diagram is a circle or an ellipse with principal axes aligned with the coordinate axes.

**Example 3.10**

Suppose that $E(t) = E_2 + E_1 e^{-t/\tau}$.

What is the relaxation strength $\Delta$?

Show that $\tan\delta = \frac{\Delta}{\sqrt{1+\Delta}} \frac{\omega\tau_m}{1+\omega^2\tau_m^2}$ with $\tau_m = \tau\sqrt{1+\Delta}$.

What is the peak $\tan\delta$?

**Solution**

The relaxation strength $\Delta$ is, by definition, the ratio of change in stiffness through the relaxation to the relaxed stiffness, so by considering limits at short time and long time, $\Delta = E_1/E_2$. By definition $\tan\delta = E''/E'$; substituting from Example 3.5,

$$\tan\delta = \frac{E_1 \frac{\omega\tau}{1+\omega^2\tau^2}}{E_2 + E_1 \frac{\omega^2\tau^2}{1+\omega^2\tau^2}}$$

$$= \frac{E_1\omega\tau}{E_2(1 + \omega^2\tau^2) + E_1\omega^2\tau^2}$$

$$= \frac{\frac{E_1}{E_2}\omega\tau}{1 + (1 + \frac{E_1}{E_2})\omega^2\tau^2}$$

$$= \frac{\frac{E_1}{E_2}}{\sqrt{1 + \frac{E_1}{E_2}}} \frac{\omega\tau\sqrt{1 + \frac{E_1}{E_2}}}{1 + \omega^2\tau^2[\sqrt{1 + \frac{E_1}{E_2}}]^2}. \tag{3.141}$$

$$\tan\delta = \frac{\Delta}{\sqrt{1 + \Delta}} \frac{\omega\tau_m}{1 + \omega^2\tau_m^2}, \tag{3.142}$$

with $\tau_m = \tau\sqrt{1+\Delta}$. So, if $E''$ is a Debye peak, $\tan\delta$ also is a Debye peak, but with a shift in frequency.

**Example 3.11**

Suppose the loss compliance is constant over a frequency range, so $J''(\omega) = J_0''$ for $0 \le \omega \le \omega_c$, with $\omega_c$ as a cut-off frequency, and $J'' = 0$ for $\omega > \omega_c$. Find the storage compliance and discuss the physical interpretation.

**Solution**

Integrating the Kramers–Kronig relation, Equation 3.43 with the aid of some identities,

$$J'(\omega) - J'(\infty) = \frac{J_0''}{\pi}[\ln(\omega_c^2 - \omega^2)] - \ln\omega^2. \tag{3.143}$$

Now, if we have $\omega \ll \omega_c$,

$$J'(\omega) - J'(\infty) \approx \frac{2J_0''}{\pi}[\ln(\omega_c) - \ln\omega]. \tag{3.144}$$

So, for this, case a constant loss compliance is associated with a linear dispersion in the storage compliance versus log frequency. Not uncommonly, the behavior of real materials is approximated by this example. In real materials, sharp cutoffs in the mechanical loss are not observed; a cutoff was incorporated in this example to achieve convergence and it was made sharp for simplicity in calculation. The singular resonant behavior in the storage compliance at the cut-off frequency is not representative of physically observed mechanical behavior of homogeneous materials.

**Example 3.12**

Suppose that the loss is concentrated at only one frequency: $J''(\omega) = k\delta(\omega - \omega_c)$ in which $\delta$ is the Dirac delta function and $k$ is a constant. Find the storage compliance and discuss the physical interpretation.

**Solution**

Again, integrating the appropriate Kramers–Kronig relation,

$$J'(\omega) - J'(\infty) = \frac{2k\omega_c}{\pi(\omega_c^2 - \omega^2)}. \tag{3.145}$$

The storage compliance here exhibits an undamped resonant behavior in which the compliance tends to infinity as $\omega_c$ is approached from below. The negative value of dynamic compliance above $\omega_c$ refers to the 180 degree-phase shift between force and displacement of a lumped system. Structural members exhibit such resonances as described in §3.5.1; the Kramers–Kronig relations make no distinction between continuum properties and structural ones. Homogeneous materials do not exhibit resonances in their mechanical properties. In composite materials (which are heterogeneous), there is the possibility of microresonance of the structural elements; given the typical size scales involved, such behavior could only occur at ultrasonic frequencies above about 1 MHz.

Optical properties of materials (specifically the electric polarizability) are also governed by the Kramers–Kronig relations. Materials can exhibit resonant absorption of light due to microresonance on the atomic scale. For example, transparent materials, such as glass and water, exhibit resonant absorption of ultraviolet light due to resonance of electrons in the material. These materials are transparent (they have a small loss term) to visible light, which has lower frequencies than that of the resonant absorption. The dispersion of visible light manifests itself as the rainbow and in the spectral colors seen in light refracted through diamonds.

## 3.12 Problems

3.1. Find $\tan\delta$ if the relaxation function is $E(t) = At^{-n}$.

3.2. Why do tires become warm after you have driven on the highway?

3.3. Estimate roughly the loss tangent of your ear.

3.4. Show that $J'' = \frac{1}{E'}\frac{\tan\delta}{1+\tan^2\delta}$.      Show that $E' = \frac{1}{J'}\frac{1}{1+\tan^2\delta}$.

3.5. What would happen if engineering materials were, in fact, perfectly elastic $(\tan\delta = 0)$? What would happen if the Earth and all the solid matter on it were perfectly elastic?

3.6. Why are marching soldiers instructed to "break step" when crossing bridges?

3.7. Suppose that two sinusoids have a phase difference $\xi$. What is the phase difference between their Fourier transforms?

3.8. It is said that some opera singers are able to shatter a glass by singing the correct note. Is this possible? How? Does it matter what kind of glass is used?

3.9. Why are there tuning forks but no tuning knives or tuning spoons?

3.10. Compare the detailed shape of the resonance peak for a lumped system and for a distributed system. This can be done by plotting both curves for a given loss tangent or by performing a perturbation analysis.

3.11. Consider an underdamped lumped resonating system. The envelope of the displacement amplitude is a smooth exponential. For comparison, calculate the rate of decay of the total mechanical energy, considered as the sum of kinetic and potential energies [29]. Must the decay of mechanical energy be uniform? Discuss.

3.12. Show that $|E^*| = E'\sqrt{1 + \tan^2\delta}$.

3.13. Show that $\sigma(t) = |E^*(\omega)|\epsilon_0 e^{i\omega t + \delta}$ with $E^* \equiv E' + iE''$.

3.14. What is the shape of the stress–strain diagram for a material under harmonic load as a natural frequency is approached?

3.15. Show that a single relaxation-time process gives a semicircle in a Cole–Cole plot.

3.16. In §3.4, the principles of nonnegative stored energy and nonnegative rate of dissipation of energy were discussed in connection with inequalities for the dynamic functions. Must we in fact have $dE'(\omega)/d\omega \geq 0$? *Hint*: consider the *structural* compliance of a system that exhibits resonance. Can materials exhibit resonance on a microstructural scale?

3.17. Consider a viscoelastic spacecraft spinning in outer space and neglect any gravitational pull from other objects. Is angular momentum conserved? Is kinetic energy conserved? Suppose that the spin axis does not exactly coincide with any principal axis of inertia, so that there is some wobble.

3.18. What conclusions can be drawn about a system if oscillations of decreasing magnitude are observed following a step load to a system?

3.19. Determine $\tan\delta$ from the stress–strain loop in Figure 3.3, (a) from the width of the loop, and (b) from the ratio of dissipated to stored energy density. Compare the results.

## BIBLIOGRAPHY

[1] Irwin, J. D., *Basic Engineering Circuit Analysis*, New York: Macmillan, 1989.

[2] Ferry, J. D., *Viscoelastic Properties of Polymers*, 2nd ed., New York: John Wiley, 1970.

[3] Kronig, R., On the Theory of Dispersion of X-rays, *J Opt Soc Am*, 12, 547, 1926.

[4] Kronig, von R. de L., and Kramers, H. A., Absorption and Dispersion in X-ray Spectra (Zur Theorie der Absorption und dispersion in den R ontgenspektren, in German), *Z Physik*, 48, 174–179, 1928.

[5] Kubo, R., and Ichimura, M., Kramers–Kronig Relations and Sum Rules, *J Math Phys*, 13, 1454–1461, 1972.

[6] Post, E. J., *Formal Structure of Electromagnetics*, Amsterdam: North Holland, 1962.

[7] Nowick, A. S., and Berry, B. S., *Anelastic Relaxation in Crystalline Solids*, New York: Academic, 1972.

[8] Hu, B., Kramers–Kronig in Two Lines, *Am J Physics*, 57, 821, 1989.

[9] Golden, J. M., and Graham, G. A. C., *Boundary Value Problems in Linear Viscoelasticity*, Berlin: Springer Verlag, 1988.

[10] McCrum, N. G., Read, B. E., and Williams, G., *Anelastic and Dielectric Effects in Polymeric Solids*, New York: Dover, 1991.

[11] Graesser, E. J., and Wong, C. R., The Relationship of Traditional Damping Measures for Materials with High Damping Capacity: A Review, in $M^3D$: *Mechanics and Mechanisms of Material Damping*, V. K. Kinra and A. Wolfenden, eds., ASTM (1916 Race St., Philadelphia, PA), 1992. Volume ASTM STP 1169.

[12] Gaul, L., and Nitsche, K., The Role of Friction in Mechanical Joints, *Appl Mech Rev*, 54, 93–105, 2001.

[13] Nashif, A. D., Jones, D. I. G., and Henderson, J. P., *Vibration Damping*, New York: John Wiley, 1985.

[14] Lazan, B. L., *Damping of Materials and Members in Structural Mechanics*, New York: Pergamon, 1968.

[15] Ungar, E. E., and Kerwin, E. M., Loss Factors of Viscoelastic Systems in Terms of Energy Concepts, *J Acoust Soc Am*, 34, 954–957, 1962.

[16] Christensen, R. M., Restrictions upon Viscoelastic Relaxation Functions and Complex Moduli, *Trans Soc Rheology*, 16, 603–614, 1972.

[17] Gurtin, M. E., and Herrera, I., On Dissipation Inequalities and Linear Viscoelasticity, *Quart Appl Math*, 23, 235–245, 1965.

[18] Christensen, R. M., *Theory of Viscoelasticity*, New York: Academic, 1982.

[19] Gottenberg, W. G., and Christensen, R. M. An Experiment for Determination of the Material Property in Shear for a Linear Isotropic Viscoelastic Solid, *Int J Eng Sci*, 2, 45–56, 1964.

[20] Main, I. G., *Vibrations and Waves in Physics*, Cambridge, UK: University Press, 1978.

[21] Lathi, B. P., *Linear Systems and Signals*, Carmichael CA: Berkeley–Cambridge Press, 1992.

[22] Pipkin, A. C., *Lectures on Viscoelasticity Theory*, Heidelberg; Springer Verlag; London: George Allen and Unwin, 1972.

[23] Thurston, R. N., Wave Propagation in Fluids and Normal Solids, in *Physical Acoustics*, New York: Academic, pp. 1–110, 1A, 1964.

[24] Bohn, E., *The Transform Analysis of Linear Systems*, Reading, MA: Addison Wesley, 1963.

[25] Main, I. G., *Vibrations and Waves in Physics*, Cambridge, UK: Cambridge University Press, 1978.

[26] Lakes, R. S., and Katz, J. L., Viscoelastic Properties of Wet Cortical Bone–III. A Non-Linear Constitutive Equation, *J of Biomechanics*, 12, 689–698, 1979.

[27] Lockett, F. J., and Gurtin, M. E., Frequency Response of Nonlinear Viscoelastic Solids. Brown University Technical Report, NONR 562(10), NONR 562(30), 1964.

[28] Lockett, F. J., *Nonlinear Viscoelastic Solids*, New York: Academic, 1972.

[29] Karlow, E. A., Ripples in the Energy of a Damped Harmonic Oscillator, *Am J Physics*, 62, 634–636, 1994.

# 4

## Conceptual Structure of Linear Viscoelasticity

### 4.1 Introduction

The purpose of Chapter 4 is to introduce the spectra, to derive approximate interrelations between the viscoelastic functions, and to assemble all the viscoelastic functions with their interrelations, with the aim of developing physical insight regarding the structure of viscoelasticity theory.

Thus far, we have considered the creep compliance $J(t)$ and the relaxation modulus $E(t)$ as the transient functions; and the storage modulus $E'(\omega)$, the loss modulus $E''(\omega)$, and the loss tangent $\tan\delta(\omega)$ as the dynamic functions. These viscoelastic functions are directly measurable, but different types of experiments, and even different types of apparatus, are used to determine each one.

The viscoelastic functions are interrelated. Let us recapitulate the salient interrelations obtained in previous chapters. We have, in §2.4, found the exact relationship, a convolution, between the relaxation modulus $E(t)$ and the creep compliance $J(t)$:

$\int_0^t J(t-\tau)E(\tau)d\tau = \int_0^t E(t-\tau)J(\tau)d\tau = t$.

The complex dynamic modulus $E^* = E' + iE''$ is related in a simpler way (§3.2) to the complex dynamic compliance $J^* = J' - iJ''$: by an inverse, that is,

$J^* = \frac{1}{E^*}$.

The dynamic moduli $E'(\omega)$ and $E''(\omega)$ are related to the relaxation modulus $E(t)$ by one sided Fourier transforms, as developed in §3.2.2 (with $\hat{E}(t) = E(t) - E_e$):

$E'(\omega) \equiv E(\infty) + \omega \int_0^\infty \hat{E}(t')\sin\omega t' dt', \qquad E''(\omega) \equiv \omega \int_0^\infty \hat{E}(t')\cos\omega t' dt'$.

The real part and the imaginary part of a complex viscoelastic function are related by the Kramers–Kronig relations. For the compliance, these are as follows:

$J'(\omega) - J'(\infty) = \frac{2}{\pi}[\wp \int_0^\infty \frac{\varpi J''(\varpi)}{\varpi^2 - \omega^2} d\varpi], J''(\omega) = \frac{2\omega}{\pi}[\wp \int_0^\infty \frac{J'(\varpi) - J'(\infty)}{\omega^2 - \varpi^2} d\varpi]$.

Interrelations among the viscoelastic functions are of use in converting experimental results from one form to another. The integral forms can be evaluated with the aid of a computer, however, the approximate relations developed below continue to be useful for a quick examination of data, for some special situations, and particularly in the development of physical insight.

## 4.2 Spectra in Linear Viscoelasticity

### 4.2.1 Definitions $H(\tau)$, $L(\tau)$ and Exact Interrelations

Materials governed by a simple relaxation mechanism may be approximated by a single exponential or a unique sum of exponentials. However, such cases are rare; most real materials exhibit behavior that covers many decades of time scale and that cannot be described in this way. If the governing mechanisms are nevertheless considered to give rise to exponential components, one may envisage a relaxation function $E(t)$ composed of a distribution of exponentials, as follows:

$$E(t) - E_e = \int_{-\infty}^{\infty} H(\tau)e^{-t/\tau}d\ln\tau = \int_{0}^{\infty} \frac{H(\tau)}{\tau}e^{-t/\tau}d\tau, \tag{4.1}$$

in which $H(\tau)$ is called the relaxation spectrum and $E_e = \lim_{t\to\infty} E(t)$ is the equilibrium modulus, the stiffness after an infinite time of relaxation. The above equation *defines* the spectrum $H(\tau)$. Similarly, the creep compliance $J(t)$ may be written as

$$J(t) = J(0) + \frac{t}{\eta} + \int_{0}^{\infty} \frac{L(\tau)}{\tau}(1 - e^{-t/\tau})d\tau, \tag{4.2}$$

in which $\eta$ is the asymptotic viscosity (which is infinite in solids) and $L(\tau)$ is the retardation spectrum. These definitions follow those of Ferry [1]; some authors, for example, Christensen [2] define the spectra which differ from the above by a factor of $\tau$ in the integrand.

The dynamic functions are given in terms of the relaxation spectrum as follows, as presented in the classic works of Ferry [1] and Nowick and Berry [3]:

$$E'(\omega) = E_e + \int_{0}^{\infty} \frac{H(\tau)}{\tau} \frac{\omega^2\tau^2}{1 + \omega^2\tau^2}d\tau, \tag{4.3}$$

$$E''(\omega) = \int_{0}^{\infty} \frac{H(\tau)}{\tau} \frac{\omega\tau}{1 + \omega^2\tau^2}d\tau. \tag{4.4}$$

The spectra are related to each other as follows [1, 3], with the integrals considered as principal parts. For a viscoelastic liquid [1], $\eta$ represents an asymptotic viscosity; for solids, the inverse of this viscosity vanishes,

$$L(\tau) = \frac{H(\tau)}{[E_e - \int_{-\infty}^{\infty} \frac{H(u)}{\frac{\tau}{u}-1}d\ln u]^2 + \pi^2 H^2(\tau)} \tag{4.5}$$

$$H(\tau) = \frac{L(\tau)}{[J(0) + \int_{-\infty}^{\infty} \frac{L(u)}{1-\frac{u}{\tau}}d\ln u - \frac{\tau}{\eta}]^2 + \pi^2 L^2(\tau)}. \tag{4.6}$$

If $H(\tau)$ can be written as a sum of delta functions, then $L(\tau)$ also consists of a sum of delta functions, albeit with different time constants. This can be demonstrated by applying a delta function method to the exact transformation equations [4] however, it is more straightforward to convert $H(\tau)$ to a sum of exponentials in $E(t)$, convert $E(t)$ to $J(t)$, then to $L(\tau)$.

### 4.2.2 Particular Spectra

*Lognormal Spectrum*

A variety of particular spectra have been developed for the description of various materials. One that has been useful is the *lognormal spectrum* [5]

$$H(\tau) = \frac{b}{\sqrt{\pi}} e^{-b^2 z^2} \tag{4.7}$$

in which $z = ln(\tau/\tau_m)$. It describes materials in which a molecular relaxation process operates in a range of atomic environments distributed in a Gaussian fashion about a mean value. The damping peak due to this spectrum is broader than a Debye peak and the breadth depends inversely on the value of the parameter $b$. Measurable functions for this spectrum are calculated numerically rather than analytically.

*Box Spectrum*

The *box* spectrum [6] is constant over a range of time values:

$$H(\tau) = H_0, \qquad \tau_1 \leq \tau \leq \tau_2,$$

$$H(\tau) = 0, \qquad \tau > \tau_2, \tau < \tau_1. \tag{4.8}$$

The box spectrum describes certain polymers and is useful in that analytic expressions are available for both the transient and dynamic functions associated with it. Specifically,

$$E(t) = E_e + H_0[Ei(-t/\tau_1) - Ei(-t/\tau_2)], \tag{4.9}$$

in which $Ei(y) \equiv - \int_y^\infty e^{-x}/x \, dx$ is the exponential integral function.

$$E'(\omega) = E_e + \frac{H_0}{2} \ln \frac{1 + \omega^2 \tau_2^2}{1 + \omega^2 \tau_1^2}, \tag{4.10}$$

$$E''(\omega) = H_0(\tan^{-1} \omega \tau_2 - \tan^{-1} \omega \tau_1). \tag{4.11}$$

A box spectrum covering four decades of time and measurable modulus functions are plotted in Figure 4.1. $E''$ is nearly constant over a wide range of frequency; $E'$ and $E(t)$ vary as the log of frequency and of time, respectively, over a wide range.

*Wedge Spectrum*

The *wedge* spectrum [6] is wedge-shaped in a log–log plot. It has been used by Tobolsky [6] to model the glass-rubber transition of polyisobutylene. $H(\tau) = k\tau^{-1/2}$,

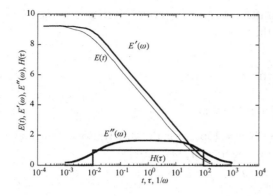

Figure 4.1. Box spectrum $H(\tau)$ covering four decades of time $t$ and measurable modulus functions $E(t)$, $E'(\omega)$, $E''(\omega)$. Arbitrary units of stiffness on the ordinate; arbitrary units of time on the abscissa.

for $a \leq \tau \leq b$, $H(\tau) = 0$, for $\tau > b, \tau < a$. For the wedge relaxation spectrum, integration of the exact interrelation gives the following retardation spectrum:

$$L(\tau) = \frac{1}{H(\tau)} \frac{1}{[\ln \frac{\sqrt{b}+\sqrt{\tau}}{\sqrt{b}-\sqrt{\tau}} \frac{\sqrt{\tau}-\sqrt{a}}{\sqrt{\tau}+\sqrt{a}}]^2 + \pi^2}, \quad a \leq \tau \leq b$$

$$L(\tau) = 0, \quad \tau \leq a, \tau \geq b. \tag{4.12}$$

If the original wedge spectrum covers many decades, $L(\tau) \approx \frac{1}{\pi^2 H} = \frac{\tau^{1/2}}{\pi^2 k}$ except over about half of a decade of time scale near the spectrum edges. Here, $L(\tau)$ increases with time over most of the range in contrast to $H(\tau)$, which decreases, again illustrating the emphasis upon longer times in the retardation spectrum in comparison with the relaxation spectrum.

### Delta Function Spectrum: Standard Linear Solid

We may also consider a *delta function* spectrum,

$$\frac{H(\tau)}{\tau} = A\delta(\tau - \tau_1), \tag{4.13}$$

in which $A$ is a constant. The measurable functions are then those of the standard linear solid. We have

$$E(t) = E_e + Ae^{-t/\tau_1}, \tag{4.14}$$

$$E' = E_e + A\frac{\omega^2\tau_1^2}{1 + \omega^2\tau_1^2}, \tag{4.15}$$

$$E'' = A\frac{\omega\tau_1}{1 + \omega^2\tau_1^2}. \tag{4.16}$$

Plotting these, assuming $E_e = 1$ and $A = 1$ in units of stress versus log time (and $\log(1/\omega)$ to view the dynamic functions on the same scale) for $\tau_1 = 1$ unit of time, we obtain Figure 4.2. Compare with Figure 4.1. Observe also that most of the change

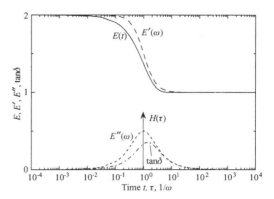

Figure 4.2. Viscoelastic functions for a delta function relaxation spectrum at time $\tau = 1$. Arbitrary units of time. These functions correspond to a standard linear solid. $E''$ and $\tan\delta$ have Debye peak functional dependence.

in the measurable functions for a standard linear solid occurs within one decade of time or frequency.

## 4.3 Approximate Interrelations of Viscoelastic Functions

Many approximate interrelations have been developed, [7, 9]; they display the connection between properties in an accessible way.

### 4.3.1 Interrelations Involving the Spectra

#### *Relaxation Spectrum from Relaxation Modulus*

The exact relation is the definition of the spectrum, $E(t) - E_e = \int_{-\infty}^{\infty} H(\tau)e^{-t/\tau}\,d\ln\tau$ but one may desire to calculate the spectrum from the relaxation modulus, because the latter is measurable. Because no exact analytical inversion formula is available, approximate analytical forms or numerical methods are used. The box spectrum may be used to obtain useful approximations. Differentiating $E(t)$ with respect to $t$ in Equation 4.9 and making use of the Leibnitz rule,

$$t\frac{dE(t)}{dt} = \frac{dE(t)}{d\ln t} = H_0(e^{-t/\tau_1} - e^{-t/\tau_2}). \tag{4.17}$$

If the box spectrum is so broad that $\tau_1 \ll \tau_2$, and if $\tau_1 \gg t \gg \tau_2$, then $e^{-t/\tau_1} \approx 0$, and $e^{-t/\tau_2} \approx 1$, then,

$$H_0 \approx -\frac{dE(t)}{d\ln t}, \tag{4.18}$$

with the approximation improving as the spectrum becomes arbitrarily broad. If the spectrum is not of the box form but is slowly varying, we have

$$H(\tau) \approx -\frac{dE(t)}{d\ln t}\Big|_{t=\tau}. \tag{4.19}$$

To obtain an approximation by a different approach, plot $e^{-t/\tau}$ versus $\tau$ on a logarithmic scale. Observe that $e^{-t/\tau} \approx 1$, for $\tau \gg t$, $e^{-t/\tau} \approx 0$, for $\tau \ll t$. The transition in $e^{-t/\tau}$ covers about one logarithmic decade; if the spectrum $H(\tau)$ varies much more slowly than $e^{-t/\tau}$, the latter may be approximated by a step function. The symbol $H$ is used for the relaxation spectrum; the Heaviside step function is distinguished by use of the script letter $\mathcal{H}$. So according to Ferry [1],

$$E(t) - E_e \approx \int_0^\infty \frac{H(\tau)}{\tau} \mathcal{H}(\tau - t) d\tau = \int_t^\infty \frac{H(\tau)}{\tau} d\tau, \qquad (4.20)$$

so, by the Leibnitz rule,

$$\boxed{H(\tau) \approx -t \frac{dE(t)}{dt} = -\frac{dE(t)}{d\ln t}\Big|_{t=\tau}.} \qquad (4.21)$$

This is known as the Alfrey approximation [10].

### Relaxation Spectrum from Dynamic Storage Modulus

The box spectrum may be used to approximately relate $E'$ to $H(\tau)$. Let us calculate $\omega\, dE'/d\omega = dE'/d\ln\omega$, since a semilog plot of $E'$ versus $\omega$ shows a straight line far from the spectrum edges in Figure 4.1. The condition $\tau_1 \ll t \ll \tau_2$ corresponds to $\omega\tau_1 \ll 1 \ll \omega\tau_2$, so $\omega\, dE'/d\omega \approx H_0$. Again, we recognize that the assumption of $t = 1/\omega$ far from the spectrum edges corresponds to a slowly varying spectrum. So, the final approximation is

$$H(\tau) \approx \omega \frac{dE'}{d\omega} = \frac{dE'}{d\ln\omega}\Big|_{\omega=\tau}. \qquad (4.22)$$

To develop this approximation by direct construction, the portion $\frac{\omega^2\tau^2}{1+\omega^2\tau^2}$ of the integrand in Equation 4.3 is approximated as a step function in that it varies versus log time much more rapidly than does the spectrum, which is assumed to vary slowly. To visualize this, plot $\frac{\omega^2\tau^2}{1+\omega^2\tau^2}$ on a log scale covering many decades; compare Figure 4.2 with Figure 4.1:

$$E(t) - E_e = \int_0^\infty \frac{H(\tau)}{\tau} \frac{\omega^2\tau^2}{1+\omega^2\tau^2} \approx \int_0^\infty \frac{H(\tau)}{\tau} \mathcal{H}(\tau - \frac{1}{\omega}) d\tau = \int_{1/\omega}^\infty \frac{H(\tau)}{\tau} d\tau \quad (4.23)$$

$$\boxed{H(\tau) \approx \omega \frac{dE'(\omega)}{d\omega} = \frac{dE'(\omega)}{d\ln\omega}\Big|_{\omega=1/\tau}.} \qquad (4.24)$$

### Relaxation Spectrum from Dynamic Loss Modulus

The box spectrum may be used to approximately relate $E''$ to $H(\tau)$. The condition $\tau_1 \ll t \ll \tau_2$ gives $E''(\omega) \approx \pi H_0/2$. For a slowly varying spectrum, we make the correspondence $\tau = 1/\omega$, and the approximation:

$$\boxed{H(\tau) \approx \frac{2}{\pi} E''(\omega)\big|_{1/\omega=\tau}.} \qquad (4.25)$$

$E''$ may be related to $H(\tau)$ by direct construction. The kernel in Equation 4.4 has the form of a hump covering about one decade. If the spectrum varies sufficiently slowly, the hump may be regarded in comparison as a delta function [7]. The area of the hump, that is, the integral of $[\omega\tau/(1 + \omega^2\tau^2)]$, is $\pi/2$. Hence,

$$E''(\omega) = \int_0^\infty \frac{H(\tau)}{\tau} \frac{\omega\tau}{1 + \omega^2\tau^2} d\tau \approx \frac{\pi}{2} \int_0^\infty \frac{H(\tau)}{\tau} \tau\delta(\tau - \frac{1}{\omega}) d\tau \qquad (4.26)$$

$$= \frac{\pi}{2} H(\tau = \frac{1}{\omega}),$$

$$H(\tau) \approx \frac{2}{\pi} E''(\omega)|_{\omega=1/\tau}. \qquad (4.27)$$

Referring to Figure 4.1, which shows the viscoelastic functions derived from exact interrelations for a box spectrum, this approximation is seen to be accurate away from the spectrum edges. If the spectrum contains a sharp peak, as shown in Figure 4.2, $E''$ is a poor approximation for the spectrum $H(\tau)$. Because a slowly varying spectrum was assumed in obtaining the approximation, there should be no surprise.

### Retardation Spectrum from Creep or Dynamic Compliance

Procedures similar to the above give approximate relations involving the creep behavior of solids [3]:

$$L(\tau) \approx \frac{dJ(t)}{d\ln t}|_{t=\tau}, \qquad (4.28)$$

$$L(\tau) \approx -\frac{dJ'(\omega)}{d\ln\omega}|_{\omega=1/\tau}, \qquad (4.29)$$

$$L(\tau) \approx \frac{2}{\pi} J''(\omega)|_{\omega=1/\tau}. \qquad (4.30)$$

### Approximate Interrelations among Spectra

Smith [7] obtains the following approximate relations between the spectra with $\tau = \omega^{-1}$:

$$L(\tau) \approx \frac{H(\tau)}{\{E'(\omega) - E''(\omega) + 1.37H(\tau)\}^2 + \pi^2 H^2(\tau))}, \qquad (4.31)$$

$$H(\tau) \approx \frac{L(\tau)}{\{J'(\omega) - J''(\omega) + 1.37L(\tau)\}^2 + \pi^2 L^2(\tau))}. \qquad (4.32)$$

These relations require knowledge of the dynamic properties as well as a spectrum. Tschoegl [Ref. 5 from Appendix] obtains the following, based on the exact form (Example 4.3) for the box spectrum:

$$L(\tau) \approx \frac{1}{H(\tau)}[\frac{1}{[\frac{E_e}{H(\tau)} + \ln\frac{b-\tau}{\tau-a}]^2 + \pi^2}], a \leq \tau \leq b,$$

$$L(\tau) = 0, \tau < a, \tau > b. \qquad (4.33)$$

### *Higher Order Approximations*

Further approximate interrelationships have been given [1, 11]. It is possible to achieve improved accuracy by the use of higher order derivatives [11]. Specifically, the spectrum is written as a series expansion involving derivatives of a measurable function, then substituted into the defining equation for the spectrum.

Results include the following [12]:

$$H(\tau) \approx \{\frac{2}{\pi}[E''(\omega) - \frac{d^2 E''}{d(\ln\omega)^2}]\}|_{\omega=1/\tau},$$ (4.34)

$$H(\tau) \approx -\{\frac{d E(t)}{d\ln t} - \frac{d^2 E}{d(\ln t)^2}\}|_{t=2\tau}.$$ (4.35)

Caution is required with these, since all experimental data contain some noise, and differentiation of the data increases the noise. Spectra, which vary abruptly with time, may be extracted via numerical methods [13].

### 4.3.2 Interrelations Involving Measurable Functions

### *Transient and Dynamic Moduli*

Setting the expressions Equations 4.18, 4.24, and 4.27 for the spectra equal, one obtains approximate interrelations among the moduli,

$$\frac{2}{\pi} E''(\omega)|_{\omega=1/t} \approx \frac{d E'(\omega)}{d\ln\omega}|_{\omega=1/t} \approx -\frac{d E(t)}{d\ln t}.$$ (4.36)

The first approximate equality provides an approximation to the Kramers–Kronig relations. It is possible to provide bounds for the dispersion via finite-frequency range relations [14]. Moreover, from the last two of Equation 4.36, to the lowest order,

$$\boxed{E'(\omega)|_{\omega=1/t} \approx E(t)}.$$ (4.37)

Similarly, from Nowick and Berry [3], $J'(\omega)|_{\omega=1/t} \approx J(t)$.

This approximation (Equation 4.37) may be compared with the curves in Figure 4.1, which neglects $E_e$ that would represent a baseline. The percent error associated with this approximation becomes smaller as $E_e$ becomes larger in relation to the total dispersion. Therefore, the approximation is the most accurate for weakly viscoelastic materials, those in which the dispersion in stiffness is small compared with the stiffness.

Approximations of improved accuracy and no greater complexity can be derived with a little additional effort from the interrelations between transient and dynamic properties [2]. For example,

$$\boxed{E'(\omega)|_{\omega=2/\pi t} \approx E(t)},$$ (4.38)

as considered in Example 4.2.

The transient and dynamic moduli and the spectrum are linked by the following exact equation [7]:

$$E'(\omega)_{\omega=1/t} - E(t) = \int_{-\infty}^{\infty} H(\tau)[\frac{1}{1+t^2/\tau^2} - e^{-t/\tau}]d\ln\tau. \tag{4.39}$$

The quantity in the [ ] brackets forms a hump covering about one decade of time scale. One can approximate the relation by assuming $H(\tau) = k\tau^{-m}$ over the range of time in which this peak differs significantly from zero.

$$E(t) \approx E'(\omega)_{\omega=1/t} - H[\tfrac{\pi}{2}\csc\tfrac{m\pi}{2} - \Gamma(m)], \quad \text{for } -1 < m < 2.$$

For $m = 0$ the quantity in the [ ] brackets converges to the Euler–Mascheroni constant $\approx 0.5772$. If the spectrum is known approximately, the transient and dynamic moduli can be related. Moreover, an approximate relation for $H$ in terms of $E''$ can be obtained by substituting $H(\tau) = k\tau^{-m}$ in $E''(\omega) = \int_{-\infty}^{\infty} \frac{H(\tau)}{\tau} \frac{\omega\tau}{1+\omega^2\tau^2} d\tau$.

Integration, again approximating the hump as a delta function [7] gives $E''(\omega) \approx H\tfrac{\pi}{2}\sec\tfrac{m\pi}{2}$, for $-1 < m < 1$. The zero-order approximation corresponds to the case $m = 0$, $H \approx \tfrac{2}{\pi}E''$, as seen in Equation 4.27.

Experimental dynamic data can be converted via a numerical procedure [8] into a discrete relaxation modulus $G(t)$ and a discrete creep compliance $J(t)$. The discrete time-dependent functions, written as a series of exponentials, are calculated for a time window corresponding to the frequency window of the input data. A nonlinear regression is used to obtain a best fit for $G'$ and $G''$ by varying parameters in the discrete time functions.

### tan$\delta$ *for the Kuhn Model*

The damping corresponding to the modified Kuhn model for logarithmic creep is as follows, with constants given in §2.6.6, Equation 2.76:

$$\tan\delta = \frac{\frac{J_B}{J_A}\tan^{-1}\frac{C_\omega}{\omega}}{1 + \frac{J_B}{J_A}\ln\sqrt{1 + (\frac{C_\omega}{\omega})^2}} \tag{4.40}$$

The loss tangent is almost constant over a wide range of frequency below the characteristic frequency $C_\omega$.

### *Creep Compliance and Relaxation Modulus*

We have considered previously in §2.6 an exact interrelation for a material with a relaxation modulus of power law form $E(t) = At^{-n}$. The value $n$ is the slope of the relaxation curve plotted on a log log scale. The relation is (see §2.6)

$$J(t) = \frac{1}{A\Gamma(1-n)\Gamma(n+1)}t^n. \tag{4.41}$$

The gamma function $\Gamma(x)$ is defined for $n > 0$ as

$$\Gamma(x) = \int_0^{\infty} t^{x-1}e^{-t}dt. \tag{4.42}$$

The identity $\Gamma(x)\Gamma(1-x) = \pi/\sin\pi x$ may be used to obtain the following [1, 7].

$$E(t) = \frac{\sin n\pi}{n\pi}\frac{1}{J(t)},$$

(4.43)

with $n < 1$. The expression is exact for a power law material. For an arbitrary material, one can fit a power law to the behavior for a particular time; the fit will be approximate over a limited interval of time scale, and Equation 4.43 becomes approximate. Moreover, for small $n$, the expression converges to the elastic relation $E = 1/J$.

### The Loss Tangent from Moduli

The loss tangent is related to the slope of the log of moduli or compliances on log-frequency or log-time scales. The approximations are based on the *exact* relationship for a power-law relaxation, $E(t) = At^{-n}$, for which creep is also a power law, and the loss angle is (Example 4.1):

$$\delta = \frac{n\pi}{2}.$$

(4.44)

For an arbitrary relaxation function a power-law may be fitted for a particular value of time. The loss tangent obtained is then an approximation. The approximations are most satisfactory if the slope is small. Moreover,

$$\tan\delta = \frac{E''}{E'} \approx \frac{\pi}{2}\frac{dE'}{d\ln\omega}\frac{1}{E'}$$

(4.45)

from Equation 4.36, but $\frac{d\ln E'}{d\ln\omega} = \frac{1}{E'}\frac{dE'}{d\ln\omega}$, so [3],

$$\tan\delta \approx \frac{\pi}{2}\frac{d\ln E'}{d\ln\omega}.$$

(4.46)

Similarly, the loss tangent is related to the slope of the relaxation [3] curve by the following:

$$\tan\delta \approx -\frac{\pi}{2}\frac{d\ln E(t)}{d\ln t}\Big|_{t=1/\omega}.$$

(4.47)

A different approach to this derivation is given in Zener [15]. The loss tangent is related to the slope of the creep curve by the following:

$$\tan\delta \approx \frac{\pi}{2}\frac{d\ln J(t)}{d\ln t}\Big|_{t=1/\omega}.$$

(4.48)

The loss tangent is related to the slope of the dynamic compliance [3] by the following:

$$\tan\delta \approx -\frac{\pi}{2}\frac{d\ln J'}{d\ln\omega}.$$

(4.49)

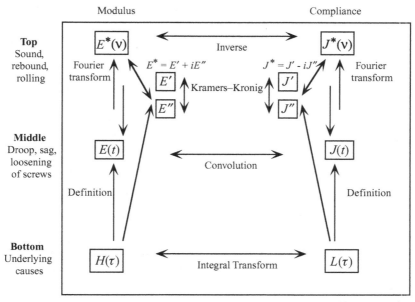

Figure 4.3. Conceptual organization of the viscoelastic functions. The dynamic modulus $E^*(\omega)$ is defined in terms of $E(t)$ in Equations 3.5 and 3.6; $E^*(\omega)$ and $J^*(\omega)$ are related in Equation 3.13. The real and imaginary parts of $J^*(\omega)$ are related in Equations 3.43 and 3.44. The relaxation modulus $E(t)$ is related to the creep compliance $J(t)$ in Equation 2.20. The spectrum $H(\tau)$ is defined in Equation 4.1 and $L(\tau)$ is defined in Equation 4.2. The spectra are related by integral transforms Equations 4.5 and 4.6.

### 4.3.3 Summary, Approximate Relations

The approximate interrelations are useful for developing our intuition concerning the connection between the viscoelastic functions, as well as for calculations. Here, we see that viscoelastic loss is always associated with creep, relaxation, and dispersion of the dynamic stiffness; the slope of the modulus or compliance curves versus log time or log frequency is approximately related to the loss.

## 4.4 Conceptual Organization of the Viscoelastic Functions

In this section, we collect and discuss the viscoelastic functions studied so far according to the conceptual view of Gross [17].

The dynamic functions $E^* = E' + iE''$ and $J^* = J' - iJ''$ are considered to be the ones most closely related to direct experience because they are directly connected with observable phenomena (Figure 4.3). The rate of decay of vibration of a bell or bar after it is struck, acoustic "deadness" or ringing (Table 4.1), and the degree of rebound of a ball, are related to the phase angle $\delta$ between stress and strain. The fundamental resonant frequency of a struck bell, bar, or tuning fork (see cover) depends on the dynamic modulus as well as the object's size and shape. Indeed, some of the first experiments done in childhood in exploring the degree to which

Table 4.1. *Perceived sound of an object with a resonant frequency in the acoustic range, following an impact giving rise to free decay of vibration (adapted from Kerwin and Ungar [16])*

| $\tan\delta$ | Cycles for decay to one-tenth of initial amplitude | Sound |
|---|---|---|
| $\leq 0.001$ | $\geq 730$ | "Clang," approaching pure tone |
| 0.01 | 73 | "Bong": clear pitch |
| 0.1 | 7.3 | "Bunk": discernible pitch |
| 0.5 | 1.5 | "Thud": almost without pitch |

round objects can be bounced or rolled, represent an informal measure of $\tan\delta$; the formal relationship is developed in Chapter 10. Therefore, the dynamic functions are therefore considered to be at the top level in this conceptual scheme. The relation between the dynamic modulus and the dynamic compliance is also the most easily grasped mathematically: it is simply an inverse as in the case of elasticity. The difference is that the viscoelastic dynamic functions are complex numbers, while the elastic moduli are real numbers.

Dynamic functions of frequency also describe several optical phenomena, which are directly experienced. The phase angle between electric field and electric displacement governs optical phenomena such as color (related to absorption of light at different frequencies). Dispersion of light (the rainbow; the rainbow effect in gemstones, referred to as "fire") arises from dependence of the refractive index or optical compliance of the material on frequency.

The transient properties, relaxation and creep are considered one level down in terms of direct perception (Figure 4.3). The associated phenomena of sag and droop can be directly perceived, but such motion is only perceived if it lies within our attention span in the time domain. Rapid creep, faster than one second, is not perceived by us as droop. Slow processes, such as the flow of glaciers and of rock, and even relaxations that occur in hours or days, are not immediately apparent. The transient functions are therefore considered to be at the middle level. The interrelation between transient modulus and compliance is more complicated than in the dynamic case: a convolution. The transient functions were introduced first as fundamental, primarily because most engineers or scientists interested in viscoelasticity initially prefer the time domain to the frequency domain.

The spectra, which relate to the underlying causes for the phenomena represent the bottom level since they are not directly observable; they are even difficult to extract accurately from experimental data. The interrelation between the spectra is a complicated integral transform. The spectra have a significance related to causes only if the fundamental relaxation process is an exponential one in time, $e^{-t/\tau}$. As we will see in Chapter 8, exponentials occur naturally in a variety of processes. Some relaxation mechanisms, however, give rise naturally to a power law relaxation function; a spectrum of exponentials would be less appropriate for materials governed by such mechanisms. Nevertheless, the spectra are useful for visualization because,

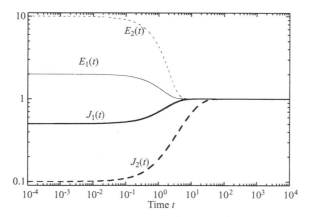

Figure 4.4. Equivalent relaxation and creep curves corresponding to a single relaxation time and a single retardation time respectively, calculated via Equations 2.40, 2.47, and 2.48. $E_1(t)$ and $J_1(t)$, represent a factors of two change in stiffness in relaxation and creep, respectively. $E_2(t)$ and $J_2(t)$, a factor-of-10 change in stiffness in relaxation and creep, respectively. The transition in the creep curve appears to occur at a longer time than in the relaxation curve according to Equation 2.48. These are arbitrary units of time, stiffness, and compliance.

in the absence of local microinertial effects, no relaxation can be more abrupt than a Debye relaxation or single relaxation time process; this corresponds to a spectrum in the form of a delta function.

Viscoelastic response is a single physical reality that can be manifested in many ways, some of which are conceptualized in terms of the viscoelastic functions. A thorough understanding of viscoelasticity entails a simultaneous appreciation of all these manifestations and their analytical representations. Each viscoelastic function, if known over the full range of time or frequency, contains complete information regarding the material's linear behavior. Nevertheless, each viscoelastic function emphasizes a different aspect of the behavior. For example, in §2.6, a material with an exponential relaxation response was seen to exhibit a creep response with a *longer* time constant than that for relaxation. Moreover, as seen Figure 4.4, the creep response of a standard linear solid emphasizes longer term processes than does the relaxation response; the corresponding spectra contain the same difference in emphasis; see Example 4.3. There is a complementary relation between time and frequency as embodied in the Fourier transform. As with the other integral equations considered, the Fourier transform is nonlocal in that the value of a function at a particular frequency depends upon the value of its transform at *all* values of time. All values of time are not equally weighted; we have considered approximations based on locality or on "weak" nonlocality associated with derivatives. In that vein, we may consider frequencies $f$ to be complementary to time $t$ according to $\omega = 2\pi f = 1/t$.

The range of time and frequency permitted by mathematics is the full range from zero to infinity. The range of time and frequency permitted by physics is bounded. Mechanical disturbances in real solids are limited to frequencies below about $10^{13}$ Hz since the wavelength of a sound wave in a material cannot be shorter than the spacing between atoms. A portion of the range of times and frequencies

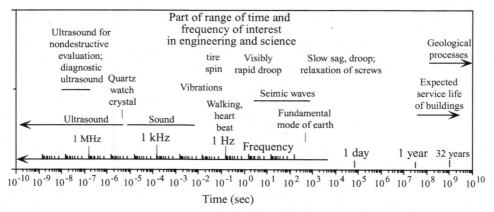

Figure 4.5. A portion of the range of time and frequency of interest in science and engineering: 20 decades.

of interest in science and engineering is shown in Figure 4.5, with some of the salient phenomena. Under some circumstances material behavior over a range of 20 decades or more may be of interest. In geological problems, and in the study of historical buildings, longer time scales than those shown (320 years) are relevant. Ultrasonic waves of frequency up to 10 or 20 MHz are commonly used in nondestructive testing of engineering materials and in medical diagnosis; higher frequencies are used in scientific research. Therefore, it is sensible to think in a logarithmic sense to appreciate the range of times and frequencies of interest.

## 4.5 Summary

The physical reality of viscoelastic behavior may be represented mathematically by a variety of functions of time or frequency. Some of these, the dynamic functions, are regarded as most closely related to direct perception. All these functions are interrelated; most of the exact relationships are integral transforms. Simpler approximate interrelations are obtained under the assumption of slowly varying viscoelastic functions. In considering the full range of viscoelastic behavior, we observe that 20 decades or more (a factor of $10^{20}$) of time and frequency may be of interest in scientific study and engineering applications of viscoelastic materials.

## 4.6 Examples

### Example 4.1

Suppose the relaxation function follows a power law: $E(t) = At^{-n}$. Determine the loss tangent.

### Solution

Substitute $E(t)$ in the transform integrals, with $E_e = 0$, and make the substitution $\omega t' = z$. $E'(\omega) = \omega A \int_0^\infty \frac{\sin \omega t'}{t'^n} dt' = \omega A \int_0^\infty \frac{\sin z}{z^n} \omega^n \frac{1}{\omega} dz$. This is a standard integral form, so $E'(\omega) = \omega^n \frac{A\pi}{2\Gamma(n)\sin\frac{n\pi}{2}}$. Similarly, $E''(\omega) = \omega^n \frac{A\pi}{2\Gamma(n)\cos\frac{n\pi}{2}}$. $\tan\delta(\omega) = \frac{E''(\omega)}{E'(\omega)} = \tan\frac{n\pi}{2}$.

$$\boxed{\delta = \frac{n\pi}{2}}. \qquad (4.50)$$

So the loss angle is proportional to the slope of the power-law relaxation function on a log–log plot. The relationship is exact for power-law relaxation, and it may be used as an approximation if a portion of a relaxation curve can be fitted by a power law.

**Example 4.2**
Show that $E(t) \approx E'(\omega)|_{\omega=2/\pi t}$.
**Solution**
The exact interrelation is an inverse one sided Fourier transform [1]. In the absence of an equilibrium modulus,

$$E(t) = \frac{2}{\pi} \int_0^\infty E'(\omega) \frac{\sin\omega t}{\omega} d\omega. \qquad (4.51)$$

The quantity $\omega^{-1}\sin\omega t$ is a peaked function (Figure 4.6), the integral of which is $\pi/2$. If $E'(\omega)$ varies sufficiently slowly, the peak will seem sharp by comparison, so the approximation,

$$\omega^{-1}\sin\omega t \approx \frac{\pi}{2}\delta(\omega - \omega_0), \qquad (4.52)$$

is sensible [2]. To find $\omega_0$ in terms of $t$, recognize that $\omega_0$ specifies the position of the delta function on the angular frequency axis. The function $\omega^{-1}\sin\omega t$ is peaked but not sharply peaked, so the position is diffuse. An effective position may be defined in terms of the first moment of this function, so,

$$\omega_0 = \frac{\int_0^\infty \omega \frac{\sin\omega t}{\omega} d\omega}{\int_0^\infty \frac{\sin\omega t}{\omega} d\omega} \qquad (4.53)$$

Figure 4.6. Plot of $\omega^{-1}\sin\omega t$.

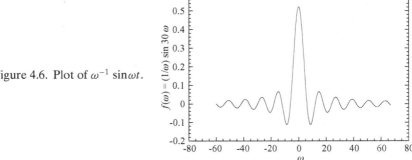

The integral in the numerator does not converge. However, multiply the integrand by $e^{-q\omega}$, which gives

$$\int_0^\infty e^{-q\omega}\sin\omega t\, d\omega = \frac{t}{q^2 + t^2}. \tag{4.54}$$

Taking the limit as $q \to 0$, gives $\omega_0 = (1/t)/(\pi/2)$ so $\omega_0 = 2/\pi t$. The sifting property of the delta function refers to the fact that $\delta(x - a)$ in an integral with respect to $x$ sifts out the value of the multiplied function at $x = a$, as discussed in Appendix §A.1.4. Using the sifting property in the exact equation gives

$$E(t) \approx E'(\omega)|_{\omega=2/\pi t} \tag{4.55}$$

as desired.

This is a better approximation than Equation 4.37 as can be seen by examining Figures 4.1–4.2.

### Example 4.3

Determine the retardation spectrum $L(\tau)$ corresponding to the "box" spectrum of relaxation times $H(\tau) = H_0$, $a \leq \tau \leq b$, and $0, \tau < a, \tau > b$. Discuss the physical implications.

### Solution

Use the exact interrelation between the spectra from Equation 4.5, to obtain

$$L(\tau) = \frac{1}{H_0} \frac{1}{[\frac{E_e}{H_0} + \ln\frac{b-\tau}{\tau-a}]^2 + \pi^2}, \quad a \leq \tau \leq b, \tag{4.56}$$

$L(\tau) = 0, \tau < a$, and $\tau > b$.

Observe in plotting this function (Figure 4.7) that while the assumed relaxation spectrum is flat, the retardation spectrum increases sharply with time [7]. Consequently, the slower viscoelastic processes are emphasized more in a creep test than in a relaxation test.

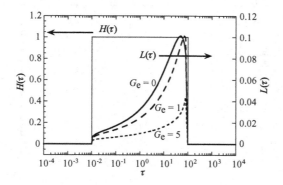

Figure 4.7. Plot of a box function relaxation spectrum $H(\tau)$ and the corresponding retardation spectrum $L(\tau)$.

**Example 4.4**

Suppose that for a material the relaxation strength is 0.1. What is the maximum tan $\delta$ (a) if a Debye relaxation is assumed and (b) if it is desired that tan$\delta$ be uniform over the acoustic range 20 Hz to 20 kHz. Discuss.

**Solution**

For the Debye form, maximum tan$\delta$ is $\tan\delta_{max} = \frac{1}{2}\frac{\Delta}{\sqrt{1+\Delta}} = 0.05$. For a constant loss from angular frequency $\omega_1$ to $\omega_2$, from Example 3.4,

$$J'(0) - J'(\infty) = \int_0^\infty J''(\varpi)d\ln\varpi \approx J''(\ln\omega_1 - \ln\omega_2). \qquad (4.57)$$

This result is approximate since if tan $\delta$ is constant, $J'$ must vary with frequency so $J''$ cannot be constant. Divide both sides by $J'$ to obtain the relaxation strength $\delta J'/J'$.

$$\tan\delta_{max} \approx \frac{1}{6.91}\frac{\delta J'}{J'} = 0.0145. \qquad (4.58)$$

For a particular given relaxation strength, the wider the range of frequency the smaller the magnitude of the damping.

**Example 4.5**

To what degree do viscoelastic materials remember the path by which they arrived at the present state of stress and strain? Does path memory depend on the details of the relaxation function?

**Solution**

Consider the stress that results from a superposition of two strain step functions of amplitude $a$ and $b$ at times $t_1$ and $t_2$,

$$\epsilon(t) = \epsilon_0[a\mathcal{H}(t - t_1) + b\mathcal{H}(t - t_2)]. \qquad (4.59)$$

The resulting stress is, by the principle of superposition and the definition of the relaxation modulus,

$$\sigma(t) = \epsilon_0[a E(t - t_1) + b E(t - t_2)]. \qquad (4.60)$$

Suppose the relaxation function is a simple exponential, $E(t) = E_0 e^{-t/\tau}$, with $\tau$ as the relaxation time. Then, the stress is

$$\sigma(t) = \epsilon_0 E_0[ae^{-(t-t_1)/\tau} + be^{-(t-t_2)/\tau}], \qquad (4.61)$$

$$\sigma(t) = \epsilon_0 E_0[ae^{-t/\tau}e^{-t_1/\tau} + be^{-t/\tau}e^{-t_2/\tau}], \qquad (4.62)$$

$$\sigma(t) = \epsilon_0 E_0[ae^{-t_1/\tau} + be^{-t_2/\tau}]e^{-t/\tau}, \qquad (4.63)$$

$$\sigma(t) = \epsilon_0 E_0[ae^{-t_1/\tau}e^{-t_2/\tau} + be^{-t_2/\tau}e^{-t_2/\tau}]e^{-t/\tau}e^{-t_2/\tau}, \qquad (4.64)$$

$$\sigma(t) = \epsilon_0 E_0[ae^{-(t_2-t_1)/\tau} + b]e^{-(t-t_2)/\tau}. \qquad (4.65)$$

In the last expression, the quantity in the [ ] brackets is constant independent of time $t$. Therefore, the response is the same as if a single step strain had been applied at time $t_2$ [19]. The material that follows a single exponential exhibits memory of transient strain, but no memory of any details of the strain history. Consider, by contrast, relaxation functions that are not exponential, such as the stretched exponential

$$E(t) = (E_0 - E_\infty)e^{-(t/\tau)^\beta} + E_\infty,$$

or sums of exponentials. The rate of relaxation of an exponential relaxation is exponential, but the rate of relaxation for the stretched exponential is $-(\beta/\tau^\beta t^{1-\beta})e^{-(t/\tau)^\beta}$, which depends on the time following perturbation.

### Example 4.6

Consider the path memory for a history in which strain is maintained constant for time $t_1$ to $t_2$, then the *stress* is forced to zero at time $t_2$ by a second step in strain followed by control of stress. How does the response to such a history depend on the relaxation function? Is there anything special about the exponential function?

### Solution

For time $t_1 < t < t_2$, the strain is $\epsilon(t) = \epsilon_0 \mathcal{H}(t - t_1)$, so the stress is $\sigma(t) = \epsilon_0 E(t - t_1)$. The stress at time $t_2$ is $\epsilon_0 E(t_2 - t_1)$, and the corresponding strain increment is $\epsilon_0 E(t_2 - t_1)/E(0)$ because we are dealing with the immediate response to the second step function. To null the stress for time $t > t_2$, the strain history must be

$$\epsilon(t) = \epsilon_0[\mathcal{H}(t - t_1) - \frac{E(t_2 - t_1)}{E(0)}\mathcal{H}(t - t_2)], \tag{4.66}$$

so,

$$\sigma(t) = \epsilon_0[E(t - t_1) - \frac{E(t_2 - t_1)}{E(0)}E(t - t_2)]. \tag{4.67}$$

Suppose again that the relaxation function is an exponential, $E(t) = E_0 e^{-t/\tau}$, $\sigma(t) = \epsilon_0 E_0[[e^{(t-t_1)/\tau} - e^{-(t-t_2)/\tau}e^{-(t_2-t_1)/\tau}]] = \epsilon_0 E_0[[e^{-t/\tau}e^{t_1/\tau} - e^{-t/\tau}e^{t_1/\tau}e^{+t_2/\tau}e^{-t_2/\tau}]] = 0$. So, for an exponential relaxation function, the action of interrupting the relaxation by applying a second step strain to momentarily force the stress to zero, results in zero stress for all time following. For other relaxation functions there is a rebound effect ($\sigma(t) > 0$) that arises from the fact that the momentarily enforced equilibrium is actually a mixture of nonequilibrium states [19].

### Example 4.7

Show that $J''(\omega) \approx -\frac{\pi}{2}\frac{dJ'(\omega)}{d\ln\omega} = -\frac{\pi}{2}\omega\frac{dJ'(\omega)}{d\omega}$. *Hint*: Use the Kramers–Kronig relation for $J'$, differentiate within the integral, and assume that $J''$ is slowly varying.

### Solution

The Kramers–Kronig relation is $J'(\omega) - J'(\infty) = \frac{2}{\pi}[\wp \int_0^\infty \frac{\varpi J''(\varpi)}{\varpi^2 - \omega^2}d\varpi]$. Differentiating, and assuming that $J''$ varies sufficiently slowly that it can be regarded as constant compared with the kernel, $\frac{dJ'(\omega)}{d\omega} \approx \frac{2}{\pi}2\omega J''\int_0^\infty \frac{\varpi}{(\varpi^2 - \omega^2)^2}d\varpi$. This is a standard integral of the following form: recall that $\varpi$ is a variable of integration, which

can be renamed, $\int \frac{x}{(x^2-a^2)^2}dx = -\frac{1}{2(x^2-a^2)}$. $\frac{dJ'(\omega)}{d\omega} \approx \frac{2}{\pi}2\omega J''\{-\frac{1}{2(\varpi^2-\omega^2)}\}^{\infty}_{\varpi=0}$. $J''(\omega) \approx$
$-\frac{\pi}{2}\omega\frac{dJ'(\omega)}{d\omega}$. Moreover, $\frac{d\ln\omega}{d\omega} = \frac{1}{\omega}$, so $J''(\omega) \approx -\frac{\pi}{2}\frac{dJ'(\omega)}{d\ln\omega}$.

## 4.7 Problems

4.1. Consider the stretched exponential form $E(t) = (E_0 - E_\infty)e^{-(t/\tau_r)^\beta} + E_\infty$, with $0 < \beta \leq 1$. Determine the $\tan\delta$ associated with this relaxation. Use an analytical approximation scheme or use a numerical approach to evaluate the transformation integrals. Discuss how the loss peak differs from a Debye peak.

4.2. For a glassy polymer, suppose that $G' = 1$ GPa at 1 Hz and $\tan\delta \approx 0.08$, with the loss approximately independent of frequency. Estimate the stiffness $G$ at 30 kHz, 1 MHz, and after 2 weeks of creep.

4.3. If you have done an informal creep test in an earlier assignment, transform your results to $|E^*|$ (or $|G^*|$) and $\tan\delta$ versus frequency and discuss.

4.4. The approximate interrelations convert nonlocal integral equations into local expressions or weakly nonlocal ones involving derivatives. Some of the approximate interrelations are exact for power law relaxation. Analytically determine the effect at time $t$ of a superposed exponential relaxation of time $\tau$ as a function of $t/\tau$.

4.5. For $H(\tau)/\tau = E_0(\delta(\tau - \tau_1) + \delta(\tau - 10\tau_1))$, and for $H(\tau)/\tau = E_0(\delta(\tau - \tau_1) + \delta(\tau - 100\tau_1))$, determine $E(t)$, $J(t)$, $L(\tau)$ and plot them versus log time. For simplicity, let $E_e = E_0$. Discuss.

4.6. Show that $H(\tau) \approx \frac{2}{\pi}(E''(\omega) - \frac{d^2 E''}{d(\ln\omega)^2})|_{\omega=1/\tau}$.

4.7. If $J(t) = At^n$, what is the retardation spectrum? *Hint*: consider the definition of the gamma function as an integral representation [18]. Write $J(t)$ in terms of the definition of the retardation spectrum and perform a change of variables to make contact with the gamma integral representation.

4.8. To further illustrate path memory effects, calculate the stress from Equation 4.60 assuming a relaxation function consisting of a sum of two exponentials of the same amplitude but different time constants, plot the response, and discuss.

4.9. Sketch as a *space curve* the dependence of stress and strain as they depend on time during dynamic loading of a viscoelastic material. Illustrate that the projection of this curve on the stress–strain plane is an ellipse, and that the projection on the stress–time plane is a sinusoid. What does the space curve look like in the case of an elastic material?

### BIBLIOGRAPHY

[1] Ferry, J. D., *Viscoelastic Properties of Polymers*, 2nd ed., New York: John Wiley, 1970.
[2] Christensen, R. M., *Theory of Viscoelasticity*, Academic, New York: 1982.
[3] Nowick, A. S., and Berry, B. S., *Anelastic Relaxation in Crystalline Solids*, New York: Academic, 1972.

[4] Gross, B., On Creep and Relaxation, *J Appl Phys*, 18, 212–221, 1947.

[5] Nowick, A. S., and Berry, B. S., Lognormal Distribution Function for Describing Anelastic and Other Relaxation Processes, *IBM J Research and Development*, 5, 297–312, 1961.

[6] Tobolsky, A. B., *Properties and Structure of Polymers*, New York: John Wiley, 1960.

[7] Smith, T. L., Approximate Equations for Interconverting the Various Mechanical Properties of Linear Viscoelastic Materials, *Trans Soc Rheol* 2, 131–151, 1958.

[8] Baumgaertel, M., and Winter, H. H., Determination of Discrete Relaxation and Retardation Time Spectra from Dynamic Mechanical Data, *Rheologica Acta*, 28, 511–519, 1989.

[9] Marvin, R. S., The Linear Viscoelastic Behavior of Rubberlike Polymers and Its Molecular Interpretation, in *Viscoelasticity, Phenomenological Aspects*, Academic, New York: 1960.

[10] Alfrey, T., *Mechanical Behaviour of High Polymers*, New York: Interscience, 1948.

[11] Schwarzl, F. R., and Struik, L. C. E., Analysis of Relaxation Measurements, in *Advances in Molecular Relaxation Processes*, Vol. 1, Amsterdam: Elsevier, 1967, pp. 201–255.

[12] Schwarzl, F., and Staverman, A. J., Higher Approximation Methods for the Relaxation Spectrum from Static and Dynamic Measurements of Visco-Elastic Materials, *Appl Sci Res*, A4, 127–141, 1953.

[13] Wiff, D. R., RQP Method of Inferring a Mechanical Relaxation Spectrum, *J Rheology*, 22, 589–597, 1978.

[14] Milton, G. W., Eyre, D. J., and Mantese, J. V., Finite Frequency Range Kramers–Kronig Relations: Bounds on the Dispersion, *Phys Rev Lett*, 79, 3062–3065, 1997.

[15] Zener, C., *Elasticity and Anelasticity of Metals*, Chicago: University of Chicago Press, 1948.

[16] Kerwin, E. M., Jr., and Ungar, E. E., Requirements imposed on polymeric materials in structural damping applications, in *Sound and Vibration Damping with Polymers*, R. D. Corsaro and L. H. Sperling, eds., Washington DC: American Chemical Society, 1990.

[17] Gross, B., *Mathematical Structure of the Theories of Viscoelasticity*, Paris: Hermann, 1968.

[18] Findley, W. N., Lai, J. S., and Onaran, K., *Creep and Relaxation of Nonlinear Viscoelastic Materials*, Amsterdam: North Holland, 1976.

[19] Rekhson, S. M., Viscoelasticity of Glass, Ch. 1 in *Glass: Science and Technology*, D. R. Uhlmann and N. J. Kreidl, eds., Vol. 3, Viscosity and Relaxation, New York: Academic, 1986.

# Viscoelastic Stress and Deformation Analysis

## 5.1 Introduction

In this section, we consider problems of determining stress, strain, or displacement fields in viscoelastic bodies. First, the constitutive equations are generalized from one dimension to three dimensions. Secondly, a problem in simple beam theory is then solved for a viscoelastic material, illustrating the approach of direct construction. It is simpler in this case to use the correspondence principle, which simplifies solutions for certain boundary-value problems in viscoelastic materials. As input to the correspondence principle, one uses a corresponding solution for an elastic material. Several examples involving use of the correspondence principle are developed. Finally, problems not amenable to the correspondence principle are examined.

## 5.2 Three-Dimensional Constitutive Equation

In stress and deformation analysis, constitutive equations are used. For example, in three dimensions, Hooke's law of linear *elasticity* is given by

$$\sigma_{ij} = C_{ijkl}\epsilon_{kl}, \tag{5.1}$$

with $C_{ijkl}$ as the elastic modulus tensor. The usual Einstein summation convention is used when we sum over repeated indices [1, 2]. There are 81 components of $C_{ijkl}$. As discussed in §2.8, the number of independent constants is reduced to 21 by considering symmetry in the stress and strain, and a strain-energy density function. Isotropic materials, which have properties independent of direction, are describable by two independent elastic constants.

To obtain the constitutive equation for *viscoelastic* materials, one repeats, for each component of strain and stress, arguments identical to those given in §2.2 for one dimension. The following is the result:

$$\sigma_{ij}(t) = \int_0^t C_{ijkl}(t - \tau)\frac{d\epsilon_{kl}}{d\tau}d\tau. \tag{5.2}$$

This constitutive equation can accommodate any degree of anisotropy. Each independent component of the modulus tensor can have a different time dependence.

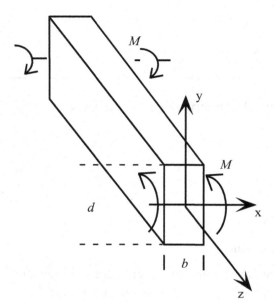

Figure 5.1. Pure bending of a bar.

Many practical materials are approximately isotropic (their properties are independent of direction). For isotropic elastic materials, the constitutive equation [1] is written in §2.8. Engineering constants [3] such as Young's modulus $E$, shear modulus $G$, and Poisson's ratio $\nu$ are frequently used rather than tensorial constants. In isotropic materials, there are two independent viscoelastic functions, which can have different time dependence. As in the case of elasticity, we may consider engineering functions rather than tensorial ones. The viscoelastic time-dependent functions are Young's modulus $E(t)$, shear modulus $G(t) = \mu(t)$, Poisson's ratio $\nu(t)$, and bulk modulus $B(t)$.

## 5.3 Pure Bending by Direct Construction

The problem addressed here is the determination of the relation between the moment and the curvature under the assumptions of elementary beam theory, for a linearly viscoelastic beam. The assumptions are (a) plane sections remain plane; (b) effects of shear are neglected; (c) the material has identical properties in tension and compression. Consider a beam of isotropic viscoelastic material, of rectangular cross-sectional dimensions $b$ and $d$ uniform along the beam axis (the $z$ direction), bent by moments $M(t)$ which are applied at each end (Figure 5.1).

The relation between the curvature $\Omega_c$ and strain $\epsilon$ (in the $z$ direction) is obtained from (a).

$$\epsilon(y, t) = -\Omega_c(t)(y - y_n(t)) = -\Omega_c(t)y', \qquad (5.3)$$

in which $y_n$ is the position of the neutral axis. Substitute Equation 5.3 in the Boltzmann superposition integral, which, by virtue of linearity, entails assumption (c).

Stress is in the $z$ direction: $\sigma(y, t) = \sigma_{zz}(y, t)$ and the absence of other stresses follows from the above assumption that we neglect shear,

$$\sigma(y, t) = -y' \int_0^t E(t - \tau) \frac{d\Omega_c}{d\tau} d\tau. \tag{5.4}$$

The equilibrium conditions $\Sigma F = 0$, $\Sigma M = 0$ give

$$\int \int \sigma(y, t) dx dy = 0, \tag{5.5}$$

$$\int \int \sigma(y, t) y dx dy + M(t) = 0. \tag{5.6}$$

By substituting Equation 5.4 into Equation 5.5,

$$b \int_{-d/2}^{d/2} -y' \int_0^t E(t - \tau) \frac{d\Omega_c}{d\tau} d\tau dy = 0. \tag{5.7}$$

Interchanging the order of integration with the aim of locating the neutral axis,

$$\int_0^t E(t - \tau) \frac{d\Omega_c}{d\tau} d\tau [-b \int_{-d/2}^{d/2} (y - y_n(t)) dy] = 0, \tag{5.8}$$

$$-b \int_{-d/2}^{d/2} (y - y_n(t)) dy = -b[\frac{y^2}{2} - y_n y]_{-d/2}^{d/2} = 0 \tag{5.9}$$

So, $y_n = 0$ so that the neutral axis coincides with the centroid of the cross-section. To examine the moments, substitute Equation 5.4 into Equation 5.6:

$$b \int_{-d/2}^{d/2} -y^2 \int_0^t E(t - \tau) \frac{d\Omega_c}{d\tau} d\tau dy = M(t), \tag{5.10}$$

$$\frac{bd^3}{12} \int_0^t E(t - \tau) \frac{d\Omega_c}{d\tau} d\tau = M(t), \tag{5.11}$$

$$\frac{M(t)}{I} = \int_0^t E(t - \tau) \frac{d\Omega_c}{d\tau} d\tau. \tag{5.12}$$

This is the desired relationship between the curvature history $\Omega_c(t)$ and moment in the bending of a viscoelastic beam; $I$ is the area moment of inertia, which is $\frac{1}{12} bd^3$ for a rectangular section. Equation 5.12 gives the moment history in terms of the curvature history. If we know the moment and wish to calculate the curvature, an appropriate relation can be obtained by taking a Laplace transform of Equation 5.12,

$$s E(s) \Omega_c(s) = \frac{M(s)}{I}, \tag{5.13}$$

but from Laplace transformation of the Boltzmann integrals in the compliance and modulus formulations, Equation 2.18,

$$s J(s) = \frac{1}{s E(s)}, \tag{5.14}$$

so,

$$\Omega_c(s) = \frac{1}{I} M(s) s J(s), \tag{5.15}$$

so, taking the inverse transform,

$$\Omega_c(t) = \frac{1}{I} \int_0^t J(t - \tau) \frac{dM}{d\tau} d\tau. \tag{5.16}$$

This last expression gives the curvature history in terms of the moment history for the bent viscoelastic beam.

Observe that most of the effort in obtaining Equation 5.12 is identical to one which was already expended (in earlier studies of elastic response of bent beams) in obtaining the corresponding solution for an elastic material. One may surmise from the similarity in the elastic and viscoelastic analyses that there should be an easier way to solve-boundary value problems for viscoelastic materials. For a substantial class of such problems, this turns out to be the case, and is discussed in §5.4.

## 5.4 Correspondence Principle

The correspondence principle [5–7] states that if a solution to a linear elasticity problem is known, the solution to the corresponding problem for a linearly viscoelastic material can be obtained by replacing each quantity, which can depend on time by its Laplace transform multiplied by the transform variable ($p$ or $s$), and then transforming back to the time domain. There is the restriction that the interface between boundaries under prescribed load and boundaries under prescribed displacement may not change with time, although the loads and displacements can be time dependent.

The correspondence principle is demonstrated as follows: in solving problems in mechanics, one uses the equilibrium equations,

$$\frac{\partial \sigma_{ij}}{\partial x_i} + F_j = 0, \tag{5.17}$$

and the strain-displacement relations,

$$\epsilon_{ij} = \frac{1}{2} [\frac{\partial u_j}{\partial x_i} + \frac{\partial u_i}{\partial x_i}]. \tag{5.18}$$

One also has the boundary conditions on part of the surface of the object $\Gamma_T$, for which the surface tractions are prescribed,

$$T_j = \sigma_{ij} n_i \tag{5.19}$$

with $n$ as the unit normal. On another part of the surface the surface displacements are prescribed,

$$u_j = U_j. \tag{5.20}$$

Finally, one makes use of the constitutive equation that relates stress to strain. In some elasticity problems, compatibility conditions for the strains must be used.

Here, the standard notation is used in which indices can have values from one to three, repeated indices are summed over. The relations given are for the three dimensional theory of elasticity; simplified counterparts may be used in analyses at the level of elementary mechanics as was done in §5.3. Among the above equations, only the viscoelastic constitutive equation contains a product of time dependent quantities. Therefore, following a Laplace transform only the viscoelastic constitutive equation will differ from its elastic counterpart, and the difference is a factor of the transform variable $s$ (sometimes called $p$). In one dimension, the Boltzmann integral in the Laplace plane is $\sigma(s) = sE(s)\epsilon(s)$. Thus, as the governing equations are carried through to obtain a solution, the distinction between the elastic and viscoelastic cases in the Laplace domain consists of merely the extra factors of $s$, which then appear in the final solution. So, if we have an elasticity solution, the corresponding solution for a viscoelastic material may be obtained by replacing each quantity, which can depend on time by its $s$- (or $p$-) multiplied Laplace transform, and transforming back to the time domain. In order that the elastic and viscoelastic solutions correspond, the interface between boundary regions under specified displacement and under specified stress must not change with time.

There is also a correspondence principle involving Fourier transforms [4, 7]. If a solution to a linear elasticity problem is known, the solution to the corresponding problem for a linearly viscoelastic material can be obtained by replacing each quantity which can depend on frequency given by its Fourier transform. This principle is simpler conceptually than the above Laplace principle since the Fourier transform of a transient viscoelastic response function is a dynamic response function which is itself *measurable experimentally*. The Fourier transform of the stress history is given by

$$\sigma(\omega) = \int_{-\infty}^{\infty} \sigma(t)e^{-i\omega t}dt = \int_{-\infty}^{\infty}\int_{-\infty}^{t} E(t-\tau)\frac{d\epsilon(\tau)}{d\tau}d\tau e^{-i\omega t}dt. \tag{5.21}$$

Using the shift, convolution, and derivative theorems for Fourier transforms discussed in Appendix §A.2.2, the constitutive equation becomes

$$\sigma(\omega) = E^*(\omega)\epsilon(\omega). \tag{5.22}$$

There is no extra multiple of the transform variable in this case because the $\omega$ is absorbed in the definition of the dynamic functions. As in the case of the Laplace correspondence principle described above, Fourier transformation of the other equations (strain displacement, equilibrium, boundary conditions, etc.) used in obtaining an elasticity solution leads to no change in the appearance of these equations. To apply this correspondence principle, begin with a known elasticity solution. Wherever an elastic constant (such as $G$) appears, replace it with the complex dynamic viscoelastic function (such as $G^*$). The result will be valid for linearly viscoelastic materials. The same restrictions associated with the Laplace-type correspondence principle apply here.

## 5.5 Pure Bending by Correspondence

We have the solution relating curvature $\Omega_c$ to moment $M$ for an elastic beam,

$$\Omega_c(t) = \frac{M(t)}{EI}. \tag{5.23}$$

Apply the correspondence principle to obtain the solution of the bending problem for a linearly viscoelastic material. In the Laplace domain, with $E(s)J(s) = 1/s^2$ from the Boltzmann integral,

$$\Omega_c(s) = \frac{M(s)}{s E(s)I} = \frac{1}{I} s M(s) J(s). \tag{5.24}$$

Performing the inverse transform,

$$\Omega_c(t) = \frac{1}{I} \int_0^t J(t - \tau) \frac{dM}{d\tau} d\tau, \tag{5.25}$$

which is identical to the relation obtained earlier by direct construction but is obtained with less effort. We may also consider the relation $\sigma(y) = -\frac{My}{I}$. No elastic constants are present. By the correspondence principle, $\sigma(y, s) = -\frac{M(s)y}{I}$. Therefore, in the viscoelastic case, by an inverse transform, $\sigma(y, t) = -\frac{M(t)y}{I}$. Relaxation has no effect on the relationship between stress and the applied bending moment. If this is counterintuitive, consider that in the elastic case a bar of steel and a bar the same shape of rubber have the same relationship between stress and moment; consider the reason why.

## 5.6 Correspondence Principle in Three Dimensions

In this section, we consider several examples in which stress distributions are taken as given from classical elasticity. In several cases of practical interest, such as stress concentrations and Saint Venant's principle, viscoelasticity can have a significant effect on stress distribution. The correspondence principle is also of use in the study of composite materials as examined in Chapter 9. For reasons explained in §5.7, the time-dependent Poisson's ratio is assumed to be of the relaxation form.

### 5.6.1 Constitutive Equations

One may obtain the viscoelastic constitutive equations in three dimensions by using the elastic versions and the correspondence principle rather than repeating the direct demonstrations as suggested in §5.3. For example, in the elastic case, $\sigma_{ij} = C_{ijkl}\epsilon_{kl}$; in the Laplace plane, $\sigma_{ij}(s) = sC_{ijkl}(s)\epsilon_{kl}(s)$, transforming back, $\sigma_{ij}(t) = \int_0^t C_{ijkl}(t - \tau)\frac{d\epsilon_{kl}}{d\tau} d\tau$, as was argued above in §5.2.

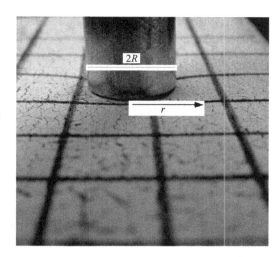

Figure 5.2. Indentation of a semi-infinite solid by a cylindrical indenter of radius $R$.

### 5.6.2 Rigid Indenter on a Semi-Infinite Solid

*Normal Force*

Consider a rigid circular cylindrical indenter of radius $R$ pressed with load $F$ on a semi-infinite viscoelastic solid substrate (Figure 5.2). This could represent a building erected on compliant earth, or an industrial press operation, or an experiment to determine material properties. A solution for an elastic solid of Young's modulus $E$ and Poisson's ratio $v$ is available. The indenter displacement [2] is

$$w = F\frac{1 - v^2}{2RE}.$$ (5.26)

The pressure distribution $q(r)$ as a function of radial coordinate $r$ is as follows: the singularity at $r = R$ occurs as a result of the sudden end of the rigid constraint,

$$q(r) = \frac{F}{2\pi R\sqrt{R^2 - r^2}}.$$ (5.27)

These elastic solutions are valid if $F$, $q$, and $w$ are slowly varying functions of time; the time scale of variation is much longer than the period associated with any resonance. If rapid variations occurred, a new solution containing inertial terms would be required. In the viscoelastic case, which allows similar slow time dependence, the stiffness $E$ and Poisson's ratio $v$ depend on time as well. In the Laplace domain,

$$q(r, s) = \frac{F(s)}{2\pi R\sqrt{R^2 - r^2}};$$ (5.28)

transforming back, the time-dependent pressure follows the time dependence of the force $F$,

$$q(r, t) = \frac{F(t)}{2\pi R\sqrt{R^2 - r^2}}.$$ (5.29)

This case is simple because there are no elastic constants present. As for the displacement $w$,

$$w(s) = \frac{F(s)}{2RsE(s)}(1 - s^2 v(s)v(s)) = \frac{1}{2R}F(s)(sJ(s) - s^3J(s)v(s)v(s)). \quad (5.30)$$

Transforming back, the time-dependent displacement is

$$w(t) = \frac{1}{2R}\int_0^t J(t-\tau)\frac{dF}{d\tau}d\tau$$

$$-\frac{1}{2R}\{\int_0^t J(t-\tau)[\int_0^\tau v(\tau-\eta)[\int_0^\eta v(\eta-\xi)\frac{d^3F(\xi)}{d\xi^3}d\xi]d\eta]d\tau\}. \quad (5.31)$$

The second term is obtained by repeatedly applying the convolution theorem to $s^3J(s)v(s)v(s)$, peeling off one expression at a time. These are cascaded convolutions. We remark that if Poisson's ratio is independent of time, the expression becomes much simpler because

$$w(s) = \frac{1}{2R}F(s)sJ(s)(1 - v^2), \quad (5.32)$$

so,

$$w(t) = \frac{1}{2R}(1 - v^2)\int_0^t J(t-\tau)\frac{dF(\tau)}{d\tau}d\tau. \quad (5.33)$$

If the force $F$ is a step function in time, the indentation displacement is proportional to the creep compliance in compression, understood as $J_E(t)$. Therefore, for the case of constant and known Poisson's ratio, a time-dependent indentation test can be used to determine the creep compliance. If the Poisson's ratio is not constant, a separate experiment is required to determine it.

### Torsion

For torsion of a rigid cylinder of radius $R$ upon a semi-infinite elastic region [8] the shear stress is $\sigma = 4G\phi[\pi\sqrt{((R/r)^2 - 1)}]^{-1}$ for radial coordinate $r < R$. As in the case of normal force there is a singularity in stress at $r = R$. The angular displacement $\phi$ is, in terms of the twisting moment $M$ and shear modulus $G$,

$$\phi = M\frac{3}{16}\frac{1}{R^3 G}. \quad (5.34)$$

Writing $J = 1/G$ and using the correspondence principle as above gives the viscoelastic solution in terms of the creep compliance in shear, $J_G(t)$. Since Poisson's ratio does not appear, interpretation of this sort of experiment is simpler than for indentation by a normal force,

$$\phi(t) = \frac{3}{16}\frac{1}{R^3}\int_0^t J_G(t-\tau)\frac{dM(\tau)}{d\tau}d\tau. \quad (5.35)$$

### 5.6.3 Viscoelastic Rod Held at Constant Extension

Does a stretched viscoelastic rod held at constant extension get fatter or thinner with time [10]? Consider first polymer materials, then arbitrary materials. When an elastic rod is stretched, it gets thinner provided it is made of an "ordinary" material for which $v > 0$. We do not consider here materials for which $v < 0$ [9]. As for a viscoelastic rod, consider the interrelations between Poisson's ratio and the stiffnesses. For an isotropic elastic material, the elastic constants are related [1] as follows (see also Appendix §A.5), $v = \frac{1}{2} - \frac{E}{6B}$, in which $E$ is Young's modulus, $G$ is the shear modulus, $B$ is the bulk modulus, $v$ is Poisson's ratio, and $\lambda$ and $\mu$ are the Lamé constants given above. The last equation for Poisson's ratio is the easiest to convert into viscoelastic form in the context of the stated problem for polymers. Consider the bulk compliance $\kappa = 1/B$, then the Poisson's ratio may be written as follows:

$$v = \frac{1}{2} - \kappa \frac{E}{6}. \tag{5.36}$$

If we consider a polymer material, there is much more relaxation in the shear and axial properties (typically a factor of 1,000 or more through the glass–rubber transition) than in the bulk properties (possibly a factor of two [15]). For this case, approximate $B(t)$ as a constant. Using the correspondence principle and passing to the Laplace domain,

$$sv(s) = \frac{1}{2} - \frac{1}{6}s\kappa E(s), \tag{5.37}$$

since the bulk compliance $\kappa$ is assumed to be constant. Transforming back,

$$v(t) = \frac{1}{2} - \kappa \frac{E(t)}{6}. \tag{5.38}$$

Since $E(t)$ decreases monotonically, $v(t)$ increases for this case (in a typical polymer $v \approx 1/3$ in the glassy region, $v \approx 1/2$ in the rubbery region) so the viscoelastic polymer rod under constant tensile strain becomes *thinner* with time.

As for a general material for which both Young's and the bulk moduli are time dependent, again applying the correspondence principle, $sv(s) = \frac{1}{2} - \frac{1}{6}s^2\kappa(s)E(s)$. Dividing by $s$ and transforming back,

$$v(t) = \frac{1}{2} - \frac{1}{6}\int_0^t E(t-\tau)\frac{d\kappa(\tau)}{d\tau}d\tau. \tag{5.39}$$

It is not immediately obvious from this expression whether there is now any restriction on whether Poisson's ratio increases or decreases; see Problem 5.3.

### 5.6.4 Stress Concentration

The stress concentration factor is the ratio of the maximum stress near an heterogeneity to the stress far from it. In three dimensional elasticity theory, expressions for stress concentration factors usually contain Poisson's ratio. We may consider the effect of relaxation in Poisson's ratio of viscoelastic materials upon the stress concentration factor (SCF). For example, for a spherical cavity in biaxial tension [11],

$$SCF = \frac{\sigma_{\max}}{\sigma_0} = \frac{12}{(7 - 5v)}. \tag{5.40}$$

So $\sigma_{\max}(7 - 5v) = 12\sigma_0$. Applying the correspondence principle, $\sigma_{\max}(s)(7 - 5sv(s)) = 12\sigma_0(s)$. Transforming back,

$$7\sigma_{\max}(t) - 5 \int_0^t v(t - \tau)\frac{d\sigma_{\max}(\tau)}{d\tau}d\tau = 12\sigma_0(t). \tag{5.41}$$

An explicit relation can be obtained given an explicit functional form for $v(t)$. However, suppose, for the sake of simplicity, that for asymptotically short times the material behaves elastically with $v = 1/3$, and at asymptotically long times also behaves elastically with $v \approx 1/2$. The stress concentration factor goes from 2.25 to 2.667, a 19 percent increase. The functional form of the time dependent stress concentration factor is not obtained by an asymptotic argument; it can only be obtained by solving the convolution equation. In a polymer, such a change in Poisson's ratio would likely be accompanied by a relaxation of several orders of magnitude in shear modulus. In many cases, relaxation in Poisson's ratio is neglected because its effect is often much less than that of the relaxation in stiffness. A counterexample is given in §5.6.5.

### 5.6.5 Saint Venant's Principle

Saint Venant's principle is important in the application of elasticity solutions in many practical situations in which boundary conditions are satisfied in the sense of resultants rather than pointwise. For example, a bending moment may be applied to a beam via a complex array of bolted joints, which generate a locally complex stress pattern. In view of Saint Venant's principle, one expects to observe bending type stresses far from the ends.

Saint Venant's principle states that a localized self-equilibrated load system produces stresses that decay with distance more rapidly than stresses due to forces and moments. It is applicable in many situations of interest in engineering. There are some counterexamples (see, e.g., Fung [3]). Consider a sandwich panel with rigid face sheets and an elastic material of Poisson's ratio $v$ sandwiched between them. For Poisson's ratios in the vicinity of 0.5, stresses applied to the end will decay [12] with distance $z$ as $\sigma(z) \propto e^{-\gamma z}$. The decay rate is

$$\gamma \propto \sqrt{\frac{3(1 - 2v)}{3 - 4v}}. \tag{5.42}$$

The form of $\gamma(v)$ makes it difficult to proceed with a transform solution unless a specific functional form of $v(t)$ is chosen. We may assume the material is a viscoelastic polymer in which Poisson's ratio changes from one-third at asymptotically

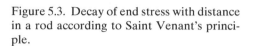

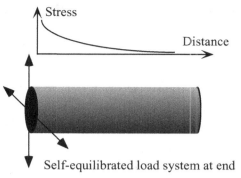

Figure 5.3. Decay of end stress with distance in a rod according to Saint Venant's principle.

short times to nearly one-half at asymptotically long times. The decay rate $\gamma$ for end stresses tends to zero as relaxation proceeds. Therefore, the distance $1/\gamma$, over which there is significant stress, diverges as Poisson's ratio relaxes to one-half. As relaxation progresses in this example, stress propagates further and further into the core material. Practical sandwich panels incorporate as core cellular solids such as anisotropic honeycombs or relatively stiff foam (with a Poisson's ratio of about 0.3) rather than soft polymers. So, this example serves to show that dramatic effects can result from relaxation in Poisson's ratio, rather than as a practical case study.

As for unconstrained rods, a self-equilibrated load system applied to a rod (Figure 5.3) with free lateral surfaces gives rise to stress that decays with distance according to Saint Venant's principle. The distance required for substantial decay is comparable to the rod radius for normal values of Poisson's ratio, but decay can be slow [13] for Poisson's ratio is approaching $-1$, a situation that is possible in foams and composites. If the Poisson's ratio changes with time in a viscoelastic material with negative Poisson's ratio, the spatial rate of stress decay can also change with time.

## 5.7 Poisson's Ratio $v(t)$

In viscoelastic materials Poisson's ratio may be defined in several ways. Some definitions are problematical [16]. Moreover the assumption of a Poisson's ratio constant in time, while helpful in simplifying analysis, is inconsistent with experiment for most materials, and can involve unrealistic theoretical elements. Care is needed in the use of the correspondence principle with Poisson's ratio. In this segment, time-dependent Poisson's ratio is considered [17]. It is shown that Poisson's ratio for tensile relaxation has a different time-dependence than Poisson's ratio in tensile creep.

### 5.7.1 Relaxation in Tension

Apply uniaxial tension to an elastic bar with elastic modulus $E$ and Poisson's ratio $v$. First, consider the elastic case, then apply the correspondence principle to the moduli to obtain the solution for the viscoelastic case. The longitudinal strain $\epsilon_L$ is

$$\epsilon_L = \sigma_L/E. \tag{5.43}$$

The transverse strain $\epsilon_{\text{tr}}$ is, in terms of Poisson's ratio $\nu$,

$$\epsilon_{\text{tr}} = -\nu\epsilon_L = -\nu\sigma_L/E. \tag{5.44}$$

To relate the Poisson's ratio $\nu$ with relaxation in the moduli, express it using the relation for an isotropic elastic solid in terms of the bulk modulus $B$ or the bulk compliance $\kappa = 1/B$ as follows:

$$\nu = \frac{1}{2} - \frac{E}{6B} = \frac{1}{2} - \frac{1}{6}\kappa E. \tag{5.45}$$

Using Equations 5.44 and 5.45, the transverse strain is written as

$$\epsilon_{\text{tr}} = -[\frac{1}{2} - \frac{1}{6}\kappa E]\epsilon_L. \tag{5.46}$$

This relation between strains and moduli for elastic materials is converted to one for viscoelastic materials by use of the correspondence principle. Assume, to simplify analysis, that the bulk compliance is constant in time. The assumption of a constant bulk compliance is a good approximation for polymers in the glass–rubber transition, but it does not apply in general. Apply the correspondence principle to the moduli in Equation 5.46 to obtain

$$\epsilon_{\text{tr}}(s) = -[\frac{1}{2} - \frac{1}{6}\kappa s\, E(s)]\epsilon_L(s). \tag{5.47}$$

Transform back to the time domain via the convolution theorem and use the derivative theorem of the Laplace transform, to obtain the following solution for the transverse strain in a viscoelastic material:

$$\epsilon_{\text{tr}}(t) = -\frac{1}{2}\epsilon_L(t) + \frac{1}{6}\kappa \int_0^t E(t-\tau)\frac{d\epsilon_L(\tau)}{d\tau}d\tau. \tag{5.48}$$

Equation 5.48 expresses the transverse strain for any history of longitudinal strain in terms of the (constant) bulk modulus and the relaxation function in tension. In stress relaxation, the longitudinal strain is $\epsilon_L(t) = \epsilon_0\mathcal{H}(t)$ with $\mathcal{H}(t)$ as a Heaviside step function of time $t$. The step function is chosen to begin an infinitesimal time after zero to avoid a surface term. Substituting in Equation 5.48 gives

$$\nu_r(t) = -\frac{\epsilon_{\text{tr}}(t)}{\epsilon_0} = \frac{1}{2} - \frac{1}{6}\kappa E(t). \tag{5.49}$$

This ratio of transverse to longitudinal strains is a Poisson's ratio; call it $\nu_r(t)$, a Poisson's ratio in relaxation. Observe that $\nu_r(t)$ is less time-dependent than $E(t)$

because, for a typical value $v = 0.3$, the second (time varying) term in Equation 5.49 is about half of the total.

### 5.7.2 Creep in Tension

Consider creep in uniaxial tension: apply a stress of magnitude $\sigma_0$ as a step function of time: $\sigma_L(t) = \sigma_0 \mathcal{H}(t)$. The longitudinal strain is time dependent in a viscoelastic material:

$$\epsilon_L(t) = \sigma_0 J_E(t), \tag{5.50}$$

with $J_E(t)$ as the creep compliance for axial deformation. Substitute the Laplace transform of Equation 5.50 into Equation 5.47. This gives the relation,

$$\epsilon_{tr}(s) = -\frac{1}{2}\epsilon_L(s) + \frac{1}{6}\kappa\sigma_0 s\, E(s) J_E(s) \tag{5.51}$$

The interrelation between creep and relaxation in the Laplace domain is

$$s^2 E(s) J_E(s) = 1. \tag{5.52}$$

Use this result, and the correspondence principle, and apply an inverse Laplace transform to obtain the following relation in the time domain:

$$\epsilon_{tr}(t) = -\frac{1}{2}\epsilon_L(t) + \frac{1}{6}\kappa\sigma_0 \mathcal{H}(t). \tag{5.53}$$

Divide the above transverse strain by the longitudinal strain and recognize the ratio as $v_c(t)$, the Poisson's ratio in creep, to obtain the following:

$$v_c(t) = -\frac{\epsilon_{tr}(t)}{\epsilon_L(t)} = \frac{1}{2} - \frac{1}{6}\kappa\sigma_0 \frac{\mathcal{H}(t)}{\epsilon_L(t)}. \tag{5.54}$$

Recall that the creep compliance is $J_E(t) = \epsilon_L(t)/\sigma_0$. Use the relation in Equation 5.50 and recognize that $t > 0$; the Poisson's ratio in creep is

$$v_c(t) = -\frac{\epsilon_{tr}(t)}{\epsilon_L(t)} = \frac{1}{2} - \frac{1}{6}\kappa\frac{1}{J_E(t)}. \tag{5.55}$$

Observe that the Poisson's ratio $v_c(t)$ in creep differs from the Poisson's ratio $v_r(t)$ in relaxation given in Equation 5.49, since $E(t) \neq 1/J_E(t)$. Specifically, $E(t)J_E(t) \leq 1$ in which the equality corresponds to an elastic solid. From Equations 5.49 and 5.55, the interrelation between $v_c(t)$ and $v_r(t)$ is as follows in terms of $E(t)$ and $J_E(t)$:

$$v_c(t) = v_r(t) + \frac{1}{6}\kappa E(t)[1 - \frac{1}{E(t)J_E(t)}]. \tag{5.56}$$

With a constant bulk modulus assumed in Equation 5.49 and 5.55, $v_r(t)$ and $v_c(t)$ are both increasing functions of time. Moreover, since $E(t)J_E(t) \leq 1$,

Equation 5.56 implies $v_c(t) < v_r(t)$. Poisson's ratio need not increase for all materials. Several counterexamples are given in Chapter 8.

Although the Poisson's ratio in creep differs from the Poisson's ratio in relaxation, the difference is negligible for $\tan\delta = 0.05$, and is small for $\tan\delta = 0.5$ [17]. Only for large relaxations corresponding to the glass–rubber transition in a polymer is the difference substantial. The correspondence principle applied to the Poisson's ratio in the time domain is interpreted as the relaxation type [17].

In the frequency domain, Poisson's ratio is complex. The dynamic correspondence principle gives rise to simple algebraic equations in that case.

## 5.8 Dynamic Problems: Effects of Inertia

It is usually expedient to treat dynamic problems in the frequency domain. One may proceed ab initio or make use of the dynamic correspondence principle to convert an elastic solution [18–20] into a viscoelastic one.

### 5.8.1 Longitudinal Vibration and Waves in a Rod

A uniform rod of length $L$, cross-sectional area $A$, dynamic modulus $E^*$, and density $\rho$ is rigidly fixed at one end and has the other end free. A one-dimensional analysis is performed on the rod; neglect of the Poisson effects is warrantable if the wavelength of stress waves in the rod is much greater than the rod diameter. In §3.7, the stress–strain relation was combined with Newton's second law and the viscoelastic stress–strain relation to obtain the following *wave equation*:

$$\frac{\partial^2 u}{\partial t^2} = \frac{E^*}{\rho}\frac{\partial^2 u}{\partial x^2}. \tag{5.57}$$

This admits traveling wave solutions, such as

$$u(x,t) = u_0\exp[i\omega(\frac{x}{c} - t)]\exp[-x\frac{\omega}{2c}\tan\delta], \tag{5.58}$$

for small loss, with the wave speed

$$c = \sqrt{\frac{E'}{\rho}}. \tag{5.59}$$

The real exponential in Equation 5.58 represents wave attenuation. If we are given a wave equation for elastic materials,

$$\frac{\partial^2 u}{\partial t^2} = \frac{E}{\rho}\frac{\partial^2 u}{\partial x^2}, \tag{5.60}$$

the corresponding wave equation for viscoelastic materials is obtained by replacing the elastic modulus $E$ with the complex modulus $E^*$. The wave speed and attenuation from this wave equation are then determined as in §3.7.

Standing wave solutions corresponding to vibrations are also admissible. A standing wave can be formed as a superposition of two traveling waves, one going right and the other going left. For example, for an elastic material [18],

$$u_n = A_n \cos \frac{\omega_n x}{c} + B_n \sin \frac{\omega_n x}{c}. \tag{5.61}$$

The coefficients $A_n$ are determined from the boundary condition that displacement is zero at one end, $x = 0$. Substituting in the wave equation, $A_n = 0$. At the other end, there is no force, so $\partial u/\partial x = 0$ at $x = L$. So $(\omega_n/c)B_n\cos(\omega_n L/c) = 0$, and $\omega_n = \frac{c\pi n}{2L}, n = 1, 3, 5 \ldots$ Here $n$, an integer, is the mode number. So for fixed-free vibration of a rod, the odd modes are present. If both ends are free, all modes are present. Several experimental methods make use of vibrating rods as presented in §6.9. For viscoelastic materials, the dynamic compliance at the resonant frequencies becomes finite, and the resonance peaks acquire finite width proportional to tan$\delta$, as presented in §3.5.1.

### 5.8.2 Torsional Waves and Vibration in a Rod

In this section, a detailed solution is obtained for an elastic material, then the correspondence principle is applied to obtain a viscoelastic solution. The equation of motion in linear elasticity is

$$\frac{\partial \sigma_{ij}(x_j, t)}{\partial x_j} + F_i(x_j, t) = \rho \frac{\partial^2 u_i(x_j, t)}{\partial t^2}, \tag{5.62}$$

in which $\rho$ is the material density and $u_i$ is the particle displacement [3]; the Einstein summation convention for repeated indices is used. For torsional oscillation, there is only one component of displacement (in the circumferential direction), $u_\theta$, in cylindrical coordinates. The constitutive equation is

$$\sigma = 2G\epsilon, \tag{5.63}$$

with $\epsilon$ as the tensorial shear strain. So, using the strain-displacement relation from elasticity theory,

$$G\nabla^2 u_\theta = \rho \frac{\partial^2 u_i}{\partial t^2}. \tag{5.64}$$

In cylindrical coordinates,

$$\nabla^2 \mathbf{A} = \mathbf{a}_r [\nabla^2 A_r - \frac{A_r}{r^2} - \frac{2}{r^2} \frac{\partial A_\theta}{\partial \theta}] \tag{5.65}$$

$$+ \mathbf{a}_\theta [\nabla^2 A_\theta + \frac{A_\theta}{r^2} + \frac{2}{r^2} \frac{\partial A_r}{\partial \theta}] + \mathbf{a}_z \nabla^2 A_z,$$

in which $\mathbf{A}$ is a vector and $\mathbf{a_r}$, $\mathbf{a}_\theta$, $\mathbf{a_z}$ are unit vectors. $z$ is along the rod axis, $r$ is radial, and $\theta$ is in the circumferential direction. Also,

$$\nabla^2 A_\theta = \frac{1}{r} \frac{\partial}{\partial r} \{r \frac{\partial A_\theta}{\partial r}\} + \frac{1}{r^2} \frac{\partial^2 A_\theta}{\partial \theta^2} + \frac{\partial^2 A_z}{\partial z^2}. \tag{5.66}$$

So,

$$\frac{\partial^2 u_\theta}{\partial r^2} + \frac{1}{r}\frac{\partial u_\theta}{\partial r} + \frac{\partial^2 u_\theta}{\partial z^2} - \frac{u_\theta}{r^2} = \frac{\rho}{G}\frac{\partial^2 u_\theta}{\partial t^2}. \tag{5.67}$$

This was the starting point of the analysis of Christensen [4]. This is also a *wave equation*. Consider a solution of the form:

$$u_\theta = rf(z)e^{i\omega t}. \tag{5.68}$$

This represents torsional waves in that planar cross sections rotate with respect to each other but do not deform. The first term vanishes and the second and fourth terms of the wave equation cancel. The following remains:

$$\frac{\partial^2 u_\theta}{\partial t^2} = \frac{G}{\rho}\frac{\partial^2 u_\theta}{\partial z^2}. \tag{5.69}$$

This *wave equation* is similar to Equation 5.60 for longitudinal waves in one dimension. Other dynamic motions are possible in shear. Radial functions containing Bessel functions are also admissible. The following constitutes a solution for traveling waves. Here, the angular frequency is $\omega = 2\pi\nu$ with $\nu$ as frequency, and the wave number is $k = 2\pi/\lambda$ with $\lambda$ as the wavelength. $u_\theta(x,t) = u_{\theta_0}e^{i(kx-\omega t)}$. Substitute in the wave equation to obtain the *dispersion relation*. $\omega^2 = k^2\frac{G}{\rho}$. Again, $\omega/k = \lambda\nu$ represents the phase velocity or wave speed $c_T$, which for torsional waves depends on the shear modulus $G$ rather than Young's modulus. $c_T = \sqrt{\frac{G}{\rho}}$. The treatment is purely elastic so far. The wave speed for torsional waves is independent of frequency in the elastic case, so these waves propagate nondispersively. The wave propagation solution for a viscoelastic material is obtained by using the correspondence principle to replace $G$ by $G^*$ in the dispersion relation. Then, the dispersion relation becomes $k\sqrt{\frac{G'}{\rho}}\sqrt{1 + i\tan\delta} = \omega$. Substituting in the wave equation, with $\tan\delta \ll 1$,

$$u_\theta(x,t) = u_{\theta_0}\exp[i\omega\{\frac{x}{c_T} - t\}]\exp\{-x\frac{\omega}{2c_T}\tan\delta\}, \tag{5.70}$$

which shows wave propagation and attenuation.

Consider now standing waves, which correspond to vibrations. Returning to Equation 5.68 in the elastic case, the $z$ dependence of the displacement $u_\theta$ is

$$f(z) = A\sin\frac{\Omega z}{L} + B\cos\frac{\Omega z}{L}, \tag{5.71}$$

with

$$\Omega = \omega L\sqrt{\frac{\rho}{G}}. \tag{5.72}$$

To calculate the moment necessary to drive these vibrations, the shear stress is obtained from the displacement field by the strain–displacement relation and stress–strain relation and is integrated over the cross-section. The cross-section is assumed here to be a solid circle of radius $R$,

$$M(z,t) = \frac{\pi}{2}R^4\frac{\rho\omega^2 L}{\Omega}[A\sin\frac{\Omega z}{L} - BA\cos\frac{\Omega z}{L}]e^{i\omega t}. \tag{5.73}$$

The values of $A$ and $B$ come from the end conditions. Consider a rod free at one end and fixed at the other end. The dynamic rigidity (torque $M$ divided by angular displacement $\theta$) in torsional vibration of an elastic right circular cylinder of radius $R$, density $\rho$, length $L$, and shear modulus $G$ fixed at one end and dynamically driven at the other end is given by the following [4]:

$$\frac{M}{\theta} = \frac{\pi R^4}{2} [\rho \omega^2 L] \frac{\mathrm{ctn}\Omega}{\Omega}. \tag{5.74}$$

The development is elastic so far. For a viscoelastic cylinder, the Fourier correspondence principle yields

$$\frac{M^*}{\theta} = \frac{\pi R^4}{2} [\rho \omega^2 L] \frac{\mathrm{ctn}\Omega^*}{\Omega^*}, \tag{5.75}$$

in which

$$\Omega^* = \omega L \sqrt{\frac{\rho}{G^*}}. \tag{5.76}$$

and $G^*$ is the complex shear modulus. These vibrations represent standing waves. If the loss tangent is small, considerable simplification is possible, as has been done in §3.5.1.

Results for torsional vibration of a circular cylindrical rod for materials with several values of tan$\delta$ are plotted in Figure 3.8 (see also [21]). For low loss, resonance peaks occur in the structural compliance as anticipated by the analysis in §3.5.1; the higher harmonics occur at progressively smaller amplitude. The phase angle passes through 90 degrees at the resonances. When tan$\delta = 1$ the dynamic structural compliance displays no resonance peak due to the heavy damping. The dynamic compliance $\theta/M$ differs from the normalized compliance $\Gamma = G'k\theta/\tau$ considered in §3.5.1. Because the dynamic storage modulus $G'$ depends on frequency by virtue of the Kramers–Kronig relations, the compliance $\theta/M$ rises at low frequency.

Kolsky [22] obtains the torsional wave equation Equation 5.69 by more elementary means as follows, by considering the torque $M$ on a thin differential slice of length $dz$ of the bar. The slice experiences a twisting moment $M$ on one side and $M + dM$ on the other side, since the moment varies with position $z$ in the dynamic case. Newton's second law for the slice is written in terms of its angle of rotation $\theta$ and its mass moment of inertia $I = \frac{1}{2}\pi\rho r^4 dz$, with $\rho$ as density and $r$ as radius. $\frac{\partial M}{\partial z} dz = I \frac{\partial^2 \theta}{\partial t^2}$. The moment and angular displacement are related by the shear modulus $G$ of the bar material. $M = \frac{1}{2}\pi G r^4 \frac{\partial \theta}{\partial z}$. Substituting, $\frac{G}{\rho} \frac{\partial^2 \theta}{\partial z^2} = \frac{\partial^2 \theta}{\partial t^2}$, which is the wave equation.

Torsional vibration is simpler than longitudinal vibration since in the latter there is the following effect of Poisson's ratio. For longitudinal of long wavelength, Poisson expansion and contraction occur freely. This is the case considered above. As the wavelength approaches the rod diameter, Poisson contraction, because it varies with axial position, is accompanied by shear. For sufficiently high frequency, the wavelength of longitudinal waves becomes smaller than the cylinder size. In that case, the Poisson expansion and contraction are suppressed, leading to dispersion in which wave speed depends on frequency.

### 5.8.3 Bending Waves and Vibration

For bending, analyze wave motion by the correspondence principle. For bending of an elastic rod, the relationship between moment and curvature is given by

$$M(z, t) = -EI\frac{\partial^2 u}{\partial z^2}, \tag{5.77}$$

in which $z$ is a coordinate along the rod axis, $u$ is a displacement in that direction, and $I$ is the area moment of inertia of the cross-section. Consider now a differential element of length $dz$ along the rod. Applying Newton's second law, and considering that for small displacements the angular acceleration of the element may be neglected, the sum of moments [18] on the elements vanishes,

$$M(z + dz, t) - M(z, t) = V dz, \tag{5.78}$$

so $\frac{\partial M}{\partial z} = V$, with $V$ as the shear force. As for the transverse motion,

$$\rho A dz \frac{\partial^2 u}{\partial t^2} = V(z + dz, t) - V(z, t) + f(z, t) dz \tag{5.79}$$

Dividing by $dz$ and assuming the forcing function $f(z, t)$ to vanish, $\rho A \frac{\partial^2 u}{\partial t^2} = \frac{\partial V}{\partial z}$. Using the moment shear relation,

$$-\frac{\partial^2 u}{\partial t^2} = \frac{E}{\rho}\frac{I}{A}\frac{\partial^4 u}{\partial z^4}. \tag{5.80}$$

This is also a wave equation, however the wave speed depends on frequency or wavelength even in the elastic case. To demonstrate this, substitute a trial wave solution of the form $u(z, t) = \sin(kz - \omega t)$. Then, the dispersion relation is

$$\omega^2 = \frac{E}{\rho}\frac{I}{A}k^4. \tag{5.81}$$

The wave speed [3], specifically the phase velocity, for bending waves in an elastic rod is

$$c_B = \lambda \nu = \frac{\omega}{k} = k\sqrt{\frac{E}{\rho}\frac{I}{A}}. \tag{5.82}$$

Recall that for the angular frequency, $\omega = 2\pi\nu$ and for the wave number $k = 2\pi/\lambda$. The wave speed for bending waves is inversely proportional to the wavelength, so there is *dispersion*, in contrast to the case of torsional waves.

The wave speed $c_B$ is the *phase velocity*, the speed at which the wave crests, or other features of constant phase, propagate [22]. In a dispersive medium such as this one, the energy of a pulse travels at a different speed, called the *group velocity*. For bending waves, both the phase and group velocity given by the elementary formula for a slender bar tend to infinity for short wavelength. This is not realistic.

Therefore, when the wavelength becomes comparable to the bar thickness, more sophistication must be incorporated into the analysis. Specifically, the rotary motion and shear deformations of segments of the bar must be taken into account when the wavelength of flexural waves is small. Moreover, higher order modes can occur [19], not accounted for in the elementary analysis.

The wave propagation solution for a viscoelastic material is obtained as was done above (§5.8.2) for torsion by using the correspondence principle to replace $E$ by $E^*$ in the dispersion relation. The complex modulus depends on frequency, therefore, the wave speed depends on frequency. The imaginary part of the complex modulus gives rise to wave attenuation. Then, the dispersion relation becomes

$$\omega^2 = \frac{E^*}{\rho} \frac{I}{A} k^4. \tag{5.83}$$

By substituting in the wave equation, one finds that the loss tangent gives rise to attenuation, and the frequency dependent real part of the modulus alters the pre-existing dispersion of waves. Standing waves can be excited in a bar of finite length. Such waves are useful in experimental settings [14]. For a bar free at both ends, the natural frequencies go approximately as $3^2, 5^2, 7^2$. For a bar fixed at one end and free at the other, the natural frequencies, go approximately as $(1.194)^2, (2.988)^2, 5^2, 7^2$.

As in the case of torsion, the loss tangent can be extracted from the width of the resonance curve under driven conditions or from the time of decay of vibrations.

### 5.8.4 Waves in Three Dimensions

Waves in one dimension have been considered in §3.7. In three dimensions, the field equation for classical elasticity is Navier's equation, with $G$ (also called $\mu$) as the shear modulus, $\lambda$ as the second Lamé elastic constant (no relationship to the wavelength $\lambda$), $F_i$ as the body force, $\rho$ as the material density and $u$ as the particle displacement:

$$G\frac{\partial^2 u_i}{\partial x_j^2} + (\lambda + G)\frac{\partial^2 u_j}{\partial x_j \partial x_i} + F_i = \rho\frac{\partial^2 u_i}{\partial t^2}. \tag{5.84}$$

Navier's equation is obtained from the elastic constitutive equation for *isotropic* materials, $\sigma_{ij} = \lambda\epsilon_{kk}\delta_{ij} + 2\mu\epsilon_{ij}$ in which $\lambda$ and $\mu$ are the two independent Lamé elastic constants, $\delta_{ij}$ is the Kronecker delta (1 if $i = j$, 0 if $i \neq j$) and $\epsilon_{kk} = \epsilon_{11} + \epsilon_{22} + \epsilon_{33}$, and from the equation of motion $\rho\frac{\partial^2 u_i}{\partial t^2} = \frac{\partial \sigma_{ij}}{\partial x_j} + F_i$. The equation of motion is the continuum representation of Newton's second law.

In three dimensions, one can have *longitudinal* waves, also called *dilatational* or *irrotational waves* [2], in which the particle displacement is in the same direction as the wave propagation, as follows:

$$u_x = A\sin\frac{2\pi}{\lambda}(x - c_L t), u_y = 0, u_z = 0, \tag{5.85}$$

with $\lambda$ as the wavelength. This is a plane wave solution in which surfaces of constant phase, for example, the wave crests, form parallel planes. The solution satisfies Navier's equation provided the wave speed is taken as [2]

$$c_L = \sqrt{\frac{\lambda + 2G}{\rho}}. \tag{5.86}$$

Here, $\lambda$ is the Lamé elastic constant. An alternative form [3] is

$$c_L = \sqrt{\frac{E(1 - v)}{\rho(1 + v)(1 - 2v)}}. \tag{5.87}$$

As for interpretation, the modulus tensor element $C_{1111}$ is given for isotropic materials by [2] as shown in Example 5.9 governs the velocity of longitudinal waves in the one direction,

$$C_{1111} = E\frac{1 - v}{(1 + v)(1 - 2v)}, \tag{5.88}$$

This tensor element represents stiffness in the one direction, when there is no strain in the two or three directions; it is a constrained modulus. Deformation in longitudinal plane waves is constrained in that there is no room for transverse motion.

In three dimensions, it is also possible to have *transverse* (or shear) waves in which the particle displacement is orthogonal to the direction of wave propagation, as follows. Transverse waves are characterized by *polarization*, or the plane containing the propagation direction and the direction in which the particles move. For example, for particle motion in the $y$ direction,

$$u_y = A\sin\frac{2\pi}{\lambda}(x - c_T t), u_x = 0, u_z = 0, \tag{5.89}$$

with $\lambda$ as the wavelength. The solution satisfies Navier's equation provided the wave speed is taken as

$$c_T = \sqrt{\frac{G}{\rho}}. \tag{5.90}$$

The ratio of wave speeds depends on Poisson's ratio. For $v = 0.25$, $c_L = c_T\sqrt{3}$; for Poisson's ratio approaching 0.5, the longitudinal wave-speed is much greater than the transverse wave-speed, $c_L \gg c_T$.

The above development is for elastic solids. In three-dimensional viscoelastic bodies, we may apply the dynamic correspondence principle and replace the elastic constants that appear in the particle displacement equations via the wave speeds, with the corresponding complex viscoelastic functions. Attenuations are obtained as in the one dimensional case considered in §3.7. The attenuations for longitudinal and transverse waves can differ, depending on the material, because an isotropic material has two independent stiffness values and two independent loss tangents.

## 5.9 Noncorrespondence Problems

There are problems of practical interest that cannot be treated by the correspondence principle. These include extending cracks, moving loads upon a surface, and ablation (or evaporation) of a solid. The correspondence principle cannot be applied in these cases because the interface between boundary regions under specified displacement and under specified stress moves with time, contradicting one of the assumptions used in deriving the correspondence principle. Solutions, nonetheless, can be obtained albeit with more difficulty [23].

### 5.9.1 Solution by Direct Construction: Example

In the direct construction approach, one proceeds ab initio, as was done in §5.3. The stress or strain history of each differential element of the material is examined explicitly with no appeal to any elastic solutions.

As a simple example consider a weight $w$ hanging from a viscoelastic rope of cross sectional area $A$, length $L$, and weight that is negligible compared with $w$. The rope is initially supported at its upper end, and, of the total length $L_{\text{total}} = L + L_1 + L_2$, the length $L_1 + L_2$ is initially embedded in the support. The weight is applied at time $t = 0$. At a later time $t_1 > 0$, a section of depth $L_1$ of the support crumbles (or ablates), exposing more rope; at a later time $t_2 > t_1$, a section of depth $L_2$ of the support crumbles. Determine the downward displacement of the weight.

For an elastic rope, the displacement $u$ of the weight is, for time $t > t_2$,

$$u = \frac{w}{EA}(L + L_1 + L_2). \tag{5.91}$$

The correspondence principle cannot be applied in this case since regions of the rope pass from a condition of specified displacement to a condition of specified load. We may consider the deformation of each region of the rope separately so that in the elastic case,

$$u(t) = \frac{wJ}{A}(L\mathcal{H}(t) + L_1\mathcal{H}(t - t_1) + L_2\mathcal{H}(t - t_2)). \tag{5.92}$$

In the viscoelastic case, via superposition,

$$u(t) = \frac{w}{A}(LJ(t) + L_1 J(t - t_1) + L_2 J(t - t_2)). \tag{5.93}$$

This solution reflects the fact that different regions of the rope are loaded at different times.

This problem is not amenable to the correspondence principle; one may attempt a "solution" obtained by that method with the actual solution and compare. Again, consider Equation 5.91 for the displacement of the end of an elastic rope for time $t > t_2$. Incorporating the $s$-multiplied Laplace transform of the compliance $J = E^{-1}$, $u(s) = sJ(s)\frac{w(s)}{A}(L + L_1 + L_2)$. Transforming back, $u(t) = \frac{L+L_1+L_2}{A}\int_0^t J(t - \tau)\frac{dw(\tau)}{d\tau}d\tau$. To proceed further, the load history $w(t)$ is required, however different regions of the rope are loaded at different times. Therefore, a single functional form for the load history is not available.

### 5.9.2 A Generalized Correspondence Principle

A generalized correspondence principle was developed by Graham [25] and discussed by Christensen [4]. One assumes that an elastic solution is known in which a displacement, strain or stress $c^{el}(\mathbf{r}, t)$ is known, which is separable into a product of a function $\xi(\mu, B)$ of the elastic constants $\mu$ and $B$, and a function $C^e(\mathbf{r}, t)$ of the spatial coordinates $\mathbf{r}$ and time $t$,

$$c^{el}(\mathbf{r}, t) = \xi(\mu, B)C^e(\mathbf{r}, t). \tag{5.94}$$

Suppose, further, that $C^e(\mathbf{r}, t) = 0$ upon a region $B_2$ (part of the boundary) and the region $B_2$ is monotonically decreasing. Then, with $s$ as the Laplace transform variable,

$$c^{visc}(\mathbf{r}, t) = K(t)C^e(\mathbf{r}, 0) + \int_0^t K(t - \tau)\frac{dC^e(\mathbf{r}, \tau)}{d\tau}d\tau \tag{5.95}$$

with ($K$ has nothing to do with the bulk modulus)

$$K(t) = \mathcal{L}^{-1}\{[\frac{1}{s}\xi(s\mu(s), s B(s))]\}. \tag{5.96}$$

So, a subset of problems not amenable to the correspondence principle can be solved by the generalized correspondence principle.

### 5.9.3 Contact Problems

Many noncorrespondence type problems and their solutions are given in Christensen [4] and Graham [29]. An interesting example is that of the winding of tape or paper onto rolls [26] in which the tape is under tension and is added one layer at a time. As each layer is added, it alters the stress state in the hub and in the layers present at that time. If the tape is viscoelastic, the stress state at the end of the winding phase depends on the winding speed. Storage or pausing of the tape will result in stress redistribution.

#### *Rolling*

Rolling of a rigid cylinder [27] on a viscoelastic substrate is a noncorrespondence problem because regions of the substrate experience different boundary conditions as time progresses and they come into contact and lose contact with the roller. Viscoelasticity of the substrate gives rise to rolling friction as discussed in §10.8. Similarly, rolling resistance occurs in the rolling of a stiff sphere or cylinder on a stiff surface.

### *Hertz Problem: Sphere Contact*

For two spheres in contact, both the displacement $u$ and the radius of contact $a$ are nonlinear functions of the force $F$ pressing the spheres together. This is the classic Hertz problem treated by Timoshenko [2] §140. The displacement is, in terms of the radii $R_1$, $R_2$, Young's moduli $E_1$, $E_2$, and Poisson's ratios $v_1$, $v_2$,

$$u = (\frac{9}{16})^{1/3} \pi^{2/3} F^{2/3} [\frac{1 - v_1^2}{\pi E_1} + \frac{1 - v_2^2}{\pi E_2}]^{2/3} (\frac{R_1 + R_2}{R_1 R_2})^{1/3}. \tag{5.97}$$

This relation between force and displacement manifests structural nonlinearity; the material assumed to be linearly elastic gives rise to a nonlinear force-displacement characteristic from the changing area of contact. Special cases of interest in sports and in experimental testing include a rigid ball pressed into a substrate with a flat surface, corresponding to one radius tending to infinity, and a compliant ball compressed between rigid flat surfaces.

Although the solution for an elastic solid is simple, the solution is not amenable to the correspondence principle due to the change of boundary condition as the contact advances or recedes. Indentation of a substrate by a curved indenter also gives rise to a noncorrespondence problem [28, 29] for which alternate methods are used for the solution.

### 5.10 Bending in Nonlinear Viscoelasticity

Problems considered thus far have been within the linear theory. We consider here a simple problem of bending of a beam of material, which undergoes steady state (secondary) creep [30] governed by the following nonlinear relation:

$$\frac{d\epsilon}{dt} = A\sigma^n. \tag{5.98}$$

Although this relation between stress and strain rate is nonlinear, it deals only with creep and offers no prediction of the response to stress which is not constant in time. Consider $z$ to be along the beam axis and $y$ to be along the beam depth so that the beam is bent in the plane $zy$. The beam has a depth $d$ in the $y$ direction and a width $b$ so that $I = bd^3/12$ is the area moment of inertia as in Figure 5.1. The "initial" response, considered to be elastic, is that the curvature $\Omega_c$ is $\Omega_c = \frac{M}{EI} = \frac{MJ}{I}$, and the stress is $\sigma = \frac{My}{I}$. The bending moment is

$$b \int_{-d/2}^{d/2} y\sigma \, dy = M(t). \tag{5.99}$$

The creep strain rate is, from the strain-curvature relation,

$$\frac{d\epsilon}{dt} = y\frac{d\Omega_c}{dt} = A\sigma^n. \tag{5.100}$$

Combining the above,

$$M = [\Omega_c \frac{b^n d^{2n+1}(n)^n}{2A(4n+2)^n}]^{1/n}. \tag{5.101}$$

The stress, obtained from the Equation 5.99 is

$$\frac{\sigma(y)}{\sigma_{\max}} = \frac{2}{3}(1 + \frac{1}{2n})|\frac{2y}{d}|^{1/n}, \tag{5.102}$$

for $0 < \frac{2y}{d} < 1$

$$\frac{\sigma(y)}{\sigma_{\max}} = -\frac{2}{3}(1 + \frac{1}{2n})|\frac{2y}{d}|^{1/n}, \tag{5.103}$$

for $-1 < \frac{2y}{d} < 0$.

The stress distribution is no longer linear across the depth of the beam as it is in linear elasticity and in linear viscoelasticity. Instead, the stress is an antisymmetric function of the coordinate $y$ and the creep exponent $n$, but not on time. The stress would redistribute itself with time during the primary phase of nonlinear creep, not considered here.

## 5.11 Summary

Many problems in determining stress distributions in viscoelastic objects can be treated simply via the correspondence principle, provided a solution for an elastic object of the same geometry is available. In problems involving changing boundary conditions or nonlinear behavior, the correspondence principle is not applicable; direct construction or other methods may be applied in those cases.

## 5.12 Examples

**Example 5.1**
An elastic bar (of negligible damping) has a Young's modulus $E_1$ and dimensions $b$ and $h$. A thin layer of thickness $\xi \ll h$ of a linearly viscoelastic material of stiffness $E_2^*$ is applied to two lateral surfaces of width $b$, as shown in Figure 5.4.

Show that the effective loss tangent of the composite bar in bending is

$$\tan\delta_{eff} \approx 6\frac{\xi}{h}\frac{E_2'}{E_1}\tan\delta_2. \tag{5.104}$$

**Solution**
Use the correspondence principle. Because the loss tangent is required, the Fourier version of the correspondence principle is appropriate. First, obtain the elastic solution for the bending rigidity. The bar's area moment of inertia is $\frac{1}{12}bh^3$. The added layers are also rectangular in cross-section with area $A = b\xi$ each and are displaced from the neutral axis by a distance $d$. Their contribution to the total area moment is given by the parallel axis theorem:

$$Ad^2 = 2b\xi\frac{(h + \xi)^2}{4}. \tag{5.105}$$

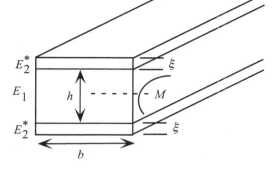

Figure 5.4. Elastic bar with viscoelastic layers.

The bending rigidity, assuming *elastic* layers is, therefore,

$$\Gamma_b = \frac{M}{\Omega_c} = E_1[\frac{1}{12}bh^3 + 2b\xi \frac{E_2}{E_1}\frac{1}{4}(h+\xi)^2]. \tag{5.106}$$

Apply the correspondence principle, and take the ratio of real and imaginary parts:

$$\frac{M^*}{\Omega_c} = E_1[\frac{1}{12}bh^3 + 2b\xi \frac{E_2^*}{E_1}\frac{1}{4}(h+\xi)^2]. \tag{5.107}$$

$$\tan\delta_{\text{eff}} = \frac{\Im(\frac{M^*}{\Omega_c})}{\Re(\frac{M^*}{\Omega_c})} = \frac{2b\xi \frac{E_2''}{E_1}\frac{1}{4}(h+\xi)^2}{\frac{1}{12}bh^3 + 2b\xi \frac{E_2'}{E_1}\frac{1}{4}(h+\xi)^2}. \tag{5.108}$$

Multiplying both numerator and denominator by $12/bh^3$,

$$\tan\delta_{\text{eff}} = \frac{6\frac{\xi}{h}\frac{E_2''}{E_1}\frac{(h+\xi)^2}{h^2}}{1 + 6\frac{\xi}{h}\frac{E_2'}{E_1}\frac{(h+\xi)^2}{h^2}}. \tag{5.109}$$

For a thin layer, $\xi \ll h$, so, with the definition of $\tan\delta$,

$$\tan\delta_{\text{eff}} \approx \frac{6\frac{\xi}{h}\frac{E_2'}{E_1}\tan\delta_2}{1 + 6\frac{\xi}{h}\frac{E_2'}{E_1}}. \tag{5.110}$$

This may be simplified further if the layer material is not too stiff:

$$\tan\delta_{\text{eff}} \approx 6\frac{\xi}{h}\frac{E_2'}{E_1}\tan\delta_2. \tag{5.111}$$

Consequently the effective damping of the composite beam is maximized if the added layer is made as thick as possible, and of a stiff material with the largest possible value of loss tangent. Polymer layers are applied to metallic structural elements to reduce vibration and noise as discussed in §10.6.

**Example 5.2**
A concentrated load $F$ is suddenly applied orthogonal to the surface of a semi-infinite viscoelastic region with a horizontal boundary, then the load is held constant.

What is the vertical motion of points on the surface as a function of time? Suppose the Poisson's ratio is constant in time; ignore inertial effects.

**Solution**

Use the correspondence principle. The quasistatic solution for the displacement field $u$ of an elastic region [2] under a concentrated load $F$ is as follows:

$$u_x = -F\frac{(1-2v)(1+v)}{2\pi\,Er}, u_z = \frac{F(1-v^2)}{\pi\,Er}. \tag{5.112}$$

The coordinate $z$ is down, perpendicular to the surface; $r$ is the radial distance from the point of load application. Applying the correspondence principle for the time domain, and allowing the load to be time dependent,

$$u_z(s,r) = \frac{(1-v^2)}{\pi r}sF(s)J(s). \tag{5.113}$$

Transforming back,

$$u_z(r,t) = \frac{(1-v^2)}{\pi r}\int_0^t J(t-t)\frac{dF}{d\tau}d\tau. \tag{5.114}$$

Because the load history is a step function in time, the displacement history follows the creep function,

$$u_z(r,t) = \frac{(1-v^2)}{\pi r}FJ(t). \tag{5.115}$$

**Example 5.3**

A viscoelastic rope of density $\rho$ radius $r$, and length $L$ is suddenly lowered off a cliff at time zero, and immediately a climber of mass $m$ begins to climb down at speed $v$. Write expressions for the stress, strain, and displacement of the rope as a function of a vertical coordinate $y$ (in the down direction) and time $t$. Neglect thrashing oscillations of the rope after it is lowered.

**Solution**

Proceed by direct construction. The stress is

$$\sigma(y,t) = \rho gy\mathcal{H}(t) + \frac{mg}{\pi r^2}\mathcal{H}(t-\frac{y}{v}). \tag{5.116}$$

The first term represents the stress due to the weight of the rope and is simply the weight of the rope divided by its cross sectional area. The step function $\mathcal{H}(t)$ reflects the fact that the gravitational stress begins at time zero. The second term represents the stress due to the weight of the climber. The step function has a value of unity for $vt > y$ or $y < vt$. Only the rope above the climber experiences gravitational load due to his weight, but as he descends, more rope is stressed.

Each differential element of rope, as it receives a step stress, exhibits a strain response that follows the creep function. Therefore, the strain in the $y$ direction is

$$\epsilon(y,t) = \rho gyJ(t) + \frac{mg}{\pi r^2}J(t-\frac{y}{v}). \tag{5.117}$$

The displacement $u$ is obtained by integrating the strain with respect to $y$. The upper limit in the second term is the $y$ position of the climber, $y_m = vt$,

$$u(y, t) = \rho g \frac{y^2}{2} J(t) + \frac{mg}{\pi r^2} \int_0^{vt} J(t - \frac{y}{v}) dy. \tag{5.118}$$

Consider as a special case an elastic rope, for which $J(t) = J_0$ is constant and equivalent to the inverse of Young's modulus $E$. Then,

$$u(y, t) = \rho g \frac{y^2}{2} \frac{1}{E} + \frac{mg}{\pi r^2} \frac{1}{E} \int_0^{vt} \mathcal{H}(t - \frac{y}{v}) dy. \tag{5.119}$$

The integral of a step function is a slope function,

$$u(y, t) = \rho g \frac{y^2}{2} \frac{1}{E} + \frac{mg}{\pi r^2} \frac{1}{E} y \mathcal{H}(t - \frac{y}{v}). \tag{5.120}$$

**Example 5.4**

Consider, again, the viscoelastic rope in Example 5.3. Is the generalized correspondence principle applicable? If so, use it to obtain a solution. Consider only the deformation due to the weight of the climber.

**Solution**

The elastic solution in Equation 5.119 can be decomposed into a product of a function $\xi$ of the elastic constant $E = J^{-1}$, and a function $C^e(y, t)$ of the spatial coordinate $y$ and time $t$, as follows:

$\xi = J \frac{mg}{\pi r^2}$,

$C^e(y, t) = \int_0^{vt} \mathcal{H}(t - \frac{y}{v}) dy$.

The deformation due to the weight of the climber occurs only for $y < vt$, and that deformation is zero in the monotonically decreasing region of rope below the climber, as required in the statement of the generalized correspondence principle. Following Equation 5.96, evaluate $K$,

$K(t) = \mathcal{L}^{-1}\{(\frac{1}{s}\xi(sJ(s)))\} = J(t)\frac{mg}{\pi r^2}$.

Following Equation 5.95, the viscoelastic solution for the displacement is

$u^{\text{visc}}(y, t) = K(t)C^e(y, 0) + \int_0^t K(t - \tau)\frac{dC^e(y,\tau)}{d\tau} d\tau$.

Substituting,

$u^{\text{visc}}(y, t) = J(t)\frac{mg}{\pi r^2}\int_0^0 \mathcal{H}(t - \frac{y}{v})dy + \frac{mg}{\pi r^2}\int_0^t J(t - \tau)\frac{d}{d\tau}\int_0^{v\tau} \mathcal{H}(\tau - \frac{y}{v})dy d\tau$.

Interchanging the order of integration in the second term and carrying out the differentiation,

$u^{\text{visc}}(y, t) = 0 + \frac{mg}{\pi r^2}\int_0^{vt}\int_0^t J(t - \tau)\delta(\tau - \frac{y}{v})d\tau dy$. With the sifting property of the delta function,

$u^{\text{visc}}(y, t) = \frac{mg}{\pi r^2}\int_0^{vt} J(t - \frac{y}{v})dy$.

This is identical to the second term of Equation 5.118, which was obtained by direct construction. In this example, the generalized correspondence principle does not offer much simplification since most of the effort in solving this problem is in keeping track of the relationship between deformation, time, and position, and that effort is required to obtain the elastic solution.

**Example 5.5**

Consider bending of a circular thin plate of thickness $2h$ and radius $a$, clamped at the edge. For an elastic plate of Young's modulus $E$ and Poisson's ratio $v$ the deflection $w$ in response to a uniformly applied force $F$ is as follows [31], provided $w \ll h$.

$w = \frac{F}{64\pi a^2 D}(a^2 - r^2)^2$ , with $D = \frac{2}{3}Eh^3 \frac{1}{1-v^2}$ as the flexural rigidity.

How is the deflection related to the force for a viscoelastic plate, if Poisson's ratio is constant in time?

**Solution**

Apply the correspondence principle to the elastic solution. Poisson's ratio has been assumed constant, so it is not transformed,

$w(s) = \frac{F(s)(1-v^2)}{\frac{2}{3}64\pi a^2 s E(s)h^3}(a^2 - r^2)^2.$

Converting to the compliance formulation to obtain a product of transforms, which can be handled with the convolution theorem,

$w(s) = \frac{(1-v^2)}{\frac{2}{3}64\pi a^2 h^3}(a^2 - r^2)^2 F(s)s J(s).$

Transforming back to obtain the time dependence of the deflection,

$w(t) = \frac{(1-v^2)}{\frac{2}{3}64\pi a^2 h^3}(a^2 - r^2)^2 \int_0^t J(t-\tau)\frac{dF}{d\tau}d\tau.$

This result consists of the product of a geometrical factor with a Boltzmann integral as was the case with beam bending examined in §5.5. In the present example, such a simple form is obtained only for Poisson's ratio as a constant.

**Example 5.6**

Determine the viscoelastic Poisson's ratio $v$ of a material from its shear and tensile properties.

**Solution**

Referring to the interrelations among elastic constants given in §5.6.3,

$$v = \frac{E}{2G} - 1. \tag{5.121}$$

Applying the correspondence principle,

$$sv(s) = \frac{sE(s)}{2sG(s)} - 1. \tag{5.122}$$

Since the convolution theorem is to be used in evaluating the inverse transform, write this as a product in terms of the shear compliance $J_G = G^{-1}$.

$$sv(s) = \frac{1}{2}sE(s)sJ_G(s) - 1. \tag{5.123}$$

Dividing by $s$ and transforming back,

$$v(t) = \frac{1}{2}\int_0^t E(t-\tau)\frac{dJ_G(\tau)}{d\tau}d\tau - 1. \tag{5.124}$$

Here, the time-dependent Poisson's ratio (which is of relaxation form, §5.7) is expressed in terms of a convolution of the axial relaxation modulus and the derivative of the shear creep compliance.

If *dynamic* data are available, the dynamic, frequency dependent Poisson's ratio may be obtained by applying the Fourier version of the correspondence principle as follows:

$$v^* = \frac{E^*}{2G^*} - 1 = \frac{E' + iE''}{2(G' + iG'')} - 1, \tag{5.125}$$

$$v^* = \frac{1}{2}(E' + iE'')\{\frac{G'}{G'^2 + G''^2} - i\frac{G''}{G'^2 + G''^2})\} - 1. \tag{5.126}$$

Referring to Example 3.6, since $G^* = 1/J_G^*$,

$$v^* = \frac{1}{2}(E' + iE'')(J_G' - iJ_G'') - 1. \tag{5.127}$$

The dynamic relationship is simpler to obtain since the algebraic manipulation of complex numbers is in this case simpler than calculation of a convolution integral.

**Example 5.7**

Must the phase angle in Poisson's ratio $v^*$ be positive?

**Solution**

Referring to Equation 5.127, consider the following cases. For $E^* = 2.6(1 + 0.1i)$ GPa and $J_G = 1$ GPa$^{-1}$, the loss tangent is 0.1 in tension and zero in torsion, we obtain $v^* = 0.3 + 0.13i$.

For $E = 2.6$ GPa, $J_G^* = 1(1 - 0.1i)$ GPa$^{-1}$, the loss tangent is 0.1 in torsion and zero in tension,

we obtain $v^* = 0.3 - 0.13i$.

So, for allowable, positive, phase angles in the axial relaxation modulus and the shear compliance, the phase angle in Poisson's ratio can be positive or negative. As for interpretation, observe that Poisson's ratio is defined as the ratio of the transverse contraction strain to the longitudinal extension strain. Therefore, there is no energy associated with Poisson's ratio or its phase angles. This is in contrast with the phase angle between stress and strain, which, as presented in §3.4, is associated with work done per cycle per unit volume.

**Example 5.8**

Explicitly derive Equation 5.31, showing all of the steps.

**Solution**

Consider Equation 5.30,
$w(s) = \frac{1}{2R}[sF(s)J(s) - J(s)v(s)v(s)s^3F(s)]$.
To apply the convolution theorem, which deals with a product of two functions in the Laplace plane, define $v(s)v(s)s^3F(s) = \Xi(s)$. So, following an inverse Laplace transform,
$w(t) = \frac{1}{2R}\int_0^t J(t-\tau)\frac{dF(\tau)}{d\tau}d\tau - \frac{1}{2R}\int_0^t J(t-\tau)\frac{d\Xi(\tau)}{d\tau}d\tau$.
Again, define $\Lambda(s) = v(s)s^3F(s)$, so, $\Xi(s) = \Lambda(s)v(s)$, so, with another inverse Laplace transform,
$\Xi(\tau) = \int_0^\tau v(\tau - \eta)\frac{d\Lambda(\eta)}{d\eta}d\eta$.

Since $\Xi$ is a function of $\tau$, a new variable of integration is named $\eta$. Similarly, transforming back $\Lambda(s)$ gives the following. Different equivalent forms are possible depending on how one collects the multiples of $s$, for example, $v(s)\{s^3 F(s)\} = \{sv(s)\}\{s^2 F(s)\}$. $\Lambda(\eta) = \int_0^\eta v(\eta - \xi)\frac{d^3 F(\xi)}{d\xi^3}d\xi$.

Combining these gives Equation 5.31, $w(t) = \frac{1}{2R}\int_0^t J(t - \tau)\frac{dF}{d\tau}d\tau - \frac{1}{2R}\{\int_0^t J(t - \tau)[\int_0^\tau v(\tau - \eta)[\int_0^\eta v(\eta - \xi)\frac{d^3 F(\xi)}{d\xi^3}d\xi]d\eta]d\tau\}$.

**Example 5.9**

Analyze the behavior of a thin layer of rubbery material compressed between rigid platens. Discuss the implications both for an elastic layer and for a viscoelastic layer.

**Solution**

Proceed first to consider elastic materials via the tensorial version of Hooke's law, then the elementary three dimensional form; then consider viscoelastic materials. In three dimensions, Hooke's law of linear elasticity is given in tensorial form by Equation 2.81, $\sigma_{ij} = C_{ijkl}\epsilon_{kl}$. The index notation, with $i$, $j$, $k$, and $l$ each able to assume values from one to three is used as is the Einstein summation convention in which repeated indices are summed over.

Consider simple tension or compression in the 1 (or $x$) direction: $\sigma_{11} \neq 0$, $\sigma_{22} = 0$, $\sigma_{11} = 0$. Expand the sum, assuming orthotropic symmetry.

$\sigma_{11} = C_{1111}\epsilon_{11} + C_{1122}\epsilon_{22} + C_{1133}\epsilon_{33}$.

Use the definition of Poisson's ratio to write this in mixed tensorial and engineering symbols.

$\sigma_{11} = C_{1111}\epsilon_{11} + C_{1122}(-v_{12}\epsilon_{11}) + C_{1133}(-v_{13}\epsilon_{11})$.

The ratio of stress to strain in simple tension in the 1 direction is Young's modulus $E_1$ in the 1 direction.

$E_1 = \frac{\sigma_{11}}{\epsilon_{11}} = C_{1111} - v_{12}C_{1122} - v_{13}C_{1133}$.

Consider the 2 (or $y$) direction in which there is zero stress, and again incorporate the definition of the Poisson's ratios.

$0 = \sigma_{22} = C_{2211}\epsilon_{11} + C_{2222}\epsilon_{22} + C_{2233}\epsilon_{33} = \{C_{2211} - v_{12}C_{2222} - v_{13}C_{2233}\}\epsilon_{11}$.

Assume the material is isotropic, so both the stiffness and the Poisson's ratio are independent of direction and $C_{1111} = C_{2222}$. Then,

$C_{2211} - vC_{2211} = vC_{1111}$; or

$C_{2211} = \frac{v}{1-v}C_{1111}$.

Substituting above, and dropping the subscript on $E$ since in an isotropic material there is no dependence of properties on direction,

$E = [1 - \frac{2v^2}{1-v}]C_{1111}$

$C_{1111} = E\frac{1-v}{(1+v)(1-2v)}$.

Since $E = 2G(1 + v)$ and

$v = \frac{3B-2G}{6B+2G}$, the modulus tensor element is written:

$C_{1111} = B + \frac{4}{3}G$.

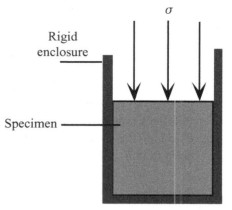

Figure 5.5. Geometry for constrained compression by a stress $\sigma$.

One may also work with the elementary isotropic form for Hooke's law.

$\epsilon_{xx} = \frac{1}{E}\{\sigma_{xx} - \nu\sigma_{yy} - \nu\sigma_{zz}\},$

$\epsilon_{yy} = \frac{1}{E}\{\sigma_{yy} - \nu\sigma_{xx} - \nu\sigma_{zz}\},$

$\epsilon_{zz} = \frac{1}{E}\{\sigma_{zz} - \nu\sigma_{xx} - \nu\sigma_{yy}\}.$

Consider simple tension or compression in the x direction: $\sigma_{xx} \neq 0, \sigma_{yy} = 0, \sigma_{zz} = 0$. Then,

$\frac{\sigma_{xx}}{\epsilon_{xx}} = E.$

Consider constrained compression, with $\epsilon_{yy} = 0, \epsilon_{zz} = 0$. Then,

$\sigma_{yy} = \nu\sigma_{xx} + \nu\sigma_{zz}.$

$\sigma_{zz} = \nu\sigma_{xx} + \nu\sigma_{yy}.$

Substituting,

$\sigma_{yy} = \sigma_{zz} = \sigma_{xx}\frac{\nu(1+\nu)}{1-\nu^2}.$

So, substituting into Hooke's law, the stress–strain ratio for constrained compression, which by definition is the constrained modulus $C_{1111}$, is

$\frac{\sigma_{xx}}{\epsilon_{xx}} = C_{1111} = E\frac{1-\nu}{(1+\nu)(1-2\nu)}.$

The physical meaning of $C_{1111}$ is the stiffness for tension or compression in the x (or 1) direction, when strain in the $y$ and $z$ directions is constrained to be zero. The reason is that for such a constraint the sum in Equation 2.81 collapses into a single term containing only $C_{1111}$. The constraint could be applied by a rigid mold, as shown in Figure 5.5, or if the material is compressed in a thin layer between rigid platens. $C_{1111}$ also governs the propagation of longitudinal waves in an extended medium, since the waves undergo a similar constraint on transverse displacement.

*Rubbery* materials have Poisson's ratios very close to 1/2, shear moduli on the order of a MPa, and bulk moduli on the order of a GPa. Therefore, the constrained modulus $C_{1111}$ is comparable to the bulk modulus and is much larger than the shear or Young's modulus of rubber.

For a viscoelastic material, apply the dynamic correspondence principle,

$C_{1111}^* = B^* + \frac{4}{3}G^*.$

Express this in terms of the loss tangents associated with each modulus,

$C_{1111}'(1 + i\tan\delta_c) = B'(1 + i\tan\delta_B) + \frac{4}{3}G'(1 + i\tan\delta_G).$

Although $B' \gg G'$ in the rubbery regime of polymers, we may have $B'' \approx G''$ in the leathery regime as discussed in §7.2.

## 5.13 Problems

5.1. Determine the relationship between the torque history and angular displacement history for torsion of a rod of circular cross section. Does relaxation in Poisson's ratio have any effect?

5.2. Consider the exact three-dimensional elasticity solution [2] for the displacement field in pure bending of a bar of rectangular cross-section,

$u_x = -\frac{1}{2R}\{z^2 + v(x^2 - y^2)\}$,

$u_y = -\frac{vxy}{R}$,

$u_z = \frac{xz}{R}$.

$R$ is the radius of curvature of bending and $v$ is Poisson's ratio.

   In what ways will the behavior of a viscoelastic bar differ from that of an elastic one? Consider in particular the effect of the viscoelastic Poisson's ratio, which is assumed to vary in time.

5.3. Does a stretched rod of *arbitrary* material held at constant extension get fatter or thinner with time? *Hint*: either make a demonstration based on continuum concepts, or envisage a material microstructure which gives rise to a decreasing Poisson's ratio.

5.4. A slender beam of uniform cross section is extruded horizontally at speed $v$ from a block of rigid material [23, 24]. The beam droops under its own weight. If the beam is linearly viscoelastic, determine the amount of droop as a function of time.

5.5. A concentrated load $F$ is suddenly applied orthogonal to the surface of a semi-infinite viscoelastic region with a horizontal boundary. Consider the load history as a step function but ignore inertial effects. What is the horizontal motion of points on the surface as a function of time? Suppose the Poisson's ratio is constant. Do points move toward or away from the load? What happens when the load is removed?

5.6. Consider bending of a circular thin plate of thickness $2h$ and radius $a$, clamped at the edge. The plate is of a linearly viscoelastic material with relaxation modulus $E(t)$ and Poisson's ratio $v(t)$, neither of which is constant in time. Find the relationship between the deflection history $w(t)$ in response to a uniformly applied force $F(t)$.

5.7. Determine the bulk properties of a viscoelastic material from its shear and tensile properties.

BIBLIOGRAPHY

[1] Sokolnikoff, I. S., *Mathematical Theory of Elasticity*, Malabar, FL: Krieger, 1983.
[2] Timoshenko, S. P., and Goodier, J. N., *Theory of Elasticity*, New York: McGraw Hill, 1982.

[3] Fung, Y. C., *Foundations of Solid Mechanics*, Prentice Hall, 1968.

[4] Christensen, R. M., *Theory of Viscoelasticity*, New York: Academic, 1982.

[5] Alfrey, T., Non-Homogeneous Stresses in Viscoelastic Media, *Quart Appl Math*, II(2), 113–119, 1944.

[6] Read, W. T., Stress Analysis for Compressible Viscoelastic Materials, *J Appl Phys*, 21, 671–674, 1950.

[7] Lee, E. H., Stress Analysis in Viscoelastic Bodies, *Quart Appl Math*, 13, 183–190, 1955.

[8] Reissner, E., and Sagoci, H. F., Forced Torsional Oscillations of An Elastic Half Space. I., *J Appl Phy*, 15, 652–654, 1944.

[9] Lakes, R. S., Foam Structures with a Negative Poisson's Ratio, *Science*, 235, 1038–1040, 1987.

[10] Pipkin, A. C., *Lectures on Viscoelasticity Theory*, Heidelberg: Springer Verlag; London: George Allen and Unwin, 1972.

[11] Goodier, J. N., Concentration of Stress Around Spherical and Cylindrical Inclusions and Flaws, *J Appl Mech*, 1, 39–44, 1933.

[12] Choi, I., and Horgan, C. O., Saint Venant End Effects for Plane Deformation of Sandwich Strips, *Int J Solids, Structures*, 14, 187–195, 1978.

[13] Lakes, R. S., Saint Venant End Effects for Materials With Negative Poisson's Ratios, *J Appl Mechanics*, 59, 744–746, 1992.

[14] Nowick, A. S., and Berry, B. S., *Anelastic Relaxation in Crystalline Solids*, New York: Academic, 1972.

[15] Ferry, J. D., *Viscoelastic Properties of Polymers*, 2nd ed., New York: John Wiley, 1970.

[16] Hilton, H. H., Implications and Constraints of Time-Independent Poisson's Ratios in Linear Isotropic and Anisotropic Viscoelasticity, *J Elasticity*, 63, 221–251, 2001.

[17] Lakes, R. S., and Wineman, A., On Poisson's Ratio in Linearly Viscoelastic Solids, *J Elasticity*, 85, 45–63, 2006.

[18] Bohn, E., *The Transform Analysis of Linear Systems*, Reading, MA: Addison Wesley, 1963.

[19] Meeker, T. R., and Meitzler, A. H., Guided Wave Propagation in Elongated Cylinders and Plates, in *Physical Acoustics*, W. P. Mason, ed., New York: Academic, 1964, pp. 1A, 111–167.

[20] Main, I., *Vibrations and Waves in Physics*, Cambridge, UK: Cambridge University Press, 1978.

[21] Chen, C. P., and Lakes, R. S., Apparatus for Determining the Properties of Materials over Ten Decades of Frequency and Time, *J Rheology*, 33, 1231–1249, 1989.

[22] Kolsky, H., *Stress Waves in Solids*, Oxford, Clarendon Press, 1953.

[23] Flügge, W., *Viscoelasticity*, Waltham, MA: Blaisdell, 1967.

[24] Wineman, A., and Rajagopal, K. R., Mechanical Response of Polymers, Cambridge, UK: Cambridge University Press, 2000.

[25] Graham, G. A. C., The Correspondence Principle of Linear Viscoelasticity Theory for Mixed Boundary Value Problems Involving Time-Dependent Boundary Conditions, *Quart Appl Math*, 26, 16, 1968.

[26] Lin, J. Y., and Westmann, R. A., Viscoelastic Winding Mechanics, *J Appl Mech*, 56, 821–827, 1989.

[27] Hunter, S. C., The Rolling Contact of a Rigid Cylinder with a Viscoelastic Half Space, *J Appl Mech*, 28, 611–617, 1961.

[28] Hunter, S. C., The Hertz Problem for a Rigid Spherical Indentor and a Viscoelastic Half-Space, *J Mech Phys Solids*, 8, 219–234, 1960.

[29] Graham, G. A. C., The Contact Problem in the Linear Theory of Viscoelasticity, *Int J Eng Sci* 3, 27–46, 1965.

[30] Goodman, A. M., and Goodall, I. W., Use of Existing Steel Data in Design for Creep, in *Creep of Engineering Materials*, C. D. Pomeroy, ed., London: Mechanical Engineering Publications, 1978.

[31] Love, A. E. H., *A Treatise on the Mathematical Theory of Elasticity*, 4th ed., New York: Dover, 1927.

# 6

# Experimental Methods

## 6.1 Introduction and General Requirements

Experimental procedures for viscoelastic materials, as in other experiments in mechanics, make use of a method for applying force or torque, a method for measuring the same, and a method for determining displacement, strain, or angular displacement of a portion of the specimen. If the experimenter intends to explore a wide range of time, and/or frequencies, the requirements for performance of the instrumentation can be severe. Elementary creep procedures are discussed first, because they are the simplest. Methods for measurement and for load application are discussed separately, since many investigators choose to assemble their own equipment from components. Many procedures in viscoelastic characterization of materials have aspects in common with other mechanical testing. Therefore, study of known standard methods [1, 2] for mechanical characterization of materials is useful. As in other forms of mechanical characterization, it is important that the stress distribution in the specimen be well-defined. End conditions in the gripping of specimens are usually not well known. The experimenter often uses elongated specimens for tension, torsion, or bending, to appeal to Saint Venant's principle in using idealized stress distributions for the purpose of analysis. The determination of viscoelastic properties as a function of frequency is at times referred to as mechanical spectroscopy.

Frequency response is an important consideration for instruments used in viscoelasticity; the transducers [3] for force generation and measurement and for measurement of deformation must respond adequately at the frequencies of interest. Resonances in these devices place an upper bound on the frequency at which meaningful measurements can be made. Manufacturers generally quote frequency response in terms of the frequency at which the amplitude $\xi$ drops by 3 dB, the "3 dB point" ($\xi(dB) = 20\log_{10}\xi$). Because dynamic viscoelastic studies involve a measurement of phase, one should be aware that significant phase shifts in the instrumentation (see §6.8) occur well below the 3 dB point. As for a lower limit on frequency, some instruments, such as electromagnetic motion detectors, only operate dynamically and have no quasistatic response. In other cases, drift in the electronics will limit the low-frequency and long-time performance. The materials used in the manufacture of the transducers in the instrumentation are real, not ideal, and, therefore, exhibit some creep [3]. This instrumental creep should be incorporated in the instrument specifications and should be taken into account in the interpretation of results. In the event that the instrumentation imposes no limit, the lowest frequencies and longest times accessible are limited by the experimenter's patience.

## 6.2 Creep

### 6.2.1 Creep: Simple Methods to Obtain $J(t)$

Creep experiments can be performed in a simple manner since a stress that is constant in time can be achieved by dead weight loading. Tensile or compressive creep tests upon a rod are the easiest to interpret since the entire specimen is under a stress that is spatially uniform, assuming uniformity at the ends. If there is nonuniformity due to gripping conditions, use of a slender specimen allows the investigator to take advantage of Saint Venant's principle in interpreting results. Uniform stress is advantageous if the investigator has any interest in examining nonlinear response. In bending and torsion creep tests, the stress is not spatially uniform within the specimen. Interpreting such tests is straightforward if the material behaves linearly; in that case the creep compliance can be extracted directly from measurements of force and displacement as developed earlier in Chapter 5. Measurement of displacement in creep can also be performed simply in the domain $t \geq 10$ sec. Among the simplest methods are the micrometer and the microscope. The microscope can be a traveling microscope, mounted on a calibrated stage and focused on a fiducial mark on the specimen or its grip; or it can be a microscope with a calibrated reticle. Other methods of displacement measurement will be described below (§6.4). Cantilever bending of a bar is an advantageous configuration in that the end displacement can be made large enough to measure easily while maintaining small strain, by choice of the bar geometry.

If tensile or compressive creep tests are to be done on relatively large specimens of a stiff material, dead weight loading becomes difficult, because large weights are required. A simple hydraulic device has therefore been developed to perform long-term compressive creep tests on concrete [4, 5]. Lever arm systems are used in commercial dead weight creep frames to amplify the load.

In polymers, one must precondition specimens by applying load cycles consisting of creep and recovery segments, to achieve reproducible results in creep, as pointed out by Leaderman [6]. Specimens of biological origin exhibit similar effects.

### 6.2.2 Effect of Risetime in Transient Tests

The definition of the creep compliance assumes a step function load history, however, in the physical world the load cannot be applied arbitrarily suddenly. To

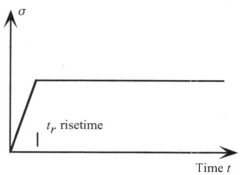

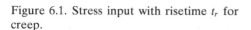

Figure 6.1. Stress input with risetime $t_r$ for creep.

determine the effect of risetime $t_r$ in transient tests, consider a stress history containing a constant load rate segment (Figure 6.1),

0 for $t < 0$,

$\sigma(t) = \frac{\sigma_0}{t_r} t$ for $0 < t < t_r$,

$\sigma_0$ for $t > t_r$.

Substituting in the Boltzmann superposition integral,

$$\epsilon(t) = \int_0^t J(t - \tau) \frac{d\sigma}{d\tau} d\tau = \frac{\sigma_0}{t_r} \int_0^{t_r} J(t - \tau) d\tau. \tag{6.1}$$

By the theorem of the mean, $\epsilon(t) = \frac{\sigma_0}{t_r} t_r J(t - t_r \zeta) = \sigma_0 J(t - t_r \zeta)$ in which $0 \leq \zeta \leq 1$. If $t \gg t_r$, then $\epsilon(t) \approx \sigma_0 J(t)$. Because the argument of the creep function differs from $t$ by an unknown but bounded amount, the analysis imposes error limits on the time and the errors are larger for shorter times. In most cases, $t \geq 10t_r$ is considered sufficient so that errors are not excessive [7]. Some authors suggest $5t_r$ is sufficient if the creep or relaxation curve is not too steep [15]. Analysis similar to the above can also be performed for relaxation. In view of the effect of risetime, it is common to begin recording transient data at a time a factor of five or ten longer than the risetime of the applied load or strain. If the stress history $\sigma(t)$ is known accurately during the initial ramp for $0 < t < t_r$, short-time data can be extracted. If the initial ramp is one of constant stress rate or constant strain rate, the analysis for interpretation of the data has been presented in §2.5.

In creep tests by dead weight loading such as described above (§6.2.1), the load would be applied in somewhat less than one second, and data collected beginning at 10 seconds. A test of 3 hours in duration ($1.08 \times 10^4$ sec) covers three decades; 28 hours is required for four decades; 11.6 days for five decades. If mechanical property data are required over many decades of time scale, either great patience in creep tests or shorter time/higher frequency data are needed, or both.

Actually (see Example 6.14), one can plot data beginning at a factor 2.5 of the risetime provided the zero of the time scale (not shown in the usual log scale for final presentation of the creep compliance) is taken as halfway through the risetime.

### 6.2.3 Creep in Anisotropic Media

Anisotropic materials have different stiffnesses in different directions. Viscoelasticity in such materials can be studied by conducting creep tests for deformation in different directions in the material. It is possible to perform creep tests in torsion and bending on beams cut in various directions with respect to the material principal axes [8]. From the results of such tests the elements of the creep compliance tensor $J_{ijkl}$ in the following Boltzmann integral can be inferred: $\epsilon_{ij}(t) = \int_0^t J_{ijkl}(t - \tau)\frac{d\sigma_{kl}}{d\tau}d\tau$.

### 6.2.4 Creep in Nonlinear Media

#### *Creep: Single Integral*

In elastic solids, one can evaluate nonlinearity from the shape of the stress–strain curve. In viscoelastic materials, the shape of the stress–strain curve is influenced both by time dependence and by strain dependence of material properties. Stress–strain curves are typically obtained at constant strain rates, so both time and strain increase simultaneously. In a linearly viscoelastic material, the stress–strain curve is curved: it is concave down because the slope depends on the relaxation modulus that decreases with time. A stress–strain curve that is concave up indicates nonlinearity because such a shape cannot arise from time dependence.

Design of experiments intended to explore nonlinear behavior is guided by the kind of constitutive equation considered for the material (see §2.12). Creep according to nonlinear superposition,

$$\epsilon(t) = \int_0^t J(t - \tau, \sigma(\tau))\frac{d\sigma}{d\tau}d\tau, \tag{6.2}$$

may be studied by performing a series of creep experiments at progressively higher stress levels, followed by recovery. The creep compliance depends on stress. If stress is sufficiently small that the experimenter is confident no irreversible changes occur, tests may be done on the same specimen after allowing sufficient time for recovery.

Quasilinear viscoelasticity (QLV) is a special case of noninear superposition in which the creep or relaxation function is separable into a function, which depends on time alone and a function, which depends on stress alone: $J(t,\sigma) = J(t)f(\sigma)$. If the material obeys QLV, then each creep curve will have the same shape (time dependence) independent of stress. If the shape of creep curves at different stress is different, then the material does not obey QLV.

#### *Creep and Recovery: Single Integral*

In linearly viscoelastic materials, recovery from creep due to the stress history $\sigma(t) = \sigma_0[\mathcal{H}(t) - \mathcal{H}(t - t_1)]$ follows: $\epsilon(t) = \sigma_0[J(t) - J(t - t_1)]$. The second term is an inverted and shifted version of the first term, so we say that recovery follows the same time dependence as creep.

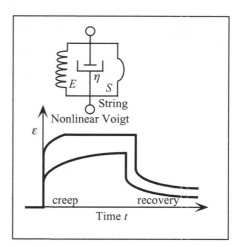

Figure 6.2. Nonlinear Voigt model.

Nonlinearity can be incorporated by allowing the creep compliance to depend on both time and stress: $J(t, \sigma)$. Interpretation of recovery in single integral models (nonlinear superposition) can be done in several ways. If one seeks to satisfy several mathematical conditions, a creep-recovery input,

$\sigma(t) = \sigma_0 \mathcal{H}(t) - [\sigma_0 - \sigma_1] \mathcal{H}(t - t_1)$

leads to a strain [13],

$\epsilon(t) = \sigma_0 J(t, \sigma_0) - \sigma_0 J(t - t_1, \sigma_0) + \sigma_1 J(t - t_1, \sigma_1),$

that is, the creep stress is turned off at time $t_1$ by the superposition of a negative step, and the recovery stress $\sigma_1$, which can be zero, is turned on. In this case, recovery to zero stress follows the same time dependence as creep. If the recovery phase is taken as [9] a result of a *change* in stress,

$\epsilon(t) = \sigma_0 J(t, \sigma_0) - [\sigma_0 - \sigma_1] J(t - t_1, (\sigma_0 - \sigma_1))$

so for recovery to zero stress, recovery strain follows creep.

If the recovery phase is taken as [10–12] a result of the *final* stress,

$\epsilon(t) = \sigma_0 J(t, \sigma_0) - [\sigma_0 - \sigma_1] J(t - t_1, \sigma_1),$

then recovery is governed by the response at $\sigma_1$, which can be zero. Recovery strain does not have to follow creep. This is the most sensible interpretation of the single integral form. Schapery [15] proposed a different single integral model in which recovery strain need not follow creep (§2.12). If a multiple integral series is used, recovery need not follow creep.

If $J(t, \sigma) = J(t) f(\sigma)$, that is, the material obeys quasilinear viscoelasticity (QLV), then recovery follows creep regardless how the single integral is interpreted.

The diagram (Figure 6.2) shows a nonlinear Voigt model, which contains a string or rope element that carries no stress when curved, then becomes stiff when straightened. Creep saturates in this case, but recovery follows the time constant $\eta/E$ corresponding to low stress; recovery does not follow creep.

The nonlinear Voigt model may be considered as the simplest approximation of the straightening of wavy fibers in tissues, such as ligament and tendon, and in string and synthetic composites that contain wavy fibers.

### Creep and Recovery: Multiple Integral

Experiments for the study of the more general nonlinearity described by the following Green–Rivlin series involve multiple steps [14] :

$$\epsilon(t) = \int_0^t J_1(t - \tau)\frac{d\epsilon}{d\tau}d\tau + \int_0^t \int_0^t J_2(t - \tau_1, t - \tau_2)\frac{d\sigma}{d\tau_1}\frac{d\sigma}{d\tau_2}d\tau_1 d\tau_2 + \cdots \quad (6.3)$$

At the simplest level, a single creep test at stress $\sigma_0$ upon a material obeying a multiple integral representation [20] gives the following:

$$\epsilon(t) = J_1(t)\sigma_0 + J_2(t, t)\sigma_0^2 + J_3(t, t, t)\sigma_0^3 + \cdots \quad (6.4)$$

A series of creep tests at different stress levels is sufficient to evaluate these kernel functions but only for the case that all time variables are equal. To evaluate the full effect of nonlinear terms such as $J_2(t - \tau_1, t - \tau_2)$ as a function of the two time variables, stress histories with two steps are applied; for example creep followed by recovery. A series of creep tests of different *durations* as well as different stress levels may be performed to determine $J_2$. The relationship between strain and the nonlinear kernel functions for a two-step stress history containing a step of magnitude $\Delta\sigma_0$ at time zero and a step of magnitude $\Delta\sigma_1$ at time $t_1$ [14] is as follows:

$$\epsilon(t) = J_1(t)\Delta\sigma_0 + J_2(t, t)(\Delta\sigma_0)^2 + J_3(t, t, t)(\Delta\sigma_0)^3 \quad (6.5)$$

$$+ J_1(t - t_1)(\Delta\sigma_1) + J_2(t - t_1, t - t_1)(\Delta\sigma_1)^2 + J_3(t - t_1, t - t_1, t - t_1)(\Delta\sigma_1)^3$$

$$+ 2J_2(t, t - t_1)(\Delta\sigma_0)(\Delta\sigma_1) + 3J_3(t, t, t - t_1)(\Delta\sigma_0)^2(\Delta\sigma_1)$$

$$+ 3J_3(t, t - t_1, t - t_1)(\Delta\sigma_0)(\Delta\sigma_1)^2 + \cdots$$

A creep and recovery test contains two steps, with $\Delta\sigma_1 = -\Delta\sigma_0$. In a linearly viscoelastic material, only the terms containing $J_1$ are present, so that recovery follows $J_1(t - t_1)(-\Delta\sigma_0)$, which has the same functional dependence as creep $(J_1(t)\Delta\sigma_0)$ only delayed in time and inverted. The terms with $J_2$ and $J_3$ show the effect of time-interaction type nonlinearity; if these are present, recovery does *not* follow the same pattern as creep.

## 6.3 Inference of Moduli

### 6.3.1 Use of Analytical Solutions

Inference of moduli from experiment usually involves relating the stress–strain equation which defines the modulus to a relation between force and displacement or torque and angular displacement as determined from an experiment. For quasi-static axial tension, bending, or torsion, exact analytical solutions for viscoelastic materials are available for this purpose, as presented in Chapter 5.

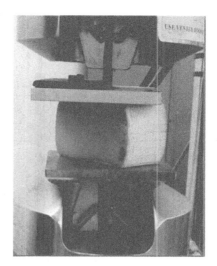

Figure 6.3. Compression of a block of rubber, showing bulge due to restraint of the Poisson effect.

### 6.3.2 Compression of a Block

The Young's modulus $E$ of a material can be determined via a tension or compression experiment upon a slender bar of material, provided the lateral surfaces are free of stress. It is not always practical to work with slender bars. For compression of a thick bar or block, the lateral deformation associated with Poisson's ratio is restrained by contact with the surfaces of the test device as shown in Figure 6.3. The specimen bulges and the modulus measured differs from the true Young's modulus. The effect has been calculated via approximate methods by several authors. The apparent Young's modulus $E_{\text{app}}$ of a compressed rubber cylindrical block [16] is given as follows, with $R$ as radius and $h$ as thickness:

$$E_{\text{app}} = E(1 + 2S^2) = E(1 + R^2/2h^2) \tag{6.6}$$

in which $E$ is the true Young's modulus and $S$ is the ratio of one loaded area to the force-free area. Similarly, for a rectangular block of square cross-section, $E_{\text{app}} = E(1 + 2.2S^2)$.

Rubber has a Poisson's ratio approaching one-half so the compressive deformation is governed by shear. If the rubber block is viscoelastic, then by the correspondence principle, the compressive modulus will have the same time dependence as the shear modulus. If the Poisson's ratio is not one-half, the correction formula differs from the above. The general relations are rather complex [17]. Some special cases are given for Poisson's ratio from zero to one-third [18]. For a cylinder with diameter equal to its length and Poisson's ratio of 0.3, then $E_{\text{app}} = 1.03E$. If the Poisson's ratio is constant in time, then both moduli have identical time dependence. The viscoelastic Poisson's ratio can depend on time [19]; in that case, interpretation is less simple. A solution for deformation of an elastic block is needed, but for this geometry only numerical results are available. To apply the correspondence principle, it would be necessary to fit curves to the tabulated values to obtain an approximate analytical form for the elastic case.

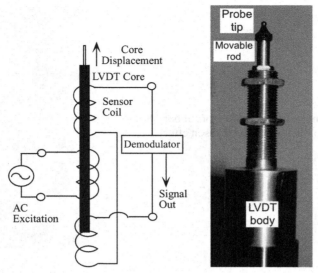

Figure 6.4. Left: Schematic diagram of an LVDT displacement transducer. Right: image of the LVDT.

## 6.4 Displacement and Strain Measurement

The micrometers and calibrated microscopes described above are usable for quasistatic measurements such as those performed in slow creep tests. In addition to simplicity, they have the advantage of maintaining their accuracy over very long periods of time. However, they are not suitable for rapid experiments. Other methods are used if a rapid response is required, or if measurements are to be made over the surface of a specimen; see, for example, [2, 21, 22].

LVDTs, or linear variable differential transformers (Figure 6.4) consist of several coils of wire surrounding a core of iron or other magnetic material. The core is usually attached to the specimen grip or a connected part of the apparatus. Devices with a spring loaded core as shown in the photograph, are only suitable for specimens so stiff that the spring force is negligible. One coil, the primary, is supplied with a sinusoidal electrical signal at several kilohertz. Electrical signals induced in the secondary coils depend upon the position of the core. These signals are demodulated, rectified and filtered to obtain an electrical signal proportional to core displacement. Similar transducers used to detect rotary motion are called RVDTs. The upper bound for frequency of the displacement signal may be as high as 1 kHz, depending on the physical size of the LVDT. Studies at low frequencies or long times are limited by drift in the electronics.

Foil strain gages (Figure 6.5) consist of an electrically conductive film, commonly a metal, on a flexible substrate. Strain in the gage results in a change in electrical resistance. Strain gages are cemented to the specimen and connected to electronics which convert the resistance change to a voltage proportional to strain. Strain gages are too stiff to be used on soft polymers, rubber, or flexible foam; errors due to gage stiffness can even be problematic for glassy polymers. The strain

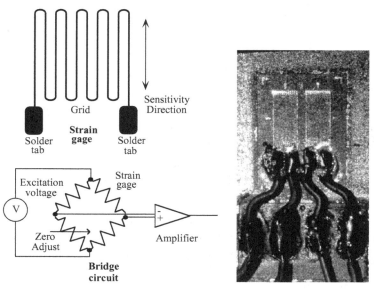

Figure 6.5. Strain gage. Top left: shape of foil strain gage grid. Bottom left: bridge circuit for strain gages. Right: photograph of pair of strain gages, grid length is about 3 mm, with wires.

gage imposes no upper limit on frequency other than that of the size of the gage in relation to the wavelength of stress waves in the material to which it is cemented. For gages of typical size, this upper limit can be on the order of 1 MHz. However, the electronic amplification system used with strain gages may impose further frequency limits. Strain gages respond to strain from thermal effects as well as strain resulting from stress. For accurate measurements of strain over long periods of time, considerable care must be exercised in the control of temperature and in matching the thermal expansion of the strain gage with that of the material to which it is cemented. Moreover, the cement used to attach the strain gage must be evaluated for possible creep, which could introduce further errors. Cyanoacrylate cement intended for strain gage use exhibits lower creep than superglue cyanoacrylate cement for general purposes.

For tests on rubbery materials or soft biological tissues, foil strain gages are not suitable. One can use a thin rubber tube filled with mercury as a strain gage for tests involving large deformation or compliant materials.

Capacitive transducers (Figure 6.6) are based on the fact that the capacitance $C$ of a capacitor depends upon the spacing between the plates. For a parallel plate, capacitor of plate is area $A$ and spacing $d$, $C = \epsilon_0 A/d$, with $\epsilon_0$ as the permittivity of free space. This variation of capacitance with plate position can be used to convert linear or angular displacement into an electrical signal. In a typical application, the capacitive transducer is excited by a high-frequency sinusoidal electrical signal and is connected to electronics which determine the capacitance, usually by determining the current which passes through the transducer. The signal is then rectified and filtered; therefore, the frequency response is limited [23].

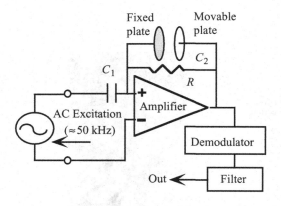

Figure 6.6. Capacitive displacement transducer and electronics.

Extensometers (Figure 6.7) are devices clipped or otherwise attached to a specimen in tension to determine the relative displacement between the two attachment points. They may be based on strain gages or LVDT's. Because they are attached off the specimen axis, response at high frequency is not to be expected; an upper bound from 30 to 100 Hz is as much as can be expected. Moreover, extensometers impose spring loads of 14 to 100 grams force (0.14 to 0.98 N) and, therefore, are suitable only for relatively stiff specimens. Extensometers serve to eliminate error due to instrument compliance in the testing of structurally stiff specimens. Specifically, the displacement signal associated with the motion of the moving grip in the test device is the sum of the instrument deformation and the specimen deformation. If the specimen is sufficiently stiff in comparison with the instrument, errors in the inferred modulus occur due to instrument compliance.

Optical methods can be extremely sensitive and offer the advantage that little or no mass need be attached to the specimen. To measure axial motion, one mirror of a Michelson interferometer (Figure 6.8) can be attached to the specimen [24]. One fringe corresponds to half a wavelength of light. The light can be converted to an electrical signal via a light sensitive diode. Fractional fringes can be readily

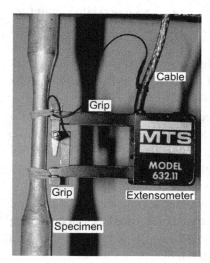

Figure 6.7. Extensometer attached to a specimen: image of device attached with the aid of rubber bands.

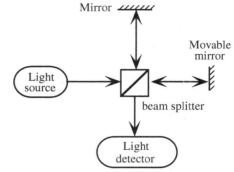

Figure 6.8. Michelson interferometer configuration for determining linear displacement. A moving mirror is attached to a specimen surface (not shown).

measured and corresponding displacement inferred via a digital counter or computer. The wavelength of light from the commonly used helium neon laser is 632.8 nm. Axial motion can also be measured using fiber optic sensors in which white light from one bundle of fibers is directed to the specimen. Reflected light from the specimen is then detected and the resulting signal is amplified electronically. This reflected light depends sensitively on displacement. Deformation of compliant materials, such as rubber or soft biological tissue, may be monitored with the aid of fiducial marks upon the specimen. For example, ink spots may be applied with a microscopic pipette, and observed via a calibrated microscope or a video system.

As for angular displacements, they are readily measured either by interferometric or "optical lever" methods [25]. An interferometer incorporating one or more right-angle prisms as reflective elements offers great sensitivity to angular displacement [26]. In the optical lever a beam of light is reflected from a mirror attached to the specimen. For creep tests, a light beam may be reflected to a distant target and its displacement measured with a caliper. For dynamic tests, it is expedient to convert the light to an electrical signal. One way of doing this involves projecting the image of a grating on another grating. The resulting light intensity signal is then converted to an electric signal by a photodiode. Light beam displacement can also be converted to an electrical signal by a split photodiode connected to a subtractor pre-amplifier. Integrated devices of this type are used in scanning probe microscopes. Semiconductor sensors sensitive to the position of an incident light beam are commercially available: an area sensor may be 1 cm or larger.

Motion can also be evaluated by electromagnetic methods: a coil or loop of wire is fixed to the specimen and immersed in a magnetic field. The voltage induced in the coil is proportional to the rate of change of magnetic flux according to Faraday's law. The induced voltage is expressed as the integral of electric field vector $\mathcal{E}$ around a closed loop which can represent the coil. The magnetic flux is the surface integral of magnetic field vector $\mathcal{B}$ over a surface spanned by the loop,

$$\int \mathcal{E} \cdot \mathbf{dl} = -\frac{d}{dt} \int \mathcal{B} \cdot \mathbf{ds}. \tag{6.7}$$

Depending on the specific configuration, this electrical signal may result from axial velocity or from angular velocity. There is no static response in this method; it is intrinsically dynamic and is suitable for resonant vibration or wave procedures.

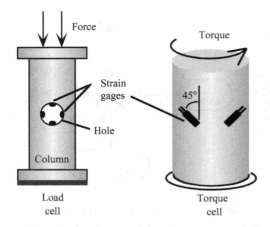

Figure 6.9. Load cell and torque cell structures.

## 6.5 Force Measurement

Force and torque transducers, known as load cells and torque cells, respectively, involve measuring the displacement or strain of a deformable substrate, typically steel. Representative arrangements of strain gages for load cells and torque cells are shown in Figure 6.9. Other forms of load cell include the proving ring which is compressed along a diameter, and configurations involving bending of plates fitted with strain gages. Torque cells are based on solid or hollow shafts, or cruciform arrangements fitted with strain gages. As with other transducers [3], load cells have characteristics of linearity, linear range, sensitivity, and frequency (or time) response. They typically contain a bar or plate of metal, usually steel, upon which strain gages are cemented. The detected strain signal on the bar is proportional to the force upon it, provided the metal and the strain gages are loaded below their proportional limit, one factor that limits the linear range or "capacity." Overload capability is limited by yield in the metal parts of the transducer. Because steel itself is not perfectly elastic but exhibits a small viscoelastic response, testing machines which use load cells of this type are not suitable for the study of materials with extremely low loss. Torque cells are similar to load cells in their use of strain gages; the metal portion and the strain gage orientation are configured differently so that the detected strain signal is a response principally to torque and not to any superposed axial load (Figure 6.9). The compliance of the load cell is important as discussed in §6.11.2. The load cell should be much stiffer than any specimen tested. It is also important that the load cell be sensitive only to the kind of load to be measured and not to other loads. In biaxial tests both tension and torsion may be applied. The torque channel should be sensitive only to torque, not axial load. This characteristic is referred to as an absence of crosstalk.

Load cells based on piezoelectric crystals are also available; they offer superior stiffness for dynamic studies. However, there is no response at zero frequency, and there are phase errors at low frequency, though the frequency range can extend below 0.001 Hz.

## 6.6 Load Application

Methods of load application for creep include dead weights as described above, electromagnetic methods, and hydraulic methods. In the electromagnetic approach, controlled electric current may be passed through a small coil of wire (which generates a magnetic dipole moment) immersed in an external magnetic field. This coil experiences a torque. Alternatively, a permanent magnet (which generates a magnetic dipole moment) is placed within a coil through which a controlled electric current passed. The magnet experiences a torque. The small coil or magnet can exert a force or torque on the specimen to which it is attached, depending on the specific geometry. The torque $\tau$ on an object of magnetic moment $\mu$ in a magnetic field vector $\mathcal{B}$ [27] is

$$\tau = \mu \times \mathcal{B}. \tag{6.8}$$

The force **F** is given by

$$\mathbf{F} = \text{grad}\{\mu \cdot \mathcal{B}\}. \tag{6.9}$$

These devices are equivalent to electric motors in which the motion is stalled by the specimen.

In hydraulic systems, an electric pump drives a fluid through a hydraulic pipe system to a movable piston which applies force to the specimen. The fluid is controlled by a valve which in turn is controlled by an electrical device.

Load application in relaxation experiments can be achieved purely mechanically by means of an eccentric cam actuated by a handle. This system provides a displacement which is constant in time after it is applied. The load is free to decrease as the specimen relaxes; it is measured independently.

## 6.7 Environmental Control

Viscoelastic properties depend on temperature, therefore some form of temperature control is usually used in viscoelastic measurements. If the method of time–temperature superposition is to be used, then capability for conducting tests at several constant temperatures is required. Alternatively, the experimenter may measure dynamic properties at constant frequency, and perform a temperature scan.

As for the precision of temperature control, polymers are very sensitive to temperature, particularly in the transition region. Therefore it is desirable to control temperature to within 0.1°C over the length of the specimen and for the duration of the experiment. To achieve such control, a stream of air can be passed through a temperature control device, and directed to the specimen through a system of baffles. In tension/compression tests, a length change can arise from creep or from thermal expansion. Excellent temperature control is required in such studies to obtain good creep or relaxation results. Since thermal expansion of an isotropic material has no shear or torsional component, torsion tests are less demanding of accurate temperature control than are tension/compression tests.

Many polymers, as well as materials (such as wood) of biological origin, are sensitive to hydration as well as to temperature. Humidity control is important for such materials; in the study of tissues such as bone, muscle or ligament, the specimen should be kept fully hydrated under physiological saline solution. The creep behavior of concrete is also dependent upon hydration.

Metals which exhibit magnetic behavior are sensitive to external magnetic fields. If electromagnetic methods are used to apply force or torque to the specimen, care must be taken that magnetic fields in the vicinity of the specimen are controlled.

## 6.8 Subresonant Dynamic Methods

### 6.8.1 Phase Determination

#### *Phase Measurement*

In subresonant dynamic methods, a sinusoidal load is applied at frequencies well below the lowest resonance of the specimen including attached inertia, if any. The test frequency must also be well below any resonances in the instrument. Under these conditions, the phase angle $\phi$ between force and displacement (or between torque and angular displacement) is the same as the loss angle $\delta$, as derived in §3.5. Direct measurement of phase is necessary in these methods.

Phase measurement can be performed by determining the time delay between the sinusoids on an oscilloscope (for high frequencies) or on a chart or digital recorder (for low frequencies). This approach is suitable for materials with a loss tangent greater than about 0.1; there is a limit in the resolution of the phase. A better method of phase determination is by evaluation of the elliptic stress–strain curve, which is called a Lissajous figure (Figure 3.3). If the phase is sufficiently large, it may be easily determined from the width of the elliptic curve. In materials with low loss, the ellipse is nearly a straight line. The middle of the ellipse can be magnified on an oscilloscope, computer display, or recorder to achieve improved phase resolution, provided that the ratio of signal to noise is sufficiently high. It is possible by such magnification, to measure a tan$\delta$ less than 0.01 in the subresonant domain.

Phase can be measured at high resolution by detecting the zero-crossing of the load and displacement signals and measuring the time delay with a high-speed timer [28]. A time accuracy of 1 $\mu$sec is claimed, which corresponds to a phase uncertainty $\delta\phi \approx 6 \times 10^{-5}$ at 10 Hz, and even better at lower frequencies. Use of such a method requires signals, which are low in noise, drift, and DC offset. To that end it is necessary to minimize electrical noise and noise from mechanical sources, such as building vibration.

An interesting approach to achieving better resolution involves subtracting a signal of zero phase to fatten the ellipse [29] and facilitate measurement (see Example 6.13). This approach is also limited by the ratio of signal to noise.

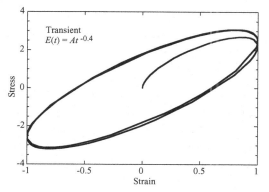

Figure 6.10. Transient behavior in response to a load history $\sigma = 0$ for $t < 0$, $\sigma = A\sin(2\pi\nu t)$ for $t \geq 0$, calculated for a material with power-law relaxation.

A two-phase lock in amplifier can be used to achieve good phase resolution ($\delta\phi \approx 10^{-3}$ to $10^{-4}$ or better), however not all models offer such resolution. Lock in amplifiers are capable of rejecting noise. It is a painstaking process to use most lock in amplifiers below about 10 Hz, however some instruments are capable of measurements down to 0.001 Hz. Good phase resolution can also be achieved by Fourier transformation of the signals for stress and strain [30]. This procedure makes use of the fact that the phase difference between two sinusoids is equal to the phase difference between their Fourier transforms, provided that the data sample contains a sufficient number of cycles [30]. A digital fast Fourier transform with 16 bit resolution gives a phase resolution of $\delta\phi \approx 2\pi/2^{16} = 9.59 \times 10^{-5}$ radians provided the signal occupies the full scale.

As for dynamic measurements at low frequency, a mathematical sinusoid has no beginning or end; by contrast, an experimentally applied sinusoid must have a starting point. This raises an issue of concern in interpreting dynamic experiments, particularly those at low frequency. An analytical solution of the problem has been presented, based on Laplace transform methods [31]. Results of a numerical analysis, in which the Boltzmann integral of the stress history was evaluated using a computer, are shown in Figure 6.10. The response to the experimental "sinusoid" beginning at time zero converges rapidly, within one or two cycles, to a sinusoid, even in a material with a high tan$\delta$. Low frequency dynamic tests are limited by the patience of the experimenter and by drift in electronics.

### Phase Error Calculation

Phase measurements are influenced by phase shifts in any electronics used to process the signals. Consider a low-pass electric filter consisting of a resistor $R$ and capacitor $C$ in series. The output voltage is taken across the capacitor. Such a filter may be used deliberately to reduce noise superposed on an electric signal, or the low-pass characteristic may be inherent in the amplifier or signal conditioning electronics used. The input voltage is $v_i$ and the current $I$ is $I = \frac{v_i}{R + \frac{1}{i\omega C}}$. The output voltage is $v_o = IZ = I/i\omega C$, with $Z$ as the electrical impedance of the capacitor and $i = \sqrt{-1}$, so the gain $\mathcal{G}$ is, with the cut-off angular frequency $\omega_0 = (RC)^{-1}$,

$\mathcal{G} = \frac{v_o}{v_i} = \frac{1}{1+i\omega RC} = \frac{1}{1+i\frac{\omega}{\omega_0}}$. Here, $RC$ is the time constant. The absolute value of the gain is $|\mathcal{G}| = \frac{1}{\sqrt{1+(\frac{\omega}{\omega_0})^2}}$. For $\omega = \omega_0$, $|\mathcal{G}| = 1/\sqrt{2}$. The phase $\phi$ introduced by the filter is given by $\tan\phi = \frac{\omega}{\omega_0}$.

If $\omega = 0.1\omega_0$, then $|\mathcal{G}| = 0.995$ and $\tan\phi = 0.1$. If $\omega = 0.01\omega_0$, then $|\mathcal{G}| = 0.99995$ and $\tan\phi = 0.01$. If this filter is used as part of a viscoelasticity test apparatus, then phase shifts occur at frequencies significantly below the cut-off frequency. If not properly taken into account, these phase shifts can cause significant error in the inference of $\tan\delta$. The AC coupling setting in an oscilloscope is a high pass filter; it can also generate large phase errors. Use the DC coupling instead.

### Consistency Checks

In experimental studies on viscoelastic materials, the Kramers–Kronig relations are useful in performing consistency checks on stiffness and damping data. The Kramers–Kronig relations are also useful when only one dynamic viscoelastic function is available, and the other one is desired.

### 6.8.2 Nonlinear Materials

### Hysteresis Curve Shape: Gradients

The stress–strain curve under dynamic loading is elliptical only if the material is linearly viscoelastic. If nonlinearity is present, the response will no longer be sinusoidal, and the stress–strain curve (hysteresis loop, Figure 3.11) is not elliptic. Energy dissipation per cycle can then be determined from the area within the hysteresis loop, and expressed as specific damping capacity (§3.4) rather than a loss tangent.

Experiments performed in torsion or bending present some complications for nonlinear materials because the strain distribution is not uniform. This nonuniformity must be taken into account to properly interpret the results of dynamic damping experiments [32, 33].

### Spectral Hole Burning

Pump-probe methods offer the intriguing possibility to distinguish between viscoelasticity which results from a distribution of exponential relaxation processes and an intrinsically nonexponential process. In this approach, first demonstrated in dielectric systems, a strong sinusoidal pump signal is applied for a period of time with the aim of saturating a particular time constant. A step function probe is then applied. By performing two experiments in which the probe is of opposite sign, it is possible to remove the nonlinear after effect. The response is equivalent to the linear response unless there is dynamic heterogeneity in material response. Such effects are manifest as a "hole" in the response for a time constant corresponding to the frequency applied in the pump signal; these effects have been observed in polymer melts [34].

### 6.8.3 Rebound Test

Materials may be rapidly screened for their effective $\tan\delta$ by measuring the height $H_1$ of rebound of a steel ball dropped from height $H_0$ on a block of material. Alternatively, the ball is made of the viscoelastic material to be tested, and the substrate on which it is dropped is made stiff and massive. For small damping the approximate relation, as follows, Equation 10.42 is sufficient,

$$\tan\delta \cong \frac{1}{\pi} \ln(\frac{H_0}{H_1}). \tag{6.10}$$

The effective frequency for this is proportional to the inverse of the impact time is $t_{\text{imp}}$. For higher damping, (Equation 10.41) is the more accurate. This interpretation is based on assuming that the dissipation of mechanical energy is entirely due to viscoelasticity rather than yield, cracking, or generation of waves in the block.

### 6.9 Resonance Methods

#### 6.9.1 General Principles

Resonance methods are eminently suitable for low loss materials. By contrast, small $\tan\delta$ values are difficult to measure in the subresonant domain. The resonant behavior of two materials with small $\tan\delta$ values $10^{-3}$ and $10^{-6}$ differs dramatically as we have seen in §3.5. The difference is used as the basis for experimental methods.

Resonances in tension/compression, torsion, or bending can be examined to characterize materials. The relation, developed in §3.5, $\tan\delta \approx \frac{1}{\sqrt{3}} \frac{\Delta\omega}{\omega_0}$ for the half-width $\Delta\omega$ of the structural compliance peak, is used to interpret the results. One may also infer damping from the free-decay time $t_{1/e}$ of vibration of period $T$, $\tan\delta \approx \frac{1}{\pi} \frac{T}{t_{1/e}}$. These relations are valid for small $\tan\delta$. As for the specimen stiffness, it is inferred from the resonance frequency $\nu_0$ in conjunction with the specimen geometry. Observe that in the resonance scheme there is no need to measure the force or torque, but if it is not measured it must have constant amplitude over the frequency range associated with the resonance peak. A drawback of this approach is that only one frequency is accessible in lumped configurations with a large inertia attached to the specimen. In the case of a specimen without attached inertia, there are, in principle, an infinite number of modes, but since their amplitude decreases rapidly with frequency, only a limited number of modes are used in practice since at sufficiently high frequency there is insufficient signal. If the attached inertia is comparable to the specimen inertia, the higher harmonics are further reduced in amplitude (Figure 6.11), and all the resonance modes occur at reduced frequencies. If the attached inertia is large compared with that of the specimen, only the lowest mode is observable.

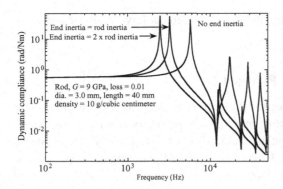

Figure 6.11. Calculated effect of end inertia on the resonance frequencies of a cylindrical rod, fixed at one end and free at the other, in torsion. The rod is of a metal alloy, with $G' = 9$ GPa, $\tan\delta = 0.01$, and it is 40 mm long, 3 mm in diameter, with a 10 g/cm$^3$ density.

The specimen may be fixed at one end and free at the other end (fixed–free configuration), or it may be free at both ends (free–free configuration). In the fixed–free approach, some of the vibration energy may be transmitted into the fixed support, leading to measured values of $\tan\delta$ which are greater than actual values. Error due to such energy loss in fixed–free systems may be quantified by considering the transmission coefficient [35] for power, which is

$$T_p = 1 - |R_w|^2 = \frac{4}{[\sqrt{\frac{Z_2}{Z_1}} + \sqrt{\frac{Z_1}{Z_2}}]^2}. \tag{6.11}$$

$Z$ is the mechanical impedance and the subscript indicates the material. $R_w$ is the reflection coefficient, $R_w = \frac{Z_1 - Z_2}{Z_1 + Z_2}$. Considering the transmission coefficient for power $T_p$ as a parasitic specific damping or energy ratio $\Psi_p$ under the assumption that all the power transmitted into the support rod is lost, the parasitic loss tangent is

$$\tan\delta_p = \frac{1}{2\pi}\Psi_p = \frac{1}{2\pi}T_p. \tag{6.12}$$

For longitudinal waves, $Z$ is the ratio of driving force to particle velocity [36]; for a cylindrical rod of radius $R$, density $\rho$, and complex Young's modulus $E^*$, $Z$ depends on the cross-sectional area,

$$Z_L = \pi R^2 \sqrt{\rho E^*}. \tag{6.13}$$

The impedance for torsion of a cylindrical rod is the ratio of torque to angular velocity; it depends on the polar moment of inertia.

$$Z_{\text{tors}} = \frac{1}{2}\pi R^4 \sqrt{\rho G^*}. \tag{6.14}$$

This is in contrast to the case of plane shear waves in which there is no mismatch in area or moment of inertia for which $Z_s = \sqrt{\rho G^*}$ represents the characteristic impedance. The difference in diameter between the specimen and the support as well as the difference in stiffness gives rise to an impedance mismatch, which reduces parasitic loss error resulting from transmission of waves into the support. In torsion, the impedance goes as the fourth power of diameter. Therefore, the desired impedance mismatch can be large in torsion. (See Examples 6.3 and 6.4.)

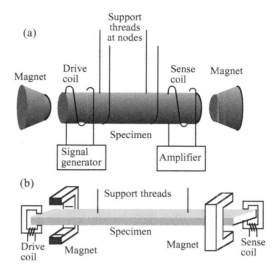

Figure 6.12. A configuration for resonance of an electrically conducting bar, free at both ends. Excitation and detection are via eddy currents. (a) Set-up for longitudinal vibration. (b) Set-up for flexural vibration.

Even so, for materials of very low loss, the free-free approach is preferable. In free–free resonance methods, the specimen is supported at one or more nodes (at which the vibration amplitude is zero) by a compliant support such as a thread. There is still some background damping due to losses in the support (which is of nonzero size), but that error is usually lower than in the fixed free approach. Specific methods are discussed below (§6.9.2).

### 6.9.2 Particular Resonance Methods

A torsion specimen with a large attached inertia can be made to resonate below 1 Hz. Some experimenters advocate changing the attached inertia to obtain different frequencies, but this is cumbersome and it is difficult to obtain a wide range of frequency. A short specimen without any attached inertia may resonate at a frequency exceeding 100 kHz. Resonance methods are commonly used in conjunction with a scan of temperature; results consist of a plot of stiffness and loss at a nearly constant frequency versus temperature. A full characterization of the material, however, entails a study over a range of frequencies.

Free–free resonance methods are suitable for materials of low loss. In the device shown in Figure 6.12, the driving force is generated by an interaction between eddy currents induced by the coils and the field of the permanent magnet. This approach is suitable for specimens which are electrically conducting. The specimen is shown suspended at two nodal points by fine threads or wires, and vibrates in its second or higher mode. It is also possible to drive the resonance via an electromagnetic shaker attached to one support wire, and detect the vibration with an electromagnetic pickup device supporting the other wire. In this method, due to Forster, the wires must be displaced slightly from the nodal points to drive and detect the vibration. However, if the support wires are too far from the nodes, their parasitic contribution to tan$\delta$ becomes excessive [37]; the error is evaluated theoretically.

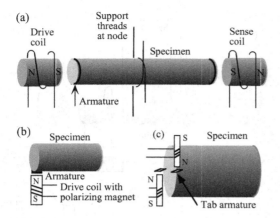

Figure 6.13. A configuration for resonance of a bar, free at both ends. Excitation and detection via magnetic interaction between external coils and a ferromagnetic armature cemented to the specimen. (a) Set-up for longitudinal vibration. (b) Set-up for flexural vibration. (c) Set-up for torsional vibration.

If the specimen is not electrically conductive, a small magnet or a ferromagnetic armature [39] may be cemented to each end, as shown in Figure 6.13. The specimen is shown suspended at the central nodal point by fine threads or wires, and vibrates in its first mode or a higher mode. The purpose of the polarizing magnet in the coil assembly is to provide a bias magnetization in the armature. As a result, the interaction between the oscillating magnetic field due to current in the coil and the bias field gives an oscillating force on the specimen at the driving frequency. This oscillating force can excite longitudinal, flexural, or torsional vibration, depending on the geometrical arrangement as shown in Figure 6.13. These magnetic approaches permit larger excitation amplitudes than the eddy current method. Coupling between specimen and driver and sensor coils leads to background loss, because some electrical energy is dissipated in the armature and magnetic pole pieces. This background loss can be minimized by using magnetic materials of low hysteresis. Both sensor and driver coils are usually encased in magnetic shielding of high permeability metal, to reduce pickup of noise and of electrical signals from the driver or heater windings. In all methods that use electromagnetic induction into coils of wire, there is no static response since the induced voltage depends on the specimen velocity.

The resonance approach can be adapted to high-loss materials for which a sharp resonance peak cannot be obtained. The procedure involves preparing a specimen of stiff, low-loss material and applying a layer of the high loss material. The properties of the high-loss layer can be extracted from resonance experiments upon the composite sandwich [38]; see §9.3 for analysis of viscoelastic laminates.

Microsamples of bone and other stiff biological materials have been studied [40] with a fixed-free electromagnetic torsion resonance device. The device provides drive torque via a miniature coil of fine wire immersed in the field of a large permanent magnet. Angular displacement is measured from the motion of a laser beam reflected from a mirror attached to the specimen. Microresonance studies in soft biological tissues have been conducted by embedding small beads of magnetic material in the tissue and oscillating them with a magnetic field [41].

The composite piezoelectric oscillator [42–44] is an interesting approach to resonance. This approach is called the *piezoelectric ultrasonic composite oscillator*

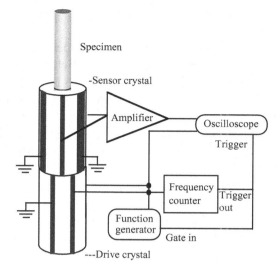

Figure 6.14. Composite torsional piezo-electric oscillator.

*technique* (PUCOT). The device consists of two piezoelectric crystals and a specimen cemented together (Figure 6.14). The fundamental resonant frequency can be from 20 kHz to about 120 kHz depending on the length of the crystals. Quartz is preferred for the crystals because it has a low tan$\delta$ on the order $10^{-6}$. One crystal is driven electrically to induce vibration and the other crystal is monitored for electrical signals induced by strain. The wires enter at the crystal midpoints, which are nodes for free–free vibration. Because the wires are of nonzero size, they allow some vibration to leak away and so represent a source of parasitic damping. Therefore the experimenter makes a measurement of damping of the supported quartz crystals alone to evaluate the parasitic damping. Tension/compression or torsion can be achieved depending on the type and orientation of crystal. Viscoelastic properties of the specimen are inferred from electrical measurements upon the sensor crystal. A correction for parasitic damping is incorporated in the calculation. The method is intrinsically dynamic, and gives no response in the subresonant or static regimes. Variations of this approach have incorporated low-temperature capability [45]. A low temperature variant of the piezoelectric composite oscillator uses a stalk, which is slender (0.5 mm diameter compared with 4 mm for the quartz crystal), and a compliant, made of copper-beryllium alloy, to support the oscillator and provide thermal contact [45]. Background loss with this system is tan$\delta_{\text{parasitic}} < 2 \times 10^{-6}$ at 2 K. Capability for high temperatures exceeding those tolerable by crystalline quartz can be achieved by attaching the specimen to the piezoelectric crystals by a long stalk of fused quartz, and placing the specimen in a furnace. The composite piezoelectric oscillator can also be used in a free decay mode (see §3.6) by modulating the drive waveform by a gate signal.

Free decay of resonant vibration (see §3.6) is also used to determine viscoelastic properties at lower frequencies. A device consisting of two large pendulums supported by ball bearings has been used to characterize PMMA at about 1.5 Hz over a range of temperatures [46]. Motion of the pendulums induces torsion in the specimen.

Resonance experiments have been applied to polymers [35], metals, fibers [38], and single crystals. Refinements include the addition of sophisticated electronics to aid in data acquisition and analysis [48]. Study of fibers by resonance methods [38] is done in an evacuated chamber [49] for thin fibers in which parasitic damping due to air viscosity is excessive. Vibration may be excited electrostatically by applying an oscillatory voltage across the plates of a capacitor, one plate of which is fixed to the specimen. Vibration amplitude is determined from electrical measurements on the capacitor.

Resonance methods can be difficult to use above 100 kHz because a multiplicity of vibrational modes can be excited. Some analysis schemes [51] have been presented to interpret experiments in the frequency range between 100 kHz and 1 MHz, above which ultrasonic wave methods become applicable.

### 6.9.3 Methods for Low-Loss or High-Loss Materials

#### *Low Loss*

For low-loss materials, resonance methods are usually superior to methods based on direct phase measurement. Small phase differences can be obscured by noise or parasitic phase shifts in the electronics and are difficult to measure directly. If it is necessary to make subresonant measurements on low-loss solids, excellent phase resolution is required but this resolution may be limited by the ratio of signal to noise. A lock-in amplifier can be beneficial in cutting through the noise and improving phase resolution. Moreover, phase errors in transducers, electronics, and filters must be calculated and minimized.

For low-loss materials ($\tan\delta < 10^{-4}$), free-decay methods are easier to use than the method of resonance half width. The reason is that the resonance peak is so sharp for low loss materials that it can be tedious to scan through it. Special experimental precautions are necessary in studying low loss materials [52, 53]. It is necessary to eliminate all spurious sources of damping other than that in the bulk matter of the specimen. Errors can arise due to radiation of sound energy into the air; these errors cause the apparent loss to be greater than the true material loss. To eliminate such errors, it may be necessary to perform the experiments in an evacuated chamber. For $\tan\delta \leq 10^{-4}$, precautions of this type are usually necessary. It is possible to calculate [53] the damping due to air friction. In particular,

$$\tan\delta_{\text{air}} = \frac{2P}{\pi\rho c}[\frac{C_P}{C_V}\frac{\mu_{\text{mw}}}{RT}]^{1/2} \tag{6.15}$$

with $P$ as the gas (air) pressure, $\mu_{\text{mw}}$, is the mean molecular weight of the gas, $\rho$ is the density of bar material, $c$ is the sound velocity in the bar, $R$ is the gas constant and $C_P$ and $C_V$ are the heat capacities of the gas at constant pressure and volume, respectively. The formula is valid if the mean free path of the gas molecules is much less than the wavelength of the sound waves in the gas. That assumption would be valid for oscillations at 10 kHz and for gas pressures greater than 0.001 Torr. For a better vacuum, a different equation applies. Bending procedures are more vulnerable to error due to air damping than torsion procedures, since the translational motion associated with bending causes more motion of air. Care in specimen preparation is also required. In particular, defects such as microcracks, spalls, absorbed molecules, and dislocations concentrated in a surface layer can result in elevated apparent damping in low-loss materials.

The resonating specimen must be supported in some way, and that support can introduce spurious losses. In fixed–free cantilever bending vibration, care is required that any grip for the specimen is not only tight but also massive [52]. In free–free vibration, specimens are commonly supported at a vibration node, where displacement is minimal, to reduce these errors. Use of a fine thread for support rather than a solid nodal mounting, serves to minimize loss due to the support. If the thread is greased to reduce transmission of shear stress, the spurious loss is reduced further [52]. Moreover, if the plane of bending vibration is horizontal rather than vertical, parasitic loss is reduced since the support threads undergo rotation rather than extension. Capacitive devices are commonly used to excite and measure vibration in low-loss materials. Although little power is available to excite vibration, little is needed. These transducers offer the advantage of contributing minimal spurious damping. Even so, a careful experimenter will calculate the spurious damping due to electrical losses in the transducer. The above experimental methods have been used to measure losses as small as $\tan\delta < 10^{-9}$ in single crystal sapphire and silicon at low temperature [53].

Parasitic damping due to the support of a vibrating cantilever of length $L$ and thickness $h$ attached to an infinitely large elastic body [55], is as follows:

$$\tan\delta_p = 0.23(\frac{h}{L})^3,\qquad(6.16)$$

so, for an aspect ratio, $L = 10h$, this parasitic damping is about $2 \times 10^{-4}$. To reduce the parasitic damping, a more slender specimen could be used, tested under vacuum.

Extra energy dissipation can occur in films adsorbed on the resonator as well as from surface damage and defects. Consequently, the surface must be kept clean and free of damage if the intrinsic damping of the material is to be approached. Specimens are polished and annealed to reduce surface and internal defects. If the specimen is small, its stored energy is minimal. Surface effects contribute substantially to the damping of small specimens, so control of surface quality must be done to obtain consistent results.

Resonant systems are commonly used to determine properties at one frequency. It is also possible to drive such a system at multiple modes. This approach was used to study wires of several low loss metals as well as sapphire and fused silica [54]. Each wire (0.1 to 0.5 mm diameter) was hung from a clamp; normal modes were excited by a capacitor plate. Deformation was determined optically via a split diode light detector. A frequency range of 1 Hz to 1 kHz was obtained. The fused silica had the lowest damping, $10^{-6}$ at the higher frequencies.

### *High Loss*

Large phase angles are easy to measure. If, however, the material tends to a fluid consistency at sufficiently low frequency or long time, the specimen may not be able to support its own weight. For such materials, a specimen can be a layer between two stiff plates, or can be studied in a cone and plate rheometer (§6.11.4). In resonant experiments, the approximate formulae for half width ($\tan\delta \approx Q^{-1} = \Delta f/\sqrt{3}f_0$) and for free-decay are not applicable for large damping. The exact solution must be used for analysis to determine large damping from resonant-peak width. Resonant peaks in high-loss materials become broader and of lower amplitude; signal strength is reduced. If the system has distributed inertia, nearby modes may overlap. For a lumped system, exact relations between $Q$ and $\tan\delta$ require a model of the viscoelastic response of the material. For example, if damping is independent of frequency, then [56], for $\tan\delta < 1$,

$$Q^{-1} = \sqrt{1 + \tan\delta} - \sqrt{1 - \tan\delta}. \qquad (6.17)$$

As for free decay of vibration the *log decrement*, for small damping, is $\Lambda \approx \pi \tan\delta$. The exact expression [56] is for a material with a complex modulus,

$$\tan\frac{\delta}{2} = \frac{\Lambda}{2\pi}. \qquad (6.18)$$

Alternatively [57],

$$\tan\delta = (1 - \exp(-2\Lambda))/2\pi. \qquad (6.19)$$

Damping cannot be obtained from a ratio of adjacent oscillations if $\tan\delta > 1/2\pi \approx 0.16$ since there are no oscillations when damping is that high.

### 6.9.4 Resonant Ultrasound Spectroscopy

Resonant ultrasound spectroscopy involves the measurement of natural frequencies for a number of a sample's normal modes of vibration [58]. Specifically, one scans the resonance structure of a compact specimen such as a cube, parallelepiped, or short cylinder to determine mechanical properties. The method is dynamic, and does not allow static or low frequency measurements; temperature dependence of properties may be studied. Elastic moduli are inferred via an inverse calculation and a knowledge of the sample shape and mass; mechanical damping is inferred from the width of the resonance curves. If the sample is a slender rod and the vibration mode is known from the excitation conditions, calculation of material properties is straightforward as described above. The mode structure of a vibrating sphere, cube, or parallelepiped is complex. With the advent of microcomputers capable of rapid

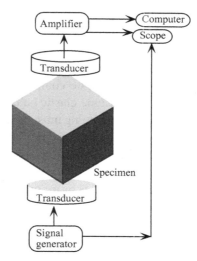

Figure 6.15. Simple configuration for resonant ultrasound spectroscopy.

computation, it has become feasible to numerically calculate moduli and damping values from the complex mode structure of single-crystal or isotropic specimens.

In a typical embodiment of the method, a rectangular parallelepiped or cubic sample is supported by transducers at opposite corners, as shown in Figure 6.15 and Figure 6.16. Corners are used since they always move during vibration (they are never nodes) and they provide elastically weak coupling to the transducers. A piezoelectric polymer film, such as polyvinylidene fluoride (PVDF) is suitable as a transducer. One transducer excites the vibration, and the other converts the resonant motion to an electrical signal. The specimen is supported by contact pressure; it does not require cementing or the use of coupling agents. The method is capable of determining, from one specimen, all the complex modulus tensor elements $C^*_{ijkl}$ of an anisotropic material, but is not capable of static or low frequency studies. As an

Figure 6.16. Resonant ultrasound spectroscopy (RUS) setup: cubical specimen between ultrasonic transducers.

example, resonant ultrasound spectroscopy, used in the study of $Ni_{80}P_{20}$ has disclosed $\tan\delta$ in the range $10^{-4}$ to $2 \times 10^{-3}$ at 320 kHz over a range of temperature [59].

In principle, one can also determine the full anisotropy of mechanical damping [60]; since resonant modes typically depend on several moduli, the numerical data analysis can be complicated. Moreover, inversion algorithms used to extract the moduli assume elastic behavior, specifically that moduli are independent of frequency. The damping must be quite small for such an assumption to be a reasonable approximation, because moduli in viscoelastic solids depend on frequency via the Kramers–Kronig relations (§3.3). If the specimen damping is sufficiently small, parasitic damping in the apparatus becomes a concern. Radiation of sound as a parasitic damping process is more problematic for compressional modes than torsion modes [61]; it can be suppressed by conducting tests in an evacuated chamber. For resonance in air of fused silica at 500 kHz, the experimental damping was $4.7 \times 10^{-5}$ about half the theoretical value. Testing under reduced pressure or under helium reduced the parasitic damping. Effects due to the point contact can be reduced by minimizing the contact force. Parasitic damping below $10^{-6}$ is possible [62]. If the specimen is isotropic or nearly so, analytical solutions may be used to extract properties, with no need for numerical inversion. After Demarest [63], the lowest mode natural frequency $f_1^{\text{cube}}$ of an elastic cube of side $L$ is a torsional mode, as follows, with $G$ as the shear modulus and $\rho$ as the density:

$$f_1^{\text{cube}} = \frac{\sqrt{2}}{\pi L}\sqrt{\frac{G}{\rho}}. \tag{6.20}$$

For a viscoelastic cube, the shear modulus $G$ is interpreted as $G'$. Higher resonant modes, for a Poisson's ratio 0.3, have frequencies in the ratios 1, 1.34, 1.38, 1.57, 1.58, 1.69 $\cdots$; Demarest also provides a plot of natural frequencies versus Poisson's ratios.

The lowest natural frequency $f_1^{\text{cyl}}$ for a cylinder of length $L$ is, for free torsional vibration in the absence of end inertia,

$$f_1^{\text{cyl}} = \frac{1}{2L}\sqrt{\frac{G}{\rho}}. \tag{6.21}$$

Higher torsional modes of the cylinder, if damping is small, occur in the ratio 1, 2, 3, 4 $\cdots$ There is no restriction on the length for this to be true, but there are other modes as well. For a long cylinder the lowest mode becomes a bending mode for which calculation is much more complicated. For RUS it is sensible to use a short cylinder with length equal to diameter. For such a cylinder, if it is isotropic, of Poisson's ratio 0.345, mode frequencies are in ratio 1, 1.27, 1.28, 1.38, 1.48, 1.5 $\cdots$ [64]. The mode structure for isotropic cubes and cylinders is summarized in Table 6.1. These known modes help the experimenter correctly identify the observed modes. The column to the right represents the sensitivity of each mode frequency (called $f$ here to distinguish it from the Poisson's ratio) to the Poisson's ratio $\nu$ of the short cylinder material.

Table 6.1. *Mode frequency ratios $f/f_1$ for isotropic solids in resonant ultrasound spectroscopy. Poisson's ratio is 0.3 for the cube and 0.345 for the cylinder. Sensitivity of cylinder mode frequencies to Poisson's ratio $v$ is shown on the right*

| Mode | Cube $f/f_1$ | Cylinder $f/f_1$ | Cyl:$\delta f/f/\delta v/v$ |
|------|------|------|------|
| 1 | 1 | 1 | 0 |
| 2 | 1.33 | 1.27 | 0.3 |
| 3 | 1.35 | 1.28 | 6.7 |
| 4 | 1.55 | 1.38 | 1.6 |
| 5 | 1.62 | 1.48 | 1.0 |
| 6 | 1.69 | 1.50 | 0.2 |
| 7 | 1.90 | 1.59 | 4.5 |
| 8 | | 1.81 | 11.6 |
| 9 | | 1.87 | 20.1 |
| 10 | | 1.98 | 5.0 |
| 11 | | 2.00 | 0 |
| 12 | | 2.01 | 25.2 |

As with other resonance methods, $\tan\delta \approx \Delta f/\sqrt{3}f_0$ is determined from the width $\Delta f$ at half of the maximum in frequency $f$ of the resonance curve. Damping in several modes can be determined, allowing a limited range of frequency, provided the damping is sufficiently small that there is no overlap of resonant peaks. RUS can be used to determine damping at the lowest (torsional) mode in a high-damping material, such as a polymer [65]. Higher modes are sufficiently close that they overlap unless the damping is small. If higher modes overlap, only the lowest mode or modes can be used for interpretation. The experiment can be done with available transducers and electronics; use of shear transducers allows a stronger signal for the easily interpreted torsion modes [66] as well as identification of modes.

## 6.10 Achieving a Wide Range of Time or Frequency

### 6.10.1 Rationale

As we discussed in Chapters 4 and 10, it is desirable to know the properties of materials over as wide a range of time or frequency as possible. Materials may be expected to support loads for periods from hours to many years, and they may be used under conditions in which acoustic behavior is important. The acoustic frequencies which can be perceived by a young human ear are 20 Hz to 20 kHz, corresponding to time scales of milliseconds or shorter. If the material is subject to ultrasonic testing, frequencies in the MHz range will be of interest. Determination of material properties over a wide range is problematic, because each experimental method typically covers no more than three or four decades, and the lumped resonance methods permit characterization at only one frequency. Specific instrumentation intended for a wide range of time and frequency is discussed in some detail in §6.11.5.

### 6.10.2 Multiple Instruments and Long Creep

A wide range of time or frequency may be achieved by performing experiments with a variety of devices, each one covering a portion of the range. This approach is cumbersome and may require preparation of several types of specimen; nevertheless, some experimenters have done such tests. Koppelmann [47] described several flexural and torsional pendulum devices and wave propagation devices for the study of polymers. The upper limit of time studied in a creep test is limited only by the patience (or lifetime) of the experimenter. Some creep tests have extended to one year, or even 20 years, as presented in §7.4.2. If, in addition, the load is applied rapidly and the response at short times is recorded electronically, a wide range can be achieved. Letherisch [67], by this approach, obtains a ten decade range of several milliseconds to one year. Extension of the range on the short time—high frequency end is achieved by the use of multiple instruments or in a single instrument by minimizing inertia and by back-calculation through specimen resonances, as discussed below in §6.11.5.

### 6.10.3 Time–Temperature Superposition

For certain types of material, it is possible to infer behavior over a wide range of equivalent time or frequency from experiments conducted at different temperatures. For a class of materials called *thermorheologically simple materials*, a change in temperature is equivalent to a shift of the behavior on the log time or log frequency axis (time–temperature superposition), as presented in §2.7. To experimentally determine if a material is thermorheologically simple, one may perform a set of creep, relaxation, or dynamic tests at different temperatures, and plot the results as illustrated in Figure 2.9. If the various curves can be made to overlap by horizontal shifts on the log-time axis, the material is considered thermorheologically simple. The curve constructed by time–temperature shifts is called a *master curve*, as illustrated in Figure 2.10. Amorphous polymers are most amenable to this approach.

Materials in which the relaxation is dominated by a thermally activated process, obeying the following Arrhenius equation, are thermorheologically simple as demonstrated in Example 6.9.

$$\nu = \nu_0 \exp -\frac{U}{RT}, \tag{6.22}$$

with $U$ as the activation energy, $\nu$ as frequency for a feature, such as a peak, and $R = 1.986$ cal/mole K as the gas constant.

If the curves do not overlap, the material is thermorheologically complex. Plazek [68–70] has pointed out that given data within a fairly narrow experimental window (3 decades or less) the test for thermorheological simplicity can only be definitive in its failure. That is, such an experiment is capable of demonstrating thermorheological complexity but it cannot demonstrate simplicity. Plazek showed that nearly perfect superposition was obtained for data for polystyrene over a restricted range of 3.4 decades, but that data over 6 decades could not be superposed.

Figure 6.17. Servohydraulic test device of 20,000-pound capacity. The lower grip is moved by a hydraulic ram below the table surface.

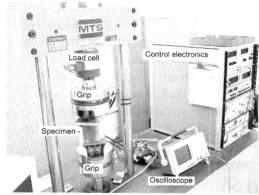

If there are multiple relaxation mechanisms, each with its own dependence on temperature, the material will not be thermorheologically simple. If the material exhibits a phase change in the temperature range considered, or if it undergoes a chemical change such as decomposition, oxidation or light-induced cross-linking, then it will not be thermorheologically simple. These issues could be problematical in inferring behavior at very long times. There is a further caveat associated with high frequencies or short times. It is possible to generate master curves for frequencies as high as $10^{28}$ Hz as discussed in §7.2. For comparison, a frequency $\nu$ of $10^{12}$ Hz corresponds, for a typical wave speed $c$ of 3 km/sec, (via $\lambda\nu = c$) to a wavelength $\lambda$ of 3 nm, or just a few atoms. Physically realizable sound waves in real materials cannot be achieved at frequencies of about $10^{13}$ Hz and higher, therefore, because the corresponding wavelength would be smaller than one atom, which is not possible because sound entails a disturbance of the atoms. Before that point is reached, new relaxation processes, such as wave scattering, become operative.

## 6.11 Test Instruments for Viscoelasticity

### 6.11.1 Servohydraulic Test Machines

Electronically controlled hydraulic systems are used to produce load in commercially available testing machines (e.g., Instron Co., Canton, MA, USA; MTS Systems Co., Eden Prairie, MN, USA). Large forces are readily generated. A test frame of 20,000 pound capacity is shown in Figure 6.17. Such machines are available for tension–compression or tension–compression and torsion tests. These machines are commonly used to conduct stress–strain testing to fracture of structural materials. Servohydraulic machines are equipped with transducers, usually LVDTs, to measure the appropriate displacements, and are fitted to accept signals from extensometers and strain gages. Signals associated with force, displacement, or strain may be used as a feedback input to a servocontroller, which causes the variable in question to follow the control signal. Experiments in creep, relaxation, constant strain rate, or dynamic loading may be conducted, depending on the control signal chosen. The upper bound on frequency depends both on the particular transducers used, as described above, and on the hydraulic system. If a large load or displacement amplitude is required, the frequency range is reduced. Ordinarily, it is difficult to exceed 10 Hz. Special servohydraulic instruments have a claimed upper bound of 200 Hz. Creep and relaxation tests are certainly possible, but since these test machines are expensive, many experimenters are reluctant to use them for lengthy creep tests. Servohydraulic test machines are versatile, however, many experimenters choose to build their own instruments for detailed viscoelasticity studies, particularly if a wide range of time or frequencies are required.

### 6.11.2 A Relaxation Instrument

A relaxation instrument intended for the study of glassy polymers over three decades of time used an eccentric cam system actuated by a handle for load application [71], and an axial load cell of the strain gage type for force measurement. Applied deformation was measured with a dial micrometer. In relaxation experiments it is necessary that the specimen strain be held constant. To evaluate the quality of the experiment it is necessary to consider the compliance of the load cell. In this example, the axial load cell had a 1.3 kN capacity at full scale and a deflection of 0.01 mm at full-scale load. Because the experiment is quasistatic, the force in the specimen is equal to the force in the load cell. A typical 76 mm long sample of stiff polymer experienced a strain of 1 percent at a load of 440 N. Here, the specimen end displacement is $0.01 \times 76$ mm $= 760\,\mu m$, while the load cell deflection is 0.01 mm (440 N/1.3 kN) $= 3.4\,\mu m$, or only 0.0045 of the total. The rigidity of the entire load frame is as important as that of the load cell; moreover for tension experiments thermal expansion of the load frame can introduce error. For that reason, the load frame was made robust and was enclosed in a temperature controlled chamber independent of the temperature control system for the specimen. The specimen was gripped at the ends with jaws of a flared collet design and serrated to enhance grip capability. The instrument also provided an independent loading system and a torque cell of the strain gage–type for torsional relaxation. Torsional and axial degrees of freedom were decoupled by an elaborate system of bearings. Signals from the force and torque transducers were processed with high frequency carrier amplifiers. The stiffness of the entire instrument was checked by measuring relaxation in steel, which is known to exhibit negligible relaxation in comparison with polymers.

### 6.11.3 Driven Torsion Pendulum Devices

#### *Basic Principles*

The torsion pendulum allows creep measurement, subresonant sinusoidal measurement, or resonance half width measurement, depending upon the input signal used. Free-decay of vibration can also be observed following a mechanical perturbation. Torsion pendulums have been made in many variants. A typical embodiment is shown in Figure 6.18. Alternatively one may attach the permanent magnet to the

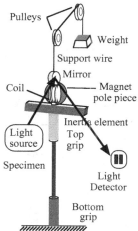

Figure 6.18. Driven torsion pendulum. Torque on the coil is generated by the action of a magnetic field supplied by a permanent magnet (one pole piece shown).

specimen and apply a controlled magnetic field via an external coil. This modality eliminates error due to stiffness and damping due to wires from a moving coil; it is facilitated by use of high-intensity permanent magnets. Angular displacement is determined via an optical lever approach. For quasistatic tests, one can project the light beam on a calibrated scale and capture data visually. For dynamic tests, and in all cases to facilitate data export, a laser beam is input to a split silicon sensor connected to a subtraction amplifier.

### Subresonant Devices

A high-resolution inverted subresonant torsion pendulum, Figure 6.18, provides a six-decade range of frequency from $10^{-5}$ Hz to 10 Hz; the natural frequency is 100 Hz to 500 Hz. It is an inverted torsion pendulum of the $K\hat{e}$ type. Torque was generated by the action of electric current in a Helmholtz coil upon a permanent magnet fixed to the specimen via an extension rod. The specimen was surrounded by furnace windings well below the driving coil. The assembly was stabilized via a weight acting through a rocker shown as a scale beam in Figure 6.19. Angular displacement was measured by reflecting a beam of light from a mirror fixed to the specimen upon a differential light-sensitive diode. Phase determination is by measuring with a timer, the time delay between the zero-crossing of the load and displacement signals. A time accuracy of 1 $\mu$sec was claimed, which corresponds to a phase uncertainty $\delta\phi \approx 6 \times 10^{-5}$ at 10 Hz, and even better at lower frequencies. Control of electronic and mechanical noise is essential in this approach; building vibration amplitude was at most 0.1 $\mu m$. Noise of mechanical origin was reduced by the suspension system that suppresses transverse vibrations. Transverse vibrations can be problematical since they occur at considerably lower frequency than torsional vibrations and so are more easily excited by movement of the supporting base. The upper frequency limit was about 10 Hz because the pendulum natural frequency was 100 to 500 Hz. Tuning through the resonance was not done in view of the

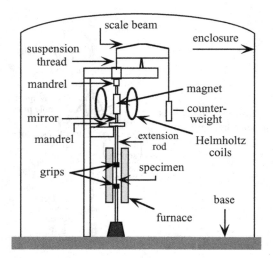

Figure 6.19. Driven subresonant torsion pendulum with environmental chamber and furnace (Adapted from Woirgard et al. [28]).

complex arrangement of stalks which would give a resonance structure difficult to interpret. The temperature range is room temperature to 800°C, under vacuum [28]. Subresonant instruments also include the device of D'Anna and Benoit [76], which allows measurement at cryogenic temperatures. The frequency range is limited to subaudio values. An automated instrument was developed for dynamic testing from $10^{-5}$ Hz to 1 Hz, and for transient studies [72]. The device was a torsion pendulum containing a magnet and mirror fixed to a long stalk attached to the top of the specimen, which is contained within a temperature-controlled furnace. Current in a Helmholtz coil was used to apply the torque. Representative results for a polymer and an oxide glass were reported [72]. A driven torsion pendulum [77] operates over an extended range of temperature, 80–1250 K, at low frequency from $10^{-4} - 10$ Hz. Torque is generated by a current through a Helmholtz coil acting on a magnet isolated from the specimen by a stalk. Phase resolution is enhanced by integrating electronic phase measurements over a number of cycles.

### Biaxial Pendulum

A biaxial driven torsion pendulum [73, 74] is shown in Figure 6.20. It permits viscoelastic measurements in torsion, in the presence of a superposed axial static load. Biaxial experiments are of interest in the study of polymers because the tensile stress alters the free volume, which alters the relaxation kinetics. The torque was generated electromagnetically: a large electromagnet provided a field for a magnesium rotor frame containing many turns of fine wire through which an amplified electric current was passed. The lower part of the specimen was mounted to a reaction torque sensor of the strain gage variety. Angular displacement was measured using an LVDT core supported by a wire wrapped around a rotor disc concentric with the torsion axis. Signals from the angular displacement and torque transducers were processed with high frequency carrier amplifiers. Transient experiments were

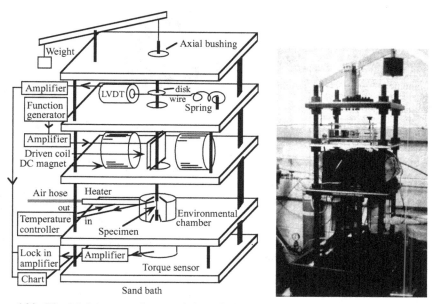

Figure 6.20. Biaxial driven torsion pendulum. Left: diagram adapted from Meyers, Camma, and Sternstein [73]. Right: photograph.

conducted using an electronic servo controller. The rise-time for relaxation tests was 60 to 100 msec, so data were taken beginning at 1 sec. The amplifiers exhibited zero drift of about 1 percent per day and sensitivity drift of 1.5 percent per day, which limited transient experiments to less than one day. Dynamic experiments employed a sinusoidal input signal to the torque motor and a lock-in amplifier to measure phase. Phase measurements at lower frequencies down to about 0.001 Hz were done using an oscilloscope to examine elliptic Lissajous figures; the analysis is shown in §3.2.2. The system permitted a resolution in tanδ of 0.001 from 0.5 Hz to 100 Hz; the lower frequency limit being due to the lock-in amplifier and the upper limit due to resonances in the LVDT assembly. Resonance of the specimen combined with added inertia occurred at 30 Hz to 60 Hz. It was possible to take measurements through this resonance because the torque at the bottom of the specimen was measured, not the driving torque upon the inertia member. Resonance due to elasticity of the torque sensor combined with the inertia of the specimen, driver, and other parts occurred at about 700 Hz. Resonance in the LVDT system occurred at about 100 Hz, which dictated the upper limit of frequency. The overall range of time and frequency accessible with this instrument was eight decades.

### Subresonant and Resonant Device

A device used for dynamic torsional measurements on polymers over a five-decade range from 0.01 Hz to 1 kHz used an electromagnetic drive system in which a coil carrying an electric current was immersed in a static magnetic field [75]. Angular

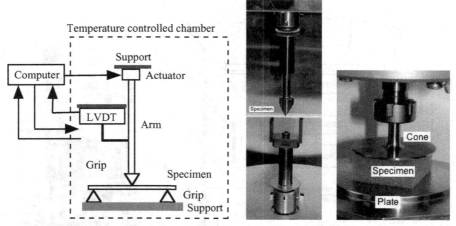

Figure 6.21. Left: schematic diagram of a commercial dynamic mechanical analyzer. Center: detail of fixture for three-point bending with a span of 25 mm. Right: detail of fixture of a cone and plate rheometer.

displacement was measured via a capacitive sensor. The specimen was attached to a lumped inertia element and the dynamic material properties were extracted from the measured structural compliance and phase via the lumped system equations such as those developed in §3.5. Resonance was typically at about 1 kHz; measurements could be made at, near, or well below resonance. The specimen was restrained by pivots supported on leaf springs to allow axial expansion with temperature changes. Spurious phase shifts were introduced by the capacitive sensor and these were corrected by subtracting the "background" associated with a low-loss metal specimen.

### 6.11.4 Commercial Viscoelastic Instrumentation

Instrumentation specifically intended for viscoelastic characterization of materials is commercially available (e.g., [78]). The "Dynamic Mechanical Analyzer" can apply flexural load to a bar shaped specimen via an electromagnetic driver coupled to the specimen by sample arms and clamps (Figure 6.21). A variety of fixtures may be used. Deflection is measured by LVDTs. Temperature and load history are controlled by a computer. Creep and stress relaxation experiments can be performed, as well as low-frequency dynamic measurements. The mass of the sample arms and gripping fixtures limits the frequency range to less than about 10 Hz. However, for thermorheologically simple materials, an extended time or frequency range can be obtained by performing experiments at different temperatures under computer control.

Rheometers are intended for materials with a fluid or soft solid consistency. The specimen is placed between parallel plates or between a shallow cone and a plate and subjected to torsion (Figure 6.21). Step, ramp, or oscillatory histories may be applied at different temperatures via computer control.

### 6.11.5 Instruments for a Wide Range of Time and Frequency

*Rationale*

Measurements over a wide range of frequency are preferable to the more common approach of varying the temperature, for the following reasons [28]:

1. Temperature related methods are restricted to activated processes. Viscoelastic spectra due to thermoelasticity or magnetic flux diffusion cannot be obtained by temperature scans.
2. The underlying viscoelastic theory is established for isothermal conditions and is not directly applicable if temperature is varied continuously.
3. Temperature variations can introduce important structural changes in the specimen during testing.

Most available experimental methods for characterizing viscoelastic materials are applicable over restricted portions of the time and frequency domains, usually no more than three or four decades. In the case of thermorheologically simple materials, such a limitation can be circumvented by performing a series of tests at different temperatures and computing a "master curve" of effective viscoelastic behavior. When a master curve is to be verified or in thermorheologically complex materials, direct measurements over many decades of time/frequency are required. Even for materials that are thought to be thermorheologically simple, a wide range of time and frequency is of use. Plazek [68–70] suggested that experimental data within a fairly narrow experimental window (3 decades or less) are capable of demonstrating thermorheological complexity but cannot demonstrate simplicity. Plazek showed that nearly perfect superposition was obtained for data for polystyrene over a restricted range of 3.4 decades, but that data over six decades could not be superposed. So, proper use of time temperature superposition involves critical analysis of the range of input data and if possible, use of input data over a moderately extended range of time or frequency.

*Instruments of Birnboim, Schrag, and Plazek*

In most reports of measurements over a wide range, many different apparati are used, each of which covering two or three decades of the time/frequency domain [38, 79]. However, several authors reported instruments covering wider ranges, as follows: the multiple lumped resonator concept of Birnboim has been applied by Schrag and Johnson [80] to viscoelastic measurements on *fluids*. In this approach, torsional oscillations are set up in a system of cylindrical inertia members joined by rods. Angular displacement is measured by reflecting light from a mirror on the oscillator and modulating the light intensity via Ronchi gratings. This is a resonant method which provides data at discrete frequencies from 100 Hz to 8300 Hz: two decades. Quasistatic experiments cannot be done with this apparatus, however modifications of the Birnboim apparatus permit low frequency studies down to

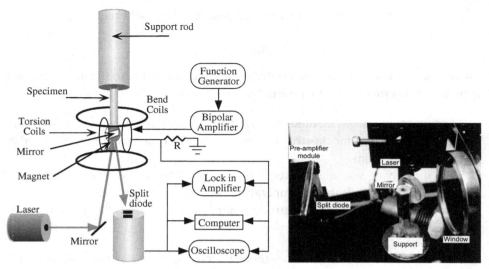

Figure 6.22. Instrument for 11 decades of time and frequency, capable of transient creep tests and dynamic subresonant and resonant tests to 100 kHz. Left: diagram. Right: photograph of light detector.

0.01 Hz [81]. Recent developments in signal processing permit good precision down to $10^{-6}$ Hz [82]. An extended frequency range exceeding seven decades has been obtained for fluids [83]. The fluid specimen is sandwiched between discs of piezo-electric ceramic, and mechanical properties inferred from electrical impedance measurements. The discs deform radially inducing shear deformation in the small sample. Since the specimen and discs are so small the lowest apparatus resonance is about 100 kHz, permitting subresonant measurements to 50 kHz.

As for solids, Letherisch [67] obtained ten decades in creep by extending the tests to one year and by using fast transducers to capture the initial transient. Moreover, six decades were obtained via an automated creep apparatus described by Plazek [109]; it uses electromagnetic drive and optical measurement of rotation. It covers a range of time from about half a second to several days. The approach of Woirgard et al. [28] provides six decades of frequency from $10^{-5}$ to 10 Hz via subresonant phase measurement. About 6.5 decades in the frequency domain has been attained using two apparati [110] one from 0.0002 Hz to 30 Hz, and a second from 20 Hz to 1 kHz. A biaxial driven torsion pendulum [73] after modification [74] provided eight decades up to 100 Hz.

### Instrument for Eleven Decades

An experimental apparatus (Figure 6.22) and analysis scheme was developed for determining the viscoelastic properties of a solid material isothermally, with a single apparatus, over more than 11 decades of time and frequency [86, 87], from 100 kHz in resonant harmonics to creep over several days. For an experimenter with the patience to extend creep tests to one year, the range extends to 13 decades. Torque is applied to the specimen electromagnetically via current through a Helmholtz coil. Specimen deformation is determined by electronic measurement of displacement of a laser beam. Resonances are eliminated from the torque and angle measuring devices by this approach. The effect of resonances remaining in the specimen itself are corrected by a numerical analysis scheme based on an analytical solution which is applicable to homogeneous cylindrical specimens of any degree of loss. Stiffness and damping are calculated by numerical solution of an exact relationship (recall Equation 3.68) for the torsional rigidity (ratio of torque $M^*$ to angular displacement $\theta$) of a viscoelastic cylinder of radius $R$ length $L$, and density $\rho$ with an attached mass of mass moment of inertia $I_{at}$ at one end and fixed at the other end [85]:

$$\frac{M^*}{\theta} = [\frac{1}{2}\rho\pi R^4][\omega^2 L]\frac{\text{ctn}\Omega^*}{\Omega^*} - I_{at}\omega^2, \tag{6.23}$$

where $\Omega^* = \sqrt{\frac{\rho\omega^2 L^2}{KG^*}}$, and $K$ is a geometrical constant (equal to 1 for a cylindrical specimen with circular cross-section). For low-loss materials the method of resonance half-widths is applicable.

$$\tan\delta \approx \frac{1}{\sqrt{3}}\frac{\Delta\omega}{\omega_0}, \tag{6.24}$$

in which $\Delta\omega$ is the full width of the resonance curve (of structural compliance vs. angular frequency) at half maximum and $\omega_0$ is a resonant angular frequency. If the magnet at the end has a sufficiently small inertia, higher order harmonics can be studied using the same method to extend the frequency range.

The apparatus is capable of creep, constant load rate, subresonant dynamic, and resonant dynamic experiments in bending and torsion. The range of equivalent frequency for torsion is from less than $10^{-6}$ Hz (limited by the patience of the experimenter in conducting slow oscillatory or long term creep tests) to about $10^5$ Hz (limited by the ratio of signal to noise in the study of resonances of a high order). Refinements to this instrument include the use of a microcomputer to control subresonant dynamic experiments and display expanded Lissajous figures, and improved isolation from mechanical and thermal perturbations from the environment. A high-resolution digital lock in amplifier has been used to achieve further improvements in phase resolution. The method is called *broadband viscoelastic spectroscopy.*

The apparatus is a modified version of a micromechanics apparatus used earlier for study of microsamples of composites and cellular solids, such as foams in bending and torsion. The specimen is driven by an electromagnetic torque acting upon a permanent magnet at one end and the angular displacement is measured optically. The rationale for the torsion geometry is that the dynamical boundary value problem for the specimen, for which all dimensions are finite [50, 85], has an exact solution. The wide range of effective frequency is obtained as follows [65, 86, 87]:

1. A fixed-free specimen geometry is used in combination with near-zero drift and zero friction methods for torque generation and angular displacement measurement. The lower bound on frequency (or equivalently, the upper bound on creep time) is dictated by the experimenter's patience, not the instrument.

2. The fundamental resonance frequency of the specimen is made as high as possible by minimizing the inertia of the torque generator (a high intensity neodymium iron boron magnet fixed to the specimen end) and of the angular displacement measuring system (a mirror 3 mm square fixed to the specimen end); and by using the shortest possible specimen.

3. Resonances in the apparatus itself are eliminated or are moved to frequencies well above the range of interest.

4. The torsion or bending configuration allows a large impedance mismatch between the specimen and support rod, to minimize parasitic damping and to reduce any coupling to instrument resonances. See Examples 6.3 and 6.4.

5. The geometry of specimen and attachment is made as simple as possible. The governing equation for the specimen resonance is therefore sufficiently simple that it can be uniquely inverted numerically. The inversion method extracts the viscoelastic properties from the measured dynamic compliance and phase in the vicinity of specimen resonance, even for high-loss materials.

Representative experimental results for indium-tin alloys over more than 11 decades up to 100 kHz are presented in §7.3.3, Figure 7.16 [88]. Creep results are shown on a time scale linked to the main scale of frequency $f$ by the relation $t = 1/(2\pi f)$. When both creep and dynamic results are available they may also be plotted on a common frequency scale using approximate or exact conversion relations as presented in Chapter 4.

### 6.11.6 Fluctuation–Dissipation Relation

In experiments of high precision or high sensitivity, limitations associated with thermal noise may become evident. Thermal fluctuations in a system are related to its dissipation via the fluctuation–dissipation theorem which predicts the spectral density as a function of frequency. A classic example is the Nyquist formula for the Johnson noise in an electrical resistor.

For sufficiently compliant mechanical systems, Brownian motion can be observed as noise in the deformation signal. For example, a torsion pendulum made with a Nylon polymer fiber 0.15 mm in diameter and 98 mm-long exhibited noise in the torsional angular displacement associated with the mechanical damping of the polymer [89]. In compliant systems, the noise can be of sufficient magnitude that it can be used to infer viscoelastic dissipation even if no external force is applied. The fluctuation–dissipation theorem can be written in terms of the real part of the admittance $Y$,

$$S(f) = 4kT\Re[Y(f)]/(2\pi f)^2, \qquad (6.25)$$

in which $S(f)$ is the power spectrum of fluctuations, $f$ is frequency, $T$ is absolute temperature, and $k$ is Boltzmann's constant. The real part represents dissipation for the following reason. The admittance in an electrical system is the ratio of current to voltage. For the torsion pendulum, the admittance is the ratio of angular velocity to torque.

$$Y(f) = \frac{i\omega}{-J_t\omega^2 + K_t(1 + i\delta)},\tag{6.26}$$

with $K_t = G'\pi d^4/32L$ as the structural stiffness (in torsion) of the fiber of diameter $d$, length $L$, loss angle $\delta$ and shear storage modulus $G'$, $J_t$ as the moment of inertia of the mass in the torsion pendulum, and $\omega = 2\pi f$. The noise therefore has a peak at the natural frequency of the system. The admittance is proportional to the imaginary part of the *structural* compliance. The root mean square value of the overall noise is given by the equipartition theorem as $(kT/J_t\omega^2)^{1/2}$, for the example given, about $10^{-7}$ rad.

To study soft materials such as gels and biological cells, one can embed a small magnetic bead that can be excited by an external field. If the material is sufficiently compliant one need not necessarily excite the bead externally; thermal motion of the bead provides enough motion from which one can infer material properties. An embedded micrometer-size bead is illuminated with an infrared laser. Viscoelastic properties are inferred from the power spectrum of Brownian motion of the bead; motion is determined from an image of the bead projected upon a split silicon-light sensor [90]. The frequency range is from 100 Hz to several kilohertz.

Noise associated with material damping is of interest in the context of experiments designed to detect gravitational waves [91]. The wires supporting interferometer mirrors generate noise proportional to their damping, so low damping materials are preferable.

### 6.11.7 Mapping Properties by Indentation

Mapping of fine scale properties at the surface has been attempted with scanning-probe microscopes, which can provide phase data as a contrast variable in images. Hard materials have been difficult to characterize in this way, owing to a highly indirect chain of inference. The shape of the probe tip is often not well known. Moreover, adhesive and contact forces can complicate the analysis. Indeed, mapping of stiffness in a graphite-epoxy composite disclosed reasonable values of moduli but damping values for graphite were too high by orders of magnitude [92, 93]. Soft materials such as gels and biological materials have been studied with modified scanning-probe microscopes [94] with better results. Rubbery materials have been studied by indentation tests with a flat end indenter of diameter 0.1 mm [95]. Results compared well with results from dynamic mechanical analysis on a sample 35 mm by 15 mm by 7 mm from 1 Hz to 50 Hz.

The indenter shape is important in interpretation. For a stiff flat end cylindrical indenter, if Poisson's ratio of the substrate that it indents is assumed constant, then the relationship between force-displacement indentation behavior and linear viscoelastic properties can be inferred via the analysis of §5.6. Indeed, indentation creep with a flat-end indenter has been suggested [96] to maintain a constant contact area during indentation; interpretation was based on secondary creep, not linear viscoelasticity. The possibility of the time-variable Poisson's ratio complicates the interpretation. For rubbery materials, this is less of a problem because the Poisson's ratio is close to 0.5. Torsion of a probe on a substrate would eliminate this complication as discussed in §5.6. For other indenter shapes, such as cones, pyramids, and spheres, some analytical solutions (of a noncorrespondence type) are available for interpretation, but they are more complex than for a flat ended indenter. The relation between force and displacement is nonlinear even if the material is linear, owing to geometrical nonlinearity. The Hertz solution for contact of spheres or a sphere and a flat block, discussed in §5.9, reveals this nonlinearity.

## 6.12  Wave Methods

Wave methods are suitable for studies at high frequencies above those attainable by resonance methods. Surveys of wave methods are given in McSkimin [36] and Kolsky [106]. The stiffness is inferred from the wave speed $c$ and the density $\rho$. The speed $c_T$ of shear or transverse waves in an unbounded isotropic elastic medium is given in Thurston [97] and Gorelik [98].

$$c_T = \sqrt{\frac{G}{\rho}}, \qquad (6.27)$$

and the speed $c_L$ of longitudinal or dilatational waves is

$$c_L = \sqrt{\frac{(B + \frac{4}{3}G)}{\rho}}, \qquad (6.28)$$

in which $G$ is the shear modulus and $B$ is the bulk modulus of the material. Observe that $B + \frac{4}{3}G = C_{1111}$, which is the elastic modulus tensor element for constrained axial compression. In viscoelastic solids, the moduli are complex numbers, $G^*$ and $B^*$. Interpretation of the modulus elements is discussed in Example 5.9. The storage modulus is inferred from the wave-speed and the density, and the loss tangent is inferred from the attenuation $\alpha$ as presented in §3.7. Specifically, $\alpha \approx (\omega/2c)\tan\delta$, and the exact version is

$$\alpha = \frac{\omega}{c}\tan\frac{\delta}{2}. \qquad (6.29)$$

Ultrasonic methods are suitable for frequencies from about 0.5 MHz to 20 MHz and beyond. Special techniques permit study to more than 100 GHz.

The attenuation $\alpha$ is measured in nepers per unit length or in decibels (dB) per unit length. One neper is a decrease in amplitude of a factor of $1/e$. This corresponds to $\alpha z = 1$ in Equation 3.108. The word "neper" is a degradation of the name, Napier [99], of the inventor of the natural logarithm.

Attenuation of ultrasonic waves can be measured by comparing the signal amplitude transmitted through two samples of different length [100], as shown in Figure 6.23. The electrical signal is a group of sinusoidal oscillations known as a tone

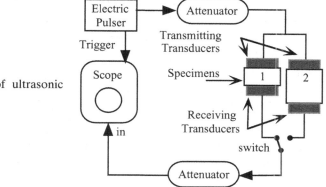

Figure 6.23. Measurement of ultrasonic attenuation.

burst. The attenuator is used to isolate the signal generator from the reflection of the waveform at the transducer. The velocity $c$ is determined from the difference $\Delta t$ in transit times of a particular zero-crossing in the signal, and the known lengths $l_1$ and $l_2$ of the specimens,

$c = \frac{(l_1 - l_2)}{\Delta t}$.

The attenuation $\alpha$, in units of nepers per unit length is determined from the magnitudes of the signals: $A_1$ through a specimen of length $l_1$, $A_2$ through a specimen of length $l_2$.

$\alpha = \frac{\ln(A_1/A_2)}{(l_1 - l_2)}$.

One may also use a single transducer on one face of a specimen of parallelepiped shape to emit a waveform which reflects off the back surface of the specimen [99, 101]. The transducer is excited with bursts of electrical oscillations at its resonant frequency. The bursts must be short enough that successive echoes can be distinguished, but they must have at least several cycles so that the frequency is well defined. The waveform is received by the same transducer. A series of echoes is observed, and the attenuation is inferred from the decrease in amplitude of the echoes. The experimenter must take care that spurious sources of energy loss do not influence the results. In this echo method, quartz is favored for the transducer material. Quartz has a relatively weak electromechanical coupling, therefore little energy from the sound wave is converted back into electricity and lost at the transducer interface. The quartz plate is bonded to the specimen or, if only longitudinal waves are to be used, coupled to it with a thin layer of oil or grease. Alternatively, the quartz transducer may be bonded to a slab-shaped specimen with salol (phenyl salicylate), which melts at 42°C. Moreover, the transducer should not be impedance-matched to the driving and detection circuitry since any energy absorbed by the circuit will increase the apparent attenuation. Commercial transducers for nondestructive testing (NDT) are not appropriate in this method because they exhibit strong coupling. Therefore, the transducer extracts considerable energy from the sound wave at each echo. Moreover, these transducers are heavily damped to achieve broadband response. In a single transducer echo method, these transducers are inappropriate due to the parasitic damping.

In attenuation measurements, a long buffer rod may be interposed between the transducer and the specimen [99]. The rationale for this approach is to eliminate parasitic energy loss from sound waves entering the transducer. A broadband NDT type transducer can be used in the buffer rod approach. Attenuation is inferred from the magnitude of subsequent echoes $A$, $B$, and $C$. The reflection coefficient $R_c$ for the rod-specimen interface must be known. It may be calculated from the echoes as follows. First normalize the echo amplitudes, retaining their sign, $\underline{A} = \frac{A}{B}$, $\underline{C} = \frac{C}{B}$. The reflection coefficient is $R_c = \sqrt{\frac{AC}{AC-1}}$. The reflection coefficient can also be calculated from the acoustic impedance values (§6.9.1) of each material. Finally the attenuation is $\alpha = \frac{1}{2}\frac{1}{l}ln[-\frac{R_c}{C}]$ with $l$ as the specimen thickness.

Waves can also be transmitted through long rods or plates [102, 103]. As for rods, bending waves and longitudinal waves are dispersive (the wave speed depends on frequency) even for an elastic material. For longitudinal waves in a rod, the wave speed $c_L$ is

$$c_L = \sqrt{\frac{E}{\rho}}, \qquad (6.30)$$

provided that the wavelength is much larger than the rod diameter. For shorter wavelengths smaller than the rod diameter, the Poisson effect cannot readily occur, giving rise to dispersion. Torsional waves of lowest order are nondispersive in an elastic medium, and their velocity is the shear wave velocity given above; in a viscoelastic medium the dispersion results from viscoelastic effects only. Again, in viscoelastic materials the elastic modulus is a complex quantity, so that the displacement is

$$u(x, t) = u_0 \exp i\omega\{\frac{x}{c} - t\}\exp\{-x\frac{\omega}{2c}\tan\delta\} = u_0 \exp i\omega\{\frac{x}{c} - t\}\exp\{-\alpha x\}, \qquad (6.31)$$

and the loss tangent is inferred from the attenuation $\alpha$; Equation 6.29.

Ultrasonic methods for determining the complex bulk and shear moduli were presented by Waterman [104]. The method involves transmitting longitudinal ultrasonic waves through a slab (Figure 6.24) of specimen material. The slab is immersed in a fluid, which serves as the medium for wave propagation. The transducers emit longitudinal waves, the only kind that propagate through a fluid. The slab is then rotated through known angles. Oblique incidence converts part of the incident longitudinal wave to a transverse wave, so that the speed and attenuation of both waves can be measured. A variant of this approach [105] involving comparison of amplitudes $A_1$ and $A_2$ transmitted through otherwise identical specimens of different thicknesses $L_1$ and $L_2$ allows simpler interpretation because the loss due to surface wave reflection need not be considered because it is the same for both specimens. When the specimen is held perpendicular to the ultrasound beam, only longitudinal waves are generated in the specimen. The attenuation $\alpha_L$ of longitudinal waves, in decibels per centimeter, is given by

$$\alpha_L = \alpha_{\text{liquid}} + 20(L_2 - L_1)^{-1}\log(A_1/A_2). \qquad (6.32)$$

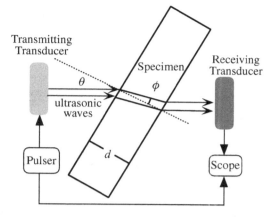

Figure 6.24. Ultrasonic method for determining complex bulk and shear moduli. The transducers and specimen are immersed in a fluid, such as water.

The wave speed is calculated by allowing for the time delay in the liquid. If the ultrasound beam is incident at an oblique angle $\theta$, shear waves are generated in a beam displaced from the longitudinal beam. If the angle exceeds a critical angle, the longitudinal wave is totally reflected in the specimen and only a shear wave is propagated. For oblique incidence, allowance is made for refraction of the ultrasonic wave, to calculate the speed $c_T$ and attenuation $\alpha_T$ (in decibels per centimeter) of shear (transverse) waves as follows:

$$c_T = c_{\text{liquid}}[(\cos\theta - c_{\text{liquid}}\frac{\Delta t}{L})^2 + \sin^2\theta]^{-1/2} \tag{6.33}$$

$$\alpha_T = 20(L_2 - L_1)^{-1}\{1 - [c_T\sin\theta/c_{\text{liquid}}]^2\}^{1/2}\log(\frac{A_1}{A_2}),$$

with $\Delta t$ as the time delay and $L$ as the specimen thickness.

Kolsky [106] has reviewed some wave methods and adduced articles by Nolle [107, 108] who studied rubbery materials from 0.1 Hz to 120 kHz using five different experimental methods. From 0.1 to 25 Hz, the specimen provided a restoring force for a beam rocking on a knife edge. From 10 to 500 Hz a vibrating reed type resonance method was used. In this method the frequency can be changed only by changing the size or shape of the specimen. Longitudinal wave methods in thin strips were used for higher frequencies from 1 to 40 kHz. For 12 kHz to 120 kHz, a compound magnetostrictive resonator was used, in which a rubber–metal sandwich was examined. In Buna rubber over a range of frequency and temperature, the $\alpha$ transition in modulus, and the corresponding $\alpha$ peak in loss versus temperature, was demonstrated to be sharpest at low frequency and progressively less sharp at higher (ultrasonic) frequency. Kolsky [109] also demonstrated the excitation of pulsed waveforms via an explosive charge on the end of a rod, and the interpretation of results for nonsinusoidal waveforms.

Ultrasonic waves are usually generated and detected by piezoelectric transducers; these are commonly used for ultrasonic studies and are available to generate longitudinal or shear waves. The thickness of the piezoelectric element governs its natural frequency. Piezoelectric transducers are available for frequencies

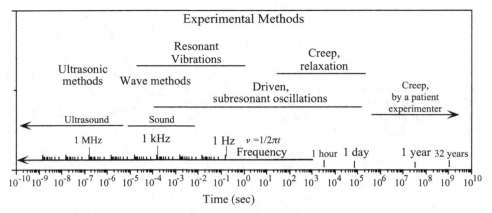

Figure 6.25. Summary of experimental methods in the time and frequency domains.

between about 0.5 MHz and 20 MHz, the frequency range most commonly used for nondestructive evaluation of machine parts and for diagnostic ultrasonic diagnosis of disorders in the human body. Electromagnetic methods are also used, particularly at lower frequencies.

Piezoelectric transducers may be prepared as thin layers for ultrasonic frequencies as high as 1 GHz. Ultrasonic attenuation measurements at frequencies as high as 440 GHz have been performed as follows [110]. A pulsed laser generates a short light pulse of picosecond duration. This light is directed at a solid surface where it is absorbed, raising the temperature of the surface a few degrees. The sudden temperature change induces a pulse of thermal stress, which propagates into the solid as an ultrasonic wave. This wave is detected as it is reflected back to the surface by an optical method which uses a probe pulse of laser light to determine the strain-induced change in reflectivity due to the ultrasonic wave. The method has been used to characterize the properties of thin films and small-scale structures with features as small as 200 nm.

## 6.13 Summary

A brief review of some experimental methods has been presented. Although experimental methods are available over 20 decades of time and frequency (Figure 6.25), most individual methods permit study of three decades or less for a given specimen. Study of an extended range requires multiple apparati, use of time–temperature superposition, or special techniques.

## 6.14 Examples

**Example 6.1**
A durometer is a device used to measure the stiffness of rubbery materials. The durometer has a flat surface that the user presses to the rubber specimen. A spring-loaded protruding probe causes an indentation in the rubber. The observer reads the

value of the indentation from a dial gage, which reads from zero to 100 (representing percent of the deformation taken by the spring), linked to the probe. A spring within the durometer provides the indenting force exerted by the probe. If the rubber is viscoelastic, the dial reading changes with time after the durometer is pressed to the rubber specimen.

Calculate the relationship between the time-dependent dial reading and the creep function.

Remark: The analysis is applicable to a variety of indentation methods, including nanoindentation.

Hint: first calculate the relationship between the dial gage reading and the stiffness of an elastic material considered as a spring of unknown stiffness.

Hint: the sum of the deflections of the specimen and the internal spring as an experiment is conducted may be approximated as a step function in time. The internal spring deflection is observable as the dial reading. The specimen deflection is not directly observed.

**Solution**

For the durometer, the sum of the spring displacement $X_1$ and the specimen displacement $X_2$ is a step function in time: $X_1 + X_2 = X_{tot}\mathcal{H}(t)$. The force in the spring equals the force in the specimen since inertial effects are neglected in the quasistatic regime: $F_1 = F_2$.

For an elastic material, $F_1 = F_2$ implies, with $k_1$ as the stiffness of the spring and $k_2$ as the *structural* stiffness of the specimen, $k_1 X_1 = k_2 X_2$ so $\frac{X_1}{X_{tot}} = \frac{k_2}{k_1+k_2}$.

For a viscoelastic material, with $k_1$ as the stiffness of the spring and $k_2(t)$ as the *structural* stiffness of the specimen,

$$k_1 X_1(t) = \int_0^t k_2(t-\tau)\frac{dX_2}{d\tau}d\tau. \tag{6.34}$$

The structural stiffness of the specimen can be related to its modulus via the rigid indenter solution analyzed in Chapter 5, §5.6.2. Take the Laplace transform of Equation 6.34:

$$k_1 X_1(s) = sk_2(s)X_2(s) = sk_2(s)(X_2(s) + X_1(s)) - sk_2(s)X_1(s). \tag{6.35}$$

Incorporate

$$X_1(t) + X_2(t) = X_{tot}\mathcal{H}(t), \tag{6.36}$$

and take the Laplace transform.

$$k_1 X_1(s) = sk_2(s)X_{tot}\frac{1}{s} - sk_2(s)X_1(s). \tag{6.37}$$

Define a new (dimensionless) $x_1$ as $\frac{X_1}{X_{tot}}$. Then,

$$k_1 x_1(s) = k_2(s)(1 - sx_1(s)), \tag{6.38}$$

so,

$$x_1(s)(k_1 + sk_2(s)) = k_2(s), \tag{6.39}$$

so,

$$x_1(s) = \frac{k_2(s)}{k_1 + sk_2(s)}. \tag{6.40}$$

Rewrite this in a form amenable to power series expansion (because it is difficult to achieve an inverse transform) and recognize from Equation 2.19 that $J(s)k_2(s) = 1/s^2$, with $J$ as the specimen compliance and $k_2$ as its stiffness,

$$x_1(s) = \frac{\frac{1}{s}}{1 + \frac{k_1}{sk_2(s)}} = \frac{1}{s}\{1 - sk_1 J(s) + s^2 k_1^2 J(s)J(s) - \cdots\} \tag{6.41}$$

Transforming back to the time domain,

$$x_1(t) = 1 - k_1 J(t) + k_1^2 \int_0^t J(t - \tau)\frac{dJ}{d\tau}d\tau - \cdots \tag{6.42}$$

The result is an exact series representation thus far. If $k_2 \gg k_1$ for all time, then the series can be approximated by retaining only the first two terms. Here the spring is much more compliant than the specimen material, so most of the deformation occurs in the spring and little in the specimen so that the durometer reading approaches 100 percent. In that case, the normalized spring displacement $x_1(t)$ follows the time dependence of the creep function since the spring force in this regime is nearly constant in time.

If the indenter is flat-ended, the analysis of §5.6.2 can be applied to extract the creep compliance from the time dependent indentation displacement.

**Example 6.2**
Suppose experimental data are available for both creep and dynamic properties of a material. Suppose the creep results are converted to dynamic form with the relations given in Chapter 4, and that reasonable agreement is observed in the region of overlap. Suppose moreover that the real and imaginary parts of the dynamic compliance are found to obey the Kramers–Kronig relations. Can one conclude that the experimental results are valid?
**Answer**
The experimental results are not necessarily valid. If phase angles of electrical origin occur in the transducers or the preamplifier circuits, they will also obey the interconversion formulae (since parasitic phase angles are also due to causal processes) but will generate errors in the results. Independent calibration of the phase behavior of the instrumentation is required.

**Example 6.3**
A specimen of diameter $d_1$, length $L_1$, shear modulus $G_1$ and damping $\tan\delta_1$ is supported by a rod of diameter $d_2$, length $L_2$, shear modulus $G_2$ and damping $\tan\delta_2$ in a fixed–free torsional configuration. How much error is introduced in the measured stiffness and damping by the fact the support rod is not infinitely rigid? Assume that the frequency is well below any specimen or structural natural frequency.

**Solution**

The structural compliance defined as the ratio of the angle $\theta$ to the torque $\tau$ of specimen rod and the support rod is the sum of the compliances. Consider the elastic case first, in terms of the compliances $J_1 = 1/G_1$ and $J_2 = 1/G_2$. At low frequency, the static relations are appropriate. The structural compliance is as follows:

$$\frac{\theta}{\tau} = \frac{32}{\pi}\{J_1 \frac{L_1}{d_1^4} + J_2 \frac{L_2}{d_2^4}\}.$$

Apply the dynamic correspondence principle.

$$\frac{\theta^*}{\tau} = \frac{32}{\pi}\{J_1'(1 + i\,\tan\delta_1)\frac{L_1}{d_1^4} + J_2'(1 + i\,\tan\delta_2)\frac{L_2}{d_2^4}\}.$$

The fractional error $\xi_s$ in stiffness is the real part of the second term divided by the real part of the first term. $\xi_s = \frac{G_1'}{G_2'}\frac{d_1^4}{d_2^4}\frac{L_2}{L_1}$. The observed phase angle $\varphi$ between torque and angular displacement is as follows:

$$\tan\varphi = \frac{\Im[\frac{\theta^*}{\tau}]}{\Re[\frac{\theta^*}{\tau}]} = \frac{J_1'\frac{L_1}{d_1^4}\tan\delta_1 + J_2'\frac{L_2}{d_2^4}\tan\delta_2}{J_1'\frac{L_1}{d_1^4} + J_2'\frac{L_2}{d_2^4}},$$

$$\tan\varphi = \frac{\tan\delta_1 + \frac{G_1'}{G_2'}\frac{d_1^4}{d_2^4}\frac{L_2}{L_1}\tan\delta_2}{1 + \frac{G_1'}{G_2'}\frac{d_1^4}{d_2^4}\frac{L_2}{L_1}}.$$

Since $\xi_s$ is small for a specimen sufficiently thinner than the support rod,

$$\tan\varphi \approx [\tan\delta_1 + \frac{G_1'}{G_2'}\frac{d_1^4}{d_2^4}\frac{L_2}{L_1}\tan\delta_2]\{1 - \frac{G_1'}{G_2'}\frac{d_1^4}{d_2^4}\frac{L_2}{L_1}\}.$$

The quantity in the {} brackets represents a multiplicative error in the loss. The term containing $\tan\delta_2$ in the [ ] brackets is a parasitic damping, $\tan\delta_p$. The fractional error $\xi_d$ associated with this parasitic damping is as follows:

$$\xi_d = \frac{\tan\delta_p}{\tan\delta_1} = \frac{G_1'}{G_2'}\frac{d_1^4}{d_2^4}\frac{L_2}{L_1}\frac{\tan\delta_2}{\tan\delta_1}.$$

Use of a stiff material such as steel or tungsten for the support rod entails $\tan\delta_2 \approx 10^{-3}$. In view of the limited resolution for specimen damping in the subresonant regime, and the ratio of specimen diameter, the parasitic damping error is usually negligible.

For example, if $d_1 = 3.18$ mm, $d_2 = 12.7$ mm, $L_1 = 30$ mm, $L_2 = 120$ mm, $G_1 = 15.4$ GPa (a metal alloy specimen), $G_2 = 154$ GPa (a tungsten support rod), then $\xi_s = 1.56 \times 10^{-3}$, which is a small fractional error in stiffness. Moreover, the parasitic damping, at $\tan\delta_p = 1.56 \times 10^{-6}$, is too small to easily be resolved in the subresonant domain. If $\tan\delta_1 = 10^{-3}$ and $\tan\delta_2 = 10^{-3}$, then $\xi_d = 1.56 \times 10^{-3}$, which is a small fractional error in damping. In the subresonant regime (no inertial effects or waves), phase uncertainty due to electrical and mechanical noise is more of a limitation. These errors in modulus and phase are much smaller in importance in the study of polymers, which are more compliant and higher in damping than the alloy considered above.

By contrast, in the higher frequency regime, wave transmission into the support rod can be problematical, particularly in view of the fact that at resonance one can resolve small values of $\tan\delta$ by the method of resonance half-width or the method of free decay of vibration. An illustration is given in the following example.

**Example 6.4**

Consider parasitic damping at high frequency due to transmission of waves into the support rod in a fixed–free torsion device. Calculate the parasitic damping using the following data. *Case 1*: the support rod is steel, with $G_2 = 78$ GPa, $\rho_2 = 7.9\,g/cm^3$, $r_2 = 6.2$ mm, and material 1, and a specimen is aluminum alloy with $G_1 = 27$ GPa, $\rho_1 = 2.7\,g/cm^3$, $r_1 = 0.8$ mm. *Case 2*: the support rod is tungsten, with $G_2 = 154$ GPa, $d_2 = 12.7$ mm, $L_2 = 120$ mm, $\rho_2 = 19.3\,g/cm^3$, and a specimen is an alloy with $G_1 = 15.4$ GPa, $d_1 = 3.18$ mm, $L_1 = 30$ mm, $\rho_1 = 7\,g/cm^3$.

**Solution**

For case 1, the transmission coefficient for power is $T_p = 4.7 \times 10^{-3}$ based on Equation 6.11. Considering this as a parasitic specific damping or energy ratio $\Psi_p$ under the assumption that all the power transmitted into the support rod is lost, the parasitic loss tangent is $\tan\delta_p = \frac{1}{2\pi}\Psi_p = 7.5 \times 10^{-4}$.

This is comparable to damping seen in pure aluminum and is larger than damping seen in many aluminum alloys. The parasitic loss can be reduced by using tungsten in the support rod, since it is stiffer ($E = 400$ GPa) and denser ($\rho = 19.3\,g/cm^3$) than steel.

For case 2, we have the same alloy specimen considered in the prior example and the same tungsten support rod, $\tan\delta_p = 4.8 \times 10^{-4}$.

Here, in the resonant modality, the parasitic damping is much larger than the value found above under quasistatic, subresonant conditions. Moreover at a resonance, the experimenter can resolve much smaller values of $\tan\delta$ by the method of resonant halfwidth or by the method of free-decay of vibration. Therefore, the fixed-free configuration can be problematical in the study of low-loss stiff materials.

One may use a larger diameter support rod to increase the impedance mismatch and so reduce the parasitic loss in the fixed-free configuration. Even so, a free–free configuration is recommended for low-loss materials, though such methods are not amenable to static measurements. The fixed–free approach is adequate for high-loss metals which are usually less stiff than aluminum. For such metals the error would be correspondingly small, and a negligible percent of the total damping. Polymers have even lower stiffness and higher damping, so the parasitic loss in a fixed–free configuration is negligible if the specimen is slender.

**Example 6.5**

Discuss the trade-off between the upper frequency attainable and the phase resolution of the torsion apparatus of Woirgard et al. [28] shown in Figure 6.19? Refer to parts of the apparatus shown in the diagram.

**Answer**

Phase resolution is improved by the suppression of noise due to mechanical and electrical causes. Mechanical noise is reduced by constraints that suppress parasitic bending vibration. Support of the specimen at both ends reduces parasitic vibration but also adds inertia that reduces the resonant frequencies. Resonant frequencies are also lowered by the fact that the specimen is isolated from the driving magnet

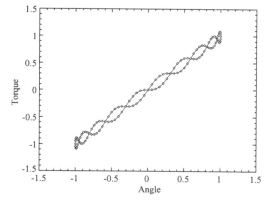

Figure 6.26. An observed Lissajous figure of unusual shape.

by a long stalk, which allows the specimen to be heated without heating the magnet. The stalks and grips contribute a considerable inertia, as well as a complex resonance structure. Therefore measurements are conducted in the subresonant regime, well below the lowest resonance frequency. Restriction of the frequency range has a side effect of reducing electrical noise, hence, improving phase resolution.

## Example 6.6

Ultrasonic attenuation is presented as 10 dB/$\mu$sec at 10 MHz, and the velocity is 6 km/sec. What is the attenuation in nepers/mm, and what is tan$\delta$?

### Solution

To convert units divide the attenuation by the velocity, so $\alpha = 10\frac{dB}{\mu\,\sec}\frac{1}{6\frac{km}{\sec}} = 1.67 \times 10^3 \frac{dB}{m}$. But from §3.7, $\alpha(dB/cm) = 8.68\alpha(neper/cm)$. $\alpha = 192\frac{neper}{m}$. Also, $\alpha = \frac{\omega}{c}\tan\frac{\delta}{2}$. With $\omega = 2\pi \times 10^6$ Hz, and $c = 6$ km/sec, then tan$\delta = 0.037$.

## Example 6.7

During dynamic subresonant torsional testing of a material, the Lissajous figure shown in Figure 6.26 suddenly appears following a small change in driving frequency. What might be the cause?

### Answer

Because the figure is not elliptical, some form of nonlinearity is present. Specifically, a high order Fourier harmonic of the driving frequency is present. One possibility is that a resonance has been excited by a weak Fourier harmonic of the driving signal, which is in the subresonant domain. If the material is of low loss, the magnification, proportional to $(\tan\delta)^{-1}$ of any input, is large. In that case even a small nonlinearity in any part of the system can excite the resonance. If such a pattern only appears at frequencies an integer fraction of a natural frequency of the specimen, then the resonant phenomenon is identified with the specimen itself. The nonlinearity could be in the specimen, the transducers, the signal generator, or the electronics.

## Example 6.8

During dynamic torsional testing of a material, the Lissajous figure shown in Figure 6.27 is observed. What might be the cause?

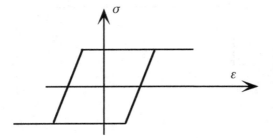

Figure 6.27. A stress–strain plot inferred from observations of torque and angular displacement.

## Answer

This is clearly a nonlinear response of the saturation type. Hard saturation is shown with sharp corners on the curve; saturation may also give rise to rounded corners. There are several possible causes:

1. Transducers can saturate if driven beyond their range of linear behavior. Examine the specifications or calibration curve for the load transducer to see if this is the case.
2. If the specimen is deformed sufficiently it may yield. Calculate the strain level and consider the type of material.
3. Some kinds of material, particularly those with magnetoelastic coupling can exhibit nonlinear hysteresis behavior.
4. If the specimen is gripped in the apparatus, a stick–slip phenomenon in the grip can give the impression of material nonlinearity. Tighten the grips and observe whether the nonlinear behavior persists.

## Example 6.9

How does one obtain the activation energy for thermally activated processes, experimentally?

## Answer

Do experiments at different temperatures. The shift factor for a thermally activated process can be obtained as follows [20]. Consider that at temperatures $T_1$ and $T_2$ the peak loss occurs at frequencies $\nu_1$ and $\nu_2$, respectively. Then, suppose an Arrhenius equation (§2.7) applies, which arises when the rate limiting step of the relaxation consists of movement of an atomic or molecular process over an energy barrier,

$$\nu = \nu_0 \exp -\frac{U}{RT}, \tag{6.43}$$

with $U$ as the activation energy and $R = 1.986$ cal/moleK as the gas constant used if $U$ is to be expressed in calories per mole. When expressed in energy per unit temperature one speaks of the Boltzmann constant $k = 1.38 \times 10^{-16}$ erg/°K. Boltzmann's constant may also be written in units of electron volts (eV), defined as the energy acquired by an electron in traversing a potential difference of one volt, $k = 8.64 \times 10^{-5}$

eV/K. Considering different frequencies and taking a ratio,

$$\frac{\nu_1}{\nu_2} = \frac{\exp(-U/RT_1)}{\exp(-U/RT_2)}, \tag{6.44}$$

$$ln[\frac{\nu_1}{\nu_2}] = \ln a_T = \frac{U}{R}[\frac{1}{T_2} - \frac{1}{T_1}]. \tag{6.45}$$

So, the activation energy $U$ is given by

$$U = R\ln[\frac{\nu_1}{\nu_2}][\frac{1}{T_2} - \frac{1}{T_1}]^{-1}. \tag{6.46}$$

One may also perform experiments at more than two temperature values. Supposing the temperature at which a loss peak occurs is called $T_p$, one may plot $\ln \nu$ versus $1/T_p$, and compute a slope. If a straight line is observed, the slope is interpreted as $U/R$.

**Example 6.10**

What is the appearance of a Debye peak of a thermally activated material when plotted as a function of temperature?

**Solution**

Consider the Debye peak,

$$\tan\delta = A\frac{\omega\tau}{1 + \omega^2\tau^2}, \tag{6.47}$$

with the time constant

$$\tau = \tau_0 \exp\frac{U}{RT}. \tag{6.48}$$

in which $\tau_0$ is a characteristic time. Here the experimentalist maintains the angular frequency $\omega$ constant in Equation 6.47 and varies the temperature $T$. The time constant may also be written $\tau^{-1} = \nu_0 \exp(-U/RT)$ with $\nu_0$ as a characteristic frequency. Because the exponential dependence on temperature, the time constant may be varied over a wide range by changing the temperature. This aspect facilitates the experimental characterization of this class of materials. Multiplying both sides of Equation 6.48 by $\omega$ and taking the natural log,

$$\ln\omega\tau = \ln\omega\tau_0 + \frac{U}{R}\frac{1}{T}. \tag{6.49}$$

At the maximum of a Debye peak, $\omega\tau = 1$, so for the peak at temperature $T_{peak}$, $\ln\omega\tau_0 + \frac{U}{R}\frac{1}{T_{peak}} = 0$.

The full width at half maximum of that peak on an inverse temperature scale is [39]:

$$\Delta(T^{-1}) = 1.144(2.303R/U) = 2.635R/U. \tag{6.50}$$

The width can be used to infer the activation energy provided that the peak is in fact a Debye peak.

**Example 6.11**
How might the viscoelastic Poisson's ratio of an isotropic material be determined experimentally from shear and axial tests?
**Answer**

In Example 5.6, the correspondence principle was used to obtain the following, in the time domain and in the frequency domain, respectively:

$$v(t) = \frac{1}{2} \int_0^t E(t - \tau) \frac{dJ_G(\tau)}{d\tau} d\tau - 1, \tag{6.51}$$

$$v^* = \frac{1}{2}(E' + i E'')(J_G' - iJ_G'') - 1. \tag{6.52}$$

Experimental data do not cover the full range of times from zero to infinity. Therefore, a portion of the range of integration in Equation 6.51 is inaccessible. If the material under consideration were reasonably well-understood, the investigator could attempt to write bounds upon the strength of relaxation in the short time region, and calculate the effect upon the integral. A better choice would be to work in the frequency domain, following Equation 6.52. The calculation is done pointwise, for each frequency, and so is simpler and less demanding of input data.

**Example 6.12**
Determine the $\tan\delta$ from the graph in Figure 6.10. Compare with the value expected from the transient data from which this plot was calculated.
**Solution**
Referring to Figure 3.3, use $\sin\delta = \frac{A}{B}$ and measure the corresponding dimensions on the elliptic figure. So, making measurements with a millimeter scale or with a micrometer, $\delta = \sin^{-1}\frac{38.5\text{mm}}{56\text{mm}} = 0.63$, so $\tan\delta = 0.74$. The relaxation function was given as a power law, $E(t) = At^n$, with $n = 0.4$. So, the inferred phase is $\delta = n\frac{\pi}{2} = 0.628$, so that $\tan\delta = 0.73$; a satisfactory agreement in view of the nature of the measurement.

**Example 6.13**
Calculate the bulk properties of an isotropic viscoelastic material from the shear and tensile properties. What is the effect of a 5 percent error in the measured stiffness? What is the effect of a 5 percent error in the measured loss tangent? Discuss the results.
**Solution**
The relation between the bulk modulus $B$, shear modulus $G$, and Young's modulus $E$ for an isotropic elastic material is as follows: $B = \frac{GE}{3(3G-E)}$. Passing to the compliance formulation, with $J_E = E^{-1}$ and $J_G = G^{-1}$, to simplify inversion of the Laplace transform. $\kappa = B^{-1} = \frac{3(3G-E)}{GE} = 3(3J_E - J_G)$. Applying the correspondence principle, $s\kappa(s) = 3(3sJ_E(s) - sJ_G(s))$. Inverting to obtain the time domain behavior, $\kappa(t) = 3(3J_E(t) - J_G(t))$. As for the *bulk modulus* in the time domain, since $B\kappa = 1$, with the correspondence principle, $B(s)\kappa(s) = s^{-2}$, transforming back, the following time-domain convolution relation between bulk creep compliance and bulk relaxation modulus is obtained. $\int_0^t \kappa(t - \tau)B(\tau)d\tau = t$. In the frequency domain,

applying the dynamic correspondence principle, $\kappa^*(\omega) = 3(3J_E^*(\omega) - J_G^*(\omega))$. Moreover, $\kappa^* = \{B^*\}^{-1}$.

Consider now errors in the input data, for the frequency domain case. Suppose the Poisson's ratio $\nu$ is 0.3, so with $E = 2G(1 + \nu)$, for an elastic material, if $G = 1$ GPa, then $E = 2.6$ GPa, and $J_G = 1$ GPa$^{-1}$, and $J_E = 0.3846$ GPa$^{-1}$. So the bulk compliance is $\kappa = 0.4615$ GPa$^{-1}$. Suppose there is a $-5$ percent error in $J_G$, then there is a 32 percent error in $\kappa$. Suppose there is a $-5$ percent error in $J_E$, then there is a $-37$ percent error in $\kappa$. For a viscoelastic material, we have $\kappa(1 - i\tan\delta_\kappa) = 3(3J_E(1 - i\tan\delta_E) - J_G^*(1 - i\tan\delta_G))$. Suppose $\tan\delta_E = \tan\delta_G = 0.01$. Then, for a $-5$ percent error in $\tan\delta_G$, there is a 32 percent error in $\tan\delta_\kappa$. Errors in the input data are magnified because the bulk compliance is written in terms of a difference of terms of similar magnitude. This situation becomes even more severe if the Poisson's ratio of the material is larger. Consequently, input data of considerable precision and accuracy in both magnitude and phase are required to infer bulk properties from shear and axial data.

## Example 6.14

Suppose that electrical signals are available, proportional to stress and to strain. The stress and strain are sinusoidal in time; the material is viscoelastic, so there is a phase shift. Suppose further that the phase shift is small enough that it is difficult to determine via a Lissajous figures (see Chapter 3). Develop a method to subtract a portion of the in-phase component to obtain a wider loop which is easier to analyze.

## Solution

Consider a quantity $\Gamma_f$ formed by subtracting a fraction $f_s$ of one signal from the other. The signals are assumed to be normalized to unity; $\phi$ is the observed phase shift,

$$\Gamma_f = \sin(\omega t + \phi) - f_s \sin\omega t \tag{6.53}$$

so, $\Gamma_f = \sin\omega t \cos\phi + \cos\omega t \sin\phi - f_s \sin\omega t = \sin\omega t \cos\phi + \cos\omega t \sin\phi - \frac{f_s}{\cos\phi} \times \sin\omega t \cos\phi$

$$\Gamma_f = (1 - \frac{f_s}{\cos\phi}) \sin\omega t \cos\phi + \cos\omega t \sin\phi. \tag{6.54}$$

But $\sin(\omega t + \xi) = \sin\omega t \cos\xi + \cos\omega t \sin\xi$. Set $\Gamma_f = a\sin(\omega t + \xi)$, and solve for $a$ and $\xi$.

$$(1 - \frac{f_s}{\cos\phi}) \sin\omega t \cos\phi + \cos\omega t \sin\phi = a\sin\omega t \cos\xi + a\cos\omega t \sin\xi. \tag{6.55}$$

So, $a\cos\xi = (1 - \frac{f_s}{\cos\phi}) \cos\phi$, $a\sin\xi = \sin\phi$,
so, $a^2\cos^2\xi + a^2\sin^2\xi = a^2(\cos^2\xi + \sin^2\xi) = a^2 = (1 - \frac{f_s}{\cos\phi})^2\cos^2\phi + \sin^2\phi$.
So,

$$a = \sqrt{[(1 - \frac{f_s}{\cos\phi})^2\cos^2\phi + \sin^2\phi]} \tag{6.56}$$

$$\sin\xi = \frac{1}{a} \sin\phi.$$

Here, $\xi$ is the phase angle in the new signal $\Gamma_f$, which is obtained by subtraction of the normalized stress and strain signals. Because $\xi$ is larger than $\phi$, it is easier to measure. Consider several special cases. For example, if $\xi = 90°$, then $\sin \xi = 1$ and $a = \sin \phi$, so,

$$(1 - \frac{f_s}{\cos \phi})^2 \cos^2 \phi = 0, \quad (1 - \frac{f_s}{\cos \phi}) \cos \phi = 0. \tag{6.57}$$

Since $\cos \phi \neq 0$ (otherwise we would not magnify it),

$f_s = \cos \phi$, $a = \sin \phi$, if $\xi = 90°$, that is, the phase is substantially magnified. Now consider the case $\phi \ll 1$.

$$a = \sqrt{\{1 - \frac{2 f_s}{\cos \phi} + \frac{f_s^2}{\cos^2 \phi}\} \cos^2 \phi + \sin^2 \phi}, \tag{6.58}$$

but $(\cos^2 \phi + \sin^2 \phi) = 1$, so $a = \sqrt{1 - 2 f_s \cos \phi + f_s^2}$. This is still exact. Now for $\phi \ll 1$, $a \approx \sqrt{1 - 2 f_s + f_s^2} = 1 - f_s$. Then,

$$\sin \xi = \frac{\sin \phi}{(1 - f_s)} \approx \frac{\phi}{(1 - f_s)}. \tag{6.59}$$

If the phase angle $\xi$ in the processed signal is also small, then the subtraction process magnifies the phase between the input signals as follows:

$$\xi \approx \frac{\phi}{(1 - f_s)}. \tag{6.60}$$

To summarize, one can magnify small phase differences by a subtraction process. This procedure has been used by geologists who are interested in the behavior of rocks and minerals as low loss materials, at low frequencies of interest in seismology and geophysics [111]. At low frequencies, resonance methods are impractical, so a subresonant method is used. In this subresonant method a standard sinusoid is subtracted from the desired signal to fatten up the hysteresis loop, improving phase resolution. Loss tangents of 0.001 can be measured this way provided the ratio of signal to noise is sufficient.

**Example 6.15**
Show that creep results are meaningful beginning at a factor 2.5 of the risetime provided the zero of the time scale is taken as halfway through the risetime.
**Solution**
The definition of the creep compliance assumes a step function load history, however in the physical world the load cannot be applied arbitrarily suddenly. To determine the effect of rise time $t_r$ in transient tests, consider a stress history containing a constant load rate segment (Figure 6.1), also (Figure 6.28). Limits on the creep function error due to rise time were obtained in §6.2.2. Though some authors conservatively suggest to begin plotting data at a time $10 t_r$, Schapery [15] suggests $5 t_r$ is sufficient if the creep or relaxation curve is not too steep. In the following, it is demonstrated how shorter time data may be used.

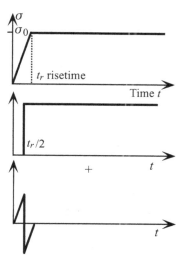

Figure 6.28. Step in stress with risetime $t_r$ (top) analyzed as a superposition of an ideal step delayed by time $t_r/2$ and a shaped pulse that approximates a doublet.

To examine more closely the effect of rise time, and to quantitatively determine the error, consider the stress history to be a superposition of an ideal step shifted by half of the risetime and a shaped pulse. The shifted step stress history is $\sigma(t) = \sigma_0 \mathcal{H}(t - t_r/2)$. The strain response to the shifted step stress is $\epsilon(t) = \sigma_0 J(t - t_r/2)$.

The shaped pulse, which provides the slope of the initial portion of the step,
$$\sigma_p(t) = [\sigma_0 t/t_r, \text{ for } 0 < t < t_r/2, \sigma_0 t/t_r - \sigma_0 \text{ for } t_r/2 < t < t_r],$$
can be approximated as a doublet $\Psi(t)$ for $t \gg t_r$.
Since the delta (which has integral one) is the integral of the doublet,
$$\sigma_p(t) \approx (\sigma_0 t_r{}^2/24)\Psi(t).$$

By the sifting property of the doublet, we write,
$$\epsilon(t) \approx \sigma_0 J(t - t_r/2) + (\sigma_0 t_r{}^2/24)\frac{d^2 J(t - t_r/2)}{dt^2}.$$
In the interpretation of experiments one may neglect the second term and as above, and begin taking creep data at a time some multiple of the risetime. In the following, we determine how soon after the rise time one can begin taking data without excessive error from the fact the step function used is not an ideal one.

As an example suppose, (with $J_0$ a constant),
$J(t) = J_0(t/t_r)^n$, then $d^2 J(t)/dt^2$
$= J_0(n(n-1)/t_r^2)(t/t_r)^{n-2} =$
$J_0(n(n-1)/t_r^2)(t/t_r)^n(t/t_r)^{-2} = J(t)(n(n-1)/t_r^2)(t/t_r)^{-2}$. So, for this case (actually a power law can fit any creep curve over a range of time),
$$\epsilon(t) \approx \sigma_0 J(t - t_r/2)[1 + \tfrac{1}{24}(n(n-1)/2)((t - t_r/2)/t_r)^{-2}].$$

If we suppose Andrade creep in which $n = 1/3$,
$$\epsilon(t) \approx \sigma_0 J(t - t_r/2)[1 - (0.0046)((t - t_r/2)/t_r)^{-2}],$$
so, even for $t = 5t_r/2$, the deviation from the delayed step is less than 1 percent, so we can judiciously obtain creep data earlier than anticipated above, $t = 10t_r$, in §6.1.

In creep tests by dead weight loading such as described above, if data were collected beginning at 10 seconds, a test of 3 hours duration ($1.08 \times 10^4$ sec) covers three decades; 28 hours is required for four decades; 11.6 days for five decades. If mechanical property data are required over many decades of time scale, either great patience in creep tests or shorter time or higher frequency data are needed, or both. In this context it is helpful to use the above approach to obtain data at shorter times than the usual rule to begin plotting data at a time a factor of 10 greater than the rise time. Specifically, one can plot data beginning at a factor 2.5 of the risetime provided the zero of the time scale (not shown in the usual log scale for final presentation of the creep compliance) is taken as halfway through the rise time.

**Example 6.16**
How does one determine the instantaneous modulus in a creep or a relaxation test?
**Answer**
It is tempting to pick off the intercept on the ordinate, when time is plotted on a linear scale. The difficulty with such an approach is that viscoelasticity occurs at short times as well as long times. The modulus may change significantly over time scales shorter than the risetime of the transient experiment. Indeed, for many materials, the modulus changes with time at the shortest times accessible in experiments as is seen clearly in plots of properties versus log time.

**Example 6.17**
How does one distinguish a Debye peak from a resonance?
**Answer**
Resonance is a consequence of inertia in the system. At the lowest natural frequency, the phase angle between cause and effect (e.g., force and displacement) is 90 degrees. For sufficiently simple systems, natural frequencies of resonances can be calculated. A Debye peak is about a factor ten in frequency wide at half maximum. That is much wider than the most heavily damped resonances. If the experiment is repeated with a different inertia, resonant frequencies change but Debye peaks, because they depend on the material itself, do not change.

**Example 6.18**
Does testing at a slow strain rate eliminate the effect of viscoelasticity?
**Answer**
Viscoelasticity can occur over a wide range of rates, including values too slow to be accessible in experiments by humans. Therefore use of a slow rate does not eliminate the effect of viscoelasticity.

## 6.15 Problems

6.1. A vibrating bar of a continuous material has an infinite number of resonance frequencies, yet it is usually possible to use only the lowest few in experiments. Why?

6.2. (a) Determine the creep properties of a foam earplug (see §10.2) using available materials and methods. If earplug foam is unavailable, obtain an alternate flexible foam. Design the experiment so that data are obtained over as many decades of time scale as possible, and so that the error bars are not excessively large. If you have already done a crude creep experiment in Problem 2.4, refine your technique here. It is not necessary to have sophisticated laboratory electronics to do creep tests. Be creative. Also write down what you do and make diagrams of your apparatus. Prepare error estimates for your creep results. Is your specimen sensitive to changes in relative humidity?

(b) Determine the creep properties of a foam earplug at several temperatures. If your apparatus is reasonably compact, it can be placed in a refrigerator. Does time–temperature superposition apply to the material? If so, prepare a master curve.

6.3. Consider phase measurement in relation to the problem of phase resolution. How does the signal subtraction method described in Example 6.13 differ in effectiveness from a measurement based on the width of an amplified Lissajous figure.

6.4. Grip a thin rod of plastic or other inert material between your teeth and tweak it into "cantilever" bending. Describe your observations. Discuss implications regarding the effect of grip conditions on measurements of free decay of vibration in cantilever bending.

6.5. A durometer is a device used to measure the hardness of materials. Durometers used for compliant materials, such as rubbers, contain a spring-loaded probe connected to a dial gage. The experimenter presses the device upon the material to be tested and reads the indentation from the dial. If a durometer is available, observe the dial reading as a function of time following the application of the instrument to a viscoelastic rubber. Write down the data. Prepare a computation outline. Can these results be interpreted as creep?

BIBLIOGRAPHY

[1] Dally, J. W., and Riley, W. F., *Experimental Stress Analysis*, 2nd ed., McGraw Hill, New York: 1978.

[2] Whitney, J. M., Daniel, I. M., and Pipes, R. B., *Experimental Mechanics of Fiber Reinforced Composite Materials*, Englewood Cliffs, NJ: SASE/Prentice Hall, 1982.

[3] Norton, H. N. *Handbook of Transducers*, Englewood Cliffs, NJ: Prentice Hall, 1989.

[4] L'Hermite, R., What Do We Know about the Plastic Deformation and Creep of Concrete?, *Bull RILEM*, 1, 22–51, 1959.

[5] Bazant, Z., *Mathematical Modeling of Creep and Shrinkage of Concrete*, New York: John Wiley, 1988, p. 30.

[6] Leaderman, H., in *Rheology*, F. R. Eirich, ed., Vol. II, Academic, New York: 1958.

[7] Turner, S., Creep in Glassy Polymers, in *The Physics of Glassy Polymers*, R. H. Howard, ed., New York: John Wiley, 1973.

[8] Johnson, A. F., Bending and torsion of anisotropic beams, *Int J Solids, Structures*, 9, 527–551, 1973.

[9] Ward, I. M., and Wolfe, J. M., The Non-Linear Mechanical Behaviour of Polypropylene Fibres under Complex Loading Programmes, *J Mech Phys Solids*, 14, 131–140, 1966.

[10] Sternstein, S. S., private communication, 1973.

[11] Drugan, W. D., private communication, 2008.

[12] Kinder, D. F., and Sternstein, S. S., A path Dependent Variable Approach to Nonlinear Viscoelastic Behavior, *Trans Soc Rheology*, 20, 119–140, 1976.

[13] Pipkin, A. C., and Rogers, T. G., A non-linear representation for viscoelastic behaviour, *J Mech Phys Solids*, 16, 59–72, 1968.

[14] Findley, W. N., Lai, J. S., and Onaran, K., *Creep and Relaxation of Nonlinear Viscoelastic Materials*, Amsterdam: North Holland, 1976.

[15] Schapery, R. A., On the characterization of nonlinear viscoelastic materials, *Polymer Engineering and Science*, 9, 295–340, 1969.

[16] Gent, A. N., and Lindley, P. B., The Compression of Bonded Rubber Blocks, *Proc Inst Mech Engrs*, 173, 111–117, 1959.

[17] Filon, L. N. G., On the Equilibrium of Circular Cylinders under Certain Practical Systems of Load, *Philosophical Trans. Royal Society of London, England, Ser. A*, 147–233, 1902.

[18] Chau, K. T., Young's modulus interpreted from compression tests with end friction, *J Eng Mech, ASCE*, 123, 1–7, 1997.

[19] Lakes, R. S., and Wineman, A., On Poisson's Ratio in Linearly Viscoelastic Solids, *J Elasticity*, 85, 45–63, 2006.

[20] Ward, I. M., and Hadley, D. W., *Mechanical Properties of Solid Polymers*, New York: John Wiley, 1993.

[21] Kobayashi, A. S., ed., *Manual of Engineering Stress Analysis*, 3rd ed., Englewood Cliffs NJ: Prentice Hall, 1982.

[22] Holman, J. P., *Experimental Methods for Engineers*, 6th ed., New York: McGraw Hill, 1994.

[23] Doebelin, E. O., *Measurement Systems: Application and Design*, New York: McGraw Hill, 1975.

[24] Steel, W. H., *Interferometry*, 2nd ed., Cambridge, UK: Cambridge University Press, 1986.

[25] Jones, R. V., Some Developments and Applications of the Optical Lever, *J Sci Instr*, 38, 37–45, 1961.

[26] Shi, P., and Stijns, E., New Optical Method for Measuring Small Angle Rotations, *Appl Opt*, 27, 4342–4344, 1988.

[27] Jackson, J. D., *Classical Electrodynamics*, New York: John Wiley, 1962.

[28] Woirgard, J., Sarrazin, Y., and Chaumet, H., Apparatus for the Measurement of Internal Friction as a Function of Frequency between $10^{-5}$ and 10 Hz, *Rev Sci Instr*, 48, 1322–1325, 1977.

[29] Brennan, B. J., Linear Viscoelastic Behaviour in Rocks, in *Anelasticity in the Earth*, F. D. Stacey, M. S. Paterson, and A. Nicholas, eds., *Am Geophysical Union*, 1981.

[30] Kinra, V. K., and Wren, G. G., Axial Damping in Metal-Matrix Composites. I: A New Technique for Measuring Phase Difference to $10^{-4}$ Radians, *Experimental Mechanics*, 32, 163–171, 1992.

[31] Gottenberg, W. G., and Christensen, R. M., Prediction of the transient response of a linear viscoelastic solid, *J Appl Mech*, 33, 449, 1966.

[32] Lazan, B., Effect of Damping Constants and Stress Distribution on the Resonance Response of Members, *J Appl Mech*, 75, 201–209, 1953.

[33] Graesser, E. J., and Wong, C. R., Analysis of Strain Dependent Damping in Materials via Modeling of Material Point Hysteresis, DTRC-SME-91-34, David Taylor Research Center, U.S. Navy, 1991.

[34] Shi, X., and McKenna, G. B., Spectral Hole Burning Spectroscopy: Evidence for Heterogeneous Dynamics in Polymer Systems, *Phys Rev Let*, 94, 157801, 2005.

[35] Cremer, L., and Heckl, M. *Structure Borne Sound*, 2nd ed., Berlin: Springer Verlag, 1986.

[36] McSkimin, H. J. Ultrasonic Methods for Measuring the Mechanical Properties of Liquids and Solids, in *Physical Acoustics*, E. P. Mason, ed., 1964, pp. 1A, 271–334.

[37] Wachtman, J. H., Jr., and Tefft, W. E., Effect of Suspension Position on Apparent Values of Internal Friction Determined by Forster's Method, *Rev Sci Instr*, 29, 517–520, 1958.

[38] Ferry, J. D., *Viscoelastic Properties of Polymers*, 2nd ed., New York: John Wiley, 1970.

[39] Nowick, A. S., and Berry, B. S., *Anelastic Relaxation in Crystalline Solids*, New York: Academic, 1972.

[40] Frasca, P., Harper, R. A., and Katz, J. L., Micromechanical oscillators and techniques for determining the dynamic moduli of microsamples of human cortical bone at microstrains, *J. Biomechanical Eng*, 103, 146–150, 1981.

[41] Amblard, F., Yurke, B., Pargellis, A., and Leibler, S., A Magnetic Manipulator for Studying Local Rheology and Micromechanical Properties of Biological Systems, *Rev Sci Instr*, 67, 818–827, 1996.

[42] Quimby, S. L., Experimental Determination of viscosity of vibrating solids, *Phys Rev*, 25, 558–573, 1925.

[43] Marx, J., Use of the Piezoelectric Gauge for Internal Friction Measurements, *Rev Sci Instr*, 22, 503–509, 1951.

[44] Robinson, W. H., Carpenter, S. H., and Tallon, J. L., Piezoelectric Method of Determining Torsional Mechanical Damping between 40 and 120 kHz, *J Appl Phys*, 45, 1975–1981, 1974.

[45] Cahill, D. G., and Van Cleve, J. E., Torsional Oscillator for Internal Friction Data at 100 kHz, *Rev Sci Instr*, 60, 2706–2710, 1989.

[46] Iwayanagi, S., and Hideshima, T., Low Frequency Coupled Oscillator and Its Application to High Polymer Study, *J Phys Soc Japan*, 8, 365–368, 1953.

[47] Koppelmann, V. J., Über das dynamische elastische Verhalten Hochpolymerer Stoffe, *Kolloid Zeitschrift*, 144, 12–41, 1955. In German.

[48] Garrett, S. L., Resonant determination of elastic moduli, *J Acoust Soc Am*, 88, 210–221, 1990.

[49] Simpson, H. M., and Fortner, B. E., The Dynamic Modulus and Internal Friction of a Fiber Vibrating in the Torsional Mode, *Am J Phys*, 55, 44–46, 1987.

[50] Christensen, R. M., *Theory of Viscoelasticity*, New York: Academic, 1982.

[51] Tverdokhlebov, A., Resonant Cylinder for Internal Friction Measurement, *J Acoust Soc Am*, 80, 217–224, 1986.

[52] Bishop, J. E., and Kinra, V. K., Some Improvements in the Flexural Damping Measurement Technique, in *M³D: Mechanics and Mechanisms of Material Damping*, V. K. Kinra, and A. Wolfenden, eds., Philadelphia, PA: ASTM STP 1169, 1992.

[53] Braginskii, V. B., Mitrofanov, V. P., and Panov, V. I., *Systems with Small Dissipation*, Chicago: University of Chicago Press, 1985.

[54] Cagnoli, G., Giammatoni, L., Kovalik, J. Marchesoni, F., and Punturo, M., Low Frequency Internal Friction in Clamped Free Thin Wires, *Phys Let A*, 255, 230–235, 1999.

[55] Hosaka, H., Itao, K., and Kuroda, S., Damping Characteristics of Beam Shaped Micro-Oscillators, *Sensors and Actuators A-Physical*, 49, 87–95, 1995.

[56] Graesser, E. J., and Wong, C. R., The Relationship of Traditional Damping Measures for Materials with High Damping Capacity: A Review, in *M³D: Mechanics and Mechanisms of Material Damping*, V. K. Kinra, and A. Wolfenden, eds., Philadelphia, PA: ASTM STP 1169, 1992.

[57] Zhu, X., Shui, J., and Williams, J. S., Precise Linear Internal Friction Expression for a Freely Decaying Vibrational System, *Rev Sci Instrum*, 68, 3116–3119, 1997.

[58] Maynard, J., Resonant Ultrasound Spectroscopy, *Physics Today*, 49 (Jan.), 26–31, 1996.

[59] Kuokkala, V. T., and Schwarz, R. B., The Use of Magnetostrictive Film Transducers in the Measurement of Elastic Moduli and Ultrasonic Attenuation of Solids, *Rev Sci Instrum*, 63, 3136–3142, 1992.

[60] Leisure, R. G., and Willis, F. A., Resonant Ultrasound Spectroscopy, *J Phys Condensed Matter*, 9, 6001–6029, 1997.

[61] Zhang, H., Sorbello, R. S., Hucho, C., Herro, J., Feller, J. R., Beck, D. E., Levy, M., Isaak, D., Carnes, J. D., and Anderson, O., Radiation Impedance of Resonant Ultrasound Spectroscopy Modes in Fused Silica, *J Acoust Soc Am*, 103, 2385–2394, 1998.

[62] Migliori, A., and Maynard, J. D., Implementation of a Modern Resonant Ultrasound Spectroscopy System for the Measurement of the Elastic moduli of Small Solid Specimens, *Rev Sci Instr*, 76, 121301, 2005.

[63] Demarest, R. D. Jr., Cube-Resonance Method to Determine the Elastic Constants of Solids, *J Acoustical Soc of Am*, 49, 768–775, 1971.

[64] Senoo, M., Nishimura, T., and Hirano, M., Measurement of Elastic Constants of Polycrystals by the Resonance Method in a Cylindrical Specimen, *Bull. JSME (Japan Soc. Mech. Engineers)*, 27, 2239–2346, 1984.

[65] Lee, T., Lakes, R. S., and Lal, A., Resonant Ultrasound spectroscopy for measurement of mechanical damping: comparison with broadband viscoelastic spectroscopy, *Rev Sci Instr*, 71, 2855–2861, 2000.

[66] Wang, Y. C., and Lakes, R. S., Resonant Ultrasound spectroscopy in shear mode, *Revi Sci Instr*, 74, 1371–1373, 2003.

[67] Letherisch, W., The Rheological Properties of Dielectric Polymers, *Brit J Appl Phys*, 1, 294–301, 1950.

[68] Plazek, D. J., Temperature Dependence of the Viscoelastic Behavior of Polystyrene, *J Phys Chem*, 69, 3480–3487, 1965.

[69] Plazek, D. J., Oh, Thermorheological Simplicity, Wherefore Art Thou?, *J Rheology*, 40, 987–1014, 1996.

[70] Plazek, D. J., Magnetic Bearing Torsional Creep Apparatus, *J Polymer Sci*, A-2, 6, 621–638, 1968.

[71] Sternstein, S. S., and Ho, T. C., Biaxial Stress Relaxation in Glassy Polymers, *J Appl Phys*, 43, 4370–4383, 1972.

[72] Etienne, S., Cavaille, J. Y., Perez, J., and Salvia, M., Automatic System for Micromechanical Properties Analysis, *Journal de Physique*, Colloq. C5, suppl. 10, Tome 42, C5-1129-C5-1134, 1981.

[73] Myers, F. A., Cama, F. C., and Sternstein, S. S., Mechanically Enhanced Aging of Glassy Polymers, *Ann NY Acad Sci*, 279, 94–99, 1976.

[74] Lakes, R. S., Katz, J. L., and Sternstein, S. S., Viscoelastic Properties of Wet Cortical Bone-I. Torsional and Biaxial Studies, *J Biomechanics*, 12, 657–678, 1979.

[75] Wetton, R. E., and Allen, G., The Dynamic Mechanical Properties of Some Polyethers, *Polymer*, 7, 331–365, 1966.

[76] D'Anna, G., and Benoit, W., Apparatus for Dynamic and Static Measurements of Mechanical Properties of Solids and of Flux Lattice in Type II Superconductors at Low Frequency ($10^{-5} - 10$ Hz) and temperature (4.7–500 K), *Rev Sci Instr*, 61, 3821–3826, 1990.

[77] Gutierrez-Urrutia, I., Nó, M. L., Carreno-Morelli, E., Guisolan, B., Schaller, R., and San Juan, J., High Performance Very Low Frequency Forced Pendulum, *Mat Sci Eng*, A 370, 435–439, 2004.

[78] TA Instruments, Inc., 109 Lukens Drive, New Castle, DE 19720. This instrument was formerly made by DuPont.

[79] Koppellmann, V. J., Uber die Bestimmung des dynamischen Elasticitatsmoduls und des dynamischen Schubmoduls im Frequenzbereich $10^{-5}$ bis $10^{-1}$ Hz, *Rheologica Acta*, 1, 20–28, 1958.

[80] Schrag, J. L., and Johnson, R. M., Application of the Birnboim Multiple Lumped Resonator Principle to Viscoelastic Measurements of Dilute Macromolecular Solutions, *Rev Sci Instr*, 42, 224–232, 1971.

[81] Schrag, J. L., and Ferry, J. D., Mechanical Techniques for Studying Viscoelastic Relaxation Processes in Polymer Solutions, *Faraday Symposia of the Chemical Society*, 6, 182–193, 1972.

[82] Winther, G., Parsons, D. M., and Schrag, J. L., A High-Speed, High Precision Data Acquisition and Processing System for Experiments Producing Steady State Periodic Signals, *J Polymer Sci, Part B: Polymer Physics*, 32, 659–670, 1994.

[83] Christensen, T., and Olsen, N. B., A Rheometer for the Measurement of a High Shear Modulus Covering More Than Seven Decades of Frequency below 50 kHz, *Rev Sci Instr*, 66, 5019–5031, 1995.

[84] Plazek, D. J., Magnetic Bearing Torsional Creep Apparatus, *J Polymer Sci*, Part A-2, 6, 621–638, 1968.

[85] Gottenberg, W. G., and Christensen, R. M. An Experiment for Determination of the Material Property in Shear for a Linear Isotropic Viscoelastic Solid, *Int J Eng Sci*, 2, 45–56, 1964.

[86] Chen, C. P., and Lakes, R. S., Apparatus for Determining the Properties of Materials over Ten Decades of Frequency and Time, *J Rheology*, 33(8), 1231–1249, 1989.

[87] Brodt, M., Cook, L. S., and Lakes, R. S., Apparatus for Determining the Properties of Materials over Ten Decades of Frequency and Time: Refinements, *Rev Sci Instr*, 66, 5292–5297, 1995.

[88] Lakes, R. S., and Quackenbush, J., Viscoelastic Behaviour in Indium Tin Alloys Over a Wide Range of Frequency and Time, *Philosophical Magazine Letters*, 74, 227–232, 1996.

[89] Gonzales, G. L., and Saulson, P. R., Brownian Motion of a Torsion Pendulum with Internal Friction, *Phys Let A*, 201, 12–18, 1995.

[90] Helfer, E., Harlepp, S., Bourdieu, L., Robert, J., MacKintosh, F. C., and Chatenay, D., Microrheology of Biopolymer-Membrane Complexes, *Phys Rev Let*, 85, 457–460, 2000.

[91] Gretarsson, A. M., Harry, G. M., Penn, S. D., Saulson, P. R., Startin, W. J., Rowan, S., Cagnoli, G., and Hough, J., Pendulum Mode Thermal Noise in Advanced Interferometers: A Comparison of Fused Silica Fibers and Ribbons in the Presence of Surface Loss, *Phys Let A*, 270, 108–114, 2000.

[92] Lakes, R. S., Viscoelastic Measurement Techniques, *Rev Sci Instr*, 75, 797–810, 2004.

[93] Syed Asif, S. A., Wahl, K. J., Colton, R. J., and Warren, O. L., Quantitative Imaging of Nanoscale Mechanical Properties using Hybrid Nanoindentation and Force Modulation, *J Appl Phys*, 90, 1192–1200, 2001.

[94] Mahaffy, R. E., Shih, C. K., MacKintosh, F. C., and Käs, J., Scanning Probe-Based Frequency-Dependent Microrheology of Polymer Gels and Biological Cells, *Phys Rev Let*, 85, 880–883, 2000.

[95] Herbert, E. G., Oliver, W. C., and Pharr, G. M., Nanoindentation and the Dynamic Characterization of Viscoelastic Solids, *J Phys D*, 41, 074021, 2008.

[96] Chu, S., and Li, J., Impression Creep, a New Creep Test, *J Mat Sci*, 12, 2200–2208, 1977.

[97] Thurston, R. N., Wave Propagation in Fluids and Normal Solids, in *Physical Acoustics*, E. P. Mason, ed., 1964, pp. 1A, 1–110.

[98] Gorelik, G. S., *Oscillations and Waves*, Fizmatgiz, Moscow, 1959.

[99] Papadakis, E., The measurement of Ultrasonic Attenuation, in *Physical Acoustics*, R. N. Thurston and A. D. Pierce, eds., XIX, New York: Academic Press, 1990, pp. 107–155.

[100] Truel, R., Elbaum, C., and Chick, B., *Ultrasonic Methods in Solid State Physics*, New York: Academic, 1966.

[101] Chung, D. H., Silversmith, D. J., and Chick, B. B., A Modified Ultrasonic Pulse Echo Overlap Method for Determining Sound Velocities and Attenuation of Solids, *Rev Sci Instr*, 40, 718–720, 1969.

[102] Meeker, T. R., and Meitzler, A. H., Guided Wave Propagation in Elongated Cylinders and Plates, in *Physical Acoustics*, E. P. Mason, ed., 1964, pp. 1A, 111–167, Vol 1A.

[103] McSkimin, H. J., Notes and References for the Measurement of Elastic Moduli by Means of Ultrasonic Waves, *J Acoust Soc Am*, 33, 606–615, 1961.

[104] Waterman, H. A., Determination of the Complex Moduli of Viscoelastic Materials with the Ultrasonic Pulse Method (Part I), *Kolloid Zeitschrift. & Z für Polym*, 192, 1–16, 1963.

[105] Hartmann, B., and Jarzynski, J., Immersion Apparatus for Ultrasonic Measurements in Polymers, *J Acoust Soc Am*, 56, 1469–1477, 1974.

[106] Kolsky, H. *Stress Waves in Solids*, Oxford: Clarendon Press, 1953.

[107] Nolle, A. W., Methods for Measuring the Dynamic Mechanical Properties of Rubber-Like Materials, *J Appl Phys*, 19, 753–774, 1948.

[108] Nolle, A. W., Dynamic Mechanical Properties of Rubberlike Materials, *J Polymer Sci*, 5, 1–54, 1949.

[109] Kolsky, J., The Propagation of Stress Pulses in Viscoelastic Rods, *Philosophical Magazine*, 1, 693–711, 1957.

[110] Lin, H., Maris, H. J., Freund, L. B., Lee, K. Y., Luhn, H., and Kern, D. P., Study of Vibrational Modes of Gold Nanostructures by Picosecond Ultrasonics, *J Appl Phys*, 73, 37–45, 1993.

[111] Brennan, B. J., Linear Viscoelastic Behaviour in Rocks, in *Anelasticity in the Earth*, F. D. Stacey, M. S. Paterson, and A. Nicholas, eds. Washington, DC: American Geophysical Union, 1981.

# Viscoelastic Properties of Materials

## 7.1 Introduction

### 7.1.1 Rationale

The purpose of Chapter 7 is to present the viscoelastic behavior of representative real materials so that the reader can gain a sense of orders of magnitude of the effects. Engineers and scientists who deal with elastic behavior of materials are aware of the moduli of various common materials. Similarly, knowledge of the viscoelastic properties of particular materials is essential to rationally apply them. Further examples, with analysis of the physical causes of the viscoelasticity, are provided in Chapter 8. Materials presented here are classified as polymers, metals, ceramics, biological composites and synthetic composites (Chapter 9). In a survey, damping properties of metals, ceramics, and metal matrix composites are compared [1]. Structural metals tend to be low damping, with tan$\delta$ on the order of $10^{-3}$ or less. Experimental results for the same material can differ substantially depending on purity and permanent deformation. Most ceramics also exhibit low damping at ambient temperature, but some exhibit modest damping at elevated temperature.

### 7.1.2 Overview: Some Common Materials

An overview of modulus and damping of selected classes of materials at small strain is shown in Figure 7.1. At high strain amplitude, some metals exhibit higher damping [2] in such maps.

The loss tangents of some well-known materials at various temperatures and frequencies are presented in Table 7.1. Most of the data are at room temperature, denoted rt in Table 7.1 except as noted, and at audio or subaudio frequencies. Further comparison of materials [3–14] is given in the stiffness-loss map in Figure 7.2. Since viscoelastic properties depend on frequency, each material at a given temperature is associated with a *curve*, not a point, in a stiffness-loss map. If only a narrow range of time or frequency is available the curve may appear to degenerate into a point, particularly if the damping is small. Properties of many materials have also

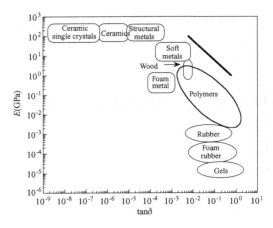

Figure 7.1. Stiffness loss map of Young's modulus versus damping tan$\delta$ for several classes of materials at ambient temperature. The diagonal line represents $E'' = E' \tan\delta = 0.6$ GPa. Most materials occupy the region to the left of that line.

been compiled by Ashby [16] including charts of damping versus stiffness, as well as stiffness and strength versus density.

## 7.2 Polymers

### 7.2.1 Shear and Extension in Amorphous Polymers

For illustrative purposes we consider only a few of the many available polymers, particularly PMMA (polymethyl methacrylate), for which many different methods of characterization have been applied. The properties of many others are described in Ferry [15]. The general features of the viscoelastic response of a polymer are shown

Table 7.1. *Loss tangent of common materials at various temperatures* T; *rt refers to room temperature*

| Material | T | Frequency | tan$\delta$ | Ref. |
|----------|---|-----------|-------------|------|
| Sapphire | 4.2K | 30 kHz | $2.5 \times 10^{-10}$ | [3] |
| Sapphire | rt | 30 kHz | $5 \times 10^{-9}$ | [3] |
| Silicon | rt | 20 kHz | $3 \times 10^{-8}$ | [3] |
| Quartz | rt | 1 MHz | $\approx 10^{-7}$ | [4] |
| Aluminum | rt | 20 kHz | $< 10^{-5}$ | [5, 7] |
| Cu-31%Zn | rt | 6 kHz | $9 \times 10^{-5}$ | [5, 7] |
| Steel | rt | 1 Hz | 0.0005 | [6] |
| Aluminum | rt | 1 Hz | 0.001 | [6] |
| Fe-V0.62% | 33°C | 0.95 Hz | 0.0016 | [5, 8] |
| Basalt | rt | 0.001–0.5 Hz | 0.0017 | [9] |
| Granite | rt | 0.001–0.5 Hz | 0.0031 | [9] |
| Glass | rt | 1 Hz | 0.0043 | [6] |
| Wood | rt | 1.5-8 Hz | 0.0083 | [10] |
| Wood | rt | $\approx$1 Hz | 0.02 | [182] |
| Bone | 37°C | 1–100 Hz | 0.01 | [11] |
| Lead | rt | 1–15 kHz | 0.029 | [10] |
| PMMA | rt | 1 Hz | 0.1 | [12, 13] |
| PMMA | 135°C | 0.5 Hz | 1.7 | [14] |

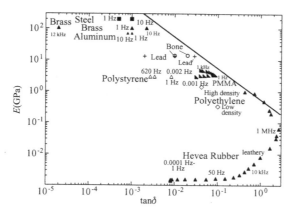

Figure 7.2. Stiffness-loss map for some materials. Temperature is near room temperature. Data points are adapted from the following sources: polymethyl methacrylate (PMMA), $\beta$ peak, various frequencies, [12]; lead, 1–15 kHz, [10]; single-crystal lead, 64 kHz, [54]; bone, 1–100 Hz [11]; steel, 1 Hz, [6]; aluminum, 1 Hz, [6]; Hevea rubber, [15, p. 50], polystyrene, 0.001 Hz to 1 kHz [15, p. 468]. The diagonal line represents $E'' = E' \tan\delta = 0.6$ GPa. Most materials occupy the region to the left of that line. The product $E' \tan\delta$ represents a figure of merit for damping layers as discussed in §10.6.

in Figure 7.3. Polymers consist of long-chain molecules. A *linear polymer* is one in which the polymer molecule is a long chain with no branch points or side appendages. In a *cross-linked polymer* the chains are linked at various points. In *amorphous polymers* there is no long range order in the molecular arrangement. Amorphous polymers have a glass transition temperature but no melting temperature. They are stiff below that temperature and rubbery or viscous above it. Glassy polymers such as PMMA and polystyrene are brittle; a second component can be added to improve toughness. Many amorphous polymers are considered to be thermorheologically simple. Polymethyl methacrylate (PMMA) is a representative amorphous linear polymer. At room temperature, $|G^*| \approx 1$ GPa and $\tan\delta \approx 0.1$ at 1 Hz [12, 13]. A plot of viscoelastic properties at constant frequency versus temperature [14] is shown in Figure 7.4. PMMA exhibits a large peak [17] (called an $\alpha$ peak) in the loss tangent at 135°C and a smaller one (a $\beta$ peak) at 20°C. A plot of properties versus frequency at constant temperature [13], shown in Figure 7.5, shows the $\beta$ peak in the frequency domain.

Figure 7.3. General form of viscoelastic behavior of a polymer.

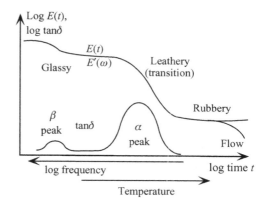

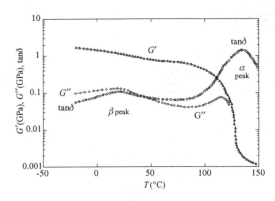

Figure 7.4. Dynamic properties of PMMA at 1 Hz versus temperature (adapted from Iwayanagi and Hideshima [14]).

A master curve derived from relaxation tests at various temperatures (Figure 7.6) [18] is shown in Figure 7.7. At ambient temperatures and short times, PMMA is comparatively stiff; loss as well as dispersion are relatively small: it exhibits the *glassy* state of an amorphous polymer. At higher temperatures or longer times, the stiffness decreases rapidly with increasing temperature or time: the *transition* or *leathery* region. In the transition region the loss tangent can attain values exceeding one. At yet higher temperatures and longer times, the stiffness is low as are the dispersion and the loss: the *rubbery* region. Polymers for which the rubbery region occurs at ambient temperatures are called rubbers; such polymers exhibit glassy behavior at sufficiently low temperatures or high frequencies. The effect of any cross-links and of molecular weight [19] generally manifests itself at long times or low frequencies.

Although amorphous polymers are ordinarily viewed as thermorheologically simple, deviations from time–temperature superposition are known [17]. For example, in a study of properties of PMMA versus frequency and temperature, the $\alpha$ peak hardly shifted as the frequency was changed from 7 Hz to 60 Hz, but the $\beta$ peak shifted markedly to higher temperature, so much as to coalesce at 60 Hz with the $\alpha$ peak [20]. This behavior is indicative of different activation energies for the peaks: on the order of 80 to 200 kcal/mole for the $\alpha$ peak and 29 kcal/mole for the $\beta$ peak [17]. In this vein, viscoelastic and dielectric relaxation phenomena are to be distinguished from other frequency dependent phenomena, such as the infrared

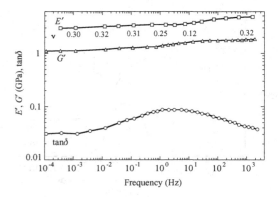

Figure 7.5. Dynamic properties (Young's modulus $E'$, shear modulus $G'$, loss tangent $\tan\delta$ for both torsion and tension) of PMMA at room temperature versus frequency, in the glassy region adapted from Koppelmann [12]. $\nu$ is the Poisson's ratio calculated from the axial and shear properties.

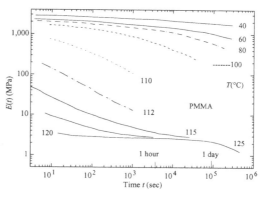

Figure 7.6. Relaxation properties of PMMA at various temperatures (adapted from McLoughlin and Tobolsky [18]).

spectra of polymers. Infrared spectra result from resonances of chemical bonds, and the frequencies at which these resonances occur are quite insensitive to temperature changes [21].

Rubbery materials (elastomers) can differ significantly in their viscoelastic behavior. Properties of several types of rubber used for vibration damping and shock isolation have been presented [22]. Natural rubber has a low-loss tangent of about 0.03 at room temperature, up to about 100 Hz increasing to 0.2 at 10 kHz; neoprene rubber has a loss tangent of 0.1 to 0.15 at room temperature, up to about 100 Hz. Shear moduli on the order of 1 MPa are representative in the plateau region of rubbery materials behavior.

High-loss rubbers (viscoelastic elastomers) can be prepared by incorporating plasticizing agents in the rubber or by a nonstoichiometric composition. It is possible to achieve a loss tangent exceeding 1.0 (Figure 7.8) over one or more decades of frequency. Further results for rubbery materials given in Wetton [24] disclose loss peaks for natural rubber at 10 GHz at room temperature, 10 kHz for polyisobutylene, and near zero to 10 kHz for plasticized polynorbornene. Since the loss peak of natural rubber is at such a high frequency, and the glass transition temperature is correspondingly low, $-73°C$, the loss tangent of natural rubber is low at room temperature and at low frequency. In many elastomers, the peak in the loss versus temperature is sharp. The loss peak of elastomers can be broadened by mixing incompatible components to generate localized fluctuations in concentration on a

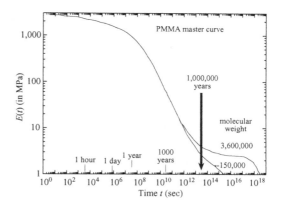

Figure 7.7. Master curve for PMMA reduced to 40°C (adapted from McLoughlin and Tobolsky [18]).

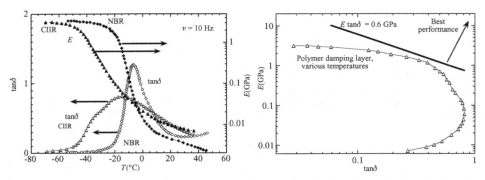

Figure 7.8. Behavior of viscoelastic elastomers used as damping layers. Viscoelastic elastomers, cured to a Shore A hardness of 55. Nitrile-epichlorohydrin rubber (ASTM designation NBR-ECO) and a chlorobutyl rubber, (CIIR). Left: temperature dependence, adapted from Capps and Beumel [23]; arrows indicate scales. Right: stiffness-loss map of CIIR.

scale of 10 to 50 Å, and a range of relaxing environments. Some specific polymeric compositions for high damping have been described in the patent literature. An energy absorbing polyurethane [25, 26] can be made by reacting a mixture of linear and branched polyols, a polyisocyanate, and optionally, a blowing agent (to make foam). The isocyanate index should be from 65 to 85. Such viscoelastic elastomers have been used in shoe insoles for the absorption of impacts [27]. Chlorosulfonated polyethylene [28], widely used commercially [29], allows the glass transition temperature to be varied from $-35°$C to near room temperature by control of the chlorine content. These polymers are often prepared with carbon black as a filler, which stiffens the material in the rubbery regime and reduces the peak damping. Further discussion of the control of damping in polymers is presented in §8.4.

### 7.2.2 Bulk Relaxation in Amorphous Polymers

Polymers exhibit relaxation in their bulk properties as well as in their shear properties [30]. Experimental studies of shear and bulk response have been conducted using ultrasonic wave procedures. The total change in shear stiffness $G'$ or extensional stiffness $E'$ through the $\alpha$ transition between glassy and rubbery consistency can exceed a factor of 1,000. By contrast, the corresponding change in the bulk stiffness $B'$ is about a factor of two. Master curves for shear and bulk behavior of polyisobutylene disclose loss modulus peaks of similar magnitude in shear ($G'' \approx 0.44$ GPa) and volumetric deformation ($B'' \approx 0.72$ GPa) at the same reduced frequency (about 100 MHz at 25°C). The loss peak in shear is broader than the one for bulk deformation. This difference is attributed to an extra physical mechanism involving entanglement of molecules; in shear the molecules can exhibit large relative motions not possible in volumetric deformation. Similar studies upon polystyrene and polymethyl methacrylate [31] disclose a difference in the temperature or reduced frequency associated with the peaks for $G''$ and $B''$.

The difference in the magnitude of relaxation in shear modulus and bulk modulus through the $\alpha$ transition gives rise to a corresponding change in Poisson's ratio that goes from about one-third to values approaching one-half as time increases or frequency decreases through the transition. By contrast, in the $\beta$ peak, observe that the Poisson's ratio of PMMA as inferred from the shear and uniaxial tensile properties [13] is not monotonic in frequency (Figure 7.5). Therefore Poisson's ratio is not monotonic in the time domain through the $\beta$ transition. As for comparisons between tension and torsion in relaxation, PMMA relaxes more slowly in tension than in torsion [32]. In biaxial tests, torsional relaxation rates decrease with superposed tensile strain, but tensile relaxation is unaffected by a superposed torsion strain.

Bulk properties also are important in determining the longitudinal wave speed and attenuation, because, as discussed in §5.8 and Example 5.9, the constrained modulus that governs longitudinal waves is $C_{1111} = B + \frac{4}{3}G$. For example, at ultrasonic frequencies, PMMA exhibits a longitudinal wave speed of $c_L = 2.74$ km/sec and an attenuation of $\alpha = 0.018$ neper/mm at 1 MHz and at 21.1°C [33]. The stiffness $C_{1111}$ is extracted via Equations 5.87 and 5.88 from the density $\rho = 1.184$ g/cm$^3$ and the wave speed $c_L$. $C_{1111} = c_L^2\rho = 8.89$ GPa. From the wave attenuation one may infer $\tan(\delta/2) = \alpha v/2\pi v = 7.85 \times 10^{-3}$. So $\tan\delta = 1.57 \times 10^{-2}$, which is considerably smaller than values observed for shear and uniaxial tension properties at acoustic frequencies and below. The reason is that the constrained modulus $C_{1111}$ can have a very different loss tangent than Young's modulus or shear modulus because it depends on the bulk modulus as seen in Example 5.9. Several epoxy resins exhibited larger attenuation. Interpretation of $C_{1111}$ is discussed in [35] and in §5.8. Ultrasonic data collected at a variety of temperatures [33] were used to generate master curves for frequencies as high as $10^{28}$ Hz. We remark that for a wave speed of 3 km/sec, a frequency of $10^{12}$ Hz corresponds (via $\lambda v = c$) to a wavelength of 3 nm, or just a few atoms. Frequencies of about $10^{13}$ Hz and higher, therefore, do not correspond to physically realizable sound waves in real materials, because one cannot have a wavelength smaller than one atom in a sound wave.

The Poisson's ratio of polymers has been measured directly by several authors. For example, Poisson's ratio of epoxy, measured using an optical method, moiré interferometry, increases with time from 10 to 1,200 seconds. Master curves constructed from such results showed good superposition and an increase in Poisson's ratio from 0.4 to 0.49 [37].

### 7.2.3 Crystalline Polymers

*Crystalline* polymers exhibit long range molecular order as a result of regularity and symmetry in the molecular chains. Linear polyethylene (Figure 7.9) [15, 34, 36], polytetrafluoroethylene and isotactic polypropylene are crystalline. Many biological polymers such as collagen are also highly ordered. Crystalline polymers are generally not thermorheologically simple, so that a master curve cannot be obtained from tests at different temperatures. The $\alpha$ transition in crystalline polymers tends to be broader than in amorphous polymers. Some results are given in Ferry [15], Hopkins and Kurkjian [17], and Aklonis, MacKnight, and Shen [38].

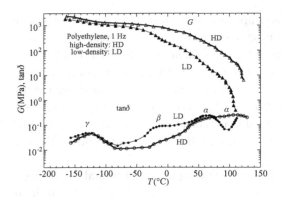

Figure 7.9. Polyethylene shear modulus and tan$\delta$ as a function of temperature (adapted from Flocke [36]).

### 7.2.4 Aging and other Relaxations

Polymers exhibit an aging behavior in their viscoelastic properties as a function of time following polymerization [39, 40]. The effect of aging time $t_a$ is to slow the characteristic retardation times $\tau_c$ by a multiplicative factor,

$$\tau_c = \tau_0 t_a^{\mu_{str}}, \tag{7.1}$$

in which $\mu_{str}$ is the Struik shift factor and $\tau_0$ is a retardation time. In such cases, the effect of aging is represented by a shift of the creep curves along the log time axis. The rate of aging in glassy polymers depends in part upon how far below the glass transition temperature $T_g$ the polymer is used.

Similarities in the *shape* of the relaxation behavior of metals and polymers have been observed [41], though polymers ordinarily exhibit much more relaxation than metals.

The thermal expansion of polymers is time dependent, as anticipated in the discussion of constitutive equations in §2.10. Following a temperature change, the strain may continue slowly to change in the same direction or may slowly change in the opposite direction [42]. In rubbery materials, the time-dependent expansion depends on the degree of prestretch [43].

### 7.2.5 Piezoelectric Polymers

Polyvinylidene fluoride (PVF$_2$) is a piezoelectric polymer: it generates an electric field when deformed; it is available in thin films. It exhibits a broad peak in tan$\delta$ of 0.09 at about 5 MHz [44]. The piezoelectric sensitivity of this polymer is less than that of piezoelectric ceramics (§7.4.5), but the damping is higher than that of most ceramics and the polymer can be prepared in flexible layers.

### 7.2.6 Asphalt

Asphalt is usually produced via vacuum distillation of petroleum crude oils [45]. The resulting residue is used as a binder or cement for mineral aggregates to form

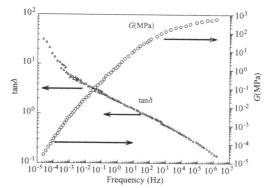

Figure 7.10. Viscoelastic properties of asphalt [45]. Master curve, $T_{ref} = 15°C$.

mixtures used in paving. Asphalt behaves as a viscoelastic liquid with a shear modulus of about $G_0 = 1$ GPa in the glassy regime of high frequency and low temperature, as shown in Figure 7.10. In the terminal regime ($10^{-6}$ to $10^{-7}$ Hz at $15°C$) the loss angle tends to $\pi/2$. The following phenomenological model has been proposed [46] for the normalized complex shear modulus:

$$\frac{G^*}{G_0} = [1 + (i\omega\tau)^{-h_a} + \xi(i\omega\tau)^{-k_a}]^{-1}, \tag{7.2}$$

with $h_a = 0.6, k_a = 0.25, \xi = 2$. The temperature dependence of the mean relaxation times follows an Arrhenius law (from $-10°C$ to $30°C$) with activation energies ranging from 125 to 258 kJ/mole. For temperatures above about $45°C$, deviations from time–temperature superposition occur and they are more noticeable in the phase angle than in the modulus. Deviations from time–temperature superposition also occur if crystalline fractions are present. Such behavior parallels are seen in structural polymers.

Asphalt behavior has been interpreted using tools, such as the WLF equation and the concept of free volume, originally developed for polymers. Asphalt also exhibits an aging effect, called *physical hardening* in this community [47]. Following aging, creep becomes progressively slower. The behavior appears to follow a shift rule in which a change in aging time results in a shift of the creep curves along the log-time axis. The aging effect may result from a progressive collapse of free volume on the molecular scale. This gives rise to slow shrinkage at low temperature; such shrinkage is implicated in the cracking of road surfaces during winter.

## 7.3 Metals

### 7.3.1 Linear Regime of Metals

At small stresses and at low temperatures far below the melting point most metals behave in a nearly elastic manner. Viscoelasticity manifests itself in the form of a small, but nonzero value of the loss tangent as shown in Table 7.1. The corresponding creep or relaxation is small; considerable precision is needed to detect it.

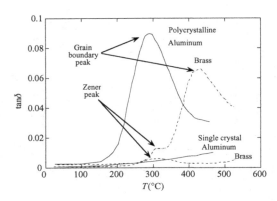

Figure 7.11. Tan$\delta$ as a function of temperature for aluminum [57] and brass [62], both polycrystalline and single crystal, at low frequency (adapted from *Kê*).

Viscoelastic effects in metals used for structural purposes are usually much smaller than those in polymers. Metals often exhibit a series of peaks in the loss tangent at various frequencies [5, 6]. A variety of causal mechanisms are responsible for the damping as discussed in Chapter 8. Results for many metals are compiled in Smithells [5].

Single crystals of metals such as copper, zinc, and lead have been examined by a number of investigators [48–55]. Part of the rationale is to eliminate damping due to the boundaries between the crystalline grains found in bulk samples of metal, in an effort to better understand the causes of damping. In single crystal lead [54], the loss tangent at 64 kHz depended on heat treatment and changed with time after mounting of the specimen; values from 0.002 to 0.01 were observed at room temperature. Damping increased with temperature.

Metals as they are used in most applications are polycrystalline. The boundaries between crystals or grains in metals give rise to additional damping as discussed in §8.5. Peaks of this type, known as *grain boundary peaks* (Figure 7.11) have been observed in a variety of metals, including aluminum [5, 56, 57]. Grain boundary peaks occur at relatively high temperatures, a significant fraction (referred to absolute zero, in Kelvins) of the melting temperature. The ratio of sample absolute temperature to melting temperature is called the *homologous temperature*. The effect of the grain boundaries differs if the grain and specimen sizes are comparable (macrocrystalline materials); the grain boundaries are called *bamboo boundaries* in this case [58–60]. In macrocrystalline materials, the loss-peak height is proportional to the number of grains; by contrast, if the grains are small, the loss-peak height is independent of grain size.

Boron fibers, which are used in high performance composites, exhibit a torsional loss tangent of about 0.001 at room temperature and a frequency of 5 Hz; the shear modulus is 100 to 160 GPa [61].

As for alloys, grain boundary effects for brass are as follows. In polycrystalline $\alpha$-brass (70 percent copper, 30 percent zinc), there is a large peak [62] of about 0.066 in the loss tangent at 430°C, at 0.5 Hz; this peak is absent in single crystal brass; under the same conditions, the loss is less than 0.0025. Polycrystalline brass at room temperature exhibits a small loss tangent of about 0.001. Alloys can also exhibit damping due to stress-induced ordering of the different atoms constituting the alloy. In $\alpha$-brass, a peak in the loss tangent [62] due to this cause occurs at 300°C and at 0.5 Hz: tan$\delta$ is 0.013 in polycrystalline brass and 0.006 in single crystal brass.

Some aluminum alloys exhibit very low damping. Aluminum alloy 2090 exhibits $Q \approx 3.1 \times 10^5$ (tan$\delta \approx 3.2 \times 10^{-6}$) in torsion at 1 kHz at room temperature (300 K) [63] and $5 \times 10^6$ (tan$\delta = 2 \times 10^{-7}$) at 0.5 K.

Alloy 2090 is a strong aluminum lithium alloy which has been developed for aircraft applications. Aluminum alloy 6061, commonly used for pipelines and structural applications, exhibits $Q \approx 2.8 \times 10^5$ (tan$\delta \approx 3.6 \times 10^{-6}$) in torsion at room temperature. Aluminum alloy 5056, commonly used for rivets and window screen, exhibits $Q \approx 2.2 \times 10^5$ (tan$\delta \approx 4.5 \times 10^{-6}$) in torsion at room temperature. Young's moduli for these alloys are $E = 78$ GPa for 2090, 69 GPa for 6061, and 72 GPa for 5056. Aluminides [64] consisting of 51–58 atomic percent iron and the balance aluminum exhibit Young's moduli (at kHz frequencies) from 260 GPa at 300 K to 200 GPa at 988 K; loss tangents are from $3 \times 10^{-4}$ to $10^{-2}$ at high temperature. For a uranium 0.75 weight % Ti alloy [65] at 40 to 80 kHz, and at 25°C, $E = 193$ GPa, $G = 81$ GPa, and Poisson's ratio $\nu = 0.17$; moduli decrease with temperature and Poisson's ratio increases to about 0.5 at 650°C.

Resonance dispersion has been reported in polycrystalline metals such as lead and aluminum [66] in the range 100 Hz to 5 kHz. The sharp resonance-type peaks observed in the loss were considered to be due to microinertial effects in the metal itself rather than due to structural resonance. Crystalline polymers have also been reported to exhibit such microresonance effects [15, 67]. Measurements were made with an electromagnetic device in which material properties are inferred from electrical measurements upon coils of wire. The anomalies occur in specimens with residual stress or under static tensile or compressive stress; they vanish upon annealing or removal of external load. Anomalous resonances were confirmed using a different apparatus [68]. It was suggested that these were due to parametric resonance, a nonlinear effect.

### 7.3.2 Nonlinear Regime of Metals

#### Dynamic

The dynamic mechanical loss in metals often depends on strain amplitude. In the presence of nonlinearity, the response to a sinusoidal input is no longer sinusoidal. Nevertheless because the losses are usually small in metals, the usual terminology associated with linear materials is still used by most writers. One source of nonlinearity is that loss due to dislocation motion increases with strain amplitude, even at small strains. In single-crystal zinc [55], the damping increased with strain along one path, then decreased with strain along a different path. Moreover, the damping increased following application of a static stress. Loss peaks due to grain boundary motion also are strain dependent [69]. Magnetoelastic coupling also gives rise to nonlinear behavior. In 0.42 percent carbon steel [70], $\tan\delta$ was less than $10^{-3}$ for a peak strain less than $10^{-4}$, rising to $8.4 \times 10^{-3}$ for a peak strain of $10^{-3}$.

*Hysteresis* in general refers to a lag between cause and effect, however, in the context of metals, some authors, such as Smithells [5], in the metals field use the following specific definition. Metals may exhibit *hysteresis damping* in which the stress–strain curve for oscillatory loading has cusps rather than being elliptical as in linear behavior. Values for hysteresis damping are normally quoted for a specific, high-stress amplitude level, for example, 34.5 MPa shear stress [5]. Hysteresis damping is usually independent of frequency, but is strain dependent and achieves a relative maximum for a particular strain amplitude. Metals exhibiting considerable hysteresis damping include cast irons, nickel titanium alloys, ferromagnetic alloys, and manganese copper alloys. Smithells also defines *anelastic damping* as damping, which is independent of strain amplitude and that exhibits a peak with respect to frequency. Recall that anelastic materials are viscoelastic materials that exhibit full recovery following creep, a definition rather more general than that of Smithells [5].

Damping in metals depends on whether they have been subjected to plastic deformation prior to the damping test. Pure iron [71] exhibits an effective torsional $\tan\delta$ below about 0.003 for peak strains below $10^{-4}$. A prior plastic deformation strain of $10^{-3}$ gives rise to increased amplitude-dependent damping, up to $\tan\delta = 0.03$ for a peak strain of $6 \times 10^{-5}$. Plastic deformation at higher temperatures gives rise to increased damping after that deformation. The effect is attributed to dislocations.

### Creep

Metals exhibit significant creep at high temperatures and high stresses. For strain between $10^{-3}$ and $10^{-2}$ at ambient temperature, metals with hexagonal crystal structure (e.g., Mg, Zr) creep more than metals with cubic structure (e.g., Al, Fe, Ti) [73], a fact attributed to differences in dislocation mobility. The behavior is nonlinear at the stress levels associated with structural members. In an effort to simplify matters so that realistic problems become tractable, many workers have adopted an equation of state approach [72] in which the creep strain depends only on the conditions of temperature $T$ and stress $\sigma$ at the present time $t$

$$\epsilon(t, T, \sigma) = \sum_{i=1}^{n} f_i(T) g_i(\sigma) h_i(t). \tag{7.3}$$

In this modeling approach, memory of past events is not incorporated, in contrast to the Boltzmann superposition integral of linear viscoelasticity that we have thus far emphasized. A specific equation of state commonly used [72] is the Bailey–Norton law intended to model primary and secondary creep via power laws,

$$\epsilon(t, \sigma) = A\sigma^m t^n, \tag{7.4}$$

with $A$, $m$, and $n$ dependent on temperature and $m > 1$; $n < 1$. Other functions [74] of stress include $g_i(\sigma) = A\sinh(\sigma/\sigma_0)$ or $g_i(\sigma) = A\exp(\sigma/\sigma_0)$. Polynomial functions of time are also used: $h_i(t) = t^n + Ct + Dt^p$ with $n$ as a fraction, often about one-fourth and $p$ an integer often considered to be 3. Creep in metals becomes substantial at temperatures above about 0.3 $T_m$ in which $T_m$ is the absolute melting temperature (Kelvin); above 0.5 $T_m$, the material is at a high homologous temperature, and creep can be rapid [74]. The specific dependence on temperature is usually assumed to be

$$f_i(T) = \exp -\frac{U}{RT}, \tag{7.5}$$

in which $U$ is the activation energy, $R$ is the gas constant, and $T$ is the absolute temperature. The state variable approach is only an approximation to the true behavior, but it is sufficient for many purposes.

As for specific metals, most structural metals such as steel and aluminum do not exhibit much creep at room temperature and at low stress. Metals with a low melting point, such as solders, creep rapidly at ambient temperature [75]. Common lead-tin solders have eutectic compositions or 60% Sn–40% Pb. These solders exhibit primary, secondary, and tertiary creep at room temperature. The secondary creep is thermally activated according to Equation 7.5.

### 7.3.3 High-Damping Metals and Alloys

The highest values of tan$\delta$ in solids are found in polymers near the glass transition and in polymers that lack cross-links, in the flow regime. Such materials are not very stiff. For damping layer applications, and for materials expected to play a structural role as well as to damp vibration, higher stiffness combined with some damping capacity is available in crystalline materials, as follows.

We consider here some special materials that exhibit relatively high effective loss tangents combined with high stiffness. Such materials are used to absorb vibration. In many of these metals, damping is inherently nonlinear even for strain below $10^{-4}$. Most authors do not attempt a full nonlinear analysis of these materials. In view of the nonlinearity, some authors refer to the specific damping capacity $\Psi = \Delta W/W$, the energy dissipated per cycle divided by the maximum energy stored per cycle as a measure of damping. Recall (§3.4) that for linear materials, tan$\delta = \Psi/2\pi$.

A variety of high damping metals are available [76–79]. Materials in the nonlinear regime have been compared by tabulating data obtained at an oscillatory stress amplitude one tenth of the tensile yield stress. Effective loss tangents under these conditions ranged from $0.5 \times 10^{-3}$ for 1100°F aluminum alloys, to about 0.1 for magnesium alloys [76]. Early work on high damping alloys dealt with ferromagnetic materials [82] in which the damping occurs via coupling between stress and motion of magnetic domain walls.

A class of manganese-copper alloys is of interest in applications which require high damping. Copper manganese alloy composition is typically 32 to 42 percent by weight manganese, 2 to 4 percent by weight aluminum, and the balance copper [80].

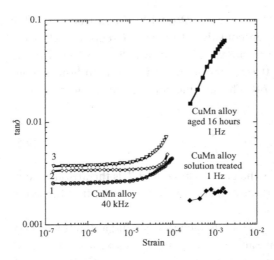

Figure 7.12. Strain-dependent damping in copper-manganese alloys at 20°C, at 40 kHz, given various heat treatments and aging (adapted from Ritchie and Pan [87], and at 1 Hz Laddha and Van Aken [89].

The damping properties of Cu-Mn alloys is enhanced by subjecting the alloy in the hot or cold worked condition to a heat treatment, which involves annealing the alloy at a temperature in the range from 1,200°F to 1,400°F, then quenching the alloy to room temperature (preferably in water), and then aging the alloy by reheating to a temperature in the range from 400°F to 900°F for a period of time from 1.5 to 24 hours, followed by cooling to room temperature (by simple air cooling). These alloys are given the trade name "Sonoston" and are used for ship propellers used in naval applications. Their response is strain dependent [87] as shown in Figure 7.12; $\tan\delta$ can exceed 0.01 at a few Hz if the strain is sufficiently large, $3 \times 10^{-4}$ [89]. The Young's modulus is 96 to 98 GPa and softens by a few percent at the highest strains in the diagram. Loss tangents as large as 0.014 occur at a stress amplitude of 4,000 psi (28 MPa) [76]. Mn-Cu alloys are nonlinear, and the loss is smaller at small stress amplitude. Hysteresis damping [5] in the nonlinear domain in these alloys can correspond to an equivalent loss tangent ($\Psi/2\pi$ in terms of the specific damping capacity $\Psi$) as large as 0.067 for alloys of 60 to 70 percent Mn at a surface shear stress of 34.5 MPa. Moreover, there is an aging effect in which loss decreases over a period of weeks [81]. Nevertheless, the damping of Mn-Cu is high for a structural metal (Figure 7.13).

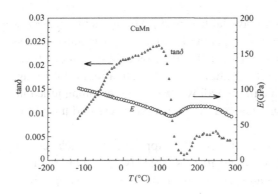

Figure 7.13. Tan$\delta$ versus temperature $T$ for a copper-manganese alloy at 1 Hz and a peak strain of $4.3 \times 10^{-4}$, aged for 16 hours [81]. Arrows indicate scales.

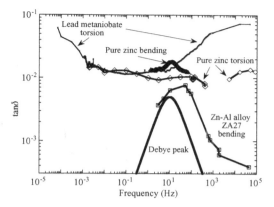

Figure 7.14. Tan$\delta$ of zinc and zinc alloy at 20°C, as a function of frequency (adapted from [87, 90]; and lead metaniobate [84]).

Zinc alloys, by contrast, show no variation in damping with strain in this range. For example, ZA27 alloy, which contains 27 percent aluminum, 2 percent copper, and 71 percent zinc, exhibits a peak tan$\delta$ of almost 0.01 near 100 Hz, decreasing to 0.001 at 1 kHz at 20°C. Young's modulus of ZA27 is about 78 GPa in contrast to about 100 GPa for pure zinc. Frequency dependent damping of ZA27 alloy [87] and pure cast zinc [90] is compared in Figure 7.14. The peaks observed in bending are due to thermoelastic effects (damping due to stress-induced heat flow), discussed in §8.2. The difference in the peak frequency is, in part, due to the difference in thickness of the specimens – 3.3 mm for pure zinc and 1 mm for the alloy. The damping in cast zinc, which is polycrystalline, is orders of magnitude greater than that of single crystal zinc [88], for which tan$\delta$ < $10^{-5}$ at small strain, at kHz frequencies. The difference is attributed to dislocations associated with grain boundaries. Shown for comparison is damping versus frequency [84] of lead metaniobate, a piezoelectric ceramic, which is used in transducers. It has a high viscoelastic damping tan$\delta$ = 0.09 at ultrasonic frequency.

Damping in all metals tends to increase with stress, and large values of damping occur as the stress approaches the fatigue limit, as noted by Lazan [78]. It would be risky to exploit this phenomenon to achieve high damping since at such stress levels, damage occurs in the metal.

Pure magnesium exhibits high damping, but it is comparatively weak. Cast commercially available magnesium [76] exhibited a loss tangent (extracted from log decrement measurements in flexural vibration) of 0.15 at a maximum strain of $6 \times 10^{-4}$, to 0.03 at a strain of $5 \times 10^{-5}$. Alloying elements can be added to improve the strength, but they usually reduce the damping. A cast AZ 81 magnesium alloy [77] exhibited a loss tangent of 0.016 at a strain of 0.001, down to about 0.0015 at a strain of $1.3 \times 10^{-4}$. Special alloys of magnesium have been prepared to achieve high damping. Magnesium-zirconium alloys with about 0.5 percent zirconium exhibit damping comparable to that of pure magnesium, but with higher strength [91]. Forged Mg 4.5 Ce exhibits an effective loss tangent (extracted approximately from reported specific damping capacity) of 0.07 in torsion near 10 Hz at a stress of 14 MPa. The effect is nonlinear so the effective loss tangent is only 0.011 at a stress of 0.86 MPa [92]. The crystals were favorably aligned for high damping in shear in the

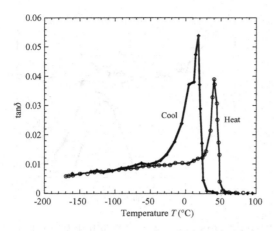

Figure 7.15. Damping versus temperature in NiTi alloy, at 1 Hz, during cooling and heating (adapted from Hasiguti and Iwasaki [86]).

forged samples. Alloys of magnesium with nickel, copper, aluminum, and tin have been studied in flexural vibration [93]. The highest loss tangent observed at small strain in these alloys was less than 0.002, but damping increased with amplitude.

Shape memory materials [85] are a class of solids which exhibit martensitic phase transformations (Chapter 8). They can be deformed while in their low-temperature phase; when heated, they change phase and revert to their original shape. The phenomenon was first reported in a Au-Cd alloy, but is best known in Ni-Ti alloys known as Nitinol. Ni-Ti alloys can exhibit high damping under some circumstances. For example, in a temperature scan of 51 percent atomic alloy, [86], peaks are observed in the damping as shown in Figure 7.15. The peaks are associated with the phase transformation (Chapter 8); they increase in magnitude with heating or cooling rate (not given in this study) and with strain amplitude ($1.8 \times 10^{-5}$ here, sufficient for some nonlinear effects). Shape memory effects are also known in In-Tl and Cu-Al-Ni alloys as well as in crystalline polymers and crystalline ceramics that undergo martensitic phase transformations. As for ceramics, Ce doped zirconia as well as ferroelectric materials such as barium titanate and lead titanate zirconate (PZT) exhibit shape memory effects. Nickel titanium alloys exhibit a large hysteresis damping at high stresses. Nitinol, which consists of 55 percent Ni and 45 percent Ti, has an equivalent loss tangent of 0.04 at a surface shear stress of 69 MPa [5]. This alloy is also called *memory metal* in view of its ability to return to a prior configuration after it is heated or cooled. A variety of magnetic alloys exhibit high damping, and some of these have been made commercially. Pure polycrystalline nickel exhibits a Young's modulus of about 200 GPa and a peak loss tangent of 0.046 at 100 kHz [94]. If a magnetic field is applied, the loss tangent decreases. A Co-35%Ni alloy at 1 Hz in torsion exhibited a loss of $\delta = 0.075$ at $10^{-4}$ shear strain $\gamma$, rising to a peak of 0.18 at $\gamma = 2.3 \times 10^{-4}$. An alloy of Co 20%Fe exhibited $\delta = 0.048$ at $\gamma = 10^{-4}$ and a peak $\delta = 0.11$ at $\gamma = 3.5 \times 10^{-4}$. More recently, an alloy commercially known as Vacrosil, containing 12%Cr, 3%Al, 85%Fe, exhibited a loss tangent [95] (defined as $Q^{-1}$) of 0.006 at 2 kHz and $\epsilon = 10^{-6}$ over a wide range of temperatures from $-200°$C to more than $300°$C. A strain-dependent peak in the loss, up to 0.1, was observed in this material at 20 Hz at $\epsilon = 7 \times 10^{-5}$. Other alloys, Fe10Cr and Fe 12Cr, exhibited similar peaks of lower amplitude; magnetoelastic alloys based on FeCo exhibited qualitatively similar behavior, tan$\delta$ to 0.075 at $\epsilon = 1.3 \times 10^{-4}$ and about 1 Hz for Fe25%Co [96]. FeMo [97] and FeMoCr [98] alloys are in the same category: tan$\delta$ up to 0.059 in Fe6%Mo. A superposed static tensile stress reduces damping in these alloys [99]. In these alloys, the loss depends on the ambient magnetic field strength, heat treatment, and the strain amplitude of vibration.

In certain compositions of CdMg (33 atomic percent Mg), high damping has been observed at room temperature [100, 101]. The peak loss is tan$\delta = 0.14$ at 1 Hz. The peak in tan$\delta$ was of magnitude 0.14 at 20°C and 0.75 Hz; it was verified to be a Debye peak in the frequency domain [102]. The peak for CdMg was about 20°C wide at constant frequency and the activation energy $U$ was inferred to be 19,000 cal/mole [101]. The loss appears to be due to a stress induced ordering of the different atoms. However this has not to the writer's knowledge been exploited practically. In $Cu_{0.81} Pd_{0.19}$ alloy, tan$\delta \approx 0.1$ to 0.24 at 2 Hz was claimed near and below room temperature [103]. Following annealing, the loss was reduced to about 0.003 at room temperature and even less at lower temperature. The loss was attributed to a Hasiguti effect in which interaction between dislocations and point defects gives rise to loss in cold worked fcc metals. In alloys of titanium and vanadium (Ti-20V) a peak in tan$\delta$ of 0.022 at 10 kHz and at 130°K is observed [104]. At room temperature, tan$\delta \approx 0.001$ and $E \approx 97$ GPa. Alloys with more vanadium exhibit less loss.

Metals of low melting point are at a high homologous temperature $T_H = T/T_m$ at ambient room temperature $T = T_r$, consequently they are expected to exhibit substantial viscoelastic response. For elements such as Cd, In, Pb, and Sn and alloys of low melting point, a high homologous temperature occurs at room temperature. Lead is popularly viewed as a high-loss metal, but it exhibits a relatively small peak tan$\delta = 0.015$ in bending [105] and tan$\delta = 0.005$ to 0.016 in the audio range in torsion [106]. Cadmium exhibits a substantial loss tangent of 0.03 to 0.04 over much of the audio range of frequencies. Indium is the softest metal which is stable in air [107]. The primary reason for its softness at room temperature is its low melting point $T_m = 429$ K, so the homologous temperature at room temperature is high: $T_H = T_r/T_m = 0.69$. Young's modulus is $E = 12.6$ GPa at $T = 300$ K and 18.4 GPa at 80 K. Indium has a yield point of only about 3 MPa, so that the strain at yield is $2.4 \times 10^{-4}$. Indium exhibits substantial creep, even at low temperature [108]. Lead is commonly thought to be a high damping metal, however compared with other metals of low melting points, such as indium, tin, and cadmium, it is not the highest in damping. Indium also exhibits high loss, particularly at low frequency [106]. Low melting point alloys, including PbSn and InSn, are used in soldering. Such alloys can exhibit substantial viscoelasticity (Figure 7.16). The effective performance of solders in electronic devices is related to their viscoelastic behavior (creep); flow or cracking of a solder joint can lead to failure of the device containing that joint.

Aluminum-zinc alloys have been studied for high damping applications [109, 110]. These alloys have from 3.5 percent to 28 percent aluminum and small amounts of copper and magnesium. At 20°C, damping is independent of strain from

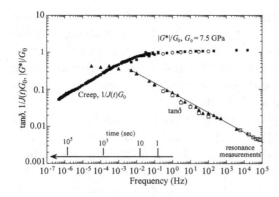

Figure 7.16. Viscoelastic behavior of an alloy of low melting point, a eutectic indium-tin alloy at room temperature (adapted from Lakes and Quackenbush [279]). Results are for a peak strain less than $10^{-5}$, measured directly with no appeal to time–temperature superposition.

$10^{-6}$ to $10^{-4}$, but increases with temperature up to about 270°C [109]. The highest damping at 20°C and at 10 Hz is 0.009 for SPZ alloy (78%Zn, 22%Al) and 0.005 for ZA27 (25–28%Al) alloy.

Gray cast iron, as described by Millett et al. [111] exhibits a loss tangent of about 0.011 over a range of frequencies, and it contains platelet shaped inclusions of graphite with a loss tangent of 0.015. The graphite gives rise to most of the loss; the underlying mechanism is stated to be dislocation loop motion (see Chapter 8).

### 7.3.4 Creep-Resistant Alloys

Metals used at high temperature, specifically at an absolute temperature which is greater than about 0.6 of the melting point, tend to creep noticeably. Aluminum alloys [113] are available for use up to about 200°C, titanium alloys to 600°C, and stainless steels to 850°C. In many applications, creep is to be minimized. To that end, a variety of creep resistant alloys has been developed [114]. The creep resistance of aluminum [115] is improved by incorporating particles of alumina ($Al_2O_3$) or of aluminum carbide. Although this alloy creeps less at high stress than pure aluminum, it has a higher $\tan\delta$ at subaudio frequencies. Nickel-based super-alloys are used at higher temperatures than aluminum alloys. A representative super-alloy has the composition, by weight given in Table 7.2. Such an alloy [116] as used in turbine

Table 7.2. *Composition by weight of a typical creep-resistant alloy (adapted from Ashby and Jones [108])*

| Element | Content | Element | Content |
|---------|---------|---------|---------|
| Ni | 59% | Mo | 0.25% |
| Co | 10% | C | 0.15% |
| W | 10% | Si | 0.1% |
| Cr | 9% | Mn | 0.1% |
| Al | 5.5% | Cu | 0.05% |
| Ta | 2.5% | Zr | 0.05% |
| Ti | 1.5% | B | 0.015% |
| Hf | 1.5 | S | <0.008% |
| Fe | 0.25 | Pb | <0.0005% |

Figure 7.17. Tan$\delta$ of low damping materials: steel (open squares), niobium (inverted triangles), tungsten (open diamonds) and fused silica (triangles) wires and fibers at ambient temperature, 300 K, as a function of frequency (adapted from Cagnoli et al. [112]). Shown for comparison is damping in bulk quartz (adapted from Berlincourt, Curran, and Jaffe [160]), bulk aluminum alloy (adapted from Duffy) [63]) (solid circles); single-crystal silicon and single crystal sapphire, [3].

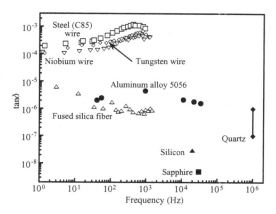

blades is expected to withstand a stress of about 250 MPa, and to creep no more than 0.1 percent over 30 hours at 850°C. Because these materials melt at $T_m = 1,280$°C, the homologous temperature is high: $T/T_m = 0.72$. By contrast, most ordinary materials creep a great deal at such temperatures. Materials that may be used at temperatures up to 1,000°C include nickel based super-alloys, refractory metals such as W-Re alloys, which have the disadvantage of being dense; ceramics based on TiC, ZrC and $Al_2O_3$, which are brittle, and ceramic composites, which are also brittle.

### *Low Damping Metals*

Properties of wires of several low-damping metals as well as fused silica [112] as a function of frequency are shown in Figure 7.17. The peak in the damping of the wires is attributed to a thermoelastic relaxation, as discussed in §8.2. The peak frequency is not definitive for study of this mechanism in the case of wires because they are not sufficiently uniform in diameter to allow accurate calculation of the peak frequency. The peak in tan$\delta$ of bulk aluminum alloy is attributed to stress induced thermal currents between crystals. The tan$\delta$ of low-damping materials is highly sensitive to crystal defects and surface quality, so a range of values is given for quartz.

### 7.3.5 Semiconductors and Amorphous Elements

Selenium is an unusual viscoelastic material in that it is a pure element, which can be prepared in the glassy state by rapid cooling from the melt; amorphous selenium exhibits dramatic viscoelastic behavior at room temperature and at accessible times. A master curve derived from relaxation experiments was obtained and is representative of polymeric behavior [117]. In its crystalline form, selenium is piezoelectric as described in §7.3.6. The flow region can be modified to obtain a rubbery plateau by the addition of arsenic or sulfur with arsenic to the selenium [117].

As for sulfur, it has long been known that fibers produced by rapidly quenching molten sulfur in cold water exhibit a rubber-like elasticity (see, e.g., [17, 117]). Amorphous sulfur becomes transformed to the crystalline form at room temperature. Adding arsenic [118, 119] or phosphorus cross-links the chains of polymeric sulfur and retards the crystallization.

### 7.3.6 Semiconductors and Acoustic Amplification

Piezoelectric materials produce an electric polarization when stressed mechanically, as discussed in §2.10. Most practical piezoelectric materials are ceramics (§7.4.5), but piezoelectric polymers and semiconductors are known.

Selenium in its crystalline form is a piezoelectric semiconductor [120] that exhibits dielectric relaxation of about 35 percent over two decades from 1 MHz to 100 MHz. The shear modulus $c_{44}$ is 18.3 GPa at ultrasonic frequency.

One can obtain amplification rather than attenuation of ultrasonic waves in piezoelectric semiconductors such as cadmium sulfide [121, 122]. This corresponds to $\tan\delta < 0$, which represents a gain tangent rather than a loss tangent. The energy required for amplification is derived from an externally applied DC electric field. The amplification can be as much as 4 percent per wavelength of the ultrasonic wave; attenuations of comparable magnitude can be achieved depending on the electric field applied; for interpretation in terms of $\tan\delta$ see §3.8. Amplification of ultrasonic waves at 15 MHz has also been observed in bismuth [123] in the presence of electric and magnetic fields. The amplification corresponds to a negative attenuation or a negative damping, $\tan\delta$, which is permissible because there is an external energy source. Similarly, ultrasonic amplification at 9 GHz was observed in germanium cooled to 4 K [124] carrying an electric current and was attributed to coupling between electrons and phonons. Consequently the positivity of the loss tangent as ordinarily expected depends on the characteristics of the material. It is not logically necessary, and is not always true; thus it is not appropriate as a mathematical postulate. The law of conservation of energy is not violated, because, in these cases for which $\tan\delta < 0$, an external source of energy, such as electrical energy, is provided. One may speak more generally of "excitable media" as physical, chemical, or biological systems in which energy dissipation is compensated by an energy supply [125, 126]. Examples include nerve tissue, heart muscle, the retina of the eye, and the Belousov–Zhabotinsky chemical reaction.

### 7.3.7 Nanoscale Properties

The microscale is usually considered to be the size scale accessible via the light microscope, 1 mm to 1 $\mu$m; the nanoscale, also called the ultrastructural scale, corresponds to dimensions below 100 nm; some authors consider the scale below 1 $\mu$m. Nanoscale properties are of interest since surface effects which are predominant at small size allow new degrees of freedom for the design of materials and systems.

Fine grain structure in materials influences material properties including viscoelasticity. For example, in iron prepared by a ball-mill process [127], to have crystal structure on the nanoscale, $Q^{-1}$ was elevated to $2 \times 10^{-3}$ at a temperature of 400°C and at 2 kHz, about three times the value seen in a reference sample of iron. The damping at ambient temperature, about $3 \times 10^{-4}$, was unchanged. Fine-grain structure also allows effects to occur such as the intercrystalline Gorsky effect (§8.6.3), in which atoms diffuse in response to stress gradients.

## 7.4 Ceramics

### 7.4.1 Rocks

Rocks ordinarily exhibit relatively small loss tangents at room temperature. For basalt, $\tan\delta = 0.0017$ and for granite $\tan\delta = 0.0031$ for small strain ($10^{-6}$) and at frequencies between 0.001 Hz and 1 Hz [9]. For comparison, frequencies of seismological interest are from about $3 \times 10^{-4}$ Hz to 1 Hz corresponding to periods from about one hour to one second. Other studies of rock are reviewed in Knopoff [10]; granite exhibited $\tan\delta = 0.005$ from 140 Hz to 1.6 kHz and limestone exhibited $\tan\delta \approx 0.02$ from 50 to 120 Hz. A small amount of water in porous rock can substantially increase the loss; for dry sandstone, $\tan\delta = 0.05$ independent of frequency in the audio range, but with 0.4 percent water, $\tan\delta$ increases with frequency up to 0.35 at 1.7 kHz.

Rocks at the elevated temperatures prevailing in the interior of the Earth can exhibit relatively large loss tangents in the absence of hydration. Naturally deformed peridotite has a loss tangent exceeding 0.05 at 5 Hz and small strain above 900°C [128]. Loss tangents for in situ rock have been inferred from seismic data [129], $\tan\delta = 0.0023$ for waves of period 3,000 seconds, to $\tan\delta = 0.0085$ for waves of period 100 seconds. Rock has also been studied at ultrasonic frequencies [130]. Westerly granite exhibits $\tan\delta$ rising from 0.01 near 30 kHz to 0.06 at 2.5 MHz. Other rocks such as slate and limestone exhibit similar broad loss peaks in the ultrasonic domain, and the loss is attributed to a Granato–Lücke dislocation process as discussed in §8.6.

The loss tangent of moon rocks differs from that of Earth rocks. Unusually long reverberations were recorded by a seismic station on the moon from lunar impacts of spacecraft [131]. $\text{Tan}\delta = 3 \times 10^{-4}$ was inferred for a frequency $f \approx 1$ Hz from these observations, compared with values from $3 \times 10^{-3}$ to 0.1 for most materials of the Earth's continental crust and 0.01 for the Earth's outer mantle [132]. Earth rocks such as limestone exhibit a broad peak of magnitude about 0.025 in damping near 1 MHz [132], attributed to dislocation motion. Rock in the laboratory has small cracks, which are likely to be closed by pressure deep underground. Moreover, stress within the Earth is larger than it is on the moon; the difference is considered sufficient to generate new dislocations [132] leading to higher damping. Lunar rocks studied on Earth have $\tan\delta = 0.01$ to 0.1. The difference is attributed to the outgassing of trace volatiles such as water in the dry lunar environment [133, 134].

Creep of rock has been studied by mining and civil engineers as well as by geologists [135]. The concern of the civil engineers is the stability of structures over years to tens of years, while the geologists are concerned with flow properties of

Figure 7.18. Rock, with layers of ice and snow.

rock over millions of years. Rock exhibits primary, secondary, and tertiary creep. The creep is attributed to the initiation and development of microcracks following chemical reactions at the crack tips [136]. The primary or transient component is often described by a power law in time or by more involved empirical equations. Creep can be substantial: delayed strain can be three or four times the initial strain over periods of a few hours. Creep in rock is nonlinearly dependent on stress level and on temperature. As for steady state creep, strain rate also depends on stress level. Empirical relationships derived from metallurgy have been used to model the creep. For example the value of $m$ in Equation 7.4 for rock is in the range 1 to 5.

Uplift of mountains (Figure 7.18) is attributed to flow of hot rock in the Earth's interior. Tectonic movement of the Earth's plates is associated with steady state creep at strain rates of $10^{-20}$ to $10^{-29}$ per second [136]. Creep in hot rock can occur by the collective motion of atoms by dislocation glide or climb, as discussed in §8.6; or by the diffusive transport of individual atoms. The constitutive relations associated with these processes differ; the creep rate due to dislocation motion is insensitive to grain size and increases nonlinearly with stress, but the creep rate due to diffusion increases dramatically with decreasing grain size and linearly with stress. Creep rate also depends on crystal structure.

In the Earth's mantle, creep in hotter regions of the upper mantle is dominated by dislocation movement; creep in the deeper regions in the mantle is governed by diffusion [138]. Creep due to dislocation motion is insensitive to grain size and increases nonlinearly with stress. Creep by diffusion decreases with grain size and increases linearly with stress [137]. Consequently, creep by diffusion in polycrystalline materials is associated with low stress levels and small grain size; creep by dislocation movement is associated with larger grain size and higher stress. The mantle is considered to be made of polycrystalline olivine. High-temperature creep experiments on polycrystalline $CaTiO_3$, an analog of the iron-magnesium silicate of the lower mantle, suggests diffusion creep to be important in the lower mantle [137]. If the mantle is viewed, as is commonly done, as a stratified linear viscoelastic material, an effective viscosity of $10^{21}$ Pa $\cdot$ s can be inferred. Moreover, a constitutive equation for steady flow is given as follows:

$$\frac{d\epsilon}{dt} = A(\frac{\sigma}{G})^n(\frac{b}{d})^m \exp[-\frac{U + PV}{RT}] \tag{7.6}$$

with $G \approx 80$ GPa as the shear modulus, $A$, $n$, and $m$ as a constant, $b \approx 0.5$ nm as the length of the Burgers vector, $d$ as the grain size, $U$ as the activation energy, $V$ as the activation volume, and $R$ as the universal gas constant.

Synthetic ceramics, owing to their high melting point, tend to exhibit relatively little damping and creep at room temperature, however they can exhibit significant viscoelastic effects at elevated temperature [139, 140]. Alumina [140] at 200 kHz exhibited $E = 436$ GPa and $\tan\delta = 9 \times 10^{-4}$ at room temperature (298 K) but $E = 246$ GPa and $\tan\delta = 1.05 \times 10^{-2}$ at 1,473 K. Silicon carbide at 120 kHz exhibited $E = 450$ GPa and $\tan\delta = 10^{-4}$ at room temperature (293 K) but $E = 414$ GPa and $\tan\delta = 1.25 \times 10^{-2}$ at 1,322 K.

Ionic crystals exhibit viscoelastic loss which is attributed to dislocation motion. In potassium chloride, internal friction ($\tan\delta$) at 40 kHz increased from $10^{-6}$ at about $430°C$ to $3 \times 10^{-4}$ at $700°C$ [141]. Polycrystalline LiF-22%CaF$_2$ eutectic salts (which are used for thermal energy storage) have been examined, again by the piezoelectric oscillator method [142] but at 80-150 kHz. At near room temperature, $E = 116$ GPa, $G = 42.6$ GPa and $\tan\delta = 3.9 \times 10^{-3}$, and $\tan\delta$ increased with temperature. For KCl near room temperature, a broad peak [143] in $\tan\delta$ was observed, $3 \times 10^{-3}$ at about 40 MHz; for LiF, there was a similar broad peak, $\tan\delta \approx 10^{-3}$ at about 10 MHz.

### 7.4.2 Concrete

Concrete is an aging material; its properties depend significantly on the time following its preparation [144]. Concrete is formed via a chemical reaction of water with cement paste. When first prepared, concrete is of a soft paste-like consistency, and it experiences permanent deformation under minimal force. The reaction is about 50 percent complete after seven days following mixing and continues for more than a year [145]. The "aggregate" component, sand and gravel, is considered in most models to behave elastically. Concrete progressively shrinks after it is formed, even if no load is applied. The creep in concrete is attributed to stress induced moisture transfer and to recrystallization in the cement or slip between layers. Microcracking is considered to cause creep at stresses above half of the short-term strength.

Concrete is used in structures that are expected to last a long time, however the time temperature superposition concept is inapplicable. Creep tests of very long duration are therefore conducted, up to 20 years [146] as shown in Figure 7.19.

A variety of empirical relations [145] have been used to model the creep strain as it depends on time $t - \tau_a$ under load, and on age of loading $\tau_a$. The effect of age is that creep is slower for older concrete and that effect is modeled by a factor $R_a = 1/\tau_a^{p_c}$, with $p_c$ as an empirical material parameter, which would imply (incorrectly) that old concrete does not creep at all. Fits to experimental data give $p_c \approx 0.1$, but the effect of loading age on creep does not change greatly after about 28 days [144]. The logarithmic model, shown incorporating the effect of age, fits the early creep reasonably well but overestimates later creep,

$$\epsilon_{\text{creep}} = \frac{1}{\tau_a^{p_c}}[c + d\log(1 + t - \tau_a)] \tag{7.7}$$

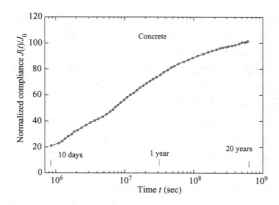

Figure 7.19. Creep of concrete for 20 years (adapted from Troxell [146]).

The power law model is a familiar one and it is successful in modeling early creep, but overestimates later creep even more than the logarithmic model,

$$\epsilon_{\text{creep}} = \frac{1}{\tau_a^{P_c}}(t - \tau_a)^m. \tag{7.8}$$

A creep function used in design [144] is as follows, with $c$ and $d$ as empirical constants.

$$\frac{\epsilon_{\text{creep}}(t)}{\epsilon(0)} = \frac{t^c}{d + t^c}\frac{\epsilon_{\text{creep}}(\infty)}{\epsilon(0)} \tag{7.9}$$

Fits to experimental data give $c = 0.4$ to $0.8$, and $d = 6$ to $30$.

An empirical kernel for the constitutive equation for creep aging (Figure 7.20) of concrete [147, 148] follows:

$$J(t, \tau_a) = \frac{1}{E(\tau_a)} + e_c^\infty[1 - (\tau_a/t)^{k_c}e^{-K_c(t-\tau_a)}], \tag{7.10}$$

with $k_c$ and $K_c$ as empirical constants, and $E(\tau_a)$ as the Young's modulus at age $\tau_a$.

Concrete under large stress can exhibit tertiary creep culminating in failure [149]. During tertiary creep, concrete accumulates damage, one consequence of which is a progressive increase in Poisson's ratio [150].

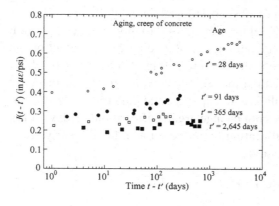

Figure 7.20. Creep as a function of age for concrete (adapted from Bazant [147]).

### 7.4.3 Inorganic Glassy Materials

Window glass, in common with glassy polymers, is also amorphous and exhibits a glass transition similar to that of a glassy polymer, only at a higher temperature, typically above 500°C. The stiffness of window glass in the glassy state is higher than that of a glassy polymer: about 70 GPa compared with 3 GPa. The loss tangent of Pyrex® glass at constant frequency increases with temperature [151]; at 1 Hz, tan$\delta$ goes from $8 \times 10^{-4}$ at 100°C to about 0.01 at 450°C to 0.1 at 600°C for annealed glass. Window glass also exhibits creep [151, 152], and significant creep may occur over long periods of time. Windows in ancient buildings such as cathedrals in Europe are often thicker at the bottom. This has been attributed to creep; however as glass manufacture in early times did not result in uniformly thick plates, the thickness of the glass is no longer attributed to creep (§10.3).

### 7.4.4 Ice

The behavior of ice is complex [153]. Stress–strain curves and other properties depend markedly on temperature and on load rate. At ambient conditions the temperature can approach the melting point, where properties depend particularly strongly on temperature. For ice under steady-state creep conditions [154], the strain rate is given by

$$\frac{d\epsilon}{dt} = A\sigma_d^n, \tag{7.11}$$

with $n \approx 3$ and $A = 2.3 \times 10^{-25}$ if $\sigma$ is in units of N/m$^2$ and $d\epsilon/dt$ is in sec$^{-1}$; this is considered valid for stresses greater than 0.1 MPa based on lab studies and 0.01 MPa based on studies of glaciers. Therefore, rheology of ice is nonlinear in this stress range. The practical significance of ice rheology is in connection with glaciers and with structures built where there is moving ice. If temperature is included in the description [153],

$$\frac{d\epsilon}{dt} = A\sigma_d^n \exp[-\frac{U}{RT}] \tag{7.12}$$

in which $U$ is the activation energy, $T$ is absolute temperature (in K), and $R$ is Boltzmann's constant.

As for dynamic studies, the damping for single crystal ice at the melting point 0°C and at small strain is tan$\delta \approx 0.13$ at low frequency [155]. Damping in ice decreases with reductions in temperature [155, 156] and increases with strain amplitude above $10^{-4}$. Pure ice also exhibits considerable damping (Figure 7.21) [156] as it approaches the melting point (273 K). Prior plastic deformation increases the damping. There is also a small peak of magnitude $10^{-3}$ below 150 K. Dislocation motion was considered as a cause for the damping. Polycrystalline pure ice at 5–8 Hz and 0°C exhibits tan$\delta \approx 0.17$ and $G = 3.3$ GPa [157]; tan$\delta$ for natural ice and compressed snow is about three times lower. At −40°C, tan$\delta$ for pure ice is $8 \times 10^{-3}$, a considerably lower loss than near the melting point. Damping in ice at −1°C has

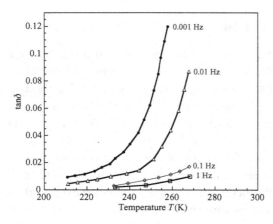

Figure 7.21. Tan$\delta$ of ice versus temperature at several frequencies at a maximum strain of $5 \times 10^{-5}$ (adapted from Tatiboët, Perez, and Vassoile [156]).

been attributed to water molecule reorientation, peaking at 10 kHz, and dislocation damping peaking at about 10 MHz [158]. Damping was inferred from ultrasonic measurements of attenuation.

Saline ice at low frequency exhibits large damping. Specifically, based on elliptic hysteresis loops, tan$\delta = 0.61$ at $-30°$C [159] in ice of 0.01 percent salinity.

### 7.4.5 Piezoelectric Ceramics

Piezoelectric materials are those which produce an electric polarization when stressed mechanically, as discussed in §2.10. They are always anisotropic. This coupling between mechanical and electrical degrees of freedom can be responsible for viscoelastic behavior if there is electrical conductivity, as presented in §8.7.2. Most practical piezoelectric materials are ceramics; piezoelectric semiconductors are also known (§7.3.6). Representative properties are given below in Table 7.3. PZT refers to lead titanate zirconate ceramic. Material properties are at ultrasonic frequency. Compliance values, such as $s_{33}^{\mathcal{E}}$, are given a superscript $\mathcal{E}$ to denote measurement at constant electric field. This compliance differs from the value measured at constant electric displacement $\mathcal{D}$ due to the coupling between fields. For example $s_{33}^{\mathcal{E}}$ is the inverse of Young's modulus in the 3 direction in the material, measured at constant electric field; $s_{44}^{\mathcal{E}}$ is the inverse of a shear modulus; the 4 direction corresponds to a shear in the 23 direction in the reduced notation used to display properties of anisotropic solids. The Curie point $T_c$ of these materials refers to the temperature at which they undergo a phase transformation to a less symmetric crystal form which lacks piezoelectricity. Piezoelectric ceramics exhibit strong piezoelectric effects; they are used in transducers and spark generators for stoves. Shear moduli from 21 to 45 GPa and loss tangents from $3 \times 10^{-3}$ to 0.02 for lead titanate zirconates to 0.09 for lead metaniobate are representative [160]. PZT ceramics are available in a variety of compositions in which different properties are maximized. Lead metaniobate, $Pb_2NbO_6$, is a piezoelectric material [83] originally notable for its high Curie temperature. In ceramic form, it has a

Table 7.3. *Piezoelectric material properties (adapted from Berlincourt, Curran, and Jaffe) [160]), at ultrasonic frequency and ambient temperature. Q refers to a mechanical quality factor d is a piezoelectric sensitivity, and s is a compliance. PZT refers to lead zirconate titanate ceramic*

| Quantity | Quartz | PZT4 | PZT5 | BaTiO$_3$ | Pb$_2$NbO$_6$ |
|---|---|---|---|---|---|
| $d_{33}$(pC/N) | 0 | 289 | 374 | 152 | 85 |
| $d_{31}$(pC/N) | 0 | −123 | −171 | −60 | −9 |
| $d_{11}$(pC/N) | 2.31 | – | – | – | – |
| $s_{33}^{\varepsilon}$ (TPa$^{-1}$) | 9.6 | 15.5 | 18.8 | 9.5 | 22 |
| $s_{11}^{\varepsilon}$ (TPa$^{-1}$) | 12.8 | 12.3 | 16.4 | 9.1 | – |
| $s_{44}^{\varepsilon}$ (TPa$^{-1}$) | 20.0 | 39 | 47.5 | 22.8 | – |
| $Q$ | 10$^6$–10$^7$ | 500 | 75 | 300 | 11 |
| $T_c$(°C) | – | 328 | 365 | 115 | 570 |

high viscoelastic damping tan$\delta = 0.09$ and a Young's modulus about 45 GPa in the longitudinal direction at ultrasonic frequency. Damping is less at audio and subaudio frequencies [84].

Quartz is a crystalline piezoelectric material used in transducers, electronic filters, and in frequency standards for radio equipment and timepieces. Its loss tangent is extremely small, $10^{-7}$ at 1 MHz increasing to $10^{-6}$ at 15 MHz [3]. Single crystal quartz, if defects are minimized, provides low damping (high $Q$), though the piezoelectric sensitivity is lower than that of the ceramics.

Viscoelastic dissipation in piezoelectric materials is relevant to their use in transducers. Resonators driven at high amplitude will overheat if the material has damping that is too high. In contrast, high damping is useful in pulsed ultrasonic transducers in which a sharp pulse, without ringing, provides accurate timing data.

## 7.5 Biological Composite Materials

Tissues are heterogeneous natural composites. They are hierarchical in their structure: structure is contained within structure. Most tissues exhibit a preconditioning effect in which the response to an initial load differs from the response to a subsequent load. In some cases, behavior does not stabilize until multiple cycles of force are applied.

Soft tissues exhibit nonlinearity at physiological strains. Tissue properties change with time. Some of the change is associated with biological aging, but a description as aging material in the context of materials science would be simplistic. Tissue in a living organism remodels so that structure and properties change in response to changes in mechanical stress, to hormones, or to changes in circulation. Some changes, such as loss of muscle or tendon strength, which are considered to be part of biological aging in humans, can be reversed by strength training. Active tissues include muscle; muscles are able to convert metabolic energy into mechanical energy. It is therefore appropriate to discuss various constitutive equations. Only the simplest ones that deal with time dependence are considered here.

## 7.5.1 Constitutive Equations

For linearly viscoelastic materials the relaxation function $E(t)$ in the following Boltzmann integral depends on time $t$ only, not on strain $\epsilon$. In many studies of tissue, the linear theory which strictly is applicable only to small strain, is sufficient. In robust experiments, tissues often exhibit a broad spectrum of relaxation times. Therefore, relaxation models which are limited to a single exponential or a few exponentials are not applicable unless a restricted window of time is used,

$$\sigma(t) = \int_0^t E(t-\tau)\frac{d\epsilon(\tau)}{d\tau}d\tau. \tag{7.13}$$

The following nonlinear relation allows for prediction of history dependence [161, 162]. This single-integral form is called *nonlinear superposition*, which allows the relaxation function $E(t, \epsilon)$ to depend on strain level. Such nonlinear effects are known in polymers [163],

$$\sigma(t) = \int_0^t E(t-\tau, \epsilon(\tau))\frac{d\epsilon(\tau)}{d\tau}d\tau. \tag{7.14}$$

A similar equation may be written in the compliance formulation (Chapter 2) based on a stress-dependent creep compliance $J(t, \sigma)$.

If the relaxation function $E(t, \epsilon)$ is separable into a time dependent function multiplied by a strain dependent function, the following nonlinear formulation is called *quasilinear viscoelasticity* (QLV):

$$E(t, \epsilon) = E(t)f(\epsilon). \tag{7.15}$$

In QLV [164, 165] the magnitude of the modulus depends on strain, but the shape (and slope) of the relaxation curve is independent of strain; similarly the shape (and slope) of the creep curve is independent of stress. QLV has been widely used in the biomechanics field. Often a specific form of $E(t)$ is used.

Effects of aging may be incorporated by allowing the modulus to depend on time after birth as well as time after application of strain. That is simplistic in that changes in biological tissue depend on the use of the tissue (e.g., exercise or inactivity; pregnancy) and other variables (e.g., disease, changes in diet) in addition to clock time after birth.

## 7.5.2 Hard Tissue: Bone

Bone is a complex composite material consisting of mineral (principally hydroxyapatite, $Ca_{10}(PO_4)_6(OH)_2$, protein (principally collagen), water, and polysaccharides. Bone is a hierarchical solid which contains structure at multiple length scales [167–169], as shown in Figure 9.12. Bone is anisotropic. Representative moduli for human compact bone at low frequency are [170] $E = 17$ GPa in the longitudinal direction, 11 GPa in the transverse direction; $G = 3.6$ GPa for torsion about the longitudinal direction. Human compact bone has hexagonal symmetry (transverse

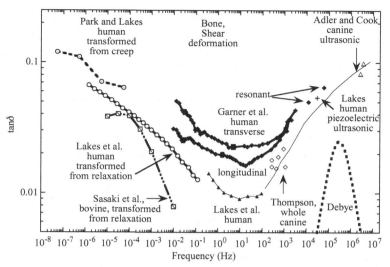

Figure 7.22. Tanδ for human compact bone (adapted from a compilation by Lakes [166]). Damping at low frequency inferred from slope of long-term creep by Park and Lakes [173], center circles. Results calculated from integration of constitutive equation of Sasaki et al. [174] for torsion relaxation in bovine bone (slant squares). Results adapted from the data of Lakes et al. [11] for wet human tibial bone at 37°C (calculated from relaxation, circles; directly measured, solid triangle). Direct measurements by Garner et al. [175] for wet human bone in torsion (diamonds). Damping data of Thompson [176] for a whole dog radius at acoustic frequencies (diamonds). Damping of wet human femoral bone by Lakes [178] via a piezoelectric ultrasonic oscillator (cross, +). Damping at ultrasonic frequency for canine bone by Adler and Cook [177] at room temperature (open triangle). Theoretical Debye peak corresponding to an exponential in the time domain, (dashed curve).

isotropy) with five independent moduli as determined by ultrasound [171]; bovine plexiform bone is orthotropic, with nine independent moduli. Moduli are higher at ultrasonic frequency than at the low frequencies associated with normal activity. Viscoelasticity can arise from multiple processes at the different scales, giving rise to a distribution of relaxation times. Viscoelastic behavior indeed occurs over a wide range of frequency [11], much wider than a Debye peak from a single relaxation time process, as shown in Figure 7.22. Observe also that the loss tangent of compact bone attains a broad *minimum* over the frequency range associated with most bodily activities. Consequently, it does not appear likely that damping in bone plays a major role in the absorption of impacts during bodily activities. The elevated damping at low frequency corresponding to long time periods in creep is due to motion at viscous like interfaces between large fibers (osteons) as discussed in §8.5.2. These fibers are sufficiently large (about 0.2 mm diameter) that they give rise to size effects [172]. Larger specimens subject to a strain gradient (as in torsion) exhibit a lower effective modulus and a higher damping than more slender ones. Therefore laboratory studies may underestimate the damping of the whole bone.

Although compact bone is a complex composite material, secondary creep is a simpler aspect of its viscoelastic behavior, and it appears to be thermally activated [179]. Both creep and fatigue in bone give rise to microscopic damage [180]. Creep under sufficiently large load, giving rise to an initial strain in the range 0.003 to 0.007, terminates in fracture [181]. Bone such as deer antler, which has less mineral than leg bone, endures higher strains without breaking [167].

Secondary creep in trabecular bone [182] follows a power law in stress: $d\epsilon/dt = 2.21 \times 10^{33}(\sigma/E_0)^{17.65}$. Creep rupture occurs at a time given by $t_{rupt} = 9.66 \times 10^{-33}(\sigma/E_0)^{-16.18}$. The strength of spongy bone can be substantially reduced if a stress approximately half the ultimate strength is applied for a few hours. Such strength reductions might play a role in the etiology of progressive, age-related collapse of vertebrae.

Living bone is an adaptive material [167]. The body gradually adds more bone tissue if a person increases physical activity, and absorbs bone tissue if physical activity is reduced. The adaptation can be rather specific: if a person uses one arm more than the other (in a sport) the bone in the exercised arm becomes thicker. The cells responsible for these changes receive stimuli associated with the mechanical stress on the bone. Some stimulus mechanisms are linked to tissue viscoelasticity. For example, fluid in the pores in bone may undergo stress-induced flow, giving rise to viscoelasticity (§8.3) and possibly to enhanced nutrition of the tissue via enhanced flow of the tissue fluid [183]. Indeed, Tami et al. [184] observed movement of fluorescent tracers in living bone to be enhanced by mechanical stress. Because viscoelasticity [185] due to stress induced fluid flow in bone appears to occur at frequency approaching 100 kHz, far above frequencies that occur in stress due to normal activity, any stimulus due to the fluid is likely to be associated with flow rather than pressure.

### 7.5.3 Collagen, Elastin, Proteoglycans

Elastin is a compliant protein: its Young's modulus is about 0.6 MPa [186]. Elastin is considered to be similar to rubber in its compliance. The stress–strain characteristic of elastin is highly linear. Elastin can be stretched to as much as 60 percent strain. Elastin in lung, blood vessels, and skin provides compliant elasticity [187]. In humans, elastin is not synthesized in adulthood. Elastin exhibits some hysteresis, for a cycle of stretch to 15% strain and release, the inferred tan$\delta$ is about 0.02. This is a small hysteresis for such a compliant material.

Proteoglycans form a matrix (ground substance) between fibers in many tissues. They are thought to contribute to tissue viscoelasticity. Solutions of proteoglycans from cartilage are highly viscoelastic [188], with tan$\delta$ exceeding 1 above 10 Hz.

Collagen is a fibrous protein found in connective tissue including tendon, ligament, skin, blood vessels, cartilage, and bone. In the fiber direction it has a Young's modulus of about 1 GPa. Collagen is a major structural protein in the body; it provides stiffness and strength where needed in tissues and organs. Tendons consist mostly of collagen [189] in a hierarchical structure of aligned fibers within fibers. The fibers have a wavy configuration that gives rise to a desirable nonlinearity as discussed below. Proteoglycans, less than 1 percent by weight, form a ground-matrix between the collagen fibers in tendon, and provide tissue hydration and separation between them. The aorta contains 50–70 percent by weight of collagen and elastin. There is about twice as much elastin as collagen in the aorta, but the proportions are reversed in the smaller arteries. The elastin in the aorta is thought to play a role in this vessel as a compliant reservoir for blood pulses during the heart beat. The elastin also maintains tension in the vessel walls during normal fluctuations in blood pressure.

Articular (joint) cartilage contains 15–20 percent collagen, 2–10 percent proteoglycans, and 70–75 percent water. The intervertebral disc contains an outer portion called the *annulus fibrosis*, which contains collagen fibers oriented obliquely within lamellae, and an inner portion called the *nucleus pulposis*, which is gelatinous in nature. The annulus fibrosis contains 50–65 percent collagen by dry weight, and 2–3 percent proteoglycans.

### 7.5.4 Ligament and Tendon

Ligaments connect bones to each other and stabilize joints; tendons connect muscles to bones. Both tendons and ligaments contain bundles of oriented collagen fibers. Tendons and ligaments are anisotropic and viscoelastic. The viscoelasticity in ligaments has been attributed to intrinsic viscoelasticity of collagen fibers, of extracellular matrix, of cross-links between molecules, or due to the fluid content, but no consensus has emerged. Dissipation of energy in the deformation of ligaments is thought to protect the tissue from injury during deformation at a high rate [190].

Stress–strain curves for ligament are concave up (Figure 7.23); they show a strain-stiffening nonlinearity, which is attributed to a straightening of initially wavy collagen fibers in the ligament. Ligament strain during normal bodily activity is typically below 4 percent [191]; from 3.6 percent for squatting, to 1.7 percent for bicycling [192]. Structural damage in laboratory specimens begins to be manifest as

Figure 7.23. Stress versus strain is nonlinear for left and right rat knee ligaments (squares) [193] and for human achilles tendon (open diamonds) and for human patellar tendon (solid diamonds). Right: ligament structure (after Provenzano et al. [194]).

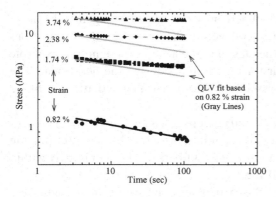

Figure 7.24. Ligament stress relaxation shows nonlinear behavior [197]. The slope depends on strain, so the relaxation function is not separable into a product. Equation 7.15 does not apply. The behavior is therefore inconsistent with QLV.

laxity upon recovery following strain exceeding 5.1 percent [193]; cellular damage is initially observed following strain below 1 percent and increases linearly with strain. A ligament sprain in a living person is painful, but some cellular damage may occur before pain is perceived.

It is not always straightforward to compare soft-tissue strain *in vivo* (in a living organism) with strain measured *in vitro* (in a laboratory preparation) since it can be difficult to define the resting length of a specimen. A preload is used *in vitro* to achieve a consistent starting point, but that reduces the initial, compliant, portion of the load-deformation curve substantially. It is considered unlikely that ligaments sustain load approaching that required for damage in normal activities.

Ligaments are viscoelastic: they exhibit creep or stress relaxation. Many studies of ligament viscoelasticity consist of a creep or relaxation test at one load or strain level, along with a stress strain curve to failure. These experiments demonstrate both nonlinearity and viscoelasticity, but they do not allow one to distinguish among constitutive models for the behavior. However, experiments [195] reveal that stress relaxation proceeds more rapidly than creep in medial collateral ligaments, a fact not explained by linear viscoelasticity theory or by QLV but which is consistent with nonlinear viscoelasticity [196]. Recent results for rat knee ligament [197] show a nonlinear behavior in which the rate of creep is dependent on stress level and the rate of relaxation is dependent on strain level, Figure 7.24. Specifically the rate of stress relaxation (slope in a logarithmic plot) decreases as the strain increases. Furthermore, relaxation proceeds faster than creep, consistent with [196].

Human knee ligaments exhibit an increased slope of relaxation at low strain [190], which confirms the observations on rat ligament. Measurements at time as short as several milliseconds indicate minimal relaxation in the short time region less than 0.1 sec. Moduli for transverse and shear deformation are several hundred times smaller than for the longitudinal direction. This is not surprising in view of the oriented fibrous structure of ligaments.

The strain-stiffening type nonlinearity of ligament allows free motion of a joint in normal activities with minimal effort required to stretch the ligament. If a joint is subject to excess load, the ligaments become stiffer by virtue of their nonlinearity, to provide stability. Ligaments in the body are subjected to repetitive loads that have a static component; the creep and relaxation behavior is pertinent to that component. Stress relaxation is relevant to athletic stretching exercises *in vivo* in which the athlete holds a limb at a prescribed angle for a period of time; creep occurs in the ligaments of a joint if the athlete applies a constant force to the limb for a while. Creep recovery occurs during periods of physical inactivity. Viscoelastic strain in the ligament recovers, provided all loads were below the threshold for damage. Therefore, there will again be a feeling of tightness after several days, so the athlete will do the stretching exercise again.

The ligamentum nuchae, at the top of the neck in cattle and horses, consists mostly of elastin. Its function differs from that of knee ligaments in that it supplies a steady force to support the head, with minimal muscular effort [187]. For this purpose, the linear stress–strain characteristic of elastin is beneficial.

*Tendon*

Tendon, similarly to collagenous ligaments, also exhibits a strain-stiffening nonlinearity [198]. Behavior of an Achilles tendon [199] from a 54-year-old woman is shown for comparison in Figure 7.23. By contrast, patellar tendon from 30-year-old male subjects [200] was obtained during surgical procedures to repair athletic injuries to the anterior cruciate ligament. This tendon is stiffer and stronger than the prior one. The difference may be due to the difference in age or physical condition of the subjects. It is known that tendons in humans become substantially stiffer and have less hysteresis following strength training [201], even for people older than 65. The rat ligament and human tendons exhibit a similar nonlinearity. Tendon properties have been explored in living human subjects by measuring the motion of bony prominences by ultrasound [202] and by measuring the force exerted by a limb. Nonlinear force-deformation characteristics, similar to those for laboratory specimens, are observed; for maximum muscle tension, the stress is 25 MPa, the strain is 0.025 and the effective modulus at that strain is 1.2 GPa. Therefore, the tendon usually operates in the toe region (the initial concave-up part of the curve) of its nonlinearity. Viscoelasticity in living tendon is also observed in which the slope of the load-deformation curve increases with rate [203].

Tendons in the leg and ligaments in the arch of the foot are sufficiently compliant that they store significant amounts of energy during walking and running [204, 205]. The total energy turnover in each stance phase of a 70 kg man running at 4.5 m sec$^{-1}$ is about 100 J. In comparison, 17 J is stored as strain energy in the compliant parts of the arch of the foot, and 35 J in the Achilles tendon. These structures store enough energy to improve the energy efficiency of running in humans [205]. The load-deformation behavior discloses nonlinearly viscoelastic behavior in that the hysteresis curve is nonlinear and encloses some area. The nonlinear behavior arises from an initial waviness in the collagen fibers in the tendon or ligament; these curves are straightened in the early stages of loading. Energy storage predominates over energy dissipation since the effective loss tangent, considered as $\Psi/2\pi$, extracted from areas in the plots, is about 0.04.

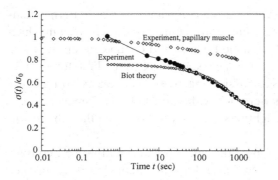

Figure 7.25. Heart ventricle muscle stress relaxation in compression (solid circles) (adapted from [206]), compared with a single relaxation time poroelastic model (open circles). Also shown is tensile relaxation in a papillary muscle (adapted from Pinto and Fung [207]).

Most land animals have compliant pads in their paws [204]. These pads cushion impacts with the ground during running. The cushioning effect arises from compliance more so than damping because the effective loss tangent is only about 0.04.

### 7.5.5 Muscle

There are three kinds of muscle: skeletal, which provides voluntary motion to the body; cardiac, in the heart; and smooth, which is involuntary. Skeletal muscle, if stimulated sufficiently by nerves, generates a maximal tension.

Heart muscle exhibits stress relaxation in which a single relaxation time process predominates [206]. This relaxation, Figure 7.25, was observed in a pig heart (left ventricle) specimen 11 mm in radius and 10-mm thick, between impermeable plates. Such behavior was interpreted in light of a poroelastic model (see §8.3) in which the viscoelasticity arises from fluid flow through pores. The longer-term portion of the relaxation in the cardiac muscle has a characteristic time which increases as the square of the sample radius, consistent with the notion of stress induced fluid flow. The short term portion of the relaxation curve is attributed to intrinsic viscoelasticity of the fibrous matrix and the biological cells. The effective Young's modulus is 4.1 kPa in the absence of fluid pressure. In contrast, rabbit papillary muscle, a cylindrical muscle in the heart, 1.3 mm in diameter and about 3.5 mm long exhibits considerably less relaxation (at body temperature) [207], as shown in Figure 7.25. Testing of a full papillary muscle is convenient because the muscle is small and it need not be cut. The difference in behavior could result from the difference in size or from the different boundary conditions for fluid flow across surfaces. Specifically, the transverse surface was cut in the case of the ventricle muscle but not in the papillary muscle. Moreover, clotting of blood in the vessels of the laboratory specimens could influence the results.

Muscle fibers from a crab [208] heart exhibit a tensile modulus of 0.01 to 2 MPa; modulus increases with strain rate and with the length of sarcomeres, which are cellular structures. These muscle fibers exhibit high damping: $\tan\delta$ of 0.1 to 0.5, increasing with frequency from 0.1 Hz to 50 Hz and with strain amplitude from 0.0036 to 0.036. The damping is considered to confer a physiological benefit of stability, in that uncontrolled resonant vibrations are prevented.

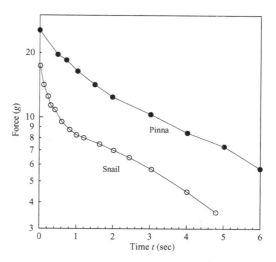

Figure 7.26. Muscle stress relaxation in tension (adapted from Abbott and Lowy [209]), for pharynx retractor of the snail Helix Pomata (open circles) and for smooth posterior adductor of Pinna Fragilis (solid circles).

Stress relaxation in resting striated and smooth muscle occurs rapidly at first, then more slowly [209]. The relaxation, shown in Figure 7.26 for muscles from a snail and from Pinna Fragilis, a large bivalve of British waters, was modeled with two exponentials. In a comparison of six species, the time constant for the slower exponential in the relaxation was inversely related to the maximum speed of shortening. For example, stress relaxation in muscle from the mussel Mytilus edulis was about 30 times slower than in muscle from a toad, but the toad muscle could contract about 40 times faster. Stress relaxation in muscle [209] in tension over a limited time scale up to about six seconds was observed and resolved into two components,

$$E(t) = E_0(a_1 e^{-t/\tau_1} + a_1 e^{-t/\tau_2}).\tag{7.16}$$

As for active muscle, the maximum tension force $F$ is related to contraction velocity $v$ from the optimal length by the following after Hill [210], for fully tensed (released from maximal isometric contraction) skeletal muscle

$$(v + b)(F + a) = b(F_0 + a)\tag{7.17}$$

in which $a, b, F_0$ are constants. This is an empirical equation based on experiment. The maximum tension force depends on how much the muscle is stretched; its peak active tension stress is about 300 kPa [214]. Since force times velocity represents power, if $a$ and $b$ are neglected (as can be done if both force and velocity are much less than maximum) then the equation approximates a condition of constant power. Active contraction of muscle is in contrast to viscoelasticity of a passive material in that force decreases with velocity for active contraction, but force increases with velocity (or strain rate) in a passive viscoelastic material. For muscle that is not fully tense, there is an additional variable: the nerve signal by which the human or animal controls the tension in the muscle. Heart muscle is never maximally tense, but it does exhibit a force-velocity characteristic similar

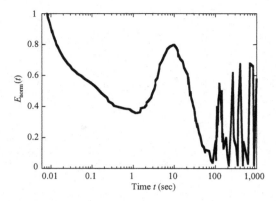

Figure 7.27. Muscle stress relaxation in tension (adapted from Price, Patitucci and Fung [215]), for smooth muscle from guinea pig cecum subjected to step tensile strain of 10 percent. Normalized modulus versus time.

to that of skeletal muscle. Also, in normal activities, a muscle may extend rather than contract under load as in lowering a weight held in the hand or in walking down stairs. More sophisticated constitutive representations are required for such conditions.

Muscle, during contraction, emits low frequency vibration and sound over a frequency range with peak amplitude near 25 Hz [211]. The first report of such sounds was made by Grimaldi in 1665. To demonstrate this sound, block each ear canal with your thumb, raise your elbows, and clench your hand in a fist. The sound is not of vascular origin as can be shown by temporarily blocking blood flow with a tight cuff. The sound is thought to be a result of pulses in individual muscle fibers. Muscle sound has been subject to correlation analysis to infer mechanical properties of muscle in living subjects [212]. Shear modulus in the frequency band 40–55 Hz increased from 100 to 220 kPa for loads between 5 and 12 pounds at the ankle; shear viscosity was about 0.5 kPa s at the highest load. By comparison, the calculated resting shear modulus of muscle at 60 Hz is about 12 kPa [213].

Stored elastic energy in tissues [214] has the benefit of substantially reducing metabolic power requirement for running humans or hopping kangaroos. Most of the stored energy in vertebrates is in the tendons rather than in the muscles. Elastic energy storage in muscle appears to be more important in insects. The dissipated energy requires metabolic input but may confer advantages in stability.

Smooth muscle exhibits spontaneous contraction that drives peristalsis in the intestine, regulatory processes in the circulation, and the pumping of lymph. In a laboratory preparation of living muscle, a step stretch (a tensile strain of 10 percent) corresponding to a stress relaxation experiment gave rise to induced oscillations in the tension stress [215]. The claimed rise time was 10 ms. Tissue response is shown in Figure 7.27. Stress has been normalized to stress (133 kPa) at 10 ms. Recall from Chapter 2 that in a passive material, the stress response to a step extension is monotonic decreasing. The muscle is not passive; it is active, and it is supplied with metabolic power. Therefore, the stress response need not be monotonic. Initially there appears to be relaxation; the muscle then exhibits a rhythmic pattern of contraction in response to the stimulus.

## 7.5.6 Fat

Adipose tissue (fat) provides chemical energy storage, buoyancy, and cushioning. Its mechanical properties are studied in part because reconstructive surgeons at times restore fat which has been removed due to disease or trauma. Because fat acts as a cushion, understanding its properties is helpful in analyzing methods to prevent pressure ulcers.

The initial elastic modulus for adipose tissue was found to be 3–24 kPa for breast and 24–180 kPa for heel pad [216]; sheep gluteal fat exhibits a relaxation modulus in confined compression of 8 kPa at 1 second, decreasing to about 4.5 kPa at 250 seconds. The initial Young's modulus of this sheep fat based on an indentation test was about 20 kPa. Human abdominal adipose tissue exhibits a shear storage modulus [217] of 1.8 to 3.4 kPa from 0.1 to 50 rad/s, with $\tan\delta$ of about 0.25. Hydrogels have been used as scaffolds for stimulation of tissue growth; their material properties have been compared with those of adipose tissue. The hydrogel exhibited a shear modulus of about 10 kPa and $\tan\delta$ of about 0.1. Fat is nonlinear; it becomes more compliant with increasing strain up to about 25 percent (strain of this magnitude occurs in normal activity), and the difference increases with frequency. Indentation creep recovery tests showed substantial nonrecoverable deformation in fat.

## 7.5.7 Brain

Brain tissue deforms dynamically during accelerations of normal activity and deforms to a greater extent in head injuries. Brain tissue also deforms slowly during creep of the head in sleeping infants. If such deformation of the head is excessive, the baby's head may become visibly misshapen. The misshapen head can be corrected by helmet therapy in which gentle pressure is applied by a spring-loaded helmet; the creep of the soft skull then results a normal head shape (§10.4.4). In some cultures, baby heads are intentionally deformed to obtain a new shape by tightly wrapping them for periods of time. Large, slow deformation of brain tissue also occur in hydrocephalus. In some patients with hydrocephalus [218], the brain volume is significantly reduced, so most of the cranium is filled with cerebrospinal fluid, but intelligence nevertheless can be average or even above average.

Laboratory studies of brain tissue under quasistatic deformation show creep, [219] hysteresis, stress relaxation, and residual strain [220]. Creep deformation of human brain is shown in Figure 7.28. The consolidation ratio $U$ is defined as the current value of the specimen compressive shortening divided by the final shortening at the end of consolidation. Brain tissue is very soft, with an effective Young's modulus $E$ of a few kPa for small deformation. The primary mechanism for delayed volumetric deformation in brain relaxation is stress-induced fluid flow according to consolidation theory. The Terzaghi [221] theory used for analysis here is a simplified special case of the Biot theory of stress induced fluid flow, discussed in §8.3. As with heart muscle and other materials that consolidate, there is a size effect in

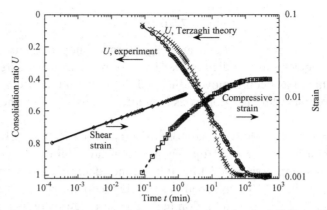

Figure 7.28. Brain compressive creep deformation under a load of 6 N in a 30 mm diameter, 7 mm thick specimen of brain, with free drainage at upper and lower surfaces (redrawn from data provided by Fraceschini et al. [220]), compared with a poroelastic model, Terzaghi theory. Creep (derived from relaxation in shear in Bilston, Liu, and Phan Thien [222]), is shown for comparison.

which the time required for relaxation scales as the square of the diffusion length corresponding to the specimen size. In these brain specimens, a 7 mm-thick specimen approached asymptotic deformation in creep after 100 min, so a 100 mm-thick segment of brain would undergo a similar deformation only after 15 days. The primary cause of the time dependent behavior in this study is consolidation. The solid phase is likely to exhibit some viscoelasticity as well. Indeed, shear relaxation of bovine brain was observed from 0.01 sec to 100 sec [222]. The relaxation curve for strain below 0.001 is consistent with a shear modulus $G(t) = At^{-n}$ with $t$ as time in seconds, $A = 2.8$ kPa and $n = 0.15$. Shear properties are not influenced by consolidation because there is no volume change. Creep derived from this relaxation in shear is shown for comparison in Figure 7.28.

Dura mater, a membrane which surrounds the brain is stiffer than brain [219]: $E$ is 7 to 14 MPa. Creep in dura mater is about a factor of two from 1 sec to 1,000 sec in a monkey, and about 50 percent over a similar time scale in a human.

### 7.5.8 Vocal Folds

Vocal fold tissue is studied in part because tissue properties influence our ability to speak; tissue properties are affected by disease and aging. The air pressure required to initiate a sound increases with the mechanical damping of the tissue. Vocal fold tissue has a shear modulus of 40 Pa to 1 kPa for male subjects and 10 Pa to 40 Pa for female subjects at 15 Hz [223]; damping is 0.1 to 0.3 over a broad range of subaudio frequencies. Tissue tends to become stiffer with age.

### 7.5.9 Cartilage and Joints

Articular (joint) cartilage serves to provide smooth mobility to the joints and to distribute stress upon the underlying bone. Friction in a normal joint is very low, 0.01

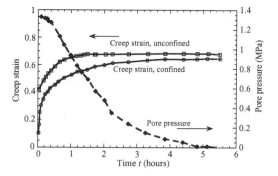

Figure 7.29. Cartilage creep in an ox patella cartilage specimen 12 mm in diameter and at least 1.6 mm thick (adapted from Oloyede and Broom [226]). Also shown is the time dependence of pore pressure.

to 0.02. Cartilage exhibits stress induced fluid flow. Such flow is considered to be important in the lubrication of joints [224, 225]. Compressive force on a joint squeezes fluid from the pores in the cartilage; this fluid provides lubrication. Creep in cartilage is associated with time dependence of pore pressure [226], which is related to stress induced fluid flow, hence to the time dependence and rate dependence of the tissue. Figure 7.29 shows creep in ox patellar cartilage. Cartilage also exhibits a strong strain rate dependence [227] with a modulus of about 1 MPa at $5 \times 10^{-5}$ /sec to about 30 MPa at $10^3$ /sec, as shown in Figure 7.30. Behavior is weakly nonlinear. At the highest rates, the behavior is nearly elastic in that rate dependence is minimal.

Spinal discs provide flexibility to the spine. Stress-induced fluid flow has long been known to be important in determining time-dependent properties of the disc [228]. Spinal discs exhibit a consolidation response [229] in which creep conducted for 50 hours in ox tail discs has a time constant of several hours and approaches an asymptote. As above, pore pressure tends to zero with the completion of creep. A nominal compressive stress of about 2 MPa gives rise to an initial strain of 0.15 and an asymptotic strain of 0.43, Figure 7.31. Similarly in human discs, the outer portion (the annulus fibrosis), confined laterally, [230] exhibits similar consolidation creep. Such behavior is pertinent to both normal motion and disease. Back aches are common. At times a disc may bulge or leak so that a spinal nerve root is compressed by bone or disc. Such nerve compression can cause pain radiating into the leg. Stretching exercises can be beneficial in such conditions, perhaps in relation to stress-induced fluid flow improving nutrition to the tissue.

Figure 7.30. Cartilage rate dependence of the stress–strain curves (adapted from Oloyede and Broom [227]).

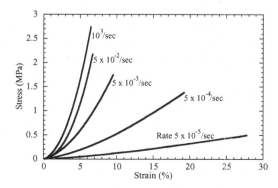

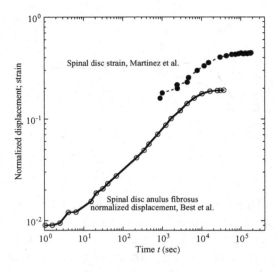

Figure 7.31. Creep in spinal discs and disc components (adapted from Martinez, Oloyede, and Broom [229] and Best et al. [230].

Synovial fluid is present in normal joints; it acts serves to lubricate the joint surfaces, also to supply nutrition to cartilage (which contains no blood vessels or nerves). Synovial fluid is highly viscoelastic [268, 269] and frequency dependent; its tan$\delta$ can exceed 1 below 0.02 Hz but becomes less than 0.3 above 1 Hz. The shear storage modulus, about 20 Pa at 1 Hz in a healthy young person, becomes less in older people, and much less in people with arthritis [270]. The viscoelastic properties are due to hyaluronic acid, a polysaccharide produced by the joint-lining cells and secreted into the fluid.

### 7.5.10 Kidney and Liver

Physical properties of these organs are studied in part for basic scientific reasons; also to provide input data in the design of pads and equipment to protect the body from trauma. Moreover, such study is done in the context of surgical phantom simulators and for diagnostic purposes. Kidney tissue exhibits dynamic viscoelastic behavior [231] in which the shear modulus of pig kidney at 1 Hz is about 4 kPa and tan$\delta$ is about 0.5. Stress relaxation plotted on a log time scale from 0.02 sec to 2000 sec shows an asymptotic modulus of about 0.6 kPa. Pig kidney tissue is linear up to a strain of about 0.002, followed by a modest shear softening. Rupture occurs at a strain greater than 1. Human cadaver kidney and liver tissues were studied using an aspiration device in which a partial vacuum caused a portion of the organ to bulge into a probe tube [232]. Stress relaxation curves were inferred, with an initial (1 sec) Young's modulus of about 40 to 50 kPa for kidney, 200 to 350 kPa for normal liver, and 50 to 120 kPa for diseased liver.

### 7.5.11 Uterus and Cervix

The uterus is an organ in which viscoelastic properties change in response to hormonal changes and to pregnancy. These changes are pertinent to its function. The

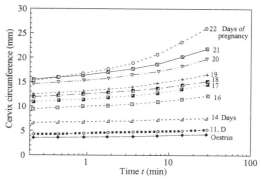

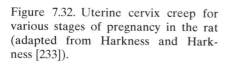

Figure 7.32. Uterine cervix creep for various stages of pregnancy in the rat (adapted from Harkness and Harkness [233]).

uterine cervix contains muscle, connective tissue, and epithelium. Tissue composition and properties change dramatically during pregnancy, to facilitate the passage of the fetus.

The cervix becomes more compliant as pregancy proceeds, and the rate of creep increases substantially [233]. The creep behavior of rat cervix under a 50 gram dilating force is shown in Figure 7.32 for the oestrus (fertile) stage and for various stages of pregnancy. Properties during the infertile (diestrous, marked D in the diagram) phase of the cycle are similar to those at 11 days pregnancy. Toward the end of pregnancy, the longer-term creep (not shown) is viscous in nature: creep is linear with time. The change in physical properties is associated with an increase in the weight of the cervix, growth of connective tissue and epithelium, and a reduction in the concentration of collagen and smooth muscle in the tissue. After the baby rats are delivered, the creep behavior reverts to a level comparable to the condition before pregnancy. The changes in the tissue cannot, therefore, be regarded as aging, though that term has been used in the wider rheology community, in reference to changes that depend on clock time rather than on time following a stress transient. In summary, the changes in the viscoelasticity of uterine tissue are essential to its function.

### 7.5.12 Arteries

The function of blood vessels depends on their mechanical properties. Knowledge of these properties in health and disease can be of use in the design of therapeutic methods. Age-related arterial stiffening is predictive of cardiovascular morbidity and mortality [236]. Artery compliance is important since if it decreases, there will be a reduction in distention due to the pulse, giving rise to increased systolic blood pressure and extra cardiac load. The role of arterial viscoelasticity in disease is less clear in this context, but it is known that increased energy dissipation [237] in vessels can increase the load on the heart.

The wall of an artery is a layered structure. Each layer consists of further structures including biological cells and collagen fibers [238]. Arteries contain smooth muscle; they are active. Moreover, they contain residual stress; if a longitudinal cut is made in a longitudinal subsection, it unrolls.

Behavior is nonlinear in the physiological range of stress. The Young's modulus of human aorta and iliac and carotid arteries increases from less than 0.5 MPa at an internal pressure of 40 mmHg to more than 1 MPa at 120 mmHg [234]; femoral arteries are more than twice as stiff. Residual stress results in a retraction strain of 10–30 percent when an artery segment is dissected from the body; this strain decreases with age. Damping tan$\delta$ is in the range 0.1 to 0.25; it tends to increase with frequency and decrease with age. Dog pulmonary artery [235] exhibits hysteresis, which increases with frequency from 0.01 Hz to 1 Hz, and increases with activation of the smooth muscle in the artery. The hysteresis was not interpreted, but we infer tan$\delta$ of about 0.15 to 0.3 at 0.1 Hz and 1 Hz, respectively. Dog aorta has a passive elastic modulus of about 0.5 MPa; hysteresis is observed in cycles of pressure versus diameter [239]. The effective tan$\delta$ is about 0.25.

Creep in arteries in organ culture has been studied [240] with the aim of developing modalities to permanently stretch arteries to increase the supply of tissue for vascular grafts. Creep occurs over several days. A threshold, axial-stretch ratio exceeding 2 was found to be necessary to obtain a permanent length change. Arteries in living organisms remodel in response to changes in blood pressure or flow to normalize mechanical loads [241]; vessels also remodel to normalize changes in longitudinal stretch.

### 7.5.13 Lung

Lung properties influence the muscular effort required for breathing. Lung tissue is viscoelastic; hysteresis in the lung is related to the amount of muscular work performed for each breath. Stress relaxation of pressure $P(t)$ following a step in volume $V_0$ in excised cat lung follows $P(t)/V_0 = A - B \log t$ from 0.1 to 50 sec, with $A$ and $B$ as constants [242]. In these experiments, a fixed-volume of gas in the lung was used to allow study of tissue properties independent of the drag due to air flow which occurs in a living organism. The area of hysteresis loops, 10 to 20 erg/ml$^2$ per cycle, is nearly independent of frequency up to 2 Hz, consistent with the relaxation results. This behavior rules out the notion of tissue viscosity in a dashpot representation. Phase angles $\phi$ [243] between pressure and volume are 4 to 9 degrees corresponding to tan$\phi$ from 0.07 to 0.16. Only about two-thirds of the lung hysteresis could be explained by linear viscoelastic inference from relaxation. It was suggested that the remaining hysteresis arises from rate-independent plastic strain. Pressure–volume characteristics of lungs in living human subjects have been determined by superposing small amplitude forced oscillations of air pressure on the normal breathing [244]. The $B/A$ ratio was inferred to be 0.11 to 0.14 in living human lungs compared with 0.16 in rat lungs (from about 0.2 to 100 seconds) [245] and 0.09 to 0.11 in above cat lungs.

### 7.5.14 The Ear

The ear is capable of fine discrimination of pitch, corresponding to the frequency of sound waveforms: a difference of 0.2–0.5 percent can be discerned. The sound signal is detected by sensitive cells in the cochlea, a coiled fluid-filled organ in the inner ear. According to Helmholtz, the cochlea contains fibers which resonate at different frequencies. These fibers protrude from hair cells which convert the motion of the fiber into a nerve signal. How can fine frequency tuning be achieved in soft tissue? Soft tissue as well as vibrating fibers immersed in fluid tend to have a low value of $Q$ (high tan$\delta$), corresponding to a broad resonant response that would impede the resolving power of a resonating fiber.

Experiments with living subjects suggest a rather high $Q$ between 200 and 350 [246] based on differences in threshold sound levels for pure tones and for pulses containing 2 to 250 oscillations. Further experiments with living subjects indicate phase memory [247]. People listened to a sequence of pulses containing at least 10 oscillations. In some tests, alternate pulses were given a phase reversal. The observers could distinguish the reversal provided the silent interval between pulses did not exceed a period of 30 oscillations. Therefore the oscillations in the ear initiated by each pulse had not died down to negligible levels prior to the arrival of the next pulse. Because a resonator decays by a factor $e \approx 2.718 \ldots$ in amplitude every $Q/\pi$ cycles, the $Q$ of sensors in the ear exceeds 100. The high $Q$ of ear structures is attributed by Thomas Gold to gain or amplification which counteracts the high damping (low $Q$) in the soft tissue [248]. This concept is by now well established [249, 250] though its acceptance took about 40 years. Indeed, pure tones can be emitted from the ear, spontaneously or in response to acoustic stimulation [251]. Active processes in the inner ear appear to give rise to negative stiffness [252] of the hair cells as well as negative damping or amplification. Metabolic energy of the living ear is necessary for such effects; the internal ear is active, not passive.

### 7.5.15 The Eye

#### *Lens*

To focus on near or far objects, muscles in the human eye deform the lens. The ability to focus (accommodation) tends to decrease with age (presbyopia), so that beyond age 45 or so, most people use eyeglasses for reading or other close work. If there is excess focusing power in the lens or cornea, the person is nearsighted or myopic; if there is too little, then farsighted or hyperopic. Myopia is prevalent in industrial societies in which people are required to focus at the near point for long periods. Properties of the lens and other eye tissues are therefore of interest.

The lens of the eye is viscoelastic in humans [253]. Lens stiffness, determined by compressing the lens axially between hard contact lenses, increases with age but by less than a factor of two (about 1.5 to 3 kPa at a few seconds after load application) from age 3 to 62 in humans [253], Figure 7.33. Similar moduli (0.8 kPa at birth to 3 kPa at age 63) were determined by spinning lenses in the laboratory [254] to mimic the radial deformation which occurs in living subjects; rabbit lenses are five times stiffer than human and cat lenses 21 times stiffer than human. This difference in modulus with age in humans seems insufficient to account for age-related changes

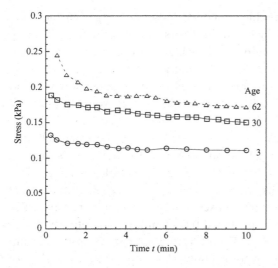

Figure 7.33. Eye lens stress relaxation in human subjects aged 3, 30, and 62 years, at a strain of 0.09 (adapted from Fukuda [253]).

in focusing ability. Continued growth of the lens through life has been suggested to cause geometric changes [255] which impair the effectiveness of the focusing muscles.

Lens creep over 5 minutes is less than 20 percent of the "initial" strain, but in lenses from young humans the creep appears not to recover fully. In contrast, recovery [257] following 3 to 5 minutes in cat and dog lens studied in more detail appears to approach completion given sufficient time. Recovery does not proceed to zero strain if a relatively large load has been applied. Creep has been observed in rabbit lens [256]. A portion of the stiffness of the lens is attributed to its capsule.

It is not, however established whether lens viscoelasticity plays a role in the development of myopia. Myopia beginning in childhood is attributed to excess growth of the posterior chamber of the eye [258]; eye growth in childhood is controlled by the retina and is linked to visual deprivation [259]. Adult onset myopia appears to be linked to change in curvature of the cornea. There is a hysteresis effect in accommodation; viscoelasticity may be involved.

### Cornea

The cornea is the transparent front portion of the eye. It provides about two-thirds of the optical focusing power of the eye, but the ability in humans to change focus is attributed to muscles that deform the lens. The cornea contains lamellae of collagen fibrils embedded in a matrix of proteoglycans, glycoproteins, and keratocytes. The cornea exhibits considerable creep, more than a factor of two over one hour [260]. Dynamic studies of the cornea [262] from 0.1 mHz to 100 Hz disclosed dependence of properties on temperature and hydration. Corneas from nearsighted people were found to be softer (by a factor of 7 at lower frequency and a factor of 4 at higher frequency) than those of emmetropic (normal) people. Corneal tissue is nonlinear in that the stress–strain curve is concave up for pig cornea [260] and it exhibits creep; the effective Young's modulus is 17 MPa at 1 second and 5 MPa at one hour. It is also

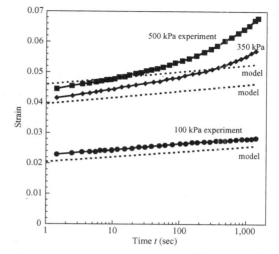

Figure 7.34. Eye cornea creep at different stresses (adapted from Boyce et al. [261]), shows nonlinear viscoelasticity that is not described by QLV (model, dashed lines).

nonlinearly viscoelastic [261]. Creep becomes more pronounced as stress increases (Figure 7.34). Therefore, the behavior does not follow quasilinear viscoelasticity (QLV).

### 7.5.16 Tissue Comparison

Table 7.4 compares properties of several tissues at low frequency. Ultrasonic properties of many kinds of tissues have been determined and compiled [272]. For example, one study of human muscle at $\nu = 1$ MHz reports the longitudinal wave velocity $c_L$ as 1,566 m/sec and the attenuation $\alpha$ as 0.16 neper/cm. The stiffness based on Equation 5.1 and 5.88, is $C_{1111} = c_L^2 \rho$, so assuming a density $\rho$ of 1 g/cm$^3$, the stiffness is 2.4 GPa. Compare with the bulk modulus, 2.18 GPa of water alone and recognize (Example 5.9) that $C_{1111}$ depends on the bulk modulus. In muscle, for that longitudinal mode of deformation, $\tan(\delta/2) = \alpha\nu/2\pi c_L = 4 \times 10^{-3}$, so, $\tan\delta = 8 \times 10^{-3}$. Liver is similar to muscle in both longitudinal wave velocity and attenuation, however, there is considerable variation in reported results; such variation is typical in biological materials. The shear wave velocity $c_T$ in dog liver at 2.2 MHz is reported

Table 7.4. *Stiffness and damping of some biological materials at 10 rad/sec (1.6 Hz); see also Chapter 9*

| Material | $|G^*|$ | $\tan\delta$ | Reference |
|---|---|---|---|
| Compact bone | 4 GPa | 0.01 | [11] |
| Spinal motion segment | 45 MPa | 0.1 | [265] |
| Articular cartilage | 0.6–1 MPa | 0.23 | [266] |
| Meniscus | 100 kPa | 0.40 | [267] |
| Nucleus pulposus | 11 kPa | 0.45 | [264] |
| Synovial fluid | 0.02 kPa | 0.29 | [269] |
| Vitreous humor | 0.002 kPa | 0.81–1.43 | [271] |

as 8.7 m/sec and the attenuation 9,200 neper/cm. So the stiffness is $C_{2323} = G = 76$ kPa and $\tan(\delta/2) = 0.58$; $\tan\delta = 1.7$. The loss is sufficiently large that the exact relation for the velocity is required, $c_T = \sqrt{\frac{|G^*|}{\rho}} \sec\frac{\delta}{2}$. The correction $\sec(\delta/2)$ is 1.16, so $|G^*| = 57$ kPa. The large difference between the longitudinal and shear properties deserves comment. Soft tissue, not unlike rubber, has a Poisson's ratio approximating 1/2 and the shear modulus is much less than the bulk modulus. Since longitudinal waves (§5.8) entail volumetric deformation, their speed depends on the bulk modulus.

### 7.5.17 Plant Seeds

Some seeds such, as those of wild wheat, have slender curved hair-like bristles called awns, which serve as wings so that the pointed end containing the seed enters the earth. The two bristles also allow the seed to drill into the soil [263]. The basis for this capability is a coupled field effect: a coupling between relative humidity and deformation. The bristles have a gradient in properties across their thickness, so that a change in humidity gives rise to bending. Daily changes in humidity cause the bristles to flex, giving rise to a slow swimming motion which drives the seed into the soil. The necessary friction is provided by tiny hairs which protrude from the bristle.

This use of coupled fields, although it occurs in a structure of biological origin, is subject to imitation by humans in designed systems. The swimming action does not involve metabolism; the energy for it is derived from periodic variations in environmental conditions, specifically humidity changes.

### 7.5.18 Wood

Wood is a cellular composite of biological origin, based on lignin, which is a natural macromolecule [273]. In addition to the results adduced in [10], numerous results ($\tan\delta$ vs. temperature at a constant frequency of 1 Hz or 1 kHz) have been obtained for dry woods, lignin, and delignified wood, as well as coal, amber, and oil shale for comparison [274]. The Young's modulus of wood along the grain is from 4 GPa to 20 GPa depending on density [16]; across the grain, the modulus is about 0.15 GPa to 2 GPa. Woods, such as spruce and beech, exhibit $\beta$ peaks corresponding to a loss tangent of 0.02 to 0.03 at a temperature of about 200K, $\tan\delta \approx 0.02$ at room temperature (about 300K), and an $\alpha$ peak of $\tan\delta \approx 0.08$ at about 400K. Bamboo [275] at room temperature is similar to wood; $\tan\delta \approx 0.02$ in torsion, 0.01 in bending for dry bamboo; wet bamboo has higher damping. Spectra for lignin, bituminous coal and amber have similar overall characteristics. The $\tan\delta$ values increase with moisture content in wood [274]. Fully wet wood (Figure 7.35) appears to be thermorheologically simple [274] as far as the stiffness, and master curves for the modulus have been produced for a frequency range $10^{-10}$ Hz ($E' = 50$ MPa) to more than 100 kHz ($E' = 400$ MPa) based on dynamic studies from 23°C to 130°C and from 0.6 Hz to 20 Hz. The properties have been associated with those of the lignin within the wood, which exhibits a glass transition temperature near 100°C.

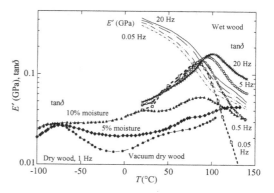

Figure 7.35. Viscoelastic properties of wood. Wet wood (adapted from Salmén [277]), dry spruce wood, several moisture contents, at 1 Hz in flexure (adapted from Kelley et al. [276]), vacuum dry spruce wood, torsion, frequency ≈ 1 Hz (adapted from Wert et al. [274]).

### 7.5.19 Soft Plant Tissue: Apple, Potato

Viscoelastic behavior of many food products has been studied. Considerable creep occurs in such materials. Power-law representations ($\epsilon = a + bt^n$) were useful in representing the creep behavior [278]. The viscoelastic behavior is nonlinear and is describable by a multiple integral representation of the Green–Rivlin type. Further examples and applications are presented in §10.3.17.

## 7.6 Common Aspects

### 7.6.1 Temperature Dependence

The phenomenology of viscoelastic response has been presented for many materials. Various materials differ substantially in their viscoelastic response. A common aspect in the viscoelasticity of materials is the role of *temperature* [116]. If the absolute temperature is small compared with the melting point of a material, it will behave nearly elastically, with minimal viscoelastic response. If the absolute temperature exceeds a significant fraction (0.3–0.5) of the melting point, (the material is at high homologous temperature) substantial creep, damping, and other manifestations of viscoelasticity may be expected. Many metals have high melting points, and these tend to exhibit little viscoelasticity at room temperature. Most polymers have low melting points, and they tend to exhibit substantial viscoelastic behavior at room temperature.

In many materials, viscoelasticity is governed by a thermally activated process for which

$$\tan\delta = f(\nu)\exp[-\frac{U}{kT}], \tag{7.18}$$

with $U$ as the activation energy, $\nu$ as frequency, $k$ as the Boltzmann's constant and $T$ as the absolute temperature in Kelvins (K). $k = 1.38 \times 10^{-16}$ erg/ K. If $U$ is expressed in calories per mole, $k$ becomes the gas constant $R = 1.986$ cal/mole K. This is an Arrhenius equation which arises when the rate limiting step of the relaxation consists of movement of an atomic or molecular process over an energy barrier.

Equation 7.18 gives rise to thermorheologically simple behavior with a shift factor $a_T$ as is shown in Example 6.9.

$$\log a_T = \frac{U}{R}[\frac{1}{T_1} - \frac{1}{T_2}]. \tag{7.19}$$

Amorphous polymers also depend sensitively on temperature. The time–temperature shift for polymers follows the empirical WLF equation (after Williams, Landel, and Ferry [15]),

$$\log a_T = -\frac{C_1(T - T_{ref})}{C_2 + (T - T_{ref})}. \tag{7.20}$$

The reference temperature $T_{ref}$ can be any temperature but is frequently taken as the glass transition temperature $T_g$. The constants $C_1$ and $C_2$ depend on the particular polymer. Even so, the behavior of different amorphous polymers is sufficiently similar in normalized coordinates that Ferry has proposed the following universal constants, for $T_{ref}$ taken as $T_g$ the glass transition temperature: $C_1 = 17.44$ and $C_2 = 51.6$. The effective activation energy associated with the WLF equation is not constant but increases rapidly with decreasing temperature below $T_g$.

Secondary creep (which is nonlinear) in a wide variety of materials is governed by the following relation for thermal activation [116]:

$$\frac{d\epsilon}{dt} = A\sigma_d^n \exp[-\frac{U}{kT}], \tag{7.21}$$

in which $n$ is a constant.

There are some exceptions to thermally activated temperature dependence. For example, thermoelastic damping, discussed in §8.2, is approximately proportional to absolute temperature. The dependence of creep rate or mechanical loss upon temperature and upon load level may be depicted in a deformation mechanism map, as discussed in §8.9.1.

### 7.6.2 High-Temperature Background

In a wide range of crystalline materials, such as metals, mechanical damping increases rapidly with temperature. The *high-temperature background* refers to such damping observed at an absolute temperature, which is a significant fraction of the melting point. The background appears to be thermally activated, with the following frequency dependence:

$$\tan\delta \propto \nu^{-n}, \tag{7.22}$$

with $n < 1$. Early examples have been reviewed by Nowick and Berry [281]. Damping in alloys of low melting point were observed to follow Equation 7.22 over a wide range of frequency [279], with $n \approx 0.2$. By contrast, an elementary dislocation model [280] predicts $n = 1$. Models are discussed in §8.6.

### 7.6.3 Negative Damping and Acoustic Emission

Materials can not only absorb sound waves, they can emit them. As discussed in §7.3.6, piezoelectric semiconductors and other semiconductors carrying electric current can exhibit acoustic amplification. Sound or ultrasound may also be emitted from materials subject to sufficiently large external force. Such *acoustic emission* occurs in tin, lead, aluminum, copper, steel, ceramics, and wood [282, 283]. The Kaiser effect refers to acoustic emission as a material is stressed to a given level; upon removal of the load, there is no further sound emission until the prior peak stress is exceeded. Emission of sound from materials has been known since early times. Ceramic pots emit noise if they crack during processing. Sufficient deformation of tin gives rise to a sound known as tin cry. Since acoustic emission can occur from flaws in a material under stress, detection of such sound forms the basis of practical methods of nondestructive testing to detect flaws [284]. Acoustic emission occurs concurrently with yield in metals and has been attributed to dislocation motion. Acoustic emission signals typically consist of a series of pulsed waveforms in a bandwidth 0.5 kHz to 100 kHz in composites and geological materials. In study of metals, detectors with a bandwidth of 100 kHz to 2 MHz may be used. Acoustic emission has been suggested [285] as a source of signals in early reports of detection of gravitational waves via resonant bar detectors.

Phase transformations can give rise to acoustic emission. Composites containing inclusions partially constrained from transformations exhibit instability manifested as slow undulations of deformation [287]. Transient negative damping also occurs in related composites that make use of negative stiffness to achieve high damping [288] or stiffness greater than that of diamond [289]. The energy source in this case is stored energy within the inclusions.

In biological materials, acoustic emission can occur. The ear can emit pure tones as discussed in §7.5.14; such emission is associated with amplification [248] or equivalently negative damping [286] in the inner ear. Muscle, as discussed in §7.5.5 emits a low-frequency sound when it contracts.

### 7.7 Summary

Virtually all materials exhibit some viscoelastic effects. The degree of viscoelasticity as quantified by the loss tangent in known solid materials differs by more than nine orders of magnitude. Some single crystal materials exhibit the lowest losses, followed by common metals. Polymers in the transition region exhibit the highest losses commonly found in solids: the loss tangent can exceed unity. For common materials, high stiffness is associated with low loss.

### 7.8 Examples

**Example 7.1**

Is the principle of dissipativity (§2.3) a real physical law? Would it be possible to create a material for which $\tan \delta < 0$? If so, what would be required?

**Answer**

In §7.4.5 piezoelectric semiconductors such as CdS were discussed. They *amplify* ultrasonic waves, hence they have tan $\delta < 0$. No thermodynamic principles are violated since the energy required for amplification is derived from an externally applied DC electric field [121, 122].

**Example 7.2**

Does a stretched rod held at constant extension get fatter or thinner in cross-section with time?

**Answer**

In §7.2, experimental results bearing upon the viscoelastic Poisson's ratio of polymers show that Poisson's ratio increases through the $\alpha$ transition but is not monotonic through the $\beta$ transition. These experimental results (assuming the specimens were in fact isotropic) provides an answer to questions raised in §5.6 and Problem 5.3. A stretched rod made of polymer held at constant extension gets thinner during the $\alpha$ transition but can get fatter through the $\beta$ transition.

**Example 7.3**

For some materials, tan$\delta > 1$. Does that mean these materials are viscoelastic liquids?

**Answer**

Polymers in the leathery regime (the $\alpha$ peak) can exhibit tan$\delta > 1$. The distinction between solids and liquids is defined (§1.3) in terms of the limiting behavior of the relaxation modulus or creep compliance as time tends to infinity. If the relaxation modulus approaches zero as time tends to infinity (or as frequency tends to zero for the storage modulus), the material is a liquid; otherwise it is a solid. For a liquid, $\lim_{\omega \to 0} \tan\delta = \infty$. So, it is not the value of tan$\delta$ at a given frequency that is relevant to the distinction between solid and liquid, but the limiting behavior at a frequency approaching zero is relevant.

**Example 7.4**

Must recovery (at zero stress) following some arbitrary stress history always be monotonic?

**Answer**

No. If the stress history contains reversals, the strain recovery after removal of the stress can contain reversals. Such reversals have been experimentally demonstrated in polymers.

## 7.9 Problems

7.1. Are there any materials that do not exhibit fading memory? Discuss your findings in connection with §2.3.

7.2. Plot the WLF and Arrhenius equations for the shift factor. Is there a regime for which the functional dependence is similar? Discuss.

7.3. Discuss applications of tin and its alloys and how they are influenced by viscoelastic properties.

7.4. In §2.6 the standard linear solid was described. Can you find any materials among the examples given in this chapter which behave in this way? How do real materials differ from the standard linear solid?

7.5. Tuning forks are usually made of a low-loss aluminum alloy. After such a fork is struck, its vibration persists for many cycles. Prepare tuning forks of several available materials, such as various polymers and woods as shown in the cover image. Describe the sounds in connection with free-decay of vibration and relate your observations to measured values of tan$\delta$.

7.6. For lung, relaxation was modeled as $A - B\log t$. For the values given, fit a power law in time. How good is the fit? Calculate the tan $\delta$ from the power law slope and compare with the phase angles quoted.

7.7. In the graph of brain relaxation, determine the final Young's modulus by applying the appropriate shape correction for the constraint of the ends.

## BIBLIOGRAPHY

[1] Zhang, J., Perez, R. J., and Lavernia, E. J., Documentation of Damping Capacity of Metallic, Ceramic and Metal-Matrix Composite Materials, *J Mat Sci* 28, 2395–2404, 1993.

[2] Ashby, M. F., *Materials Selection in Mechanical Design*, London, U.K.: Butterworths, 1992.

[3] Braginsky, V. B., Mitrofanov, V. P., and Panov, V. I., *Systems with Small Dissipation*, Chicago: University of Chicago Press, 1985.

[4] Mason, W. P., Use of Piezoelectric Crystals and Mechanical Resonators in Filters and Oscillators, in *Physical Acoustics*, E. P. Mason, ed., 1964, pp. 1A, 335–416.

[5] Smithells, C. J., *Metals Reference Book*, 5th ed., London and Boston: Butterworths, 1976.

[6] Zener, C., *Elasticity and Anelasticity of Metals*, Chicago: University of Chicago Press, 1948.

[7] Randall, B. H., and Zener, C., Internal Friction of Aluminum, *Phys Rev*, 58, 472–483, 1940.

[8] Jamieson, R. M., and Kennedy, R., Internal Friction Characteristics of Iron-Vanadium-Nitrogen Alloys, *J Iron Steel Inst*, 204, 1208–1210, 1966.

[9] Brennan, B. J., and Stacey, F. D., Frequency Dependence of Elasticity of Rock-Test of Seismic Velocity Dispersion, *Nature*, 268, 220–222, 1977.

[10] Knopoff, L., Attenuation of Elastic Waves in the Earth, in *Physical Acoustics*, E. P. Mason, ed., Vol. 3b, pp. 287–324, 1965.

[11] Lakes, R. S., Katz, J. L., and Sternstein, S. S., Viscoelastic Properties of Wet Cortical Bone: Part I, Torsional and Biaxial Studies. *J Biomechanics*, 12, 657–678, 1979.

[12] Koppelmann, V. J., Über die Bestimmung des dynamischen Elastizitätsmoduls und des dynamischen Schubmoduls im Frequenzbereich von $10^{-5}$ bis $10^{-1}$ Hz, *Rheologica Acta*, 1, 20–28, 1958.

[13] Koppelmann, V. J., Über das dynamische elastische Verhalten hochpolymerer Stoffe, *Kolloid Zeitschrift*, 144, 12–41, 1955.

[14] Iwayanagi, S., and Hideshima, T., Low Frequency Coupled Oscillator and Its Application to High Polymer Study, *J Phys Soc Japan*, 8, 365–358, 1953.

[15] Ferry, J. D., *Viscoelastic Properties of Polymers*, 2nd ed., New York: John Wiley, 1970.

[16] Ashby, M. F., On the Engineering Properties of Materials, *Acta Metall*, 37, 1273–1293, 1989.

[17] Hopkins, I. L., and Kurkjian, C. R., Relaxation Spectra and Relaxation Processes in Solid Polymers and Glasses, in *Physical Acoustics II*, W. P. Mason, ed., New York: Academic, 1965.

[18] McLoughlin, J. R., and Tobolsky, A. V., The Viscoelastic Behavior of Polymethyl Methacrylate, *J Colloid Sci*, 7, 555–568, 1952.

[19] Tobolsky, A. V., Stress Relaxation Studies of the Viscoelastic Properties of Polymers, *J Appl Phy*, 27, 673–685, 1956.

[20] Schmieder, Von K., and Wolf, K., Über die Temperatur und Frequenzabhängigkeit des mechanischen Verhaltens einiger hochpolymerer Stoffe, *Kolloid Z.*, 127, 65–78, 1952.

[21] Havrilak, S., Jr., and Havrilak, S. J., *Dielectric and Mechanical Relaxation in Materials*, Munich and Cincinnati: Hanser, 1997.

[22] Snowdon, J. C., *Vibration and Shock in Damped Mechanical Systems*, New York: John Wiley, 1968.

[23] Capps, R. N., and Beumel, L. L. Dynamic Mechanical Testing, Application of Polymer Development to Constrained Layer Damping, in *Sound and Vibration Damping with Polymers*, R. D. Corsaro and L. H. Sperling, eds., Washington, DC: American Chemical Society, 1990.

[24] Wetton, R. E., Design of Elastomers for Damping Applications, in *Elastomers: Criteria for Engineering Design*, C. Hepburn and R. J. W. Reynolds, London: Applied Science Publishers, Ltd. 1979.

[25] Hostettler, F., Energy Attenuating Polyurethanes, U.S. Patent 4,722,946.

[26] Tiao, Wen-Yu, and Tiao, Chin-Sheng, Methods for the Manufacture of Energy Attenuating Polyurethanes, U.S. Patent 4,980,386.

[27] Peoples, W. J., Shock Absorbing Device for High Heel Footwear, U.S. Patent 4,876,805.

[28] Reader, W. T., and Megill, R. W., Clorosulfonated Polyethylene: A Versatile Polymer for Damping Acoustic Waves, *Met Trans*, 22A, 633–640, 1991.

[29] Hypalon®, Dupont de Nemours, Wilmington, DE.

[30] Marvin, R., and McKinney, J. E., Volume Relaxations in Amorphous Polymers, in *Physical Acoustics*, E. P. Mason, ed., Vol. IIb, 165–229, 1964.

[31] Kono, R., The Dynamic Bulk Viscosity of Polystyrene and Polymethyl Methacrylate, *J Phys Soc Japan*, 15, 718–725, 1960.

[32] Sternstein, S. S., and Ho, T. C., Biaxial Stress Relaxation in Glassy Polymers, *J Appl Phys*, 43, 4370–4383, 1972.

[33] Sutherland, H. J., and Lingle, R., An Acoustic Characterization of Polymethyl Methacrylate and Three Epoxy Formulations, *J Appl Phys*, 43, 4022–4026, 1972.

[34] Ward, I. M., and Hadley, D. W., *Mechanical Properties of Solid Polymers*, New York: John Wiley, 1993.

[35] Sokolnikoff, I. S., *Mathematical Theory of Elasticity*, Malabar, FL: Krieger, 1983.

[36] Flocke, H. A., Ein Beitrag zum mechanischen Relaxationsverhalten von Polyäthylen, Polypropylen, Gemischen aus diesen und Mischpolymerisaten aus Propylen und Äthylen, *Kolloid Zeitschrift*, 180, 118–126, 1962.

[37] O'Brien, D. J., Sottos, N. R., and White, S. R., Cure Dependent Poisson's Ratio of Epoxy, *Experimental Mech*, 47, 237–249, 2007.

[38] Aklonis, J. J., MacKnight, W. J., and Shen, M., *Introduction to Polymer Viscoelasticity*, New York: John Wiley, 1972.

[39] Hodge, I. M., Physical Aging in Polymeric Glasses, *Science*, 267, 1945–1947, 1995.

[40] Struik, L. C. E., *Physical Aging in Amorphous Polymers and Other Materials*, Amsterdam and New York: Elsevier Scientific, 1978.

[41] Kubát, J., Stress Relaxation in Solids, *Nature*, 205, 378–379, 1965.

[42] Spencer, R. S., and Boyer, R. F., Thermal Expansion and Second Order Transition Effects in High Polymers. III. Time Effects, *J Appl Phys*, 17, 398, 1946.

[43] Thiele, J. L., and Cohen, R. E., Thermal Expansion Phenomena in Filled and Unfilled Natural Rubber Vulcanizates, *Rubber Chem and Technology*, 53, 313–320, 1980.

[44] Leung, W. P., and Yung, K. K., Internal Losses in Polyvinylidene Fluoride ($PVF_2$) Ultrasonic Transducers, *J Appl Phys*, 50, 8031–8033, 1979.

[45] Lesueur, D., Gerard, J. F., Claudy, P., Letoffe, M. M., Planche, J. P., and Martin, D., A structure related model to describe asphalt linear viscoelasticity, *J Rheology*, 40, 813–836, 1996.

[46] Sayegh, G., Variation des modules de quelques bitumes purs et bétons bitumineux, *Cahiers Rhéol*, 2, 51–74, 1966.

[47] Bahia, H. U., and Anderson, D. A., Glass Transition Behavior and Physical Hardening in Asphalt Binders, *J Association of Asphalt Paving Technologists*, 62, 93–129, 1993.

[48] Read, T. A., The Internal Friction of Single Metal Crystals, *Phys Rev*, 58, 371–380, 1940.

[49] Weertman, J., Internal Friction of Metal Single Crystals, *J Appl Phys*, 26, 202–210, 1955.

[50] Read, T. A., Internal Friction of Single Crystals of Copper and Zinc, *Trans Am Inst Mining Metal Eng*, 143, 30–44, 1941.

[51] Read, T. A., Internal Friction and Plastic Extension of Zinc Single Crystals, *J Appl Phys*, 17, 713–720, 1946.

[52] Swift, I. H., Internal Friction of Zinc Single Crystals, *J Appl Phys*, 18, 417–425, 1947.

[53] Read, T. A., Internal Friction of Zinc Single Crystals, *J Appl Phys*, 20, 29–37, 1947.

[54] Hiki, Y., Internal Friction of Lead, *J Phys Soc Japan*, 13, 1138–1144, 1958.

[55] Wert, C. A., The Internal Friction of Zinc Single Crystals, *J Appl Phys*, 20, 29–37, 1949.

[56] Kê, T. S., Cui, P., and Su, C. M., Internal Friction in High Purity Aluminium Single Crystals, *Physics Status Solidi*, 84, 157–164, 1984.

[57] Kê, T. S., Experimental Evidence of the Viscous Behavior of Grain Boundaries in Metals, *Phys Rev*, 71, 533–546, 1947.

[58] Kê, T. S., and Zhang, B. S., Contributions of Bamboo Boundaries to the Internal Friction Peak in Macrocrystalline High Purity Aluminium, *Physics Status Solidi*, 96, 515–525, 1986.

[59] Kê, T. S., Zhang, D. L., Cheng, B. L., and Zhu, A. W., On the Origin of the Macrocrystalline Internal Friction Peak (the Bamboo Boundary Peak), *Physics Status Solidi*, 108, 569–575, 1988.

[60] Zhu, A. W., and Kê, T. S., Characteristics of the Internal Friction Peak Associated with Bamboo Grain Boundaries, *Physics Status Solidi*, 113, 393–401, 1989.

[61] Firle, T. E., Amplitude Dependence of Internal Friction and Shear Modulus of Boron Fibers, *J Appl Phys*, 39, 2839–2845, 1968.

[62] Kê, T. S., Viscous Slip along Grain Boundaries and Diffusion of Zinc in Alpha Brass, *J Appl Phys*, 19, 285–290, 1948.

[63] Duffy, W., Acoustic Quality Factor of Aluminum Alloys From 50 mK to 300 K, *J Appl Phys*, 68, 5601–5609, 1990.

[64] Wolfenden, A., Harmouche, M. R., and Hartman, J. T., Mechanical Damping at High Temperature in Aluminides *J de Physique*, 46, C10-391-C10-394, 1985.

[65] Keene, K. H., Hartman, J. T., Jr, Wolfenden, A., and Ludtka, G. M., Determination of Dynamic Young's Modulus, Shear Modulus, and Poisson's Ratio as a Function of Temperature for Depleted Uranium −0.75 wt% Titanium using the Piezoelectric Ultrasonic Composite Oscillator Technique, *J Nuclear Mat*, 149, 218–226, 1987.

[66] Fitzgerald, E., Mechanical Resonance Dispersion in Metals at Audio Frequencies, *Phys Rev*, 108, 690–706, 1957.

[67] Fitzgerald, E., Mechanical Resonance Dispersion in Crystalline Polymers at Audio Frequencies, *J Chem Phys*, 27, 1180–1193, 1957.

[68] Bodner, S. R., On Anomalies in the Measurement of the Complex Modulus, *Trans Soc Rheology*, 4, 141–157, 1960.

[69] Farid, Z. M., Saleh, S., and Mahmoud, S. A., On the Grain Boundary Internal Friction Peak of $\alpha$-Brasses, *Mat Sci and Eng*, A110, L31–L34, 1989.

[70] Bratina, W. J., Internal Friction and Basic Fatigue Mechanisms in Body-Centered Cubic Metals, Mainly Iron and Carbon Steels, in *Physical Acoustics*, E. P. Mason, ed., Vol. IIIA, 1966, pp. 223–291.

[71] Fast, J. D., and Ferrup, M. B., Internal Friction in Lightly Deformed Pure Iron Wires, *Philips Re Repts*, 16, 51–65, 1961.

[72] Kraus, H., *Creep Analysis*, New York: John Wiley, 1980.

[73] Sato, E., Yamada, T., Tanaka, H., and Jimbo, I., Categorization of Ambient Temperature Creep Behavior of Metals and Alloys and Their Crystallographic Structures, *Mat Trans*, 47, 1121–1126, 2006.

[74] Fessler, H., and Hyde, T. H., Creep Deformation of Metals, in *Creep of Engineering Materials*, C. D. Pomeroy, ed., London: Mechanical Engineering Publications, Ltd., 1978.

[75] Tribula, D., and Morris, J. W., Jr., Creep in Shear of Experimental Solder Joints, *J Electronic Packaging*, 112, 87–93, 1990.

[76] Ritchie, I. G., Pan, Z. L., Sprungmann, K. W., Schmidt, H. K., and Dutton, R., High Damping Alloys–the Metallurgist's Cure for Unwanted Vibration, *Canadian Metallurgical Quarterly*, 26, 239–250, 1987.

[77] James, D. J., High Damping Metals for Engineering Applications, *Mat Sci and Eng*, 4, 1–8, 1969.

[78] Lazan, B. J., *Damping of Materials and Members in Structural Mechanics*, New York: Pergamon, 1968.

[79] Ritchie, I. G., and Pan, Z. L., High Damping Metals and Alloys, *Met Trans*, 22A, 1991, pp. 607–616.

[80] U.S. Patent 3,868,279.

[81] Ritchie, I. G., Sprungmann, K. W., and Sahoo, M., Internal Friction in Sonoston – a High Damping Mn/Cu – Based Alloy for Marine Propeller Applications, *J de Physique*, 46, C10-409-C10-412, 1985.

[82] Cochardt, A. W., High Damping Ferromagnetic Alloys, *J of Metals* (Trans. Amer. Inst. Mining Metallurgical Engineers) 206, 1295–1298, 1956.

[83] Goodman, G., Ferroelectric Properties of Lead Metaniobate, *J Amer Ceram Soc*, 36, 368–372, 1953.

[84] Lee, T., and Lakes, R. S., Damping Properties of Lead Metaniobate, *IEEE Transactions on Ultrasonics, Ferroelectrics and Frequency Control*, 48, 48–52, 2001.

[85] Otsuka, K., and Wayman, C. M., ed., *Shape Memory Materials*, Cambridge UK: Cambridge University Press, 1998.

[86] Hasiguti, R. R., and Iwasaki, K., Internal Friction and Related Properties of the TiNi Intermetallic Compound, *J Appl Phys*, 39, 2182–2186, 1968.

[87] Ritchie, I. G., and Pan, Z. L., Characterization of the Damping Properties of High Damping Alloys, in *$M^3D$: Mechanics and Mechanisms of Material Damping*, V. K. Kinra and A. Wolfenden, eds., ASTM, Phila. PA, 1992, pp. 142–157.

[88] Read, T. A., The Internal Friction of Single Metal Crystals, *Phys Rev*, 58, 371–380, 1940.

[89] Laddha, S., and Van Aken, D. C., On the Application of Magnetomechanical Models to Explain Damping in an Antiferromegnetic Copper-Manganese Alloy, *Metal and Mat Trans*, 26A, 957–964, 1995.

[90] Wang, Y. C., Ludwigson, M., and Lakes, R. S., Deformation of Extreme Viscoelastic Metals and Composites, *Mat Sci and Eng*, A, 370, 41–49, 2004.

[91] Weissmann, G. F., and Babington, W., A High Damping Magnesium Alloy for Missile Applications, *Proc ASTM*, 58, 869–892, 1958.

[92] Schwaneke, A. E., and Nash, R. W., Effect of Preferred Orientation on the Damping Capacity of Magnesium Alloys, *Metal Trans*, 2, 3453–3457, 1971.

[93] Sugimoto, K., Niiya, K., Okamoto, T., and Kishitake, K., A Study of Damping Capacity in Magnesium Alloys, *Trans JIM*, 18, 277–288, 1977.

[94] Bozorth, R. M., Mason, W. P., and McSkimin, H. J., Frequency Dependence of Elastic Constants and Losses in Nickel, *Bell System Technical J*, 30, 970–989, 1951.

[95] Schneider, W., Schrey, P., Hausch, G., and Török, Damping Capacity of Fe-Cr and Fe-Cr Based High Damping Alloys, *J de Physique*, 42, suppl 10, C5-635–C5-639, 1981.

[96] Masumoto, H., Sawaya, S., and Hinai, M., Damping Capacity of Gentalloy in the Fe-Co Alloys, *Trans Japan Inst Metals*, 19, 312–316, 1978.

[97] Masumoto, H., Sawaya, S., and Hinai, M., Damping Capacity of 'Gentalloy' in the Fe-Mo Alloys, *Trans Japan Inst Metals*, 22, 607–613, 1981.

[98] Masumoto, H., Hinai, M., and Sawaya, Damping Capacity and pitting corrosion resistance of Fe-Mo-Cr Alloys, *Trans Japan Inst Metals*, 25, 891–899, 1984.

[99] Masumoto, H., Sawaya, S., and Hinai, M., Damping Capacity of Fe-Mo Alloys, *Trans Japan Inst Metals*, 18, 581–584, 1977.

[100] Enrietto, J., and Wert, C., Anelasticity in Alloys of Cd and Mg, *Acta Metallurgica*, 6, 130–132, 1958.

[101] Lulay, J., and Wert, C., Internal Friction in Alloys of Mg and Cd, *Acta Metallurgica*, 4, 627–631, 1957.

[102] Cook, L. S., and Lakes, R. S., Viscoelastic Spectra of $Cd_{0.67}Mg_{0.33}$ in Torsion and Bending, *Metal Trans*, 26A, 2037–2039, 1995.

[103] Sobha, B., and Murti, Y. V. G. S., Low Frequency Internal Friction Spectra of $Cu_{0.81}Pd_{0.19}$ Alloy, *Bull Mater Sci*, 11, 319–328, 1988.

[104] Sommer, A. W., Motokura, S., Ono, K., and Buck, O., Relaxation Processes in Metastable Beta Titanium Alloys, *Acta Metallurgica*, 21, 489–497, 1973.

[105] Kamel, R. Measurement of the Internal Friction of Solids, *Phys Rev*, 75, 1606, 1949.

[106] Cook, L. S., and Lakes, R. S., Damping at High Homologous Temperature in Pure Cd, In, Pb, and Sn, *Scripta Metall et Mater*, 32, 773–777, 1995.

[107] Reed, R. P., McCowan, C. N., and Walsh, R. P., Delgado, L. A., and McColskey, J. D., Tensile Strength and Ductility of Indium, *Mat Sci Eng* A, 102, 227–236, 1988.

[108] Gindin, I. A., Lazarev, B. G., Starodubov, Ya. D., and Lebedev, V. P., Creep of Indium in the Normal and Superconductive States, *Fiz Metal Metalloved*, 29, 862–868, 1970.

[109] Ritchie, I. G., Pan, Z. L., and Goodwin, F. E., Characterization of the Damping Properties of Die-Cast Zinc-Aluminum Alloys, *Met Trans*, 22A, 617–622, 1991.

[110] Otani, T., Sakai, T., Hoshino, K., and Kurosawa, T., Damping Capacity of Zn-Al Alloy Castings, *J de Phy*, Colloque C10, Suppl. 12, Tome 46, c10-417–c10-420, 1985, A. V. Granato, G. Mozurkewich, C. A. Wert, eds., 8th Int'l Conf. on Internal Friction and Ultrasonic Attenuation in Solids, Urbana, IL.

[111] Millett, P., Schaller, R., and Benoit, W., Internal Friction Spectrum and Damping Capacity of Grey Cast Iron, in *Deformation of Multi-Phase and Particle Containing Materials*, Proc. 4th Riso International Symposium on Metallurgy and Materials Science, J. B. Bilde-Sorensen, N. Hansen, A. Horsewell, T. Leffers, H. Linholt, eds., Riso National Laboratory, Roskilde Denmark, 1983.

[112] Cagnoli, G., Giammatoni, L., Kovalik, J., Marchesoni, F., and Punturo, M., Low Frequency Internal Friction in Clamped Free Thin Wires, *Phys Let* A, 255, 230–235, 1999.

[113] Ashby, M. F., and Abel, C. A., Materials Selection to Resist Creep, *Phil Trans Royal Soc Lond* A, 351, 451–468, 1995.

[114] Nabarro, F. R. N., and de Villiers, H. L. *The Physics of Creep*, London: Taylor and Francis, 1995.

[115] Pichler, A., Weller, M., and Arzt, E., High Temperature Damping in Dispersion-Strengthened Aluminium Alloys, *J Alloys and Compounds*, 211, 414–418, 1994.

[116] Ashby, M. F., and Jones, D. R. H. *Engineering Materials*, Oxford: Pergamon, 1980.

[117] Eisenberg, A., and Tobolsky, A. V., Viscoelastic Properties of Amorphous Selenium, *J Polymer Sci*, 61, 483–495, 1962.

[118] Tobolsky, A. V., and Eisenberg, A., *J Am Chem Soc*, 81, 780, 1959.

[119] Tobolsky, A. V., Owen. G. D. T., and Eisenberg, A., *J Colloid Sci*, 17, 717, 1962.

[120] Royer, D., and Dieulesant, E., Elastic and Piezoelectric Constants of Trigonal Selenium and Tellurium Crystals, *J Appl Phys*, 50, 4042–4045, 1979.

[121] Hutson, A. R., McFee, J. H., and White, D. L., Ultrasonic Amplification in CdS, *Phys Rev Let*, 7, 237–239, 1961.

[122] White, D. L., Amplification of Ultrasonic Waves in Piezoelectric Semiconductors, *J Appl Phys*, 33, 2547–2554, 1962.

[123] Toxen, A. M., and Tansal, S., Ultrasonic Amplification in Bismuth, *Phys Rev Let*, 10, 481–483, 1963.

[124] Pomerantz, M., Amplification of Microwave Phonons in Germanium, *Phys Rev Let*, 13, 308–310, 1964.

[125] Holden, A. V., Markus, M., and Othmer, H. G., eds., *Nonlinear Wave Processes in Excitable Media*, New York: Plenum, 1991.

[126] Markus, M. Kloss, G., and Kusch, I., Disordered Waves in a Homogeneous, Motionless, Excitable Medium, *Nature*, 371, 402–404, 1994.

[127] Bonetti, E., Del Bianco, L., Pasquini, L., and Sampaolesi, E., Anelastic and Structural Behavior of Ball Milled Nanostructured Iron, *Nanostructured Mat*, 10, 741–753, 1998.

[128] Gueguen, Y., Woirgard, J., and Darot, M., Attenuation Mechanisms and Anelasticity, in the Upper Mantle, in *Anelasticity in the Earth*, F. D. Stacey, M. S. Paterson, and A. Nicholas, eds., Washington, DC: American Geophysical Union, 1981.

[129] Stein, S., Mills, J. M., and Geller, R. J., $Q^{-1}$ Models from Data Space Inversion of Fundamental Spheroidal Mode Attenuation Measurements, in *Anelasticity in the Earth*, F. D. Stacey, M. S. Paterson, and A. Nicholas, eds., Washington, DC: American Geophysical Union, 1981.

[130] Mason, W. P., Internal Friction at Low Frequencies Due to Dislocations: Applications to Metals and Rock Mechanics, in *Physical Acoustics*, E. P. Mason, and R. N. Thurston, eds., Vol. VIII, 347–371, 1971.

[131] Latham, G. V., Ewing, M., Dorman, J., Press, F., Toksoz, N., Sutton, G., Meissner, R., Duennebier, F., Nakamura, Y., Kovach, R., and Yates, M., Seismic Data from Manmade Impacts on the Moon, *Science*, 170, 620–626, 1970.

[132] Mason, W. P., Internal Friction in Moon and Earth Rocks, *Nature*, 234, 461–463, 1971.

[133] Tittman, B. R., Internal Friction in Lunar Rocks and Terrestrial Rocks, *1974 Ultrasonics Symposium Proceedings*, IEEE No. 74 CHO 896-ISU, 509-513, 1974.

[134] Tittman, B. R., and Housley, R. M., High $Q$ (Low Internal Friction) Observed in a Strongly Outgassed Terrestrial Analog of Lunar Basalt, *Phys Stat, Solidi* 56, K109–K110, 1973.

[135] Pomeroy, C. D., Time-Dependent Deformation of rocks, in *Creep of Engineering Materials*, C. D. Pomeroy, ed., London: Mechanical Engineering Publications, Ltd., 1978.

[136] Bassett, R. H., Time-Dependent Strains and Creep in Rock and Soil Structures, in *Creep of Engineering Materials*, C. D. Pomeroy, ed., London: Mechanical Engineering Publications, Ltd., 1978.

[137] Karato, S., and Li, P., Diffusion Creep in Perovskite: Implications for the Rheology of the Lower Mantle, *Science*, 260, 771–778, 1993.

[138] Karato, S., and Wu, P., Rheology of the Upper Mantle: A Synthesis, *Science*, 255, 1238–1240, 1992.

[139] Poirier, J. P., *Creep of Crystals*, Cambridge, UK: Cambridge University Press, 1985.

[140] Wolfenden, A., Measurement and Analysis of Elastic and Anelastic Properties of Alumina and Silicon Carbide, *J Mat Sci*, 32, 2275–2282, 1997.

[141] Robinson, W. H., and Birnbaum, H. K., High-Temperature Internal Friction in Potassium Chloride, *J Appl Phys*, 37, 3754–3766, 1966.

[142] Wolfenden, A., Lastrapes, A., Duggan, M. B., and Raj, S. V., Temperature Dependence of the Elastic Moduli and Damping for Polycrystalline $LiF - 22\%CaF_2$ eutectic salt, *J Mat Sci*, 26, 1973–1798, 1991.

[143] Robinson, W. H., Amplitude-independent Mechanical Damping in Alkali Halides, *J Mat Sci*, 7, 115–123, 1972.

[144] Branson, D., *Deformation of Concrete Structures*, New York: McGraw Hill, 1977.

[145] Illston, J. M., Creep of Concrete, in *Creep of Engineering Materials*, C. D. Pomeroy, ed., London: Mechanical Engineering Publications, Ltd., 1978.

[146] Troxell, G. E., Raphael, J. M., and Davis, R. W., Long Time Creep and Shrinkage Tests of Plain and Reinforced Concrete, *ASTM Proceedings*, 58, 1–20, 1958.

[147] Bazant, Z., *Mathematical Modeling of Creep and Shrinkage of Concrete*, New York: J. Wiley, 1988.

[148] L'Hermite, R., What Do We Know About the Plastic Deformation and Creep of Concrete? Bulletin RILEM, 1, 22–51, 1959.

[149] Neville, A., Creep of Concrete: Plain, Reinforced, and Prestressed, Amsterdam: North-Holland, 1970.

[150] Zhaoxia, L., Effective Creep Poisson's Ratio for Damaged Concrete, *Int J Fracture*, 66, 189–196, 1994.

[151] Irby, P. L., Kinetics of Mechanical Relaxation Processes in Inorganic Glasses, in *Noncrystalline Solids*, V. D. Fréchette, ed., New York: J. Wiley, 1960.

[152] Rekhson, S., Viscoelasticity of Glass, in *Glass: Science and Technology*, Vol. 3, Viscosity and Relaxation, D. R. Uhlmann and N. J. Kreidl, eds., New York: Academic, 1986.

[153] Eranti, E., and Lee, G. C., *Cold Region Structural Engineering*, New York: McGraw Hill, 1986.

[154] Jezek, K. C., Alley, R. B., and Thomas, R. H. Rheology of Glacier Ice, *Science*, 227, 1335–1337, 1985.

[155] Vassoile, R., Perez, J., Mai, C., and Gobin, P. F., Internal Friction of Ice $I_h$ due to Crystalline Defects, in *Internal Friction and Ultrasonic Attenuation in Solids*, R. R. Hasiguti and N. Mikoshiba, eds., Tokyo: University of Tokyo Press, 1977 (Proceedings of the Sixth International Confrence on Internal Friction and Ultrasonic Attenuation in Solids, June 4–7, 1977, Tokyo.

[156] Tatiboët, J., Perez, J., and Vassoile, R., Study of Lattice Defects in Ice Ih by Very Low Frequency Internal Friction Measurements, *J Physical Chem*, 87, 4050–4054, 1983.

[157] Nakamura, T., and Abe, O., Internal Friction of Snow and Ice at Low Frequency, in *Internal Friction and Ultrasonic Attenuation in Solids*, R. R. Hasiguti and N. Mikoshiba, eds., Tokyo: University of Tokyo Press, 1977.

[158] Hiki, Y., and Tamura, J., Internal Friction in Ice Crystals, *J Phys Chem*, 87, 4054–4059, 1983.

[159] Cole, D. M., and Durell, G. D., The Cyclic Loading of Saline Ice, *Philos Mag A*, 72, 209–229, 1995.

[160] Berlincourt, D. A., Curran, D. R., and Jaffe, H., Piezoelectric and Piezomagnetic Materials and Their Function in Transducers, in *Physical Acoustics*, Vol. 1A, E. P. Mason, ed., 1964, 169–270.

[161] Turner, S., The Strain Response of Plastics to Complex Stress Histories, *Polymer Engineering and Science*, 6, 306, 1966.

[162] Pipkin, A. C., and Rogers, T. G., A Non-Linear Representation for Viscoelastic Behaviour, *J Mech Phys Solids*, 16, 59–72, 1968.

[163] Ward, I. M., and Onat, E. T., Non-Linear Mechanical Behaviour of Oriented Polypropylene, *J Mech Phys Solids* 11, 217–229, 1963.

[164] Fung, Y. C., Elasticity of Soft Tissue in Simple Elongation, *Am J Physiol*, 213, 1532–1544, 1967.

[165] Fung, Y. C., Stress–Strain History Relations of Soft Tissues in Simple Elongation, in *Biomechanics, its Foundations and Objectives*, Englewood Cliffs, NJ: Prentice Hall, 1970.

[166] Lakes, R. S., Viscoelastic Properties of Cortical Bone, in *Bone Mechanics Handbook*, S. C. Corwin, ed., 2nd ed., Boca Raton, FL: CRC Press, 2001.

[167] Currey, J. D., *The Mechanical Adaptations of Bones*, Princeton NJ: Princeton University Press, 1984.

[168] Katz, J. L., Anisotropy of Young's Modulus of Bone, *Nature*, 283, 106–107, 1980.

[169] Lakes, R. S., Materials with Structural Hierarchy, *Nature*, 361, 511–515, 1993.

[170] Reilly, D. T., and Burstein, A. H., The Elastic and Ultimate Properties of Compact Bone Tissue, *J Biomechanics*, 8, 393–405, 1975.

[171] Yoon, H. S., and Katz, J. L., Ultrasonic Wave Propagation in Human Cortical Bone. II. Measurements of Elastic Properties and Microhardness, *J Biomechanics*, 9, 459–464, 1976.

[172] Buechner, P. M., and Lakes, R. S., Size Effects in the Elasticity and Viscoelasticity of Bone, *Biomechanics and Modeling in Mechanobiology*, 1 (4), 295–301, 2003.

[173] Park, H. C., and Lakes, R. S., Cosserat Micromechanics of Human Bone: Strain Redistribution by a Hydration-Sensitive Constituent, *J Biomechanics*, 19, 385–397, 1986.

[174] Sasaki, N., Nakayama, Y., Yoshikawa, M., and Enyo, A. Stress Relaxation Function of Bone and Bone Collagen, *J Biomechanics*, 26, 1369–1376, 1993.

[175] Garner, E., Lakes, R. S., Lee, T., Swan, C., and Brand, R., Viscoelastic Dissipation in Compact Bone: Implications for Stress-Induced Fluid Flow in Bone, *J Biomech Eng*, 122, 166–172, 2000.

[176] Thompson, G., Experimental Studies of Lateral and Torsional Vibration of Intact Dog Radii, PhD. thesis, biomedical engineering, Stanford University, 1971.

[177] Adler, L., and Cook, C. V., Ultrasonic Parameters of Freshly Frozen Dog Tibia, *J Acoust Soc Am*, 58, 1107–1108, 1975.

[178] Lakes, R. S., Dynamical Study of Couple Stress Effects in Human Compact Bone, *J Biomech Eng*, 104, 6–11, 1982.

[179] Rimnac, C. M., Petko, A. A., Santner, T. J., and Wright, T. M., The Effect of Temperature, Stress, and Microstructure on the Creep of Compact Bovine Bone, *J Biomechanics*, 26, 219–228, 1993.

[180] Caler, W. E., and Carter, D. R., Bone Creep-Fatigue Damage Accumulation, *J Biomechanics*, 22, 625–635, 1989.

[181] Mauch, M., Currey, J. D., and Sedman, A. J., Creep Fracture in Bones with Different Stiffnesses, *J Biomechanics*, 25, 11–16, 1992.

[182] Bowman, S. M., Keaveny, T. M., Gibson, L. J., Hayes, W. C., and McMahon, T. A., Compressive Creep Behavior of Bovine Trabecular Bone, *J Biomechanics*, 27, 301–310, 1994.

[183] Piekarski, K., and M. Munro, Transport Mechanism Operating between Blood Supply and Osteocytes in Long Bones, *Nature*, 269, 80–82, 1977.

[184] Tami, A. E., Schaffler, M. B., and Knothe Tate, M. L., Probing the Tissue to Subcellular Level Structure Underlying Bones Molecular Sieving Function, *Biorheology*, 40, 577–590, 2003.

[185] Buechner, P. M., Lakes, R. S., Swan, C., and Brand, R. A., A Broadband Viscoelastic Spectroscopic Study of Bovine Bone: Implications for Fluid Flow, *Annals of Biomedical Engineering*, 29, 719–728, 2001.

[186] Fung, Y. C., and Sobin, S. S., The Retained Elasticity of Elastin under Fixation Agents, *J Biomech Eng*, 103, 121–122, 1981.

[187] Fung, Y. C., *Biomechanics*, 2nd ed., New York: Springer, 1993.

[188] Meechai N., Jamieson A. M., Blackwell J., Carrino D. A., and Bansal, R., Viscoelastic Properties of Aggrecan Aggregate Solutions: Dependence on Aggrecan Concentration and Ionic Strength, *J Rheology*, 46(3), 685–707, 2002.

[189] Hiltner, A., Cassidy, J. J., Baer, E., Mechanical Properties of Biological Polymers, *Ann Rev Mat Sci*, 15, 455–482, 1985.

[190] Bonifasi-Lista, C., Lake, S. P., Small, M. S., and Weiss, J. A., Viscoelastic Properties of the Human Medial Collateral Ligament under Longitudinal, Transverse, and Shear Loading, *J Orthop Res*, 23, 67–76, 2005.

[191] Haut, R. C., The Mechanical and Viscoelastic Properties of the Anterior Cruciate Ligament and of ACL Fascicles, in *The Anterior Cruciate Ligaments, Current and Future Concepts*, D. W. Jackson, ed., Raven Press, 1993.

[192] Fleming, B. C., Beynonon, B. D., Renstrom, P. A., Peura, G. D., Nichols, C. E., and Johnson, R. J., The Strain Behavior of the Anterior Cruciate Ligament during Bicycling, an in Vivo Study, *Am J Sports Medicine*, 26, 109–118, 1998.

[193] Provenzano, P., Heisey, D. Hayashi, K, Lakes, R. S., and Vanderby, R., Jr., Subfailure Damage in Ligament: A Structural and Cellular Evaluation, *J Appl Physiol*, 92, 362–371, 2002.

[194] Provenzano, P., Hayashi, K., Kunz, D., Markel, M., and Vanderby, R., Jr., Healing of Subfailure Ligament Injury: Comparison between Immature and Mature Ligaments in a Rat Model, *J Orthop Res*, 20, 975–983, 2002.

[195] Thornton, G. M., Oliynyk, A., Frank, C. B., and Shrive, N. G., Ligament Creep Cannot Be Predicted from Stress Relaxation at Low Stress: A Biomechanical Study of the Rabbit Medial Collateral Ligament, *J Ortho Res*, 15, 652–656, 1997.

[196] Lakes, R. S., and Vanderby, R. Interrelation of Creep and Relaxation: A Modeling Approach for Ligaments, *ASME J Biomech Eng*, 121(6), 612–615, 1999.

[197] Provenzano, P., Lakes, R. S., Keenan, T., Vanderby, R., Jr., Non-linear Ligament Viscoelasticity, *Ann of Biomed Eng*, 29, 908–914, 2001.

[198] Rigby, J., Effect of Cyclic Extension on the Physical Properties of Tendon Collagen and Its Possible Relation to Biological Ageing of Collagen, *Nature*, 202, 1072, 1964.

[199] Abrahams, M., Mechanical Behaviour of Tendon in vitro, *Med Biol Eng*, 5, 433–443, 1967.

[200] Haraldsson, B. T., Aagaard, P., Krogsgaard, M., Alkjaer, T., Kjaer, M., and Magnusson, S. P., Region-Specific Mechanical Properties of the Human Patella Tendon, *J Appl Physiol*, 98, 1006–1012, 2005.

[201] Reeves, N. D., Narici, M. V., and Maganaris, C. N., Strength Training Alters the Viscoelastic Properties of Tendons in Elderly Humans, *Muscle Nerve*, 28, 74–81, 2003.

[202] Maganaris, C. N., and Paul, J. P., In vivo Human Tendon Mechanical Properties, *J Phys*, 521, 307–313, 1999.

[203] Pearson, S. J., Burgess, K., and Onambele, G., Creep and the in vivo Assessment of Human Patellar Tendon Mechanical Properties, *Clin Biomechanics*, 22, 712–717, 2007.

[204] Alexander, R. McN., *Elastic Mechanisms in Animal Movement,* Cambridge UK: Cambridge, University Press, 1988.

[205] Ker, R. F., Bennett, M. B., Bibby, S. R., Kester, R. C., and Alexander, R. McN., The Spring in the Arch of the Human Foot, *Nature*, 325, 147–149, 1987.

[206] Djerad, S. E., du Burck, F., Naili, S., and Oddou, C., Analyse du comportement rhéologique instationnaire d'un échantillon de muscle cardiaque, *CR Acad Sci Paris* (Comptes Rendus), II, 315, 1615–1621, 1992.

[207] Pinto, J. G., and Fung, Y. C., Mechanical Properties of the Heart Muscle in the Passive State, *J Biomech*, 6, 597–616, 1973.

[208] Meyhofer, E., Dymamic Mechanical Properties of Passive Single Cardiac Fibers from the Crab Cancer Magister, *J Exp Biol*, 185, 207–249, 1993.

[209] Abbott, B. C., and Lowy, J., Stress Relaxation in Muscle, *Proc of the Royal Soc Lond. Series B, Biological Sciences*, 146, 281–288, 1957.

[210] Hill, A. V., The Heat of Shortening and the Dynamic Constants of Muscle, *Proc Royal Soc London B*, 126, 136–195, 1938.

[211] Oster, G., and Jaffe, J. S., Low-Frequency Sounds from Sustained Contraction of Human Skeletal Muscle, *Biophys J*, 30, 119–127, 1980.

[212] Sabra, K. G., Conti, S., Roux, P., and Kuperman, W. A., Passive in vivo Elastography from Skeletal Muscle Noise, *App Phys Let*, 90, 194101, 2007.

[213] Levinson, S. F., Shinagawa, M., and Sato, T., Sonoelastic Determination of human Skeletal Muscle Elasticity, *J Biomech*, 28, 1145–1154, 1995.

[214] Alexander, R. McN., and Bennet-Clark, H. C., Storage of Elastic Strain Energy in Muscle and Other Tissues, *Nature*, 265, 114–117, 1977.

[215] Price, J. M., Patitucci, P., and Fung, Y. C., Mechanical Properties of taenia coli Smooth Muscle in Spontaneous Contraction, *Am J Physiol Cell Physiol*, 233: C47–C55, 1977.

[216] Gefen, A., and Haberman, E., Viscoelastic Properties of Ovine Adipose Tissue Covering the Gluteal Muscles, *J Biomech Eng*, 129, 924–930, 2007.

[217] Patel, P. N., Smith, C. K., and Patrick, C. W., Jr., Rheological and Recovery Properties of Poly(Ethylene Glycol) Diacrylate Hydrogels and Human Adipose Tissue, *J Biomed Mat Res*, 73A, 313–319, 2005.

[218] Lewin, R., Is Your Brain Really Necessary?, *Science*, 210, 1232–1234, 1980.

[219] Galford, J. F., and McElhaney, J. H., A Viscoelastic Study of Scalp, Brain, and Dura, *J Biomechanics*, 3, 211–221, 1970.

[220] Fraceschini, G., Bigoni, D., Regitnig, P., and Holzapfel, G. A., Brain Tissue Deforms Similarly to Filled Elastomers and Follows Consolidation Theory, *J Mech Phys Solids*, 54, 2592–2620, 2006.

[221] Terzaghi, K., *Theoretical soil mechanics*, New York: Wiley, 1943.

[222] Bilston, L. E., Liu, Z., and Phan Thien, N., Linear Viscoelastic Properties of Bovine Brain Tissue in Shear, *Biorheology*, 34, 377–385, 1997.

[223] Chan, R. W., and Titze, I. R., Viscoelastic Shear Properties of Human Vocal Fold Mucosa: Measurement Methodology and Empirical Results, *J Acoust Soc Am*, 106, 2008–2021, 1999.

[224] McCutcheon, C. W., Mechanism of Animal Joints. Sponge Hydrostatic and Weeping Bearings, *Nature*, 184, 1284, 1959.

[225] Lewis, P. R., and McCutcheon, C. W., Experimental Evidence for Weeping Lubrication in Mammalian Joints, *Nature*, 184, 1285, 1959.

[226] Oloyede, A., and Broom, N. D., Is Classical Consolidation Theory Applicable to Articular Cartilage Deformation? *Clin Biomechanics*, 6, 206–212, 1991.

[227] Oloyede, A., and Broom, N. D., Biomechanics of Cartilage Load Carriage, *Connective Tissue Research*, 34, 119–143, 1996.

[228] Virgin, W. J., Experimental Investigations into the Physical Properties of the Intervertebral Disc, *J Bone Joint Surg*, 33B, 607–611, 1951.

[229] Martinez, J. B., Oloyede, A., and Broom, N. D., Biomechanics of Load-Bearing of the Intervertebral Disc: An Experimental and Finite Element Model, *Med Eng Phys*, 19, 145–156, 1997.

[230] Best, B. A., Guilack, F., Setton, L. A., Zhu, W., Saed-Nejad, F., Ratcliffe, A., Weidenbaum, M., and Mow, V. C., Compressive Mechanical Properties of the Human Anulus Fibrosus and Their Relationship to Biochemical Composition, *Spine*, 19, 212–221, 1994.

[231] Nasseri, S., Bilston, L. E., and Phan-Thien, N., Viscoelastic Properties of Pig Kidney in Shear, Experimental Results and Modeling, *Rheol Acta*, 41, 180–192, 2002.

[232] Nava, A., Mazza, E., Kleinermann, F., Avis, N., McClure, J., and Bajka, M., Evaluation of the Mechanical Properties of Human Liver and Kidney through Aspiration Experiments, *Technology and Health Care*, 12, 269–280, 2004.

[233] Harkness, M., and Harkness, R. D., Changes in the Physical Properties of the Uterine Cervix of the Rat During Pregnancy, *J Physiol*, 148, 524–547, 1959.

[234] Learoyd, B. M., and Taylor, M. G., Alterations with Age in the Viscoelastic Properties of Human Arterial Walls, *Circ Res*, 43, 278–292 1965.

[235] Cox, R. H., Viscoelastic Properties of Canine Pulmonary Arteries, *Am J Physiol*, 246, H90–H96, 1984.

[236] Breithaupt-Grogler, K., and Belz, G. G. Epidemiology of the Arterial Stiffness. *Pathol Biol (Paris)*, 47(6), 604–613, 1999.

[237] Taylor, M. G., Wave Transmission through an Assembly of Randomly Branching Elastic Tubes, *Biophys J*, 6(6), 697–716, 1966.

[238] Vito, R. P., and Dixon, S. A., Blood Vessel Constitutive Models 1995–2002, *Ann Rev Biomed Eng*, 5, 413–439, 2003.

[239] Armentano, R. L., Barra, J. G., Levenson, J., Simon, A., and Pichel, R. H., Arterial Wall Mechanics in Conscious Dogs: Assessment of Viscous, Inertial, and Elastic Moduli to Characterize Aortic Wall Behavior, *Circ Res*, 76(3), 468–478, 1995.

[240] Davis, N. P., Han, H. C., and Wayman, B., Vito, R. Sustained Axial Loading Lengthens Arteries in Organ Culture, *Annals Biomed Eng*, 33, 867–877, 2005.

[241] Jackson, Z. S., Gotlieb, A. I., and Langille, B. L, Wall Tissue Remodeling Regulates Longitudinal Tension in Arteries, *Circ Res*, 90(8), 918–925, 2002.

[242] Hildebrandt, J. Dynamic Properties of Air-Filled Excised Cat Lung Determined by Liquid Plethysmograph. *J Appl Physiol*, 27(2), 246–250, 1969.

[243] Hildebrandt, J. Pressure-Volume Data of Cat Lung Interpreted by a Plastoelastic, Linear Viscoelastic Model. *J Appl Physiol*, 28(3), 365–372, 1970.

[244] Suki, B., Peslin, R., Duvivier, C., and Farré, R., Lung Impedance in Healthy Humans Measured by Forced Oscillations from 0.01 to 0.1 Hz., *J Appl Physiol*, 67(4), 1623–1629, 1989.

[245] Peslin, R., Duvivier, C., Bekkari, H., Reichart, E., and Gallina, C., Stress Adaptation and Low-Frequency Impedance of Rat Lungs, *J Appl Physiol*, 69, 1080–1086, 1990.

[246] Pumphrey, R. J., and Gold, T., Transient Reception and the Degree of Resonance of the Human Ear, *Nature*, 160, 124–125, 1947.

[247] Pumphrey, R. J., and Gold, T., Phase Memory of the Ear: A Proof of the Resonance Hypothesis, *Nature*, 161, 640, 1948.

[248] Gold, T., Hearing. II. The Physical Basis of the Action of the Cochlea, *Proceed Roy Soc Lond Series B, Biological Sciences*, 135, 492–498, 1948.

[249] Dallos, P., The Active Cochlea, *J Neurosci*, 12(12), 4575–85, 1992.

[250] Hudspeth, A. J., How the Ear's Works Work, *Nature*, 341, 397–404, 1989.

[251] Kemp, D. T., Stimulated Acoustic Emissions from within the Human Auditory System, *J Acoust Soc Am*, 64, 1386–1391, 1978.

[252] Iwasa, K. H., and Ehrenstein, G. Cooperative Interaction as the Physical Basis of The Negative Stiffness in Hair Cell Stereocilia, *J Acoust Soc Am*, 111, 2208–2212, Pt. 1, May 2002.

[253] Fukuda, M., Rheological Characteristics of Human Crystalline Lens, *Japanese J Ophthalmol*, 7, 47–55, 1963.

[254] Fisher, R. F., The Elastic Constants of the Human Lens, *J Physiol*, 212, 147–180, 1971.

[255] Koretz, J. F., and Handelman, G. H., How the Human Eye Focuses, *Sci Am*, 259, 92–99, 1988.

[256] Kikkawa, Y., and Sato, T. Elastic Properties of the Lens, *Exp Eye Res*, 2, 210–215, 1963.

[257] Ejiri, M., Thompson, H. E., and O'Neill, W. D., Dynamic Visco-Elastic Properties of the Lens, *Vision Res*, 9, 233–244, 1969.

[258] Bullimore, M. A., and Gilmartin, B., Aspects of Tonic Accommodation in Emmetropia and Late Onset Myopia, *Am J Opt Physio Opt*, 64, 499–503, 1987.

[259] Wallman, J. Gottlieb, M. D., Rajaram, V., and Wentzel, L., Local Retinal Regions Control Local Eye Growth and Myopia, *Science*, 237, 73–77, 1987.

[260] Nyquist, G. W., Rheology of the Cornea: Experimental Techniques and Results, *Exp Eye Res*, 7(2), 183–188, 1968.

[261] Boyce, B. L., Jones, R. E., Nguyen, T. D., and Grazier, J. M., Stress-Controlled Viscoelastic Tensile Response of Bovine Cornea, *J Biomechanics*, 40, 2637–2376, 2007.

[262] Soergel, F., Jean, B., Seiler, T., Bende, T, Mücke, S., Pechold, W., and Pels, L., Dynamic Mechanical Spectroscopy of the Cornea for Measurement of Its Viscoelastic Properties in vitro, *German J Ophthalmol*, 4, 151–156, 1995.

[263] Elbaum, R., Zaltzman, L., Burgert, I., and Fratzl, P., The Role of Wheat Awns in the Seed Dispersal Unit, *Science*, 316, 884–886, 2007.

[264] Iatridis, J. C., Weidenbaum, M., Setton, L. A., and Mow, V. C., Is the Nucleus Pulposus a Solid or a Fluid? Mechanical Behaviors of the Nucleus Pulposus of the Human Intervertebral Disc, *Spine*, 21, 1174–1184, 1996.

[265] Ohshima, H., Tsuji, H., Hirano, N., Ishihara, H., Katoh, Y., and Yamada, H., Water Diffusion Pathway, Swelling Pressure, and Biomechanical Properties of the Intervertebral Disc during Compression Load, *Spine*, 11, 1234–1244, 1989.

[266] Setton, L. A., Mow, V. C., and Howell, D. S., The Mechanical Behavior of Articular Cartilage in Shear is Altered by Transection of the Anterior Cruciate Ligament, *J Orthop Res*, 11, 228–239, 1993.

[267] Zhu, W. B., Chern, K. Y., and Mow, V. C., Anisotropic Viscoelastic Shear Properties of Bovine Meniscus, *Clin Orthop*, 306, 34–45, 1994.

[268] Ogston, A. G., Stanier, J. E., Toms, B. A., and Strawbridge, D. J., Elastic Properties of Ox Synovial Fluid, *Nature*, 165, 571, 1950.

[269] Safari, M., Bjelle, A., Gudmundsson, M., and Högfors, G. Clinical Assessment of Rheumatic Diseases using Viscoelastic Parameters for Synovial Fluid, *Biorheology*, 27, 659–674, 1990.

[270] Fam, H., Bryant, J. T., and Kontopoulo, M., Rheological Properties of Synovial Fluids, *Biorheology*, 44, 59–74, 2007.

[271] Bettelheim, F. A., and Wang, T., Dynamic Viscoelastic Properties of Bovine Vitreous, *Exp Eye Res*, 23, 435–441, 1976.

[272] Goss, S. A., Johnson, R. L., and Dunn, F., Comprehensive Compilation of Empirical Ultrasonic Properties of Mammalian Tissues, *J Acoust Soc Am*, 64, 423–457, 1978.

[273] Thomas, R. J., Wood: Formation and Morphology in *Wood Structure and Composition*, M. Lewin and I. S. Goldstein, eds., New York: Marcel Dekker, 1991.

[274] Wert, C. A., Weller, M., and Caulfield, D., Dynamic Loss Properties of Wood, *J Appl Phys*, 56, 2453–2458, 1984.

[275] Amada, S., and Lakes, R. S., Viscoelastic Properties of Bamboo, *J Mat Sci*, 32, 2693–2697, 1997.

[276] Kelley, S. S., Rials, T. G., and Glasser, W. G., Relaxation Behaviour of the Amorphous Components of Wood, *J Mat Sci*, 22, 617–624, 1987.

[277] Salmén, L., Viscoelastic Properties of *in situ* Lignin under Water Saturated Conditions, *J Mat Sci*, 19, 3090–3096, 1984.

[278] Lu, R., and Puri, V. M., Characterization of Nonlinear Creep Behavior of Two Food Products, *J Rheology*, 35, 1209–1233, 1991.

[279] Lakes, R. S., and Quackenbush, J., Viscoelastic Behaviour in Indium Tin Alloys over a Wide Range of Frequency and Time, *Philosophical Mag Let*, 74, 227–232, 1996.

[280] Schoeck, G., Bisogni, E., and Shyne, J., The Activation Energy of High Temperature Internal Friction, *Acta Metallurgica*, 12, 1466–1468, 1964.

[281] Nowick, A. S., and Berry, B. S., *Anelastic Relaxation in Crystalline Solids*, New York: Academic, 1972.

[282] Kaiser, J., Untersuchungen über das Auftreten Gerauschen beim Zugversuch, *Arkiv für das Esienhüttenwesen*, 24, 43–45, 1953.

[283] Tatro, C. A., A Welder's Introduction to Acoustic Emission Technology, 1–9, in *Acoustic Emission*, R. W. Nichols, ed., London: Applied Science Publ, 1976.

[284] Matthews, J. R., and Hay, D. R., Acoustic Emission Evaluation, in *Acoustic Emission*, J. R. Matthews, ed., New York: Gordon and Breach, 1983, 1–14.

[285] Fitzgerald, E. R., Audiofrequency Vibrations and Gravity-Wave Detectors, *Nature*, 5485, 638–640, 1974.

[286] Wilson, J. P., Mechanics of Middle and Inner ear, *British Med Bull*, 43, 821–837, 1987.

[287] Jaglinski, T., and Lakes, R. S., Anelastic Instability in Composites with Negative Stiffness Inclusions, *Philosoph Mag Let*, 84, 803–810, 2004.

[288] Lakes, R. S., Lee, T., Bersie, A., and Wang, Y. C., Extreme Damping in Composite Materials with Negative Stiffness Inclusions, *Nature*, 410, 565–567, 2001.

[289] Jaglinski, T., Stone, D. S., Kochmann, D., and Lakes, R. S., Materials with Viscoelastic Stiffness Greater than Diamond, *Science*, 315, 620–622, 2007.

[290] Leaderman, H., Elastic and Creep Properties of Filamentous Materials and Other High Polymers, Washington, DC: Textile Foundation, National Bureau of Standards, 1943.

8

---

# Causal Mechanisms

## 8.1 Introduction

### 8.1.1 Rationale

The treatment of viscoelasticity thus far has dealt with phenomena, measurement of material properties, prediction of response to various load histories, and stress analysis. In this chapter we consider the physical causes of viscoelastic response. Study of causal mechanisms is motivated by (1) the desire for scientific understanding; (2) the intention to gain ability to choose or tailor materials with specified viscoelastic properties; and (3) the utility of viscoelastic measurements as a probe of microphysical processes that are causally linked to viscoelasticity. The conceptually simplest of the causal mechanisms are developed in detail here. These processes can occur in crystalline materials including metals, ceramics as well as in polymers. Examples of damping due to fluid–solid interaction in porous materials according to the Biot theory are given in §7.5, biological tissue. Viscoelasticity in tissue also results from stress induced motion at the molecular scale. The treatment is intended to be introductory, not exhaustive.

### 8.1.2 Survey of Viscoelastic Mechanisms

There are many causal mechanisms [1–3] responsible for viscoelastic response; several of these are as follows. Mechanisms indicated by a * are called *fundamental mechanisms* because they occur even in an ideal perfect single crystal, and they are not removable, even in principle. Many effects are named for the person who discovered them or who provided a clear analysis of the physical processes responsible for them. The word relaxation is used in these names to refer to viscoelasticity in a general sense, not necessarily a specific stress relaxation experiment. An overview of representative behavior and causal mechanisms is given for polymers in Figure 8.1 and for crystalline materials including metals in Figure 8.2.

A common feature of causal mechanisms is that strain depends not only on macroscopic stress but also on an "internal variable" associated with microscopic

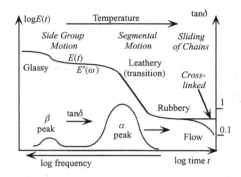

Figure 8.1. Representative viscoelastic behavior due to molecular motions in amorphous polymers.

processes or coupling to another field variable [2]. Suppose, as we have considered in §2.6 that the strain $\epsilon$ depends on the stress $\sigma$ and upon an internal variable $\xi$. The following general coupled-field formulation pertains to a variety of relaxation processes associated with internal variables.

(1) Atomic and molecular processes
  • Relaxation by motion of solute atoms (Snoek, bcc metal)
  • Relaxation by motion of solute atoms (Gorsky, strain gradient)
  • Relaxation by dislocation motion
    – Bordoni, in fcc metal; low temperature
    – Amplitude dependent, Granato–Lücke
    – Frequency dependent, Granato–Lücke
    – High-temperature "background"
  • Relaxation by molecular rearrangement (in polymers)
  • Relaxation by atom pair reorientation (Zener, alloys)
  • Relaxation by diffusion of atoms (high temperature in metals)
  • Relaxation via phase transition (sharp dependence on temperature)
  • Relaxation by electron viscosity (ultrasonic)
  • Relaxation by point-defect motion
(2) Coupled field effects
  • Thermoelastic relaxation* (in any material with thermal expansion)
  • Relaxation by fluid flow (porous materials with fluid in interstices)

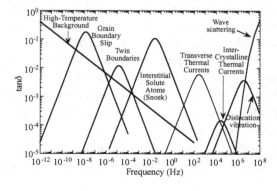

Figure 8.2. Representative viscoelastic behavior due to various processes in crystalline solids. In general, not all processes are active.

- Piezoelectric relaxation (in piezoelectric materials)
- Magnetoelastic relaxation (in magnetic materials)
- Phonon–phonon interactions* (in all materials)
- Electron–phonon interactions (in metals)

(3) Heterogeneous relaxation
  - Relaxation in composite materials (multiphase structured materials)
  - Relaxation by grain boundary slip (polycrystalline metals or ceramics)
  - Relaxation by fluid flow (porous materials with fluid in interstices)

### 8.1.3 Coupled Fields

As discussed in §2.6.2, viscoelasticity arises due to coupled fields in which stress $\sigma$ or strain $\epsilon$ is coupled to a field variable $\xi$, called an internal variable; $\Xi$ is another field variable. $J$ is a compliance and $\kappa$, $K$ and $\psi$ are material constants. Here, $K$ relates stress to an internal variable; it is not the bulk modulus,

$$\epsilon = J\sigma + \kappa\xi \tag{8.1}$$

$$\Xi = K\sigma + \psi\xi. \tag{8.2}$$

The relaxation strength is determined as follows. The compliance $J$ is considered to be at a constant value of $\xi$, so call it $J_\xi$. If the variable $\xi$ relaxes to zero or a constant with time, $J_\xi = J(\infty) = J$. Determine the compliance $J_\Xi$ at constant $\Xi$ by taking the differential $d\Xi$ and setting it equal to zero, $0 = Kd\sigma + \psi d\xi$. Take the differential of Equation 8.1, solve for $d\xi$ and substitute; solve for $d\epsilon$, and divide by $d\sigma$ to obtain $J_\Xi = J_\xi - K\kappa/\psi$. The relaxation strength $\Delta = [J(\infty) - J(0)]/J(0)$ is the change in compliance divided by the compliance at time zero (the smaller value, $J_\Xi = J(0)$).

$$\Delta = \frac{K\kappa}{J_\Xi\psi}. \tag{8.3}$$

This can be simplified if the free energy is a perfect differential with $U$ as internal energy, so $d(U - \sigma\epsilon - \xi\Xi) = -\epsilon d\sigma - \Xi d\xi$, so $\partial\epsilon/\partial\xi|_\sigma = \partial\Xi/\partial\sigma|_\xi$. So, $\kappa = K$.

Time dependent behavior enters via a diffusion process in which the rate of change of an internal variable $\xi$ is proportional to its deviation from its equilibrium value $\xi_{eq}$; $\tau$ is a time constant.

$$\frac{d\xi}{dt} = -\frac{(\xi - \xi_{eq})}{\tau}. \tag{8.4}$$

The transient solution is exponential in time. Coupling with the strain field gives rise to exponential time dependence in creep or relaxation, covering about one decade (a factor of ten) in time. This corresponds to a Debye peak in damping, about one decade (a factor of ten) in frequency. In the case of thermoelastic coupling or fluid flow, diffusion of $\xi$ to a surface leads to an equilibrium value of zero, because $\xi$ refers to stress induced temperature difference and pressure difference respectively.

Internal variables can be coupled to each other and to stress and strain in several ways.

For creep, a step stress of magnitude $\sigma_0$ is applied. Substitution of $\xi(t) = \xi_0\sigma_0(e^{-t/\tau})$, a solution of Equation 8.4, in Equation 8.1, with $\xi_0$ as a constant to be determined, gives the creep behavior,

$$\epsilon(t) = \sigma_0 J[(1 - \frac{J_\Xi}{J}\Delta e^{-t/\tau})]. \tag{8.5}$$

Rigorous analysis involves explicit treatment of boundary conditions on both the mechanical quantities and on the internal variables, leading to a discrete set of time constants depending on the shape of the object under study. The first term usually dominates in such cases. The foregoing is relevant to the large class of mechanisms for exponential relaxation. Not all mechanisms have this character.

Multiple internal variables will give rise to a discrete set of relaxation times. Some systems which exhibit a broad band of damping over many orders of magnitude in frequency are thought to be governed by mechanisms which are intrinsically nonexponential.

If a field variable (e.g., temperature, fluid pressure, electric field, magnetic field) is accessible to macroscopic influence, then mechanical behavior of the material, including damping, can be controlled by design. Some *smart materials* are based on access to coupled field variables. Examples are given in §10.7 and in Example 8.6.

## 8.2 Thermoelastic Relaxation

In this section, let us consider in some detail the relaxation due to stress induced heat flow, since it is one of the easiest to understand of the viscoelastic mechanisms. It is a coupled field effect in which strain is related to another field variable, temperature in this case, as well as to stress. The damping depends on the coefficient of thermal expansion, which is a *continuum* property having its ultimate origin in a slight nonlinearity of the interatomic force.

### 8.2.1 Thermoelasticity in One Dimension

The one-dimensional coupled field relations for thermoelasticity are as follows, with strain as $\epsilon$, stress as $\sigma$, $J$ as compliance, temperature change as $\Delta T$, entropy change as $\Delta S$, thermal expansion as $\alpha$, and $C^\sigma$ as heat capacity at constant stress,

$$\epsilon = J\sigma + \alpha\Delta T \tag{8.6}$$

$$\Delta S = \alpha\sigma + \frac{C^\sigma}{T}\Delta T. \tag{8.7}$$

Time dependence arises from thermal diffusion in which stress-generated temperature differences relax with time.

### 8.2.2 Thermoelasticity in Three Dimensions

As a result of thermoelastic coupling, the stiffness of an *elastic* material is different depending on whether loading is accomplished slowly (isothermal case) or rapidly (adiabatic case). The difference in stiffness governs the relaxation strength. It is found to depend on the coefficient of thermal expansion.

Consider [4] a unit volume of *elastic* material with strain $\epsilon_{ij}$ and entropy $S$ as dependent variables, and stress $\sigma_{ij}$ and absolute temperature $T$ as independent variables:

$$d\epsilon_{ij} = \frac{\partial \epsilon_{ij}}{\partial \sigma_{kl}}\Big|_T d\sigma_{kl} + \frac{\partial \epsilon_{ij}}{\partial T}\Big|_\sigma dT, \tag{8.8}$$

$$dS = \frac{\partial S}{\partial \sigma_{kl}}\Big|_T d\sigma_{kl} + \frac{\partial S}{\partial T}\Big|_\sigma dT. \tag{8.9}$$

$(\partial \epsilon_{ij}/\partial \sigma_{kl})$ represents elasticity, $(\partial \epsilon_{ij}/\partial T)$ represents thermal expansion, $(\partial S/\partial \sigma_{kl})$ represents the piezocaloric effect in which heat is generated in response to stress, and $(\partial S/\partial T)$ represents heat capacity. In linear materials, the elasticity equations, allowing for temperature changes, become [4],

$$\epsilon_{ij} = J^T_{ijkl}\sigma_{kl} + \alpha_{ij}\Delta T, \tag{8.10}$$

$$\Delta S = \alpha_{ij}\sigma_{ij} + \frac{C^\sigma}{T}\Delta T, \tag{8.11}$$

in which $J^T_{ijkl}$, sometimes called $S^T_{ijkl}$, is the elastic compliance tensor at constant temperature, $\alpha_{ij}$ is the thermal expansion tensor, and $C^\sigma$ is the heat capacity per unit volume at constant stress. In the isotropic case, $\alpha_{ij} = \alpha\delta_{ij}$ and Equation 8.10 is equivalent to the following elementary form [7] of Hooke's law, in engineering notation, allowing for thermal expansion:

$$\epsilon_{xx} = \frac{1}{E}(\sigma_{xx} - \nu\sigma_{yy} - \nu\sigma_{zz}) + \alpha\Delta T, \tag{8.12}$$

$$\epsilon_{yy} = \frac{1}{E}(\sigma_{yy} - \nu\sigma_{xx} - \nu\sigma_{zz}) + \alpha\Delta T, \tag{8.13}$$

$$\epsilon_{zz} = \frac{1}{E}(\sigma_{zz} - \nu\sigma_{xx} - \nu\sigma_{yy}) + \alpha\Delta T. \tag{8.14}$$

The compliance $J^T_{ijkl}$ is the isothermal compliance and is the compliance actually measured in an elastic material under deformation which is slow enough that any heat generated via the piezocaloric effect has time to flow, equalizing the temperature. The elastic compliance is different under deformation, which is sufficiently fast that this heat has no time to diffuse (adiabatic condition, $dS = 0$). To calculate the adiabatic compliance, set $dS = 0$ in Equation 8.9 and combine Equations 8.8 and 8.9 to eliminate $dT$.

$$d\epsilon_{ij} = \frac{\partial \epsilon_{ij}}{\partial \sigma_{kl}}\Big|_T d\sigma_{kl} - \frac{\left[\frac{\partial \epsilon_{ij}}{\partial T}\big|_\sigma \frac{\partial S}{\partial \sigma_{kl}}\big|_T\right]}{\frac{\partial S}{\partial T}\big|_\sigma}d\sigma_{kl}. \tag{8.15}$$

Now $(\partial \epsilon_{ij}/\partial T)_\sigma = (\partial S/\partial \sigma_{kl})_T$ since these derivatives can be expressed in terms of a thermodynamic potential function by virtue of the first and second laws of thermodynamics [4]. Dividing both sides of Equation 8.15 by $d\sigma_{kl}$ to obtain the adiabatic compliance $(\partial \epsilon_{ij}/\partial \sigma_{kl})_S$,

$$\frac{\partial \epsilon_{ij}}{\partial \sigma_{kl}}|_S - \frac{\partial \epsilon_{ij}}{\partial \sigma_{kl}}|_T = -\frac{\partial \epsilon_{ij}}{\partial T}|_\sigma \frac{\partial \epsilon_{kl}}{\partial T}|_\sigma \frac{\partial T}{\partial S}|_\sigma. \tag{8.16}$$

If the material is linear, Equation 8.16 becomes

$$J_{ijkl}^S - J_{ijkl}^T = -\alpha_{ij}\alpha_{kl}\frac{T}{C^\sigma} \tag{8.17}$$

giving a relaxation strength

$$\boxed{\Delta_{ijkl} = \frac{J_{ijkl}^T - J_{ijkl}^S}{J_{ijkl}^S} = \frac{\alpha_{ij}\alpha_{kl}}{J_{ijkl}^S}\frac{T}{C^\sigma}}. \tag{8.18}$$

The adiabatic compliance $J_{ijkl}^S$ differs from the isothermal compliance $J_{ijkl}^T$, and the difference depends on the thermal expansion and on the heat capacity. The absolute temperature is always positive. The heat capacity is positive in most systems (though it can be negative in stars and star clusters). If, as is usual, the thermal expansion coefficient is also positive, all the adiabatic compliances are smaller hence the adiabatic stiffnesses are larger than the isothermal ones. In isotropic materials, the thermal expansion is diagonal with all elements equal.

The §8.21, 8.22 deals with elastic behavior only. However, there is a difference in stiffness between fast and slow processes. The adiabatic compliance is determined in fast processes since in a material deformed rapidly there is insufficient time for heat generated by stress to flow. The isothermal compliance is determined in slow processes, such that heat generated by deformation has sufficient time to flow and equalize the temperature throughout the material. We know from the Kramers–Kronig relations discussed in §3.3 that such dispersion of the stiffness is always associated with viscoelastic loss.

### 8.2.3 Thermoelastic Relaxation Kinetics

Dynamics of thermoelastic relaxation are governed by conduction of heat generated by stress changes. Dissipation of mechanical energy in a cyclic load history with heat flow is illustrated in Figure 8.3. The history consists of three portions. First, the object is loaded slowly at constant temperature (isothermally). It is then unloaded adiabatically, too rapidly for heat flow to occur, with the slope of the stress–strain curve as the adiabatic modulus which differs from the isothermal modulus as shown in the above section. The object is then held at constant stress, and it exchanges heat with the environment. Thermal expansion occurs, so the strain changes. Mechanical energy is dissipated in this cycle, since there is a nonzero area enclosed by the load history. The loss tangent refers to sinusoidal loading that differs from the above, however the general principles are similar.

Figure 8.3. Cyclic stress–strain ($\sigma-\epsilon$) history with temperature changes allowed. Conversion of mechanical energy into thermal energy via the thermal expansion and piezocaloric effects. Compare with a similar cyclic history in poroelastic materials in Figure 8.7 and in piezoelectric materials in Figure 8.32. Energy densities $W$ are shown as shaded areas. Material is loaded slowly at constant temperature, then unloaded rapidly at constant entropy (adiabatically), which causes a temperature change, finally allowed to thermally equilibrate at zero stress. Compliance is $J$.

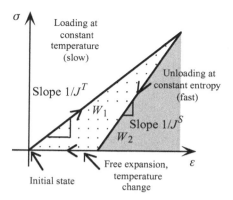

Relaxation due to thermoelastic coupling was derived by Zener [1, 5, 6]. The thermoelastic Equation 8.10 is specialized to one dimension, with the isothermal compliance tensor replaced by a number: $J^T_{ijkl} \to J_T$, as in §8.2.1.

Under adiabatic conditions, an increase in length results in a decrease in temperature, for $\alpha > 0$. Zener considers temperature changes due to adiabatic deformation and thermal diffusion separately; one can also use the approach in §8.1.3 with the one-dimensional thermoelastic equations in §8.2.1.

For a Debye peak (due to a single relaxation time process), assuming a small relaxation strength,

$$\tan\delta = \Delta\frac{f/f_0}{1 + (f/f_0)^2},\tag{8.19}$$

so that the peak loss at the characteristic frequency $f_0$ (associated with a relaxation process, not resonance) is $\tan\delta_{\text{peak}} = \Delta/2$.

$$\Delta = \frac{\alpha^2 T}{C_v J^S},\tag{8.20}$$

with $T$ as the absolute temperature and $C_v$ as the heat capacity at constant volume. This is the one-dimensional version of Equation 8.18. In most treatments of thermoelastic relaxation, the thermal expansion coefficient itself has been considered not to relax.

For a reed of thickness $d$, vibrating in bending [5, 6], the critical frequency $f_0$ is

$$f_0 = \frac{\pi}{2} D d^{-2},\tag{8.21}$$

and for a vibrating circular rod of radius $r$,

$$f_0 = 0.539 D r^{-2},\tag{8.22}$$

with $D$ as the thermal diffusion coefficient, $D = k/C_v$, with $k$ as the thermal conductivity and $C_v$ as the heat capacity per unit volume. Referring to Equation 8.5, $2\pi f_0 = 1/\tau$. More sophisticated analysis of reed vibration discloses a set of characteristic frequencies, but most of the damping is in the lowest (first) mode given above; the second mode has about 0.012 of the damping of the first. The one-dimensional

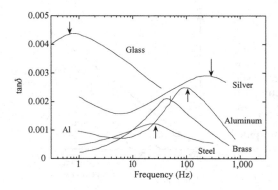

Figure 8.4. Tanδ in several materials in flexure experiments by Bennewitz and Rötger [19] (adapted from Zener [1]). The peaks correspond to the predicted peaks (frequencies indicated by arrows) for thermoelastic damping. The frequency for maximum damping depends on the specimen thickness.

analysis of Zener has been extended to three dimensions by Alblas [10, 11]. The specimen *shape* will affect the damping because specimen boundaries influence the rate of heat flow.

### 8.2.4 Heterogeneity and Thermoelastic Damping

Thermoelastic relaxation is present whenever there is heterogeneity of dilatational stress [6]. The heterogeneity can arise in the type of vibration, as in bending vibration of reeds (Figure 8.4). Heterogeneity of stress also is present if the material has cavities, discrete phases, or anisotropic crystallites with random orientation. There is also a homogeneous thermoelastic relaxation governed by heat flow between the specimen and the environment.

As for cavities [6], the heterogeneous stress around them gives rise to a thermoelastic damping of small magnitude, over a distribution of frequency. For cavities of volume fraction $v$, in a material with Poisson's ratio $\nu$, the fraction $\mathcal{R}$ of strain energy associated with fluctuations in dilatation multiplies the available relaxation strength and is given by

$$\mathcal{R} = \frac{10v}{1764} \frac{(1-2\nu)(1+\nu)}{(1-\frac{5}{7}\nu)^2}. \tag{8.23}$$

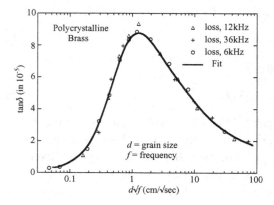

Figure 8.5. Tanδ in polycrystalline brass (adapted from Randall, Rose, and Zener [20]). The damping arises from intercrystalline thermal currents. Frequency is denoted by $f$ and grain size by $d$. Symbols: o, $f = 6$ kHz, Δ, $f = 12$ kHz; +, $f = 36$ kHz.

Table 8.1. *Predicted peak thermoelastic tan$\delta$ for inclusions in a matrix.*
*One-dimensional inclusions (adapted from Milligan and Kinra [9]*

| Inclusion | Mg | Matrix Al | In | Sn | Zn | Fe |
|---|---|---|---|---|---|---|
| $Al_2O_3$ | 0.0077 | 0.0031 | 0.0047 | 0.0093 | 0.0073 | $1.6 \cdot 10^{-5}$ |
| SiC | 0.0091 | 0.0049 | 0.0038 | 0.014 | 0.008 | 0.0002 |
| Graphite | 0.0091 | 0.0046 | 0.0063 | 0.010 | 0.0087 | 0.0005 |

For representative Poisson's ratios, the ratio involving Poisson's ratio is close to 1. The damping due to thermoelastic damping associated with cavities is small, since the volume fraction $v$ of cavities must be less than 1 and is often substantially less than 1. Cracks [12] also give rise to thermoelastic damping due to heterogeneous stress.

As for thermal currents between randomly oriented anisotropic crystals in a polycrystalline material, the corresponding values of $\mathcal{R}$ have been calculated from crystal anisotropy of several cubic crystals. Results are 0.031 for Cu, 0.0009 for Al, 0.022 for Fe, and 0.065 for Pb, therefore the associated damping is small. Even so, damping due to this source can comprise most of the total damping in some materials such as brass (69% Cu, 31% Zn), as shown in Figure 8.5.

Thermoelastic damping in *composite* materials arises due to the heterogeneity of the thermal and mechanical properties of such materials, leading to heat flow between constituents, hence, mechanical energy dissipation. The damping depends on the specific phase geometry as well as the constituents involved. The reason is that damping depends on heterogeneity of dilatational stress, and on the nature of the boundary value problem under consideration. Composites modeled as one dimensional inclusions [9], laminates [13], laminates with perfect and imperfect thermal interfaces [14], and composites with particulate inclusions, have been studied via the second law of thermodynamics. Predicted damping for one dimensional inclusions is proportional to $(\alpha_1/\rho_1 c_1 - \alpha_2/\rho_2 c_2)^2$, with $\rho$ as density and $c$ as specific heat per unit mass. Selected results are shown in Table 8.1, Table 8.2, and Table 8.3.

Thermoelastic damping has been examined in connection with the theory of thermodynamics [16–18]. Some minor corrections to the bending analysis of Zener were given as a result [16].

Table 8.2. *Predicted peak thermoelastic tan$\delta$*
*for inclusions in a matrix. Laminate,*
*uniform stress [14]*

| Inclusion | Matrix Mg | Matrix Al |
|---|---|---|
| $Al_2O_3$ | $8 \cdot 10^{-5}$ | 0.0004 |
| SiC | 0.0012 | 0.0008 |

Table 8.3. *Predicted peak thermoelastic* tanδ
*for inclusions in a matrix. Spherical
particulate inclusions [15]*

| Inclusion | Mg | Matrix Al |
|---|---|---|
| $Al_2O_3$ | 0.0035 | 0.0018 |
| SiC | 0.0067 | 0.0045 |

### 8.2.5 Material Properties and Thermoelastic Damping

For aluminum the relaxation strength is $\Delta = 0.0046$, for iron $\Delta = 0.0024$, for magnesium $\Delta = 0.005$, and for most metals $\Delta < 0.01$, so the loss due to thermoelastic effects is usually small [6]. However, the total loss in most metals used for structural purposes is also usually small. Therefore, in some frequency ranges, the thermoelastic effect can account for most of the loss in common structural metals, as has been demonstrated experimentally [8]. Table 8.4 shows a longer list of thermoelastic properties of some representative materials after Milligan and Kinra [9].

### 8.3 Relaxation by Stress-Induced Fluid Motion

#### 8.3.1 Fluid Motion in One Dimension

We remark that the Biot theory of stress-induced fluid flow has the same mathematical structure as the theory of thermoelasticity. It is a coupled field formulation. The Terzhagi [21] theory of fluid flow is a one-dimensional special case of the

Table 8.4. *Properties of some materials. Young's modulus* E, *density* $\rho$, *thermal conductivity*
k, *heat capacity per unit mass* c, *thermal expansion* $\alpha$ *at room temperature (adapted from
Milligan and Kinra [9]). Calculated relaxation strength* $\Delta$ *for thermoelastic damping*

| Solid | $E(GPa)$ | $\rho$ ($10^3$ kg/m$^3$) | $k(J/smK)$ | $c(J/kgK)$ | $\alpha(10^{-6}/K)$ | $\Delta = E\alpha^2 T/c\rho$ |
|---|---|---|---|---|---|---|
| Al | 70 | 2.7 | 222 | 900 | 23.6 | 0.0048 |
| Cu | 110 | 8.96 | 394 | 380 | 16.5 | 0.0026 |
| In | 11 | 7.51 | 23.9 | 240 | 33.0 | 0.0020 |
| Fe | 197 | 7.87 | 75.4 | 460 | 11.8 | 0.0023 |
| Mg | 44 | 1.74 | 154 | 1030 | 27.1 | 0.0054 |
| Ni | 207 | 8.9 | 92 | 440 | 13.3 | 0.0028 |
| Sn | 43 | 7.3 | 628 | 230 | 23.0 | 0.0041 |
| Ti | 116 | 4.51 | 38.9 | 519 | 8.41 | 0.0011 |
| Zn | 103 | 7.13 | 113 | 383 | 39.7 | 0.0178 |
| $Al_2O_3$ | 350 | 3.8 | 29.3 | 840 | 9.0 | 0.0027 |
| C | 379 | 2.25 | 23.9 | 691 | −0.90 | 0.00006 |
| SiC | 460 | 3.26 | 90 | 1330 | 4.30 | 0.0006 |
| TiC | 350 | 4.5 | 30.3 | 840 | 7.0 | 0.0009 |

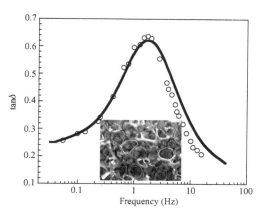

Figure 8.6. Damping in a polyurethane foam with air in the pores (adapted from Gent and Rusch [34]). The solid line represents theory, and the points are experimental. The inset shows a representative structure of a different open cell foam with communicating pores.

Biot theory, which is three dimensional. The one-dimensional version may be expressed as follows:

$$\epsilon = J\sigma + \kappa_{\text{Biot}}P, \tag{8.24}$$

$$\Xi = K\sigma + \psi_{\text{Biot}}P. \tag{8.25}$$

The variables are $\Xi = \zeta$ as increment of fluid content, $P$ as fluid pressure, $J$ as compliance at constant fluid pressure or drained compressibility, $\kappa_{\text{Biot}} = 1/H$ as volume change per pore pressure change, called poroelastic expansion coefficient, $K = 1/H_1$ relates fluid volume change to stress at constant fluid pressure, and $\psi_{\text{Biot}} = 1/\mathbf{R}$, called the specific storage coefficient, relates fluid volume to fluid pressure at constant stress. The first of the symbols for each parameter corresponds to symbols used in the above generic presentation of coupled fields; the second symbol corresponds to terminology used in geology [22].

Flow $q$ of a fluid of viscosity $\eta$ the $z$ direction in a porous medium [22] of permeability $k_p$ is governed by Darcy's law $q = (k_p/\eta)dP/dz$. Drag due to stress induced flow gives rise to time or frequency dependent behavior which is manifested as viscoelasticity in the bulk porous material.

An example of damping in a porous material, a cellular solid, due to fluid–solid interaction is shown in Figure 8.6 for a polyurethane foam of a Young's modulus about 50 kPa, with air in the pores. The theory used for comparison is simpler than the Biot theory. As for the peak damping, see Example 9.4. Soft biological tissues with fluid in the pores also exhibit damping due to stress-induced fluid flow; many examples are shown in Chapter 7.

### 8.3.2 Biot Theory: Fluid Motion in Three Dimensions

In a porous elastic material under stress, motion of fluid (such as air or water) within the pores generates viscous drag which gives rise to relaxation. This is a coupled-field type mechanism in which the stress and strain in the solid phase are coupled to the fluid pressure and fluid volume change. Biot and colleagues developed a

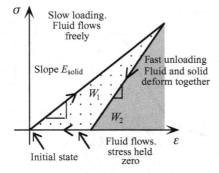

Figure 8.7. Cyclic stress–strain ($\sigma - \epsilon$) history with fluid flow, showing energy dissipation. Compare with a similar cyclic history in thermoelastic materials in Figure 8.3 and in piezoelectric materials in Figure 8.32.

substantial understanding of this process [23–28]. The relations between stress $\sigma$ and strain $\epsilon$ including the effect of fluid pressure $P$ may be written [23] as follows as a generalization of the elementary isotropic form for Hooke's law. The solid phase is in this treatment assumed to be elastic,

$$\epsilon_{xx} = \frac{1}{E}\{\sigma_{xx} - \nu\sigma_{yy} - \nu\sigma_{zz}\} + \frac{1}{3H}P, \tag{8.26}$$

$$\epsilon_{yy} = \frac{1}{E}\{\sigma_{yy} - \nu\sigma_{xx} - \nu\sigma_{zz}\} + \frac{1}{3H}P, \tag{8.27}$$

$$\epsilon_{zz} = \frac{1}{E}\{\sigma_{zz} - \nu\sigma_{yy} - \nu\sigma_{xx}\} + \frac{1}{3H}P, \tag{8.28}$$

$$2\epsilon_{xy} = \frac{\sigma_{xy}}{G}, 2\epsilon_{xz} = \frac{\sigma_{xz}}{G}, 2\epsilon_{yz} = \frac{\sigma_{yz}}{G}. \tag{8.29}$$

Here, $E$ is Young's modulus, $G$ is the shear modulus, $\nu$ is Poisson's ratio, and $H$ is a physical constant with dimensions of stress. $H^{-1}$ is a measure of the compressibility of the porous solid for a change in fluid pressure. One may write the above as follows in the tensorial formulation, with $J_{ijkl}$ as the compliance tensor:

$$\epsilon_{ij} = J_{ijkl}\sigma_{kl} + \frac{1}{3H}\delta_{ij}P. \tag{8.30}$$

The volume change $\Theta$ of fluid content is given, with $\sigma_{kk}$ as the trace of the stress tensor, by

$$\Theta = \frac{1}{3H}\{\sigma_{kk}\} + \frac{1}{\mathbf{R}}P. \tag{8.31}$$

$\mathbf{R}^{-1}$ is a physical constant representing the change in fluid content for a given change in fluid pressure. These have the same form as Equations 8.10 and 8.11 that describe thermoelastic coupling. Moreover, the treatment of damping due to fluid–solid interaction is analogous to the treatment of damping due to thermoelastic coupling, as indicated schematically in Figure 8.7.

The stress–strain relations can also be written in the modulus formulation as

$$\sigma_{xx} = 2G\{\epsilon_{xx} + \frac{\nu}{1-2\nu}(\epsilon_{xx} + \epsilon_{yy} + \epsilon_{zz})\} - AP, \tag{8.32}$$

or, in the Lamé formulation for isotropic materials,

$$\sigma_{ij} = 2G\epsilon_{ij} + (\lambda + A^2 Q_{\text{Biot}})\epsilon_{kk}\delta_{ij} - A Q_{\text{Biot}}\zeta\delta_{ij} \tag{8.33}$$

with $\zeta$ as the fluid content, $G$ as the shear modulus, and

$$A = \frac{2(1+\nu)}{3(1-2\nu)}\frac{G}{H}. \tag{8.34}$$

Since the bulk modulus $B$ of the solid skeleton without fluid is

$$B = \frac{2G(1+\nu)}{3(1-2\nu)}, \tag{8.35}$$

one can write

$$A = \frac{B}{H}. \tag{8.36}$$

The coefficient $Q_{\text{Biot}}^{-1}$ is a measure of how much fluid can be forced, into the material if it is at constant volume. It has nothing to do with the quality factor; hence the subscript Biot.

$$\frac{1}{Q_{\text{Biot}}} = \frac{1}{\mathbf{R}} - \frac{A}{H}. \tag{8.37}$$

The above coefficients have been related [24] to the following observable parameters. The *porosity* $\mathbf{f}$ is the volume fraction of void space. The *jacketed compressibility* $\kappa_{\text{Biot}}$ is the ratio of strain in the solid to the fluid pressure applied externally, with internal fluid pressure held constant. To measure $\kappa_{\text{Biot}}$, a specimen is surrounded by a thin impermeable jacket and fluid pressure in then applied. So, the compressibility $\kappa_{\text{Biot}}$ is the inverse of the bulk modulus $B$, of a dry specimen. The *unjacketed compressibility* $\delta_{\text{Biot}}$ is the ratio of strain in the solid to fluid pressure when the fluid completely penetrates the pores. The coefficient $\gamma$ of fluid content is given by an experiment in which fluid is injected into a fluid-filled chamber with and without the porous specimen. The difference in volume $\Delta V$ is given by $\Delta V = \delta_{\text{Biot}} + \gamma - c_{\text{fluid}}$, with $c_{\text{fluid}}$ as the fluid compressibility. If the porous material is isotropic, macroscopically homogeneous and fully saturated, then

$$\gamma = \mathbf{f}(c_{\text{fluid}} - \delta_{\text{Biot}}). \tag{8.38}$$

The interrelations are as follows [24, 28].

$$Q_{\text{Biot}} = \frac{\mathbf{f}(1 - \mathbf{f} - \frac{\delta_{\text{Biot}}}{\kappa_{\text{Biot}}})}{\gamma + \delta_{\text{Biot}} - \frac{\delta_{\text{Biot}}^2}{\kappa_{\text{Biot}}}}, \tag{8.39}$$

$$\mathbf{R} = \frac{\mathbf{f}^2}{\gamma + \delta_{\text{Biot}} - \frac{\delta_{\text{Biot}}^2}{\kappa_{\text{Biot}}}}, \tag{8.40}$$

$$\delta_{\text{Biot}} = (1 - A)\kappa_{\text{Biot}}, \tag{8.41}$$

$$H^{-1} = \kappa_{\text{Biot}} - \delta_{\text{Biot}}. \tag{8.42}$$

The treatment is entirely elastic thus far. Time dependent behavior is demonstrated by considering the flow of fluid in response to stress, as follows.

The flow rate vector $\mathbf{F}$ of fluid in response to a gradient of pressure $P$ is governed by Darcy's law,

$$\mathbf{F} = -k_p \nabla P, \tag{8.43}$$

with $k_p$ as the permeability. Supposing the fluid content $\Theta$ is related to the flow by

$$\frac{\partial \Theta}{\partial t} = \text{div}\mathbf{F}, \tag{8.44}$$

then combining with the generalized three-dimensional Hooke's law in Equation 8.26,

$$k_p \nabla^2 P = A \frac{\partial}{\partial t} \{\epsilon_{xx} + \epsilon_{yy} + \epsilon_{zz}\} + \frac{1}{Q_{\text{Biot}}} \frac{\partial}{\partial t} P. \tag{8.45}$$

Relaxation due to fluid flow depends on the geometry of the object and the surface boundary conditions. As a particular case [23], consider a slab of thickness $h$ of material confined on the edges, with fluid free to escape from one flat surface. To that end, the above poroelastic equations (in the modulus formulation) may be converted with the aid of the equilibrium conditions $\frac{\partial \sigma_{ij}}{\partial x_j} = 0$, into the following differential equation involving the displacement field $u_j$.

$$G \nabla^2 u_j + \frac{G}{1 - 2\nu} \frac{\partial \epsilon_{kk}}{\partial x_j} - A \frac{\partial P}{\partial x_j} = 0, \tag{8.46}$$

which, with Equation 8.45, may be specialized to one spatial dimension,

$$\frac{1}{a} \frac{\partial^2 u_z}{\partial z^2} - A \frac{\partial u_z}{\partial z} = 0, \tag{8.47}$$

$$k_p \frac{\partial^2 P}{\partial z^2} = A \frac{\partial^2 u_z}{\partial z \partial t} + \frac{1}{Q_{\text{Biot}}} \frac{\partial P}{\partial t}, \tag{8.48}$$

with $a$ as the final compressibility

$$a = \frac{1 - 2\nu}{2G(1 - \nu)}. \tag{8.49}$$

This is the inverse of $C_{1111}$ for the solid phase. Physically, $C_{1111}$ is the constrained modulus for the solid phase under compression, with the Poisson effect laterally constrained; see Example 5.9. An initial compressibility is defined in [23],

$$a_i = \frac{a}{1 + A^2 a\, Q_{\text{Biot}}}. \tag{8.50}$$

So, the relaxation strength, from the following definition (see Equation 2.52),

$$\Delta = \frac{J(\infty) - J(0)}{J(0)}, \tag{8.51}$$

with the compliance $J$ as a compressibility, is

$$\Delta = A^2 a\, Q_{\text{Biot}}. \tag{8.52}$$

If a step load of magnitude $p_0$ is applied, the *creep* displacement as a function of time $t$ is

$$u_z = hp_0\{a - \frac{8}{\pi^2}(a - a_i)\sum_{n=0}^{\infty}\frac{1}{(2n+1)^2}\exp[-(2n+1)^2\frac{t}{\tau}], \qquad (8.53)$$

with

$$\tau = \frac{4h^2}{\pi^2 k_p}\{A^2 a + \frac{1}{Q_{\text{Biot}}}\}. \qquad (8.54)$$

The time constant $\tau$ for creep depends inversely as the permeability $k_p$ and increases as the square of the slab thickness $h$. The first term in the summation is much larger than subsequent terms, so the creep is approximately of single exponential form. Therefore, the corresponding tan$\delta$ response is dominated by a Debye peak. In this example, creep results from squeezing of fluid through a surface of a compressed slab. Results will differ for other boundary conditions. For example, a uniform specimen in torsion experiences no volume change; hence no driving force for fluid flow; hence no creep due to this mechanism. This example is analogous to the case of thermoelastic damping due to exchange of heat with the environment, considered in §8.2. If the porous medium is heterogeneous on a scale larger than that of the porosity, there can be stress-induced flow of fluid between regions in the material, analogous to the intercrystalline thermal currents associated with thermoelastic coupling.

The Biot equations assume a particularly simple form for cellular solids such as low-density polymer foams for which the porosity $\mathbf{f} \approx 1$ and the unjacketed compressibility $\delta_{\text{Biot}}$ (of the solid phase) is much less than the jacketed compressibility $\kappa_{\text{Biot}}$ (of the foam). Then the relaxation strength for compression of a constrained layer is expressed in terms of the bulk moduli,

$$\Delta = \frac{B_{\text{fluid}}}{B_{\text{foam}}}\frac{1}{3}\frac{1+v}{1-v}. \qquad (8.55)$$

Poisson's ratio $v$ of the foam appears because the constrained layer compression is not a pure bulk deformation.

Introduction of anisotropy [24, 29] in the analysis permits application to problems involving bedded rock or cartilage. More sophisticated theory incorporates viscoelasticity of the solid phase, bubbles in the fluid, and better modeling of the porous structure [27, 28]. As for elastic waves, there is the possibility of dynamic relative motion between the solid and fluid phases, and this results in two kinds of compressional waves [25, 26] with different speed and different attenuation. Such waves have been observed in synthetic porous media [30, 31], and in bone [32].

Fluid flow in porous materials is important in civil engineering in understanding the consolidation of soils [23, 33]. Observe that the time scale of the creep or relaxation, expressed as the time constant $\tau$, increases as the square of the specimen size. Therefore the coupled-field equations are used directly in such problems. Fluid flow also is important in determining the mechanical and transport properties of biological materials such as cartilage, muscle, and bone, as well as of synthetic polymer foams. The size dependence of the time constant was used to infer an important role for fluid flow in the viscoelasticity of muscle, as described in §7.5.5.

Relaxation due to fluid flow has been studied [34, 35] in the context of polymer foams. The analysis, in common with that of Biot, predicts a peak in the loss tangent at a frequency, which depends on the foam slab thickness. For materials that can be idealized as an array of tubes of diameter $d$, the permeability is $k_p = \frac{d^2}{32}$. The predicted peak in the loss tangent and transition from low to high stiffness in $E'$ covers little more than one decade of frequency scale. Large loss tangents of unity or greater are possible via this mechanism. The analysis can be improved, for example, by allowing for the compressibility of the fluid as required to properly treat gases such as air. Poroelastic theory was used successfully in modeling a damping peak (Figure 8.6) of magnitude 0.6 due to air flow in a polyurethane foam.

In some polymer foams used for wheelchair cushions, bed pads, and spacecraft seats, the foam is made viscoelastic and the orifice between cells in the foam is made narrow [36]. In large blocks of foam under compression, the slow escape of air contributes significantly to the viscoelastic properties. In foams, flow of air can give rise to large values of tan$\delta$.

## 8.4 Relaxation by Molecular Rearrangement

This mechanism is the most relevant to polymers [51, 52], which consist of macromolecules; refer to observed behavior of polymers given in §7.2.

In recent studies of molecular mechanisms in polymers, regions of elevated energy states or "kinks" in long chain molecules are considered to [53, 54] oscillate at frequency $f_0$ in Brownian motion and can rearrange to overcome an energy barrier, of height $\Delta_e$. The probability of movement depends on temperature so that the transition rate is $\lambda_0 = e^{-\Delta_e/kT}$. Power law relaxation and fractional derivative representations can arise from this analysis [54].

Power law and other nonexponential relaxation relations have been attributed to quantum mechanical hopping processes [57–59]. Relaxation following the stretched exponential (discussed in §2.6) is thought to arise from interactions among a hierarchy of energy levels [60].

### 8.4.1 Glassy Region

The region of short times and low temperatures corresponds to the *glassy* region of behavior of amorphous polymers (Figure 8.1). The polymer backbones have little freedom of movement, nevertheless peaks in tan$\delta$ may be observed. In the glassy region there is a broad band of relatively small damping (much less than in the transition region) with superposed peaks in tan$\delta$. Tan$\delta$ may be as small as $3 \times 10^{-3}$, for example, in polystyrene. The peak at the lowest temperature (or highest frequency), called the $\gamma$ peak, is thought to be a result of flexing or twisting of segments of the main polymer chain. At a higher temperature, a $\beta$ peak may be observed.

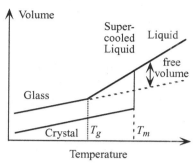

Figure 8.8. Glass transition temperature $T_g$ and melting temperature $T_m$ in polymers.

Viscoelasticity in this region arises from limited local molecular motion of side groups. Small peaks superposed on a nearly constant loss tangent in this region are called secondary maxima.

The $\alpha$ and $\beta$ peaks in glassy polymers have different activation energies. Therefore, if both peaks are present in a frequency window, thermorheological complexity will be observed [61].

### 8.4.2 Transition Region

In the *transition* region, large relaxation of a factor 100 to 10,000 occurs in the stiffness as a result of configurational rearrangement of the polymer chains. The loss tangent in this region exhibits a very large peak, called the $\alpha$ peak, of magnitude from 1 to 10.

The general shape of the viscoelastic functions in the transition region is similar among polymers, but the temperature at which it occurs differs widely. For most polymers the transition from glassy to rubbery consistency occurs within about 5 to 10°C. Viscoelastic properties in the transition region are dominated by rearrangement of molecular segments. These segments are sufficiently short that entanglements of molecules or cross-links between them play a relatively minor role. Differences in molecular weight or its distribution do not affect the behavior much. Molecules with side chains tend to exhibit the transition at progressively higher frequency for constant temperature (the glass transition temperature $T_g$ becomes less at constant frequency) as the side chain length increases. Many amorphous polymers are found to be thermorheologically simple, that is, a change in temperature is equivalent to a shift of the viscoelastic property curves along a log time or log frequency axis. This behavior implies that in similar deformations at different temperatures, the same sequence of molecular events (leading to viscoelasticity) occurs; only the speed changes with temperature [62]. The WLF equation (Equation 2.80), which describes the temperature dependence, may be obtained from concepts of free volume [63, 64]. The plot of volume per unit mass versus temperature for a typical amorphous polymer exhibits a discontinuity in *slope* at the glass transition temperature $T_g$, as shown in Figure 8.8. The additional volume within the polymer above the glass transition is considered to be free volume, or voids on the molecular scale. The fractional free volume $f_v = v_f/v$ may be written as follows:

$$f_v = f_g + \alpha_f(T - T_g), \qquad (8.56)$$

in which $T$ is temperature, $f_g$ is the fractional free volume at the glass transition temperature $T_g$, and $\alpha_f$ is the coefficient of expansion of the free volume. Suppose that each relaxation mode has a time constant $\tau = \eta/E$, in which the viscosity $\eta$ gives rise to a shift factor,

$$a_T(T) = \frac{\eta_T}{\eta_{T_g}}. \qquad (8.57)$$

Suppose, moreover that the viscosity is governed by Doolittle's equation for the viscosity of monomeric liquids, supported by experiment,

$$\eta = a \exp[bf_v/v_f], \qquad (8.58)$$

with $a$ and $b$ as constants. Combining Equations 8.57 and 8.58,

$$\ln a_T(T) = b[\frac{1}{f_v} - \frac{1}{f_g}]. \qquad (8.59)$$

Incorporating Equation 8.56,

$$\log a_T = \frac{\frac{b(T-T_g)}{2.303 f_g}}{\frac{f_g}{\alpha_f} + (T - T_g)}, \qquad (8.60)$$

which is the WLF equation.

The glass transition temperature $T_g$ is fixed for a given polymer and frequency; nevertheless, one can control $T_g$ by adding various constituents, for example, a plasticizer may be added to reduce $T_g$. Plasticizers contain small molecules which penetrate the polymer and allow molecular chain segments greater mobility. For example [65], polyvinyl chloride (PVC) is a hard plastic with $T_g \approx 80°C$; it is commonly used for water pipes. Plasticized PVC is flexible, has $T_g$ just below room temperature, and is used for shower curtains and seat covers. Particulate inclusions or "fillers" tend to raise $T_g$ slightly because free volume is reduced in the vicinity of the interface. This effect falls off with distance from the interface, therefore the transition is broadened as well as elevated in temperature. Broad transitions can be achieved by combining immiscible polymers [65] with different transition temperatures. If the free energy of mixing is close to zero, the heterogeneous structure is on a very small distance scale on the order 10 nm. In that case, much of the material is very close to an interface, and the result is a single broad transition rather than two individual transitions.

Time temperature superposition in the transition and rubbery regions may make use of vertical shifts of the modulus in addition to the usual horizontal shifts on the log time axis. According to the Rouse model (see Ferry [51]) the modulus is multiplied by a factor $b_T$:

$$b_T = \frac{T_0 \rho_0}{T \rho}. \qquad (8.61)$$

The ratio of temperature $T$ represents an effect of entropy based restoring force in the flexible chains, and the ratio of density $\rho$ represents an effect of thermal expansion. This modulus shift factor is, in some but not all cases, only weakly dependent on temperature; it is often neglected [66]. In a survey of experimental evidence of thermorheological complexity in polymers [67], a coupling model has been considered in explaining these effects in the transition and terminal regions.

### 8.4.3 Rubbery Behavior

Rubbery materials can be stretched to as much as 10 times their original length; they fully recover to the original length after the force is removed. Amorphous polymers, both natural and synthetic, behave in a rubbery fashion at temperature sufficiently higher than the glass transition point. Natural rubber is a polymer of isoprene, $(C_5H_8)_n$, derived from a kind of tree; synthetic rubbers, called elastomers, are derived from petroleum. Rubbery materials behave as solids under prolonged force provided the molecules are cross-linked. Practical rubber is vulcanized to achieve cross-linking.

In the *plateau* region of rubbery behavior, associated with long times or sufficiently high temperature, the behavior is governed by entanglements of the long chain molecules. Rearrangements on this scale occur comparatively slowly. The molecular weight of the polymer chains is important in this region as seen in the plots shown in §7.2. The asymptotic or equilibrium modulus in the *terminal* region depends on the density of cross-links and on the molecular weight. Polymers without cross-links behave as liquids for sufficiently long times, because the molecular chains can then slide over each other given enough time or thermal activation by high temperature.

### *Rubber Elasticity*

The elasticity of rubber is attributed to entropy rather than energy [68]. Analysis via thermodynamics is done in the same vein as the thermoelastic coupled field formulation presented in §8.2. Here, we consider axial stretch in one dimension in the modulus formulation with $E$ as a Young's modulus, rather than the compliance formulation. Variables are strain $\epsilon$, entropy $S$ (for a unit volume of material), stress $\sigma$ (rather than force $F$, done in [68]), and, absolute temperature $T$,

$$d\sigma = \frac{\partial \sigma}{\partial \epsilon}|_T d\epsilon + \frac{\partial \sigma}{\partial T}|_\epsilon dT. \tag{8.62}$$

The quantity $\partial \sigma / \partial \epsilon$ represents an elastic modulus. To interpret the second term, consider, for a unit volume of material, the Helmholtz energy $H_e = U - TS$ with $U$ as internal energy. The stress is given by $\sigma = \partial H_e / \partial \epsilon$. Since mixed partial derivatives are equal, $\partial^2 H_e / \partial \epsilon \partial T = \partial^2 H_e / \partial T \partial \epsilon$, then $\partial \sigma / \partial T|_\epsilon = -\partial S / \partial \epsilon|_T$. So,

$$d\sigma = \frac{\partial \sigma}{\partial \epsilon}|_T d\epsilon - \frac{\partial S}{\partial \epsilon}|_T dT. \tag{8.63}$$

The second term is the entropy contribution [68] to the stress and the first term is the energy contribution. Experiments [69] on rubber show that stress decreases with temperature at constant strain below 0.07, and stress increases with temperature for larger strains; stress at constant strain is linear with temperature and the slope increases with extension strain. The entropy term predominates at large extension strain above about 0.07. For small strain, the entropy term actually reduces the stress, hence the modulus, owing to the sign of the stress–temperature effect.

As for further interpretation, recall from §8.3.1, $\epsilon = J\sigma + \alpha\Delta T$ with $J$ as compliance and $\alpha$ as thermal expansion. If the strain $\epsilon$ is set to zero, then

$$\frac{\partial\sigma}{\partial T}|_\epsilon = -\frac{\alpha}{J} = -\alpha E. \tag{8.64}$$

Therefore, the entropy term in Equation 8.63 is related to the thermal expansion. The stress versus temperature results referred to above consequently imply a positive thermal expansion at small strain and a negative thermal expansion of larger magnitude at high strain. For example at 3 percent extension, a temperature rise of 90°C gives rise to a stress decrease of about 14 kPa, but at 370 percent extension the same temperature rise causes a stress increase of about 500 kPa [69]. Using a Young's modulus $E$ of about 1.6 MPa obtained from a stress–strain curve in the same article, the above equation gives a thermal expansion $\alpha$ of about $10^{-4}/°C$ at a 3 percent extension. The accepted thermal expansion of rubber is about $2 \times 10^{-4}/°C$ at near zero extension [71]. The comparison is reasonable in view of the nonlinearity in temperature dependent stress, hence in expansion. As for the energy term in Equation 8.63, it contains contributions due to the energy change associated with the change of molecular conformation and also due to change in volume on deformation. The volume change for small deformation is small (Poisson's ratio is close to 1/2). No term for $PdV$ with $P$ as atmospheric pressure is included since $PdV$ is much less than $Fdx$ [68].

A rapidly stretched rubber band becomes hot. Any material that is deformed sufficiently rapidly that heat has insufficient time to flow (adiabatic) undergoes a temperature change called the piezocaloric effect as presented in §8.2. With a heat change $dQ = TdS$, the temperature change [69] is $\Delta T = -\frac{T}{c_\ell}\int_{\ell_0}^{\ell}\frac{\partial S}{\partial\ell}d\ell$ with $c_\ell$ as heat capacity per unit length $\ell$; the original analysis was presented by Lord Kelvin [70]. This effect can be large in rubber, up to a 10°C increase in temperature for an extension of about 500 percent, and a corresponding decrease in temperature during sudden retraction. The temperature change increases nonlinearly with strain.

The physical origin of the large thermal effects in rubber arises from the fact that the long molecular chains are in a state of disorder in undeformed rubber, but become ordered as the rubber is stretched. This molecular order corresponds to a reduction in entropy. The number of conformations of a molecular chain decreases as the chain is extended. Since entropy is a logarithmic function of conformation number, entropy decreases with extension. Conformation changes occur by free rotation about chemical bonds; many rotational states have essentially the same energy. In rubber, the long molecules are free to rearrange, except at the cross links. A similar relation between entropy and force occurs in an ideal gas compressed by a piston in a cylinder. Physically, the force results from reversal of motion of many atoms by the constraint provided by the piston and cylinder. The corresponding force in rubber [72] arises from the constraint of thermal motion of atoms in the rubber by covalent bonds in the polymer chain backbone. These constraints prevent the atoms from moving freely. Consequently the atoms exert forces on the chain, transmitted through it to the end atoms. The entropy view and the kinetic view of the force are therefore considered equivalent.

### Rubber Viscoelasticity

Viscoelastic relaxation in rubber over short periods of time at ambient temperature is modest, about 4 percent at ambient temperature from one minute to one hour for an extension of 150 percent [71] and appears asymptotic to a constant equilibrium stress. At 70°C, relaxation has an obvious flow component. In compressive creep studies from one minute to eight years [73] at 35°C, a 15 percent initial compression increased to 60 percent to more than 200 percent depending on rubber composition, with creep continuing over the entire period.

Viscoelasticity in rubber [51] is dominated by networks of molecular entanglements in which configurational motions of neighboring molecules are coupled. For a lightly cross-linked Hevea rubber, tan$\delta$ is about 0.01 over a range of frequency, and the equilibrium Young's modulus is $E \approx 0.7$ MPa. For a styrene-butadiene amorphous copolymer, more lightly cross-linked, the equilibrium Young's modulus is $E \approx 0.15$ MPa, tan$\delta$ is about 0.06 at low frequency; there is a peak with damping greater than 0.2 due to entanglement slippage. The effects of filler (particulate inclusions) are discussed in Chapter 9.

In the *plateau* region of rubbery behavior, associated with long times or high temperature, the behavior is governed by entanglements of the long-chain molecules. Rearrangements on this scale occur comparatively slowly. The molecular weight of the polymer chains is important in this region as seen in the plots shown in §7.2.

The asymptotic or equilibrium modulus in the *terminal* region depends on the density of cross links and on the molecular weight. Polymers without cross links behave as liquids for sufficiently long times, since the molecular chains can then slide over each other given enough time or thermal activation by sufficiently high temperature. Polyethylene in the terminal zone, 130 to 190°C at 10 Hz does not obey time temperature superposition [66] as a consequence of a different temperature dependence of different molecule types.

### 8.4.4 Crystalline Polymers

The occurrence of crystallinity in polymers requires a sufficiently regular chemical structure to allow chain molecules to pack regularly in a lattice [55]. Overall, one major effect of crystallinity in polymers is to broaden the glass-transition and reduce the maximum loss tangent. The lateral forces between the oriented molecular chains are weak, facilitating torsional oscillation of the molecules [51]. Macroscopic deformation alters the distribution of oscillations, following a time lag. This gives rise to viscoelastic loss. There are domains of both ordered and amorphous material in crystalline polymers. Moreover, the viscoelastic properties within lamellae and at lamellar interfaces can differ. The polymer can therefore be regarded as a composite. Polyethylene and polytetrafluoroethylene are examples of such semicrystalline polymers.

### 8.4.5 Biological Macromolecules

Materials of biological origin generally contain macromolecules that give rise to relaxations via mechanisms of the types observed in synthetic polymers. Several examples are given by Wert [3]. Biological materials also tend to have a complex composite structure (§9.6) in which the macromolecules are highly ordered.

### 8.4.6 Polymers and Metals

Similarities in the shape (not the magnitude) of the relaxation functions of metals and polymers have been interpreted by a cooperative theory [56] in which the solid is made up of structural units that behave as two level systems. Structural units interact by exchanging phonons (quanta of sound) giving rise to cooperative behavior. Molecular dynamics simulations of stress relaxation via this theory compare well with experiments.

## 8.5 Relaxation by Interface Motion

### 8.5.1 Grain Boundary Slip in Metals

In polycrystalline metals, the boundaries between crystals or grains (Figure 8.9) exhibit viscous-like behavior under prolonged load [1, 2]. The material in the grain-boundaries is amorphous and appears to be associated with viscous flow. Grain boundary slip is responsible for significant creep in metals at high temperatures and at long times. The creep or relaxation time varies directly with the grain size. At sufficiently high temperature (a significant fraction of the melting temperature), slip can occur sufficiently rapidly that it manifests itself as a peak in the loss tangent [74]; the loss at the peak can be on the order of 0.1 (see §7.3). The effect of grain boundary slip has been studied [74] in comparisons between polycrystalline metal and the same metal in single crystal form (Figure 7.11). If the grain size is comparable to the specimen size (macrocrystalline materials), the grain boundaries are called *bamboo boundaries* (§7.3). In macrocrystalline materials, the loss peak height is proportional to the number of grains; by contrast, if the grains are small, the loss peak height is independent of grain size. One can show theoretically that grain-boundary slip can cause a total relaxation in Young's modulus of 50 percent to 75 percent of its unrelaxed value [75]. More recent analysis models creep in metals in view of viscous

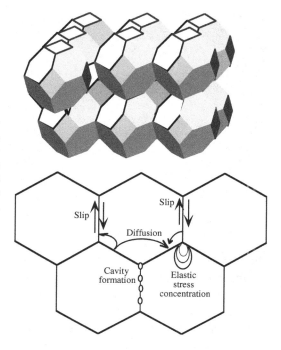

Figure 8.9. Schematic diagram of crystals (grains) in a polycrystalline material. Top: idealized tetrakaidecahedral structure of the grains. Bottom: slip motion at the grain boundaries in cross-section and how that motion can be accommodated (adapted from Crossman and Ashby [76]).

processes at the grain boundaries combined with power-law creep in the grain interior, leading to deformation mechanism maps [76] discussed in §8.9.1.

The slip between grains can be directly observed by scribing a line across a polished metal surface and then by microscopically observing offsets of the line at the grain boundaries as deformation proceeds [77]. Grain-boundary slip has been examined experimentally in eutectic Pb-62wt%Sn alloy by straining polished specimens within a scanning electron microscope [78]. As in earlier studies, grain-boundary slip was determined by measuring the offset of scribed marker lines. Grain-boundary slip along shear surfaces occurred in an heterogeneous manner, and the rate of grain-boundary slip changed with strain (at large strains) in a nonmonotonic manner. These observations were consistent with a model in which grain-boundary slip is considered a consequence of grain-boundary dislocations. Grain-boundary motion at high temperature in a magnesium alloy containing fine and coarse grains has been observed using interference optical microscopy [79]. These observations disclosed grain-boundary slip to predominate in the fine grains as a mechanism to accommodate large deformations of 7 percent. Cooperative movement of groups of grains has also been observed [80].

The grain boundaries are also important in plastic deformation of metals and in the damage that occurs under prolonged creep at high loads. Under such conditions, grain-boundary slip can be accommodated by elastic distortion of the grains, plastic flow of the grains, and by diffusion [76]. Diffusion of vacancies can lead to the formation of *cavities* at the boundaries or their junctures [78, 81]. In materials subjected to long-term high stress, the cavities can grow and coalesce, leading to fracture.

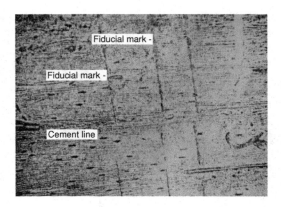

Figure 8.10. Cement line displacement in bone following long term creep in shear. Reflected light image showing offset of a fiducial mark at the cement line (horizontal) (from a specimen of the set studied in Lakes and Saha [83]).

### 8.5.2 Interface Motion in Composites

In composite materials, slip at viscous interfaces between constituents can give rise to mechanical damping. Specifically, in synthetic composites, fiber-matrix slip can cause damping. In experiments, a weak fiber-matrix bond was deliberately formed to encourage such slip [82]. In compact bone, which is a natural composite, fibers (in human bone) or laminae (in bovine bone) about 0.2 mm thick are separated by a layer of "cement substance," most likely mucopolysaccharides. Viscous-like motion at such cement lines (Figure 8.10) gives rise to significant long term creep in bovine bone [83] and in human bone [84].

### 8.5.3 Structural Interface Motion

Interface motion can also occur in rivets, bolted joints, and other interfaces in structures [85]. Effective *loss tangents* are associated with riveted or bolted thin sheet metal structures such as those used in automobiles are on the order of 0.01, even though the loss tangent of the metal itself may be 0.001 or smaller. Welded structures of thick plates, such as those used in ships exhibit effective "loss tangents" on the order of 0.001. Motion at joints may be accompanied by friction, but the amplitude of motion is so small that theories for dry friction do not apply. Much of the damping associated with joints in structures is thought to arise from the viscosity of air that is squeezed out of crevices as the structure vibrates.

## 8.6 Relaxation Processes in Crystalline Materials

Here we consider several processes that can give rise to viscoelastic loss in crystalline materials, particularly metals [2].

### 8.6.1 Snoek Relaxation: Interstitial Atoms

The *Snoek relaxation* [37, 86] refers to viscoelastic damping due to motion of interstitial solute atoms in metals with a body centered cubic (bcc) crystal structure.

Figure 8.11. Right: motion (arrow) of interstitial atom (dark circle) in a strained lattice of atoms (open circles), giving rise to a Snoek relaxation (adapted from Weller [40]). Left: motion (arrow) of an interstitial atom in a lattice with a strain gradient (shown as bending) giving rise to a Gorsky relaxation.

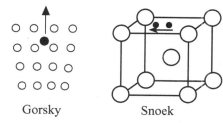

Gorsky       Snoek

As an example, carbon or nitrogen dissolved in iron can give rise to significant relaxation, up to $\tan\delta \approx 0.1$. Study of other metals and solutes disclosed that the relaxation kinetics depend strongly on temperature and on the particular alloy. Relaxation times of 0.1 to 1,000 sec are possible for temperatures between room temperature and 350°C. The effect can be observed in stress relaxation or as a phase angle in a sinusoidal experiment; it is still called the Snoek relaxation. Materials include metals such as Fe, Nb, Ta, Cr, V, and solute atoms include C, N, O. The interstitial atoms represent point defects which distort the crystal lattice. When a strain is applied to the lattice, the interstitial atoms tend to move a short distance (Figure 8.11). The motion is considered to be between sites in the lattice corresponding to octahedral symmetry, in the center of a cubic face, to a site corresponding to tetrahedral symmetry, though other short range motion may be considered [40]. Drag due to interaction of the stress-induced motion of the interstitial with thermal motion of nearby atoms gives rise to mechanical energy dissipation.

A steel tuning fork will vibrate for many cycles at room temperature, 20°C, but if the fork is warmed to about 70°C vibration is damped [38]; the damping becomes less at yet higher temperature. Snoek found that iron without any carbon or nitrogen does not exhibit this damping effect. Snoek peaks in iron with nitrogen at 0.2 Hz [37] and in niobium with 1.1 atomic percent oxygen, at a frequency of 1 Hz [39], are shown in Figure 8.12. Broadening (on the left) of the peak in niobium is attributed

Figure 8.12. Snoek damping versus temperature in iron with nitrogen adapted from [37], in niobium with oxygen (adapted from Gibala and Wert [39]), and in iron with 20 parts per million carbon (adapted from Weller [40]).

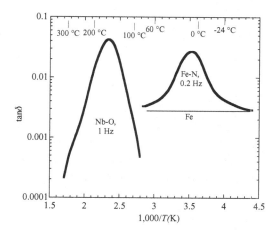

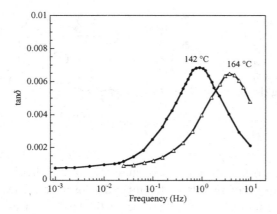

Figure 8.13. Snoek damping versus frequency, in niobium with oxygen (adapted from Woirgard, Sarrazin, and Chaumet [42]).

to contributions from clusters of several oxygen atoms. The baseline for the original Snoek study on iron is higher than that for niobium.

The theory of the Snoek relaxation is based on the analysis of point defect relaxation of Nowick and Berry [2]. The peak magnitude [40] is proportional to the concentration $c$ of solute atoms and to the inverse of the absolute temperature $T$, as follows; the frequency dependence follows a Debye peak [42], Figure 8.13, and the temperature dependence follows an Arrhenius form:

$$\tan\delta_{peak} \propto \frac{c}{T}; \qquad (8.65)$$

The time constant $\tau_r \propto a^2/D(T)$ with $D(T)$ as the (temperature dependent) diffusion coefficient and $a$ as the jump distance, which, in turn, is proportional to the interatom spacing of the crystal lattice.

Following Zener [1], the relaxation strength due to stress-induced preferential distribution of inclusions is determined as follows. Suppose that $n$ is the number of solute atoms per unit volume, and $dn$ is the excess number of such elements per volume in the preferred orientation. The strain $\epsilon$ depends on stress and $dn$. A free energy $U$ is associated with the preferential orientation. The unrelaxed modulus is $E_u$,

$$\epsilon = \frac{\sigma}{E_u} + [\frac{\partial \epsilon}{\partial (dn)}]_\sigma dn \qquad (8.66)$$

$$U = [\frac{\partial U}{\partial \sigma}]_{dn}\sigma + [\frac{\partial U}{\partial (dn)}]_\sigma dn. \qquad (8.67)$$

Here, the excess number of solute atoms $dn$ is an internal variable in the coupled field formulation. The excess of oriented atoms is considered to be thermally activated, with $T$ as absolute temperature and $k$ as Boltzmann's constant and $\beta$ a numerical coefficient of order unity, so,

$$dn = -\frac{\beta n}{kT}U. \qquad (8.68)$$

The relaxation strength $\Delta_E$ is evaluated via elimination of $dn$ and $U$.

$$\Delta_E = \frac{-E_u \frac{\partial \epsilon}{\partial (dn)} \frac{\partial U}{\partial \sigma}}{\frac{kT}{\beta n} + \frac{\partial U}{\partial (dn)}}. \tag{8.69}$$

The derivatives are estimated by consideration of the total energy $E$ per volume,

$$d(E - \epsilon \sigma) = -\epsilon d\sigma + U d(dn), \tag{8.70}$$

so,

$$[\frac{\partial U}{\partial \sigma}]_{dn} = -[\frac{\partial \epsilon}{\partial (dn)}]_\sigma. \tag{8.71}$$

Because stress and $dn$ influence $U$ via effect on strain, one may expect the changes to be of similar order of magnitude, so, with $q$ as a numerical coefficient of order of magnitude unity,

$$[\frac{\partial U}{\partial (dn)}]_\sigma = q[\frac{\partial U}{\partial \sigma}]_{dn}[\frac{\partial \sigma}{\partial \epsilon}]_{dn}[\frac{\partial \epsilon}{\partial n}]_\sigma. \tag{8.72}$$

Substituting Equation 8.71 and 8.72 into Equation 8.70,

$$\Delta_E = \frac{T_0}{T - qT_0}, \tag{8.73}$$

with (for $T > T_0$)

$$T_0 = \beta n E_u \frac{1}{k} [\frac{\partial \epsilon}{\partial (dn)}]^2 \tag{8.74}$$

as a characteristic temperature.

Equation 8.73 is analogous to the Curie–Weiss formulation of paramagnetism of magnetic materials. The equation for relaxation strength is valid provided the temperature exceeds the critical temperature for a spontaneous preferential distribution. Because $T_0$ is proportional to the concentration $n$, for small concentrations, the relaxation strength $\Delta_E$, hence the peak tan$\delta$ is proportional to $n$ of interstitial atoms, and goes as $1/T$ since $T_0$ becomes small. The Snoek relaxation satisfies these conditions.

The Snoek relaxation has been used in scientific studies of diffusion. Because the diffusion distance required to cause an observable relaxation is on the order of the lattice constant, effects can be observed at temperatures too low for macroscopic diffusion to occur over accessible time scales. Arrhenius behavior is observed over some 16 orders of magnitude in diffusion coefficient [40]. Since this relaxation is so sensitive to impurities, it has been used to determine the concentration of dissolved materials at levels too low to be studied by other techniques; the threshold concentration for detection is about one part per million [40]. The Snoek relaxation has been observed in high-temperature (77 K or greater) superconductor ceramics [41] to probe the mobility of oxygen atoms associated with the conductivity.

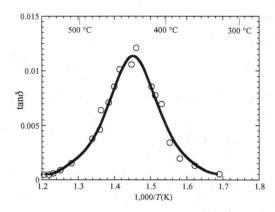

Figure 8.14. Zener tan$\delta$ in brass at 620 Hz (adapted from Zener [43]).

### 8.6.2 Zener Relaxation in Alloys: Pairs of Atoms

The *Zener relaxation* in solid solutions refers to reorientation of pairs of atoms in the alloy under stress [1, 87]. Results for brass (copper-zinc alloy), due to Zener [43], are shown in Figure 8.14. The baseline damping between 20 and 200°C is about $10^{-5}$. Zener hypothesized that under stress, pairs of atoms in the metal reorient into an energetically more favorable configuration by local diffusion. The activation energy for the anelastic relaxation is close to that for diffusion, which supports the hypothesis. The Zener relaxation and a grain-boundary peak in $\alpha$-brass is shown in Figure 7.11.

The theory assumes a relatively simple form only for a dilute concentration of solute; in that case tan$\delta$ goes as the square of the concentration. The loss peak due to the Zener relaxation is absent in pure elements. Concentrated solid solutions are more complex to understand, but experiments show a peak in tan$\delta$ as a function of concentration. The maximum loss tangent is usually less than 0.01, however some alloys, such as AgCd and AgZn, exhibit a greater loss [88]. Moreover, brass exhibits a loss peak of magnitude tan$\delta = 0.012$ near 400°C at 620 Hz. A large tan$\delta$ peak is seen in CdMg alloy described in §7.3; this alloy exhibits a loss maximum of about 0.14 at room temperature and near 1 Hz.

Solid solutions under some conditions may exhibit *lower* damping than the corresponding pure metals. Brass (CuZn) alloys with zero to 30 atomic percent zinc, at 12 kHz, near room temperature, exhibit tan$\delta$, which is approximately inversely proportional to concentration of zinc [89]. Here the temperature is too low for the Zener mechanism to be operative. The reduction in damping is attributed to the fact that dislocation mobility [90] depends on concentration.

The magnitude of the damping peak increases as the square of the concentration (Figure 8.15) [44] of alloying element, consistent with a concept based on pairs of atoms. For dilute alloys of solute concentration $c$, with [45], with $N_v$ as the number of atoms per volume and $K_z$ as a function of lattice spacings in different directions,

$$\tan\delta_{\text{peak}} \propto \frac{K_z c^2}{N_v k T}. \tag{8.75}$$

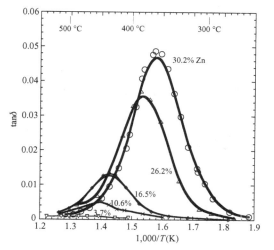

Figure 8.15. Zener tanδ in AgZn single crystal at 500 Hz as a function of temperature and of the concentration in atomic percent zinc (adapted from Seraphim and Nowick [44]).

The quadratic dependence on concentration is due to the fact that pairs of atoms are involved. The Zener relaxation has been observed in a variety of metals as well as in ionic crystalline materials. For concentrated systems, the damping peak versus frequency in an isothermal experiment increases in magnitude as one approaches a critical temperature $T_c$ for self-induced ordering [2, 46].

$$\tan\delta_{peak} \propto \frac{1}{(T - T_c)}. \tag{8.76}$$

Here, the critical temperature is proportional to the number of pairs of solute atoms, so $T_c \propto c^2$. The role of this ordering is discussed further in the context of phase transformations. For the Zener relaxation, the quantity $n$ in the above analysis of the Snoek relaxation is interpreted as the number of pairs of atoms, which may be oriented, rather than the number of atoms. Experiments with Ag-Zn alloys reveal that indeed the relaxation strength depends on temperature. Plots of inverse relaxation strength versus temperature exhibit straight lines with intercept corresponding to critical temperature at or below $-100°C$. The analysis has then been refined to allow multiple order parameters [47].

### 8.6.3 Gorsky Relaxation

The *Gorsky effect* refers to a relaxation due to motion of interstitial solute atoms (Figure 8.11) in a heterogeneous stress field such as that produced by bending [48]. The theory is analogous to that for thermoelastic damping in bending; the damping is predominantly a Debye peak. Since the effect depends on a stress gradient, the time constant for a rectangular specimen of thickness $d$ is

$$\tau_r = d^2/\pi^2 D(T), \tag{8.77}$$

with $D(T)$ as the diffusion coefficient, a function (typically an Arrhenius relation) of temperature $T$. Experimental observation over reasonable time or frequency scales

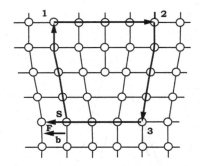

Figure 8.16. Edge dislocation (after D. S. Stone, University of Wisconsin, with permission). Atoms in the crystal lattice are represented by circles. The Burgers vector **b** points from the start S to the finish F of a path around the dislocation. The dislocation continues in the atoms in planes above and below the figure plane. Representation of dislocations as lines or loops refers to this continuation.

is facilitated by study of dissolved hydrogen, which diffuses relatively rapidly [49]; even so, a specimen 1.2 mm thick exhibited a relaxation time of 160 sec. The relaxation strength is proportional to the concentration $c$ of interstitial material and its temperature dependence goes as $(T - T_c)^{-1}$, with $T_c \propto cU$ in which $U$ as an interaction energy of the interstitial atoms. More recent experiments have used layers in which the thickness $d$ is small. Anelastic relaxation was observed [50] due to intercrystalline atomic diffusion, analogous to intercrystalline thermal diffusion, in a fine-grained polycrystalline alloy with crystals as small as 30 nm. Damping peaks of magnitude $1 - 3 \times 10^{-3}$ at frequencies in the range 150 Hz to 1 kHz and temperatures 250–500 K were observed.

### 8.6.4 Granato–Lücke Relaxation: Dislocations

*Dislocations* are line defects in crystals as shown in Figure 8.16. They experience drag as they move under stress, so they are responsible for relaxation or damping. The concept of a pinned dislocation loop oscillating under the influence of a dynamic applied stress leads to two components of loss, according to Granato and Lücke [91]. One of these depends on frequency and attains a peak in the MHz range, and the other is a hysteresis type, which is independent of frequency but dependent on strain. In the Granato–Lücke model, the material is assumed to contain a network of dislocations. There are two characteristic lengths in the model (Figure 8.17), the network length $L_n$ determined by the intersection of dislocation loops and the length $L_c$ determined by impurities. Under a small stress, the dislocation loops bow out until a break-away stress is reached at which dislocation strain increases substantially at constant stress. Under further increase of stress, new dislocation loops are created, giving rise to irreversible (plastic) strain. The frequency-dependent dislocation damping process depends upon oscillation of the dislocation loops. This is a resonant process, damped by drag upon the dislocations. Refinements of the Granato–Lücke theory include effects of point defect drag [92]. Even so, many aspects of damping in crystalline materials cannot be fully understood by this theory [93].

Damping that depends on amplitude but is independent of frequency is due to dislocations that break away from pinning points at higher stresses. Experiment shows damping to increase with amplitude [95, 96]. Representative behavior of a zinc crystal is shown in Figure 8.18. The internal friction gradually becomes less with

Figure 8.17. Diagram of dislocation loops (curved lines) impeded by impurity particles (small circles), (adapted from Granato and Lücke [91]). The curve of stress $\sigma$ versus strain $\epsilon$ shows the dislocation strain (with the elastic strain subtracted out) for a cycle of loading and unloading. If there is a distribution of loop lengths, the curve for loading becomes a smooth curve. If one continues to apply stress until new dislocation loops are formed, as in (g) in the upper diagram, the dislocation strain is irreversible and represents plastic deformation.

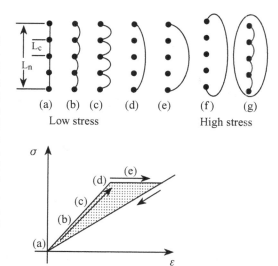

time, during room temperature annealing. This is explained as a result of mobility of dislocations resulting in their cancellation with time; positive and negative dislocations can merge. Frequency dependent damping with a peak at ultrasonic frequency is due to resonance of dislocations. An example [97] is shown in Figure 8.19. Dislocation motion is also responsible for plasticity in crystalline materials. Plasticity always shows a threshold effect, in contrast to linear viscoelasticity.

The theory of this frequency dependent relaxation [91] is based on the notion of the dislocation as a string which can vibrate with displacement $u$ in response to stress $\sigma$,

$$\frac{\partial^2 \sigma}{\partial x^2} - \frac{\rho}{G}\frac{\partial^2 \sigma}{\partial t^2} = \frac{\Lambda_L \rho a}{\ell}\frac{\partial^2}{\partial t^2}\int_0^\ell u\, dy, \tag{8.78}$$

$$A\frac{\partial^2 u}{\partial t^2} + B\frac{\partial u}{\partial t} - C\frac{\partial^2 u}{\partial y^2} = a\sigma. \tag{8.79}$$

Figure 8.18. Amplitude dependent decrement versus strain in a zinc crystal at 45 kHz, initially and after different periods of annealing at room temperature (adapted from Swift and Richardson [96]).

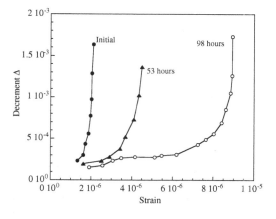

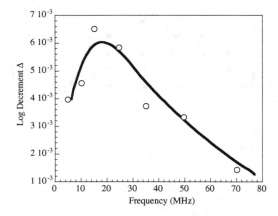

Figure 8.19. Frequency dependent decrement in copper at ultrasonic frequencies at room temperature (adapted from Alers and Thompson [97]).

The first of these is a *wave equation* in which the stress wave interacts with the dislocation motion. The second equation represents the displacement of the dislocation as that of a vibrating string, coupled to the stress field. Here, $y$ is a coordinate on the dislocation line, $\ell$ is the dislocation length, $\Lambda_L$ is a measure of the total mobile length of the dislocation line corresponding to a dislocation density, $A$ is mass per length, $B$ is damping force per length, $C$ is force per length due to the tension in a bowed out dislocation, $\rho$ is the material density, and $G$ is its shear modulus. The constants are given by $A = \pi \rho a^2$, $C = 2Ga^2/\pi(1-\nu)$, with $a$ as the Burger's vector and $\nu$ as the Poisson's ratio.

Consider a trial solution of the following form, with $k = \omega/v$, $\omega$ as angular frequency and $v = \sqrt{\dfrac{G}{\rho}}$ as a wave speed.

$$\sigma(x,t) = \sigma_0 e^{-\alpha x} e^{-i(kx-\omega t)}. \tag{8.80}$$

This leads to the following:

$$u = 4a\sigma \sum_{n=0}^{\infty} \frac{1}{2n+1} \sin\frac{(2n+1)\pi y}{\ell} \frac{e^{i(\omega t - \delta_n)}}{\sqrt{(\omega_n^2 - \omega^2)^2 + \omega^2 d^2}}, \tag{8.81}$$

$$d = \frac{B}{A}$$

$$\omega_n = (2n+1)\left(\frac{\pi}{\ell}\right)\sqrt{C/A}$$

$$\tan\delta_n = \frac{\omega d}{(\omega_n^2 - \omega^2)}. \tag{8.82}$$

This gives rise to dispersion (frequency dependence) of the velocity and attenuation of the waves. Specifically, the decrement (called $\Delta$ in the work cited, but also given the name $\Lambda$ and not to be confused with the relaxation strength) given in terms

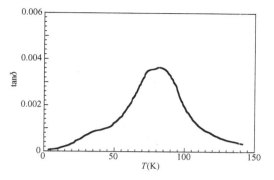

Figure 8.20. Bordoni $\tan\delta$ in copper at about 2 kHz over a range of low temperatures $T$ (adapted from Bordoni, Nuovo, and Verdini [99]).

of the attenuation $\alpha$ by $\Delta(\omega) = 2\pi\alpha/\omega$, so (neglecting the small higher terms in the series)

$$\Delta(\omega) = \frac{\Delta_0 \Lambda_L L^2}{\omega_0 (A/B)} \left[ \frac{\omega/\omega_0}{(1 - (\frac{\omega}{\omega_0})^2)^2 + (\frac{\omega}{\omega_0})^2 (\frac{B/A}{\omega_0})^2} \right]. \qquad (8.83)$$

Here, $\Delta_0 = 8Ga^2/\pi^3 C$ and $L$ is dislocation loop length. Given reasonable estimates of dislocation density, for example, $10^7$ cm/cm$^3$, and of the other parameters, the natural frequency $\omega_0$ exceeds 10 MHz. It turns out that the dislocation vibration is heavily damped, because experiments reveal a rather broad peak in overall damping.

The amplitude dependent dislocation damping is based on the concept of dislocations breaking away under applied stress, from pinning points distributed along their length. It is assumed that there is an exponential distribution of dislocation loop lengths. The resulting damping depends on strain amplitude $\epsilon_0$ as follows, with $C_1$ and $C_2$ as constants, which depend on the length and density of dislocations:

$$\tan\delta = \frac{C_1}{\epsilon_0} e^{-C_2/\epsilon_0}. \qquad (8.84)$$

More recent refinements [98] allow for movable pinning points and show that the notion of point defect drag can be considered a limiting case of the original vibrating string theory.

### 8.6.5 Bordoni Relaxation: Dislocation Kinks

The *Bordoni relaxation* [99] is due to dislocation kink motion in response to applied stress. The damping peak is observed at low temperature, as seen in Figure 8.20; the peak is several times wider than a Debye peak. The Bordoni peak appears only after cold work, and is absent in annealed material. Its magnitude increases with cold work up to about 3 percent permanent deformation and remains fairly constant for larger deformation. The Bordoni peak is essentially independent of strain amplitude, in contrast to other damping processes attributed to dislocations. Impurities tend to reduce the magnitude of the peak. The peak temperature for a fixed

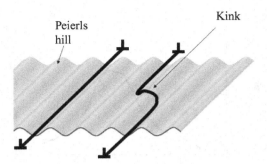

Peierls
hill

Kink

Figure 8.21. Kink in a dislocation (dark
line / curve) (after D. S. Stone, University
of Wisconsin, with permission).

frequency is independent of the amount of prestrain and of the concentration and
nature of impurities. The relaxation peak occurs at low temperature, usually one-
third of the Debye temperature (defined as the $hf_m/k$ with $h$ as Planck's constant,
$f_m$ as the maximum frequency of lattice vibration, and $k$ as Boltzmann's constant).
For example, in copper, a loss peak of width several times larger than a single relaxation time process, of magnitude $3 \times 10^{-3}$ occurred at a temperature of 72 K and at
1.9 kHz [94]. However, dislocation motion can contribute significantly to damping
at room temperature, too.

The observations are consistent with a dislocation mechanism. For example, a
dislocation segment along a Peierls valley is partly confined by the lattice but is
thought to move by nucleation of a pair of kinks (Figure 8.21), a process that is thermally activated [100]. The minimum energy configuration for a dislocation is for
it to be aligned with a symmetry axis of the crystal. The minimum stress needed to
lift a straight dislocation from one energy minimum to another is the Peierls stress
$\sigma_p$. Kinks in a dislocation carry an energy associated with the increase in potential
energy as the dislocation crosses from one valley of the energy landscape to another. A model (as a partial differential equation) of the motion of a dislocation
line was introduced by Seeger [100]. Early models predict the peak temperature to
depend on amplitude, which is not observed. More refined analysis of the Bordoni
relaxation following Marchesoni [101] models the dislocation as a string with motion $\phi(x, t)$ as a function of position $x$ and time $t$ coupled to a heat bath, such
that

$$\frac{\partial^2 \phi}{\partial t^2} - c_0^2 \frac{\partial^2 \phi}{\partial x^2} + \omega_0^2 \sin\phi = -\alpha_d \frac{\partial \phi}{\partial t} + \varphi(x, t), \qquad (8.85)$$

with $c_0$ and $\omega_0$ as parameters. The sinusoidal potential $\omega_0^2 \sin\phi$ describes the Peierls
valleys, which represent energy restraint of the lattice on the dislocation, in which
the dislocation moves and $\alpha_d$ refers to damping. The function $\varphi(x, t)$ represents
Gaussian noise with zero mean. The dislocation line is envisaged to be pinned at several points, such as $x = 0$ and $x = \ell$, with $m$ kinks. A kink, due to the perturbation
on the right side of Equation 8.85, has energy $E_0 = 8\omega_0 c_0$ subject to Brownian motion (due to coupling with heat bath) with the diffusion constant $D = (kT/E_0\alpha_d)c_0^2$.

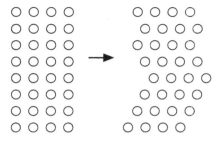

Figure 8.22. Change in crystal structure during a martensitic transformation. Circles represent atoms. Left: high-temperature phase. Right: Bands form in the low-temperature phase.

Assumption of the underdamped limit $\alpha_d \ll \omega_0$ is supported by both theory and experiment. To convert the normalized formulation to one with units appropriate to real dislocations, consider

$$\phi \rightarrow a\phi/2\pi, c_0^2 \rightarrow C_1/\pi p a^2, \tag{8.86}$$

$$\omega_0^2 \rightarrow 2\sigma_p b/\rho a^3, \alpha_d \rightarrow B_1/\pi\rho a^2, \tag{8.87}$$

with $a$ as the lattice constant, $b$ as the Burgers vector, $\rho$ as the material density, $B_1$ as the viscous force per length, $\sigma_p$ as the Peierls stress, and $C_1 = Gb^2$ as the tension force per unit dislocation length and $G$ as the shear modulus. An oscillatory external stress is applied to perform an internal friction experiment. The viscoelastic damping arises from drag of dislocation segments that have bowed out in response to the applied stress. The Bordoni peak occurs for $kT \ll E_0$ in which the Peierls potential dominates. The dislocation damping of Granato–Lücke corresponds to $kT \gg E_0$. Nucleation of thermal kink pairs and diffusion of kinks is then incorporated into the model, and the peak in internal friction is calculated.

The *Hasiguti relaxation* is attributed to interaction between point defects and dislocations [102, 103]. This relaxation occurs in metals that have been cold-worked. Loss peaks due to dislocation processes can be from 0.001 to 0.1.

### 8.6.6 Relaxation Due to Phase Transformations

*Crystal Structure and Phenomena*

Phase transformations include the conversion of a solid to a liquid during melting, as well as changes from one solid crystalline form to another. The phase transformation depends on temperature and pressure [104]. In a first-order transformation, the volume exhibits a step function and the heat capacity diverges at the critical transformation temperature $T_c$. Upon cooling through the critical temperature, the material breaks up into bands (Figure 8.22), visible as the twin crystals of the martensite, or low-temperature phase. In a second-order transformation, the volume changes continuously with temperature, and the heat capacity exhibits a finite peak. In a lambda transition, the peak in heat capacity versus temperature is symmetrical and resembles a Greek lambda.

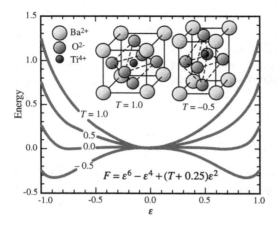

Figure 8.23. Landau [105] energy as it depends on normalized temperature $T$ and strain $\epsilon$ (adapted from Jaglinski et al. [108]). Inset, a unit cell of a crystal of barium titanate at two different temperatures. A candidate energy function $F$ is also shown.

In Landau's [105] theory of phase transformation, the energy of a crystal depends on temperature and strain. As temperature $T$ is lowered, an energy function of strain $\epsilon$ and temperature exhibits a single minimum and gradually flattens near a critical temperature $T_c$, Figure 8.23. The energy function may be expressed in terms of strain and temperature as follows: $g(\epsilon) = \epsilon^6 - \epsilon^4 + (T + T_c)\epsilon^2$ [107]. During further cooling, the curve then develops two minima or potential wells. If shear strain behaves in this way, the transformation is a martensitic one. The curvature of this energy function represents an elastic modulus. Flattening of the curve corresponds to a softening of the modulus near $T_c$. This phenomenon is observed experimentally. The energy function may also be expressed in terms of an internal variable called the *order parameter*, which is coupled to strain and represents an internal degree of freedom associated with the transformation.

Peaks in the mechanical damping occur in the vicinity of solid to solid phase transformations. Such damping is of interest in the context of applications of shape memory alloys and of piezoelectric ceramics. Also, in the context of attenuation of seismic waves in the Earth, mineral crystals in the Earth's hot interior may be in the vicinity of phase transformation temperatures.

In a martensitic transformation, a symmetric crystal structure is transformed to a less symmetric one, as illustrated in Figure 8.22. The low temperature phase (on the right) contains bands or domains. For example, in barium titanate, polycrystalline samples exhibit a complex microstructure of twin crystals, bands, or domains [106] (Figure 8.24) of width from about 10 $\mu$m to 200 nm or smaller. Single crystals can have domains as large as 0.1 mm. Single-domain single crystals on the millimeter scale can be prepared by applying a strong electric field to the crystal.

Phenomena observed in the internal friction associated with first-order martensitic transformations, are as follows:

(1) a peak in damping occurs near the transformation temperature, and the peak magnitude increases with increasing cooling rate,
(2) peak magnitude increases with decreasing test frequency, and

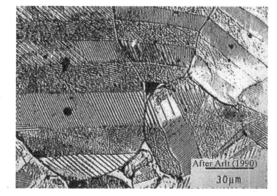

Figure 8.24. Domain structure in barium titanate; optical micrograph (from Arlt [106], with permission).

(3) hysteresis occurs between the location of the damping peak during transient heating and cooling [109, 110]. If a sufficiently large stress is applied, the material exhibits a nonlinear hysteresis in the stress–strain behavior accompanied by changes in the domain pattern [104]. Materials that exhibit such hysteresis (Figure 8.25) are called *ferroelastic*. Energy is dissipated in such a loop but it is not linearly viscoelastic or even weakly nonlinearly viscoelastic since there is a threshold stress. Similar hysteresis loops in magnetic behavior correspond to ferromagnetism, and in dielectric behavior, to ferroelectricity.

For example, barium titanate ($BaTiO_3$) is a piezoelectric, ferroelectric material used in ceramic form in capacitors and piezoelectric transducers. Barium titanate exhibits several phase transformations at different temperatures: from cubic above the Curie point 120°C to tetragonal as temperature is lowered, to orthorhombic below 5°C, to trigonal below −70°C. At each transition, there is a step change in volume, so the transitions are first order. Young's modulus [111] and damping [112, 113] are shown in Figure 8.26. Damping below the Curie point is attributed to interaction of domain walls and point defects [114].

$VO_2$ ceramic exhibits a first order transformation in which the structure changes from tetragonal to monoclinic during cooling. Experiments show a slight dip in shear modulus, and peaks in damping, up to about 0.05 (at a frequency of 0.5 Hz and a

Figure 8.25. Hysteresis of a ferroelastic crystal of $Pb_3(PO_4)_2$ (adapted from Salje [104]).

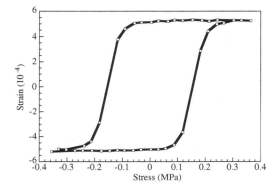

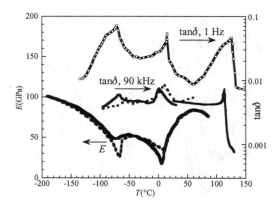

Figure 8.26. Modulus and damping for barium titanate ceramic in axial vibration at 90 kHz (adapted from [112]). Curves for heating and cooling are shown; heating is denoted by a dash curve. Damping at 1 Hz (adapted from Cheng et al. [113]), during heating at a rate 0.25°C/min is shown for comparison.

heating rate of 2.5 K/min), near the critical temperature 340 K. The peaks increase in magnitude with heating rate and decrease in magnitude with frequency [115].

*Cobalt* exhibits a first order martensitic transformation from face centered cubic to hexagonal close packed at about 420°C. Experimental studies near 1 Hz, 5 kHz, and 20 MHz reveal damping peaks near the transformation temperature [116]. The peak magnitude increases nonlinearly with temperature rate (to 0.06 at 2 K/min at 0.5 Hz) and decreases nonlinearly with frequency. An earlier model predicting linear dependence based on the notion of damping proportional to the amount of transformed phase per cycle is therefore insufficient. Even under nearly isothermal conditions, there is still a peak. The temperature-rate-dependent portion of the damping is 1,000 times smaller at 5 kHz than at 1 Hz.

As a further example, lanthanum aluminate ($LaAlO_3$), of interest in the geophysics community as a model material for minerals in the Earth, undergoes a cubic to rhombohedral phase transition below 550°C. In the low-temperature rhombohedral phase, a viscous motion of domain walls occurs. The onset of domain wall mobility in the vicinity of 200°C gives rise to a factor of 10 decrease in the storage modulus in comparison to the high-temperature cubic phase (super-elastic softening) and a large peak in damping [117], Figure 8.27. The strain in the experiments,

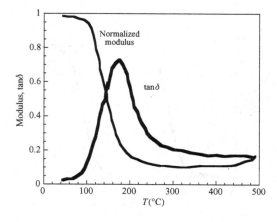

Figure 8.27. Modulus (normalized) and damping for a lanthanum aluminate crystal in bending versus temperature $T$ at 1 Hz (adapted from Harrison, Simon, and Redfern [117]). The stress amplitude is sufficiently large to unpin the domain walls.

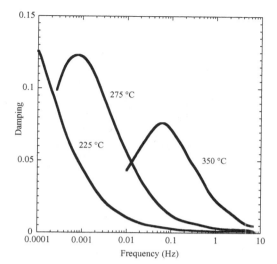

Figure 8.28. Zener tan$\delta$ in Au-Ni alloy versus frequency at versus temperatures (adapted from Mazot el al. [46]).

between 1 and $8 \times 10^{-5}$ is sufficient to release the domain walls from dislocations that restrain them; damping is much smaller at lower strain. Indeed, seismic waves develop strain in the range $10^{-8}$ to $10^{-6}$.

### Zener Relaxation: Order–Disorder

The Zener relaxation is considered to be an example of a second-order transformation called an *order-disorder transition* [109]. For example in $\beta$ brass, the body-centered cubic (bcc) lattice can be regarded as two interpenetrating simple cubic lattices. At low temperature, one of the lattices is preferentially occupied by one type of atom, for example, copper. As temperature rises, the ordering becomes gradually less; above $T_c$, the preferential order vanishes. In the theory for this class of transformation, the relaxation strength is highly sensitive to temperature, so it is necessary to measure the relaxation peak isothermally (at constant temperature), as a function of frequency [109].

For example, in a Au-Ni 30 percent alloy, the peak increases in magnitude and shifts to lower frequency as $T_c = 175°C$ is approached [46] from above, Figure 8.28. Changes in peak amplitude with temperature in frequency scans were also observed in iron-aluminum alloys [118], as shown in Figure 8.29. An order–disorder transformation is reported in this alloy at 743 K, so the results correspond to a region below and near the suggested critical temperature. The Curie point is reported as 915 K. The peak in Fe-22Al alloy shifts to higher rather than lower frequency as the critical temperature is approached from below, a fact not explained in Golovin and Rivière [118]. The glass transition in polymers has been studied as a second-order phase transition [119].

As a further example, a large damping peak in CdMg alloy with 39 percent Mg occurs at ambient temperature near 1 Hz in temperature scans [120]. In contrast to

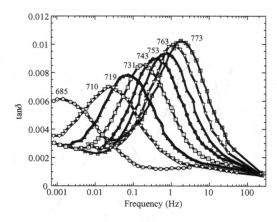

Figure 8.29. Zener tan$\delta$ in Fe-22Al alloy versus frequency at various temperatures (in K) (adapted from Golovin and Rivière [118]).

CuZn and AgZn, which exhibit the expected quadratic dependence of damping on concentration, CdMg exhibits a small maximum in damping near 15 percent Mg, a minimum near zero at 25 percent Mg, and a large peak near 30 percent. These effects are attributed to formation of a two phase mixture of a low damping intermetallic compound $MgCd_3$ and a solid solution of Mg in Cd. Frequency response of CdMg alloy at ambient temperature [121] is shown in Figure 8.30.

Time–temperature superposition fails to apply to materials that exhibit the Zener relaxation since the peak height changes with temperature; the effect of temperature is not merely a shift on the log frequency scale.

### Analysis, Second-Order Transformation

Analysis of damping associated with phase transformation is presented in the following. The Gibbs energy [2] function $g$ may be written as follows [109] for a second-order transformation, with $T$ as temperature, $\sigma$ as stress, and $\xi$ as an order parameter or internal variable; $\gamma > 0$ and $a > 0$. For a first-order transformation,

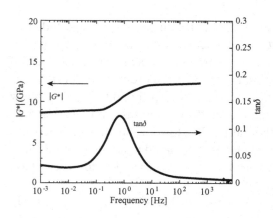

Figure 8.30. Zener tan$\delta$ in Cd-Mg alloy versus frequency at ambient temperature (adapted from Cook and Lakes [121]).

there is an additional term $\frac{1}{6}\kappa\xi^6 > 0$ [109]. For ferroelectrics, there are terms associated with electric energy.

$$g(\sigma, \xi, T) = g(0, 0, T) - \frac{1}{2}J_u\sigma^2 - \chi\sigma\xi + \frac{a}{2}(T - T_c)\xi^2 + \frac{1}{4}\gamma\xi^4. \quad (8.88)$$

Here, $J_u$ is the unrelaxed compliance. The differential form is as follows [109], with $S$ as entropy per volume and $A_\xi$ as affinity, which, as the derivative of the energy function with respect to order parameter, is a conjugate variable,

$$dg = -\epsilon d\sigma - S dT - A_\xi d\xi. \quad (8.89)$$

The strain and affinity are, from Equation 8.88,

$$\epsilon = -[\frac{\partial g}{\partial \sigma}]_{\xi, T} = J_u\sigma + \chi\xi, \quad (8.90)$$

$$A_\xi = -[\frac{\partial g}{\partial \xi}]_{\sigma, T} = \chi\sigma - a(T - T_c)\xi - \gamma\xi^3. \quad (8.91)$$

For a given stress, the equilibrium condition $dg = 0$ corresponds to $A_\xi = 0$, with $\xi_{eq}$ as the equilibrium value of $\xi$,

$$\chi\sigma = a(T - T_c)\xi_{eq} + \gamma\xi_{eq}^3. \quad (8.92)$$

For zero stress, denote $\xi_{eq} = \xi_0$, so

$$\xi_0[\gamma\xi_0^2 - a(T - T_c)] = 0, \quad (8.93)$$

so,

$$\xi_0 = 0, T > T_c;$$

$$\xi_0 = [a(T - T_c)/\gamma]^{1/2}, T < T_c. \quad (8.94)$$

The order is finite and nonzero below $T_c$, and is zero above $T_c$, with a continuous change, corresponding to a second order phase transformation. To determine the viscoelastic response of this system, apply a stress $\sigma$ at a temperature below $T_c$, so using Equation 8.92 for zero stress and a nonzero stress,

$$a(T - T_c)(\xi_{eq} - \xi_0) + \gamma(\xi_{eq}^3 - \xi_0^3) = \chi\sigma. \quad (8.95)$$

Assuming that the stress is small, the order parameter does not deviate much from its value at zero stress, $\xi_{eq} - \xi_0 \ll \xi_0$ so $\xi_{eq}^3 - \xi_0^3 \approx 3\xi_0^2(\xi_{eq} - \xi_0)$.
From Equation 8.94, for $T < T_c$,

$$(\xi_{eq} - \xi_0) = \frac{\chi\sigma}{2a(T - T_c)}. \quad (8.96)$$

The sign changes for $T > T_c$ and the expression is a factor two larger. The change in strain during the viscoelastic process as $\xi$ returns to equilibrium is, from Equation 8.90, $\chi(\xi_{eq} - \xi_0)$, so with compliance as a ratio of strain to stress,

$J_r = J(\infty)$ as the relaxed or equilibrium compliance at long time, and with Equation 8.96, (for $T < T_c$),

$$J_r - J_u = \frac{\chi^2}{2a(T_c - T)}. \tag{8.97}$$

Similarly, for $T > T_c$,

$$J_r - J_u = \frac{\chi^2}{a(T - T_c)}. \tag{8.98}$$

Since the unrelaxed compliance $J_u = J(0)$ was given as a constant, this indicates that the relaxed compliance diverges, so the modulus tends to zero as the critical temperature is approached. Such softening of particular tensorial moduli has been observed in some phase transformations.

The rate of relaxation remains to be specified. To do this, assume the following, with $L_\xi$ as a constant. For particular crystal systems, this may be justified by considering that changes in internal order exhibit drag due to coupling with thermal fluctuations.

$$\frac{d\xi}{dt} = L_\xi A_\xi. \tag{8.99}$$

The relaxation time is developed for $T < T_c$ by using the definition of $A_\xi$, Equation 8.91, and the equilibrium condition Equation 8.92 and Equation 8.93 with a small perturbation assumption $\xi_{eq} - \xi_0 \ll \xi_{eq} \approx \xi_0$,

$$A_\xi \approx 2a(T - T_c)(\xi - \xi_{eq}). \tag{8.100}$$

From Equation 8.99 and Equation 8.100 (the following form could be assumed from the outset),

$$\frac{d\xi}{dt} = -\frac{(\xi - \xi_{eq})}{\tau} \tag{8.101}$$

with, the time constant $\tau$ as follows (the time constant becomes twice as large if $T > T_c$):

$$\tau = \frac{1}{2aL_\xi(T_c - T)}, \, T < T_c$$

$$\tau = \frac{1}{aL_\xi(T - T_c)}, \, T > T_c. \tag{8.102}$$

Exponential time behavior occurs, which corresponds to a standard linear solid. In the frequency domain, (refer to §3.2), behavior is as follows: a Debye peak in damping,

$$J'(\omega) = J_u + \frac{J_r - J_u}{1 + \omega^2 \tau^2}, \tag{8.103}$$

$$J''(\omega) = [J_r - J_u]\frac{\omega \tau}{1 + \omega^2 \tau^2}, \tag{8.104}$$

$$\tan\delta = [J_r - J_u]\frac{\omega \tau}{J_r + J_u \omega^2 \tau^2}. \tag{8.105}$$

Recall that $\tan\delta = J''/J'$; in the present setting both $J_r$ and $\tau$ are functions of temperature. Both the relaxation time and the relaxation strength diverge near $T_c$. A frequency scan at constant temperature discloses a Debye peak, the height of which depends on the temperature chosen. If, by contrast, one performs a temperature scan at constant frequency, a peak of finite magnitude in damping is observed; the height of this peak is inversely proportional to the frequency. For example, in Figure 8.28, the damping peak becomes larger and shifts to lower frequency as $T_c$ is approached from above. By contrast, in ultrasonic experiments on a ferroelectric crystal, ($KD_2PO_4$) [122] attenuation increases as $\omega^2$, so $\tan\delta$ increases as $\omega$, except within $0.1°C$ of $T_c$, so $\omega\tau \ll 1$ for the entire temperature range studied. The modulus $C_{66}^{\mathcal{E}}$, at constant electric field, softens near $T_c$, but the modulus at constant polarization does not soften. In the ferroelectric $KH_2PO_4$, the modulus softens to zero within experimental resolution [123], in which temperature resolution was $0.01°C$ to resolve the cusp in the modulus. In some experiments, deviations from the ideal behavior are observed near $T_c$ due to physical effects not included in the theory. The theory assumes constant temperature, but dynamic experiments may be sufficiently rapid that the process is adiabatic. Since the distinction between adiabatic and isothermal moduli depends on heat capacity, which can become large near a phase transformation, some refinements may be called for.

*Analysis, First-Order Transformation*

In first-order transformations [109] an analysis similar to the above also predicts damping peaks. The temperature at which they occur depends on whether the material is heated or cooled: hysteresis, as shown in Figure 8.26. In such ferroelectric materials [124], there is additional energy associated with the electric field and polarization. The above analysis does not account for temperature rate effects which are seen in such materials. Specifically, the damping has three contributions [125]:

$$\tan\delta = \tan\delta_{\text{transient}} + \tan\delta_{PT} + \tan\delta_{\text{intrinsic}}. \qquad (8.106)$$

The transient term $\tan\delta_{\text{transient}}$ provides a peak at the transformation temperature; its magnitude increases with heating or cooling rate. Even as the temperature rate tends to zero, a damping peak at the transformation temperature is still observed: $\tan\delta_{PT}$. This is due to the phase transformation itself. At temperatures well above or below the transformation temperature, the intrinsic damping $\tan\delta_{\text{intrinsic}}$ of the high or low temperature phase is manifested.

In first-order transformations, the role of twin boundaries in damping of solids with martensitic structure has long been appreciated [126]; in high-damping copper-manganese alloy, annealing reduces both the visible regions of twin structure and the damping. Shape memory materials and related materials exhibit first order martensitic transformations [127, 128]. Theories for first-order transformations focusing on the creation and motion of twin boundaries during the transformation include, for example, that of Zhang et al. [110], who obtain the following relation for the magnitude of the transient internal friction peak; $q$ and $n$ are material constants,

$T$ is temperature, $t$ is time, and $\omega$ is angular frequency:

$$\tan\delta_{\text{peak}} = q\left(\frac{\partial T/\partial t}{\omega}\right)^n. \tag{8.107}$$

Theories of this type do not explain observed increases in internal friction prior to martensite formation, peak asymmetry, or the fact that polycrystalline specimens of indium-thallium alloy exhibit larger, sharper peaks than single crystals [129].

### 8.6.7 High-Temperature Background

The *high-temperature background* refers to damping observed in polycrystalline materials such as metals at an absolute temperature $T$ which is a significant fraction of the melting point $T_m$. The homologous temperature is $T_H = T/T_m$; if it exceeds 0.5, there is usually considerable viscoelasticity, manifested as creep, relaxation, or damping. The viscoelasticity is due to a variety of mechanisms, including grain boundary slip, but at sufficiently high temperature, the background predominates. Early results on the background are summarized in the work of Nowick and Berry [2]. The background depends on structure: it is smaller in single crystals than in polycrystals, it is smaller in coarse-grained polycrystals than in fine-grained polycrystals; it is enhanced in deformed and partially-recovered or polygonized samples; and it is reduced by annealing treatments at successively higher temperatures. It is generally agreed upon, however, that the background is caused by a combination of thermally-activated dislocation mechanisms.

The dependence of the background loss upon temperature and frequency has been discussed by Schoeck, Bisogni, and Shyne [130]. They considered a thermally activated dislocation-point defect mechanism. If the dislocation experiences a restoring force represented by $q$, the damping follows a Debye peak in angular frequency $\omega = 2\pi f$, with $f$ as frequency:

$$\tan\delta = \frac{Gb\gamma\Lambda_d}{q}\frac{\omega\tau}{1+\omega^2\tau^2}, \tag{8.108}$$

in which $G$ is the shear modulus. The dislocation has length $\Lambda_d$ and Burgers vector magnitude $b$; $\gamma$ is a geometrical orientation factor of order of magnitude 0.1. The time constant is $\tau = \frac{1}{pq}\exp\frac{U_0}{kT}$ in which $U_0$ is an activation energy, $T$ is the absolute temperature (in K), $k$ is Boltzmann's constant and $p$ depends on temperature only. If there is no restoring force $q$,

$$\tan\delta \propto f^{-1}. \tag{8.109}$$

Cagnoli et al. [131] think that such a model cannot account for damping at low frequency. Their recent theoretical development assuming self-organized criticality of stick-slip dislocation processes gives

$$\tan\delta \propto f^{-2}. \tag{8.110}$$

The theory is nonlinear, with frequency dependent and strain dependent expressions in multiplicative form. Cagnoli et al. point out that the Granato–Lücke theory, while successful in explaining damping at high frequency in metals fails to explain the low–frequency damping despite all later improvements in the theory.

For a *distribution* of activation energies, the following can be obtained for the case of no restoring force on the dislocations (Schoeck et al. [130]):

$$\tan\delta \propto f^{-n}. \tag{8.111}$$

Such a form has been observed in several alloys at high homologous temperature. Since it seems fortuitous that similar distributions occur in many metals, it is natural to consider mechanisms which give rise to the observed behavior without assuming special distributions. The form $\tan\delta \propto f^{-n}$ follows from a stretched exponential relaxation, which arises naturally in many complex materials with strongly interacting constituents; general aspects of such systems are discussed in §8.8. Metals of low melting point have been experimentally observed to follow the form $\tan\delta \propto f^{-n}$ with $n$ in the range 0.2 to 0.3, over many decades of frequency [132, 133]. Similar frequency dependence was inferred from *temperature dependent* damping results [134] for a nickel-aluminum intermetallic single crystal. Data were at several discrete frequencies and at high temperature, but they were fitted with $n \approx 0.6$ at frequencies above 0.1 Hz. This dependence is inconsistent with the self-organized criticality dislocation model which predicts $n = 2$. It is also inconsistent with the simple model [130], which, for a single activation energy, gives $n = 1$. A modified self-organized criticality model [135] is able to account for fractional values of the exponent $n$.

### 8.6.8 Nonremovable Relaxations

Several relaxation processes are fundamental and are not removable in principle [136]. One may envisage a single crystal without grain boundaries to slip. If the crystal is of a single constituent, processes associated with diffusion in alloys cannot occur. Crystals can be made nearly perfect, with few dislocations or point defects, hence minimal relaxation due to these causes. Therefore, many of the viscoelastic mechanisms can be made inoperative, and the associated losses removed.

There remains *thermoelastic relaxation* described above, which occurs in all materials, unless the coefficient of thermal expansion is zero. In pure shear, there is no macroscopic thermoelastic effect, although there may be loss due to intercrystalline thermal currents. In addition, *phonon–phonon interaction* gives rise to damping. A phonon is a quantum of sound. Phonons associated with the applied oscillatory stress can interact with thermal phonons, giving rise to damping. This damping tends to zero as temperature approaches absolute zero. In metals, which conduct electricity, there is also *phonon–electron interaction* which gives rise to damping. The underlying cause for these processes in the anharmonicity of the crystal lattice, which means that the interatomic force-displacement relation, hence the stress–strain relation, is slightly nonlinear. In the above processes [136] $\tan\delta \propto \omega$. For quartz, $\tan\delta \approx \omega/(6 \times 10^{16})$ due to thermoelasticity and $\tan\delta \approx \omega/10^{14}$ due to phonon–phonon interactions at room temperature (with $\omega$ in sec$^{-1}$). It is a challenge to prepare crystals of sufficient perfection to approach the fundamental limits.

### 8.6.9 Damping Due to Wave Scattering

At a very high frequency, attenuation of ultrasonic waves occurs due to scattering of the waves from heterogeneities in the material [137]. Polycrystalline materials consist of grains of the constituent material, and each grain may be anisotropic and oriented differently from nearby grains. Materials may contain inclusions, voids or other heterogeneities, and composite materials are heterogeneous by design. If the wavelength of the ultrasonic wave is much larger than the average heterogeneity size $d$, the attenuation $\alpha$ is given by $\alpha = f^4 A_{sc} d^3$, in which, $f$ is frequency and the scattering coefficient $A_{sc}$ depends on the anisotropy of the grains. This regime is known as *Rayleigh scattering* of sound waves. Rayleigh scattering of light in the atmosphere is responsible for the blue color of the sky. Scattering coefficients $A_{sc}$ have been calculated for a variety of polycrystalline materials: for longitudinal waves, $A_{sc} = 40.2$ for aluminum, $3.07 \times 10^3$ for copper, 700 for iron, $5.4 \times 10^3$ for lead, in units of dB/(cm (MHz)$^4$cm$^3$). If the wavelength of the ultrasonic wave is smaller than the heterogeneity size $d$, the attenuation $\alpha$ is given by $\alpha = f^2 B d$, with $B$ as a constant. For typical polycrystalline metals, the transition between $f^4$ frequency dependence and $f^2$ dependence occurs at frequencies between one and ten MHz. Attenuations between 1 and 10 dB/$\mu$sec at 10 MHz are representative. As for interpretation in terms of tan$\delta$, recall from Equation 3.109 that, with $c$ as the wave speed, $\alpha \approx \frac{\omega}{2c} \tan\delta$.

Damping due to scattering cannot be shifted to lower frequencies by cooling the specimen since the physical process involved is geometrical in nature and depends on the ratio of wavelength to heterogeneity size.

## 8.7 Magnetic and Piezoelectric Materials

### 8.7.1 Relaxation in Magnetic Media

Magnetoelastic relaxation is a coupled field process. Coupling between stress and magnetic field in magnetic materials can give rise to relaxation. Loss tangents from various magnetoelastic processes can be large; exceeding 0.1, as discussed in §7.3. A magnetic field that changes with time as a result of a varying strain gives rise to circulating electric currents in a conducting medium; these are called *eddy currents* [2, 138]. Eddy currents may occupy the entire volume of a magnetized specimen (macroeddy currents) or may occur locally as a result of changes in the magnetization or orientation of domains within the material (microeddy currents). All such electric currents (in a material of finite electric conductivity) involve the dissipation of energy. The damping due to macroeddy currents arises from energy dissipation from electric currents caused by changes in magnetization of the entire specimen. It depends on stress amplitude as well as on frequency and applied magnetic field. Microeddy currents can give rise to damping even if the metal as a whole is not

Figure 8.31. Damping in magnetic materials at room temperature due to stress-induced domain wall motion in nickel (adapted from Mason [140] (triangles) and Ganganna, Fiore, and Cullity [141] (diamonds). The solid curves show damping due to macroeddy currents in a nickel reed 1.52 mm thick, for two values of magnetization as a fraction of the saturation value $M_s$ (adapted from Berry and Prichet [142]).

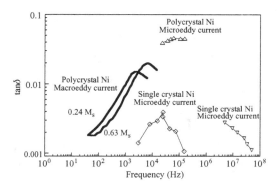

magnetized. Motion of domain walls in the material gives rise to a change in magnetic moment. This contribution depends on frequency but not on amplitude of applied stress. The above effects are linear ones. There is also hysteresis damping, associated with magnetic hysteresis, which is inherently nonlinear. Such damping depends on stress amplitude but not on frequency.

Magnetoelastic mechanisms were explored experimentally in a study of Armco iron [139] at 22.5 kHz, as a function of stress amplitude, applied magnetic field, permanent deformation, and time following vibration at high amplitude. Damping increases with vibration strain but decreases with permanent deformation. The hysteresis damping appears to be governed by bowing of the domain walls, a dynamic process which can occur at high frequency and linked to internal stress. Stress-induced motion of domain walls causes microeddy currents which dissipate energy. Since there is a range of domain sizes in these materials, there is a corresponding distribution of relaxation times [140]. Results for nickel, which has a Young's modulus of about 200 GPa, are shown in Figure 8.31. The loss is independent of strain amplitude below $2 \times 10^{-7}$ and is abolished by a sufficiently strong magnetic field, which aligns the domains [141].

Damping due to stress-induced macroscopic eddy currents [142], by contrast, occurs at lower frequency 100 Hz to 10 kHz; the effect tends to zero at full (saturation) magnetization (not shown). The peak frequency goes inversely with the specimen size $d$, $f_k = D_B k^2 / 2\pi d^2$ with $D_B \propto \rho/\mu_B$ as diffusion coefficient of magnetic flux, $\rho$ as electric resistivity, $\mu_B$ as magnetic permeability, $\mathcal{H}$ as magnetic field, $E$ as Young's modulus and $k$ as an integer. The theory is due to Zener [143]:

$$\tan\delta(f) \propto \frac{E}{\mu_B}[\frac{\partial \epsilon}{\partial \mathcal{H}}]^2 \sum_{k=1}^{\infty} \frac{(f f_k)/k^2}{1 + (f/f_k)^2}. \qquad (8.112)$$

Because the stress derivative of magnetic induction $\partial \mathcal{B}/\partial \sigma$ vanishes in both the demagnetized and saturated states, the relaxation strength also vanishes at these limits and reaches a maximum at an intermediate state.

Damping in magnetic materials can also arise in the vicinity of the phase transition between paramagnetic and ferromagnetic behavior, and due to spin reorientation. Such effects have been explored in gadolinium, which has a Curie point of

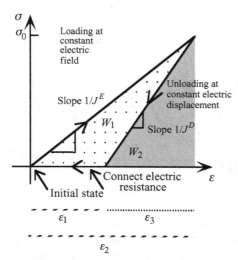

Figure 8.32. Conversion of mechanical energy to electrical energy by a piezoelectric material (adapted from IEEE [155]). Energy densities $W$ are shown as shaded areas. Material is loaded at constant electric field, unloaded at constant electric displacement, and connected to an electric load at zero stress. $W_1$ represents the work per volume of material done by the material on the electrical load, while $W_1 + W_2$ is the maximum energy per unit volume stored in the material at maximum stress. $J$ is compliance.

293 K, room temperature [144]. Polycrystalline gadolinium has a Young's modulus of about 65 GPa and a peak $\tan\delta$ of about $1.5 \times 10^{-3}$ near room temperature, while a single crystal has a peak $\tan\delta$ of about $6 \times 10^{-3}$. A smaller peak near 228 K is attributed to spin reorientation.

Nonlinear effects are also observed in magnetic materials. Damping at 1 Hz versus strain attains a maximum for a strain of $10^{-4}$ [145]. Because overall damping tends to increase with strain due to dislocation effects, the magnetic contribution is isolated by comparing the response of a magnetically saturated specimen in which magnetic damping is negligible. Magnetic damping is highest at low temperature, and tends to zero as the Curie point is approached, 360°C for nickel. The nonlinear behavior is attributed to unpinning of domain walls from crystal defects.

### 8.7.2 Relaxation in Piezoelectric Materials

#### Piezoelectric Coupling

Piezoelectricity is a coupled field effect as is thermoelasticity. In piezoelectric materials stress and strain are coupled to electrical field and polarization. Not all materials are piezoelectric; only those materials lacking a center of symmetry on the atomic scale can be piezoelectric. Examples of piezoelectric materials include single crystals of quartz, Rochelle salt, and other materials as well as polycrystalline lead titanate zirconate ceramics. A cyclic deformation with change in electrical boundary condition or with electrical dissipation can give rise to mechanical damping as shown in Figure 8.32. Damping due to piezoelectric effects can be much larger than damping due to thermoelastic effects since the strength of coupling is larger in the piezoelectric case.

Piezoelectric relaxation has been observed in many materials, including ceramics [147], composites [148], and bone [149]. Such relaxation can be represented with complex piezoelectric coefficients or by a piezoelectric loss tangent [147, 150].

Mechanical relaxation also occurs in piezoelectric materials and is important in applications: large damping is considered desirable in materials used to generate short acoustic pulses for flaw detection [151]; small damping (high mechanical $Q$) is desirable in stable resonators and high-power transducers.

### Piezoelectric Constitutive Equations

For a piezoelectric material, in the linear domain, the constitutive equations are [4]:

$$\mathcal{D}_i = [d_{ijk}]_T \sigma_{jk} + [K_{ij}]_{\sigma T}\mathcal{E}_j + [p_i]_\sigma \Delta T, \quad \text{and} \tag{8.113}$$

$$\epsilon_{ij} = [J_{ijkl}]_{\mathcal{E},T}\sigma_{kl} + [d_{kij}]_T\mathcal{E}_k + [\alpha_{ij}]_\mathcal{E}\Delta T. \tag{8.114}$$

Here, $\mathcal{D}$ is the electric displacement vector, $d$ is the piezoelectric modulus tensor at constant temperature $T$, $K$ is the dielectric tensor at constant stress $\sigma$ and temperature $T$, $\mathcal{E}$ is the electric field vector, $p$ is the pyroelectric coefficient at constant stress, $\epsilon$ is the strain, $J$ is the elastic compliance tensor at constant electric field, and $\alpha$ is the thermal expansion tensor. The usual Einstein summation convention over repeated subscripts is used. In a material described by these equations, the isothermal compliance measured at constant electric displacement $[J_{ijkl}]_\mathcal{D}$ (sometimes called $[S_{ijkl}]_\mathcal{D}$) differs [4] from the compliance measured at constant field $[J_{ijkl}]_\mathcal{E}$:

$$[J_{ijkl}]_\mathcal{D} - [J_{ijkl}]_\mathcal{E} = -d_{mij}d_{nkl}[K_{mn}]_\sigma^{-1}. \tag{8.115}$$

The derivation of this expression follows from the constitutive equations by a similar procedure to that used in obtaining the difference between adiabatic and isothermal compliances in the thermoelastic case. This difference may be regarded as the piezoelectric contribution to the compliance. Compare this expression with the thermoelastic Equation 8.17. Piezoelectric reactions also influence the apparent stiffness of a solid, under conditions in which neither $\mathcal{E}$ nor $\mathcal{D}$ is constant [152]. Similarly to the thermoelastic case, Equation 8.115 governs the relaxation strength. For materials which do relax, the coefficients in Equations 8.113 and 8.114 are complex. The sign is negative because these are compliances,

$$d^*_{ijk} = d'_{ijk} - i d''_{ijk} \tag{8.116}$$

$$K^*_{ij} = K'_{ij} - i K''_{ij}$$

$$J^*_{ijkl} = J'_{ijkl} - i J''_{ijkl}.$$

### Analysis: Mechanical Damping from Electric Phase Angles

The piezoelectric contribution to the mechanical loss tangent of a piezoelectric solid is here derived from its complex piezoelectric and dielectric coefficients [146]. This loss depends on specimen geometry as a result of differences in effects related to the electrical boundary conditions. Including a positive out-of-phase piezoelectric modulus results in reduced values of the predicted loss, which constitutes an improvement over earlier theories, which predict losses exceeding measured losses by a factor greater than two.

In this segment the connection between the dielectric, mechanical, and piezo-electric coefficients of a material which exhibits relaxation is considered. Clearly, for an ideal solid which does not relax, this connection, in the form of a piezo-electric contribution to the compliance, has been established. One can consider a piezoelectric contribution to mechanical relaxation: experimental evidence for such a contribution in quartz under quasistatic loading has appeared at least as early as 1915 [153]. A connection between dielectric loss and mechanical loss in piezoelectric solids is to be expected on heuristic grounds in that dielectric relaxation entails dissipation of electrical energy; if this energy has come from the piezoelectric conversion of mechanical energy, then mechanical relaxation or anelasticity must also occur.

The relation between electric field and electric displacement is given by the constitutive equation 8.113. For a nonpolar solid under isothermal conditions, with $K_{ij} = k_{ij}e_0$ ($e_0$ is the permittivity of free space), and with $k^* = k' - ik''$ and $d^* = d' - id''$ to describe dielectric and piezoelectric relaxation, Equations 8.113 and 8.114 become

$$\mathcal{D}_i = d^*_{ijk}\sigma_{jk} + [k^*_{ij}]_\sigma e_0 \mathcal{E}_j, \tag{8.117}$$

$$\epsilon_{ij} = [J^*_{ijkl}]_\varepsilon \sigma_{kl} + d^*_{kij}\mathcal{E}_k. \tag{8.118}$$

Suppose that the specimen in question is electrically isolated, that is, free of any attached circuit element of finite impedance and that it is subjected to a stress $\sigma_{11} \neq 0$ which is uniform within the specimen and varies sinusoidally with time: $\sigma_{11}(t) = \sigma_{11}(0)e^{i\omega t}$, $\sigma_{ij} = 0$ if $i \neq 1$ or $j \neq 1$. The supposition of uniform stress entails loading below any mechanical resonance. With these assumptions we will find that the field and the displacement are related in a way which depends on the electrical boundary conditions; therefore, several specimen geometries are considered. In all cases we assume that $d_{311} \neq 0$ and $d_{i11} = 0$ if $i \neq 3$ and that the 3 direction is a principal axis of $k_{ij}$, so that both the field and displacement will be in the 3 direction.

Consider a thin plate of piezoelectric material such that the flat surfaces are perpendicular to the 3 axis. From Gauss's law, the boundary condition on the electric displacement is $\mathcal{D}^{[\text{normal}]}_{\text{in}} - \mathcal{D}^{[\text{normal}]}_{\text{out}} = \sigma_{\text{free}}$; $\sigma_{\text{free}}$ is the density of free charge on the surface. For the thin plate geometry, $\mathcal{D}^{[\text{normal}]}_{\text{out}} = 0$. Free charge includes only charge that is not associated with processes included in the definition of $\mathcal{D}$. So, if $k''$ contains contributions from dc conductivity as well as dielectric relaxation, the free charge is zero. Then, inside of the plate,

$$\mathcal{D}_3 = 0, \tag{8.119}$$

as in the case of ideal crystals with zero conductivity [150]. The relation between electric field and electric displacement depends on boundary conditions. For example, for plates, cylinders, and spheres, within the solid, the electric field and displacement are uniform, parallel, and related by the following:

$$\mathcal{D}_3 = -\mathcal{E}_3 e_0 \Lambda_p. \tag{8.120}$$

The quantity $\Lambda_p$ is zero for the thin plate. For the cylinder, $\Lambda_p = 1$ and for the sphere, $\Lambda_p = 2$ [146].

In this section, we develop an expression for the piezoelectric contribution to the anelastic loss tangent in a solid subjected to subresonant, sinusoidal loading. Let the solid obey the constitutive equations (8.10.5) and (8.10.6), that is, let it exhibit dielectric and piezoelectric relaxation. The mechanical, dielectric, and piezoelectric loss tangents are defined as follows:

$$\tan\delta^{(s)}_{ijkl} = \frac{J''_{ijkl}}{J'_{ijkl}}, \tag{8.121}$$

$$\tan\delta^{(d)}_{ijk} = \frac{d''_{ijk}}{d'_{ijk}}, \tag{8.122}$$

$$\tan\delta^{(k)}_{ij} = \frac{k''_{ij}}{k'_{ij}}, \tag{8.123}$$

with no summation on the repeated indices.

The piezoelectric contribution to the storage compliance is formally similar to the effect of the temperature field in thermoelasticity, and the effect of the fluid pressure field in poroelasticity. The presence of piezoelectric coupling reduces the compliance and causes the material to appear stiffer. In general the stiffening effect depends also on geometry and piezoelectric phase angles. However, if we consider a flat plate specimen, $\Lambda_p = 0$, and if there is no dielectric or piezoelectric relaxation, $\tan\delta^k_{33} = 0$, $d''_{311} = 0$. Because $\Lambda_p = 0$ implies that the electric displacement is zero, we obtain

$$J'^D_{1111} - J'^{\mathcal{E}}_{1111} = -\frac{d^2_{311}}{e_0 k'_{33}}, \tag{8.124}$$

which is equivalent to Equation 8.115 in which the sum has collapsed into a single term as a result of the assumptions made. The results, therefore, reduce to those of the classical theory in which no losses or piezoelectric phase angles occur.

We remark that in the field of piezoelectric materials, dimensionless coupling factors $\mathcal{K}$ are defined for the purpose of understanding the conversion of mechanical and electrical energy [155] without damping. The range for $\mathcal{K}$ is from zero to one. The relevant energies are shown in Figure 8.32:

$$\mathcal{K}^2 = \frac{W_1}{W_1 + W_2}. \tag{8.125}$$

These coupling factors depend on elastic boundary conditions at the material surface. For example, for uniaxial stress [155],

$$\mathcal{K}_{31} = \frac{d_{31}}{\sqrt{(e_0 k_{33} J^{\mathcal{E}}_{11})}}, \tag{8.126}$$

in which $J_{11}$ is the elastic compliance in the "reduced" notation (it is equivalent to $J_{1111}$), $k_{33}$ is the dielectric constant, and $d_{31}$ is the piezoelectric modulus in the reduced notation. Compare with Equation 8.18, 8.20 for the thermoelastic case. In piezoelectric ceramics, coupling factors as large as 0.7 are possible. This is much stronger than thermo-elastic coupling.

The piezoelectric contribution to the loss tangent of the thin plate is calculated as

$$\tan\delta_{1111} = \frac{(d'^2_{311} - d''^2_{311})\tan\delta^k_{33} - 2d'_{311}d''_{311}}{\epsilon_0 k'_{33} J'_{1111}(1 + \tan^2\delta^k_{33})}. \tag{8.127}$$

$J'$ in this expression is the storage compliance for the geometry in question; this differs from the constant-field compliance $(J')_\varepsilon$ which appears in the constitutive equation. For weak coupling, the difference between these compliances is small.

The effect of the piezoelectric relaxation term $d''$ is much more pronounced in the contribution to the anelastic relaxation than in the contribution to the storage compliance. This may explain why the classical theory of linear piezoelectricity [4, 152, 154], which addresses elastic effects in the absence of dissipation, is accurate for this type of problem (prediction of piezoelectric stiffening) despite the neglect of piezoelectric phase angles.

The piezoelectric contribution to the mechanical loss of a solid in the form of a thin plate was obtained. For a different specimen *shape*, the damping may be different as a result of the electrical boundary conditions [146]. A similar shape dependence of damping was also studied in connection with thermoelastic damping [10, 11]. Phase angles in the thermal expansion coefficient [156] have been shown to affect the thermoelastic damping in a manner similar to the effect of piezoelectric phase angles.

### *Phase Transformations in Piezoelectric Materials*

Damping in piezoelectric materials can also occur as a result of motion of domains in the material as discussed in §8.6.6. For example, barium titanate is a piezoelectric, ferroelectric material; as shown in Figure 8.26, it exhibits substantial damping peaks associated with phase transformations. Above the Curie point 120°C, the piezoelectricity vanishes because the crystals lose the asymmetry required.

### *External Circuits*

An external electric resistor can be connected to the piezoelectric element to achieve mechanical damping as discussed in Example 8.6. This is done in applications to damp ultrasonic transducers and other devices as discussed in Chapter 10.

## 8.8 Nonexponential Relaxation

The stretched exponential relaxation modulus function described in §2.6,

$$E(t) = (E_0 - E_\infty)e^{-(t/\tau_r)^\beta} + E_\infty, \tag{8.128}$$

arises naturally in many complex materials with strongly interacting constituents. One may formally express such a function as a superposition of exponentials using the concept of relaxation spectrum developed in §4.2. In view of the Boltzmann superposition principle, linear macroscopic experimental measurements cannot distinguish between causes which involve a true distribution of exponential processes each with its own relaxation time, and causes that are intrinsically nonexponential.

In intrinsically nonexponential mechanisms, the stretched exponential relaxation arises from relaxation in hierarchical stages such that the constraint imposed by a faster degree of freedom must relax before a slower degree of freedom can relax [157]. The underlying feature of theories giving rise to such relaxation is the generation of a scale invariant distribution of relaxation times [158]. Stretched exponentials in slow relaxation are so widespread as to be considered "universal," [159, 160] perhaps because they represent a probability limit distribution [161]. In the frequency domain, for frequencies $f$ well above the damping peak,

$$\tan\delta \propto f^{-\beta}, \tag{8.129}$$

with $\beta < 1$. The high temperature background in metals, as discussed above in §7.6.2, follows this form. However, a copper-beryllium alloy was observed to exhibit damping of the form $\tan\delta \propto f^{-2}$ at low frequency [162], and that form was modeled theoretically under the assumption of self-organized criticality.

Andrade creep is a particular form of power law creep for which $J(t) = J_0 + At^{1/3}$. Since Andrade creep is commonly observed in soft metals as well as some polymers, particularly at large strain, it must be caused by very general effects and is not limited by specific atomic processes of flow. A mechanism based on correlation volumes has been proposed [163]. More recently [164], Andrade creep has been explained in terms of weak strain hardening associated with regeneration and exhaustion of weakly obstructed dislocation segments. If the strain hardening effect is strong, the creep becomes logarithmic in time.

Because macroscopic linear experiments do not allow one to distinguish whether nonexponential relaxation arises intrinsically or from a superposition of exponentials, there has been some debate on the nature of the causal mechanisms [165]. Recently, a nonlinear technique was used to demonstrate that a broad relaxation spectrum in a dielectric system was the result of a distribution of relaxation times [166]. The method involves brief exposure of the specimen to a sinusoidal "pump" stimulus followed by a probe stimulus of step function form. If the response to a probe stimulus is altered by the prior presence of a pump stimulus at its characteristic time, that result is interpreted as an excitation of a single exponential mode. Experiments of this type have been performed in viscoelastic materials, such as polymer melts [167].

## 8.9 Concepts for Material Design

### 8.9.1 Multiple Causes: Deformation Mechanism Maps

Deformation mechanism maps are diagrams (Figure 8.33) that display the regions of stress and temperature in which a particular mechanism of *secondary* creep or

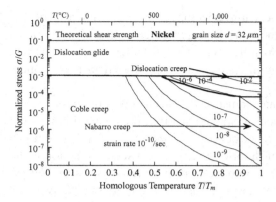

Figure 8.33. Deformation mechanism map for nickel (adapted from Ashby [168, 169]).

plastic flow dominates [168, 169]. In such a region, the particular mechanism supplies a greater strain rate in creep than any other. The stress axis is normalized to the shear modulus, and the temperature axis is normalized to the melting point. Maps may be constructed from actual creep data or from theoretical relations for each process. In the latter case, the relevant diffusion parameters must be known. The boundaries of the regions in the diagram are determined by equating strain rates associated with pairs of the constitutive equations for particular creep mechanisms.

At least six deformation mechanisms have been identified in connection with the maps. They are:

(1) flow in the absence of defects at a stress above the theoretical shear stress;
(2) flow by glide of dislocations;
(3) creep by climb of dislocations;
(4) Nabarro–Herring creep from flow of point defects through grains;
(5) Coble creep from flow of point defects along grain boundaries;
(6) twinning of crystals. Constitutive equations are as follows: dislocation glide exhibits a threshold effect,

$$\frac{d\epsilon_2}{dt} = \frac{d\epsilon}{dt}\Big|_0 \exp\{-\frac{\sigma_S - \sigma}{kT}ba\},\tag{8.130}$$

for $\sigma \geq \sigma_0$,
$\frac{d\epsilon_2}{dt} = 0,$ for $\sigma \leq \sigma_0$,

with $\sigma_S$ as a flow stress, $\sigma$ as stress, $k$ as the Boltzmann's constant, $T$ as absolute temperature, $a$ as activation area, and $\sigma_0$ as a cut off stress.

In the following, diffusional creep due to bulk transport, Nabarro–Herring creep, is represented with the subscript 3. Diffusional creep due to transport along grain boundaries, Coble creep is represented with the subscript 4,

$$\frac{d\epsilon_{3;4}}{dt} = 14\frac{\sigma\Omega_{at}}{kT}\frac{1}{d^2}D_v[1 + \frac{\pi\delta_b}{d}\frac{D_B}{D_v}],\tag{8.131}$$

in which $\Omega_{at}$ is the atomic volume, $d$ is the grain size, $D_v$ is the bulk self diffusion coefficient, $D_B$ is the boundary self diffusion coefficient and $\delta_b$ is the effective cross-section of a boundary. These mechanisms may be distinguished physically.

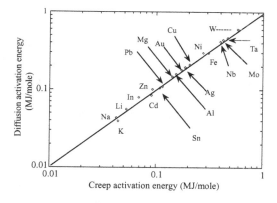

Figure 8.34. Relation between the activation energy for self diffusion and the activation energy for high-temperature, secondary creep of pure metals (adapted from Karato and Wu [171].).

Dislocation creep is nonlinear in stress: $\frac{d\epsilon_s}{dt} = AD_v\frac{Gb}{kT}(\frac{\sigma}{G})^n$, in which $G$ is the shear modulus. This process is due to aggregation of dislocations at temperatures greater than half the melting point. It is diffusion-controlled in contrast to dislocation glide.

These maps were presented initially for metals (Figure 8.33). By explicitly plotting the stress level, nonlinear behavior is allowed for, but detailed history dependence is not. Deformation mechanism maps can be useful in the design of experimental conditions for tests of particular creep processes. They are also useful in design problems involving creep. For example, a strengthening mechanism such as dispersion hardening of metals, will slow dislocation creep. Map boundaries will then shift, so that another creep mechanism dominates. Further inhibition of dislocation creep is not productive, because the metal now deforms mostly by diffusional creep, which is not inhibited by the same mechanism [168]. Ceramics [170] at sufficiently high temperature exhibit time-dependent deformation as a result of many mechanisms that operate in metals.

A further example is that multiple viscoelastic mechanisms cause flow in the "solid" interior of the Earth [171]. At relatively low stress, or for small grain size, linearly viscoelastic behavior occurs via diffusion of matter between grain boundaries. At a higher stress level, or larger grain size, nonlinearly viscoelastic behavior occurs via motion of dislocations. Creep due to dislocation motion can give rise to preferred orientation in a material. In the context of geology, this gives rise to anisotropy of seismic wave velocities. Relaxation mechanisms in the Earth have implications in understanding continental drift.

The activation energy for secondary creep at high temperature in metals is closely related to the activation energy for self-diffusion, [172] as shown in Figure 8.34. In pure metals [172] and in solid solution alloys [173], the stress dependence of steady state creep consists of three regions: a linear region, a strongly stress-dependent region, and a breakdown region. In alloys, there is an additional possibility of enhanced creep at intermediate stresses due to dragging of solute atoms by dislocations.

Deformation mechanism maps are useful for secondary (steady state) creep. As currently constituted, they do not deal with primary creep, or, in the frequency domain, peaks in the loss tangent.

### 8.9.2  Damping Mechanisms in High-Loss Alloys

In known high-damping alloys, several viscoelastic mechanisms have been identified. In zinc-aluminum alloys, thermoelastic damping contributes a noticeable peak [174] due to the high concentration of zinc which has a favorable figure of merit. The remaining damping follows a $\tan\delta \propto f^{-n}$ form at high frequency and is attributed to dislocation movement. Damping in CuMn alloys [175, 176] is attributed to the movement of twin boundaries and is dependent on the antiferromagnetic nature of the crystals in the metal. This magnetomechanical mechanism gives rise to strain dependence of damping because, at sufficiently large strain, the domains become fully aligned, giving rise to a saturation of the mechanism. Damping drops abruptly above the Néel temperature at which the material changes phase from antiferromagnetic to paramagnetic. High damping in magnesium and some of its alloys is attributed to dislocation movement [175].

### 8.9.3  Creep Mechanisms in Creep-Resistant Alloys

As temperature is increased, creep from a variety of mechanisms tends to increase. Creep-resistant alloys have been developed based on an understanding of several viscoelastic mechanisms [177], as follows:

(1) Metals of high melting point are favored since many creep processes depend on the homologous temperature. Nickel alloys are favored for turbine engine applications in which the temperature and stress are high. Cobalt alloys are used in parts for which temperature is very high, but stress is not as high. Tungsten has a higher melting point than these alloys. Tungsten is too dense for structural applications but finds use in light bulb filaments.

(2) Elements such as Co, Cr, W, Mo, and V are incorporated in a nickel matrix to form a solid solution. As dislocations move in such a material, they drag an "atmosphere" of solute atoms. The drag reduces the rate of creep. Moreover, the irregular atomic structure tends to resist the movement of dislocations in comparison with the case of a pure metal. This effect of solid solution is in contrast with the *Zener relaxation* in concentrated alloys refers to reorientation of pairs of atoms in the alloy under stress [86] discussed in §8.6.2. The Zener relaxation provides a relatively small relaxation strength and so is not responsible for large amounts of creep.

(3) Refractory particles such as $Al_2O_3$, MgO are incorporated to retard dislocation motion. This process is called *dispersion strengthening*. The particles reduce creep indirectly by suppression of grain boundary sliding during the service of the material and by the suppression of grain growth during the preparation of the material.

(4) Particles of intermetallic compounds such as $Ni_3Ti$ are incorporated. The particles exhibit the remarkable characteristic of a large reversible increase in flow stress as temperature increases.

(5) Creep due to grain boundary slip can be reduced by precipitating carbides containing Ti, Cr, W, Mo, and Zr in the grain boundaries.

(6) Creep due to grain boundary slip can be reduced by a directional solidification process which produces elongated grains aligned with the expected stress.

(7) Creep due to grain boundary slip can be eliminated by a directional solidification process which produces single crystals of metal. Then there are no grain boundaries because there is only one grain. Such crystals are used for turbine blades for jet engines. In single-crystal blades, the grain-boundary strengthening elements B, C, Zr, and Hf are omitted because they are no longer needed, and because they reduce the melting point of some alloy components.

## 8.10 Relaxation at Very Long Times

Over a sufficiently long period of time, all matter is fluid, even at zero temperature, as a result of the wave nature of matter [178]. The time scale for barrier penetration by a mass $m$ behaving as a quantum mechanical matter wave, following Gamow's analysis of radioactive decay is $\tau = \tau_0 e^{S_{act}}$, with $\tau_0$ as the natural vibration period and $S_{act}$ as the action integral

$$S_{act} = \frac{2}{\hbar} \int \sqrt{2mU(x)}dx \approx \sqrt{8mUd^2/\hbar^2}, \tag{8.132}$$

with $m$ as the particle mass, $U(x)$ as the potential barrier of thickness $d$, and $\hbar$ as the normalized Planck's constant $h/2\pi$. For atoms in a solid, $U \approx e^4 m_e/20\,\hbar^2$, and $d \approx \hbar^2/m_e e^2$. So with $A$ as the atomic weight of the atom, $m_e$ as the electron mass and $m_p$ as the proton mass, $S_{act} = \sqrt{2A\,m_p/5m_e} = 27\sqrt{A}$. For iron, $A = 56$. For atomic vibration at absolute zero temperature, $\tau_0 \approx 10^{-14}$ sec. The time scale $\tau$ is on the order of $10^{65}$ years. Because this time scale substantially exceeds estimates of the age of the universe (about $1.5 \times 10^{10}$ years) and of the remaining life of the Sun (about $5 \times 10^9$ years), it has no significance in the practical affairs of humanity. Therefore, most engineers are not concerned about such processes. Nevertheless, these arguments indicate that there are no solids.

## 8.11 Summary

Viscoelastic relaxation can occur whenever there is a delayed rearrangement of the internal structure of the material under stress. There are many ways in which this rearrangement can occur. Understanding of viscoelastic mechanisms can aid one in anticipating situations in which viscoelasticity can be expected. Moreover, by understanding the mechanisms, one can at times tailor various materials to exhibit large or small viscoelastic behavior in selected regions of time, temperature, or frequency, as required for various applications.

## 8.12 Examples

### Example 8.1

Polymers have higher thermal expansion coefficients than metals. Is thermoelastic damping an important mechanism in polymers?

**Solution**

Consider the relaxation strength, Equation 8.20 , $\Delta = \alpha^2 T / C_v J^S$, and take numerical data from Ashby's review article [179]. Suppose further that the relaxation is of the Debye form so the maximum $\tan\delta$ is approximately $\Delta/2$. Properties of high density polyethylene are:

thermal expansion coefficient $\alpha = 1.5 \times 10^{-4}/°C$,

Young's modulus $E = 0.6$ GPa,

heat capacity [180], $C = 0.55$ cal/g°C, and

density $\rho = 0.95$ g/cm$^3$.

Converting the heat capacity to SI units, and from a mass basis to a volume basis, $C = 0.55$ cal/g°C $\times 0.95$ g/cm$^3 \times 4.19$ J/cal $\times 10^6$ cm$^3$/m$^3 = 2.19 \times 10^6$ J/°Cm$^3$. Then, $\Delta = 1.85 \times 10^{-3}$ (dimensionless), so the maximum $\tan\delta$ is, for small $\Delta$, $\tan\delta_{max} \approx \Delta/2 = 0.93 \times 10^{-3}$. This is much smaller than damping observed in polymers. Most of the observed damping is due to molecular motion. Although the thermal expansion coefficient in this polymer is about 10 times as large as that of steel, the modulus is some 300 times smaller, so the thermoelastic contribution to the damping in polymers is small. By contrast in most metals, the overall damping is small, and in some frequency ranges, thermoelastic effects, though of small magnitude, can be responsible for most of the damping.

### Example 8.2

Negative values of the coefficient of thermal expansion are possible. What are the consequences in connection with thermoelastic damping?

**Solution**

The relaxation strength is given in Equation 2.52. Use the tensorial form, Equation 8.18, $\Delta_{ijkl} = \dfrac{J^T_{ijkl} - J^S_{ijkl}}{J^S_{ijkl}} = \dfrac{\alpha_{ij}\alpha_{kl}}{J^S_{ijkl}}\dfrac{T}{C^\sigma}$. Consider several components.

For example, $\Delta_{1111} > 0$ even if $\alpha_{11} < 0$, since $\alpha_{11}$ is squared, so the relaxation strength for axial deformation is always positive. Observe that $J_{1111}$ is the inverse of Young's modulus for stress in the 1 direction, and must be positive for the material to be stable.

Consider $i, j = 1; k, l = 2$. If the material is isotropic, $\alpha_{11} = \alpha_{22}$, so again a product of two negative factors appears on the right with a positive result.

If the material is anisotropic, we can have $\alpha_{11} = -\alpha_{22}$. For common materials, $J_{1122} < 0$, since its physical meaning is $-\nu/E$, and Poisson's ratio is usually positive. The possibility of $\tan\delta_{1122} < 0$ does not present physical difficulties since this 'loss tangent' represents a phase between a stress in one direction and a strain in an orthogonal direction. The angle $\delta_{1122}$ is therefore a phase in a cross property (not a modulus or compliance) with no energy content. Similarly the Poisson's ratio is a cross property, with no energy content. Therefore, not only can Poisson's ratio be negative, but phase angles in the Poisson's ratio can be positive or negative. Negative damping (or gain) in a modulus or compliance can occur only if an external source of energy is supplied.

**Example 8.3**

Consider a bar of aluminum vibrating in bending. What is the frequency $f_0$ of the damping peak if the bar is 1 mm thick? What if it is 1 cm thick? How large is the peak loss?

**Solution**

From Equation 8.21, and using data from Table 8.1,

$$f_0 = \frac{\pi}{2} D d^{-2} = 1.57 \frac{222 J/s \cdot m \cdot K}{900 J/kg \cdot K} \frac{1}{2.7 \times 10^3 kg/m^3} \frac{1}{(10^{-3}m)^2} = 144 \text{ Hz.}$$

Observe that the density has been incorporated since the equation contains heat capacity per volume, but the table refers to heat capacity per mass. A bar 1-cm thick would have a peak loss at 1.44 Hz, because the thickness $d$ is squared in Equation 8.21. The peak $\tan\delta$ is approximately $\Delta/2$. Taking $\Delta$ from Table 8.1, $\tan\delta \approx 2.4 \times 10^{-3}$, at the peak. Compare with the peak for aluminum in Figure 8.4.

**Example 8.4**

Consider thermally activated relaxation in aluminum. Following Kê [74], the grain boundary slip process gives rise to a peak in loss tangent at 0.8 Hz and 280°C. The activation energy is 34,000 cal/mole. At what frequency will this peak occur at 25°C and at 100°C? What are the corresponding relaxation times?

**Solution**

From the Arrhenius equation discussed in §2.7 and its representation in Equation 6.43, $\ln \frac{f_2}{f_1} = \frac{U}{R}[\frac{1}{T_1} - \frac{1}{T_2}]$.

The temperatures are absolute temperatures. Substituting $T_2 = 280 + 273 = 553$ K, $T_1 = 25 + 273 = 298$ K, $R = 1.98$ cal/moleK, $\ln \frac{f_2}{f_1} = 26.6$, so, $\frac{f_2}{f_1} = 3.47 \times 10^{11}$, so, for $f_1$ at 25°C, and $f_2$ at 280°C,

$f_1 = 2.31 \times 10^{-12}$ Hz.

The time constant is

$\tau = (2\pi f_1)^{-1} = 6.89 \times 10^{10}$ sec, or $2.19 \times 10^3$ years.

For $T_1 = 100$°C or 373 K, $\ln \frac{f_2}{f_1} = \frac{34000}{1.98} 8.73 \times 10^{-4} = 14.98$, so,

$f_1 = 2.49 \times 10^{-7}$ Hz.

The time constant is

$\tau = (2\pi f_1)^{-1} = 6.41 \times 10^5$ sec, or 7.41 days, a rather more accessible time than at 25°C.

**Example 8.5**

Does a stretched rod held at constant extension get fatter or thinner with time? Suggest the use of viscoelastic mechanisms to control the rate and direction of lateral expansion or contraction.

**Solution**

Recall in Example 7.2 and §7.2 that Poisson's ratio increases through the $\alpha$ transition of polymers. In that case the rod gets thinner. Consider a case in which the bulk modulus relaxes, but the shear modulus does not relax much. Envisage a porous material with fluid in the pores. Squeezing the fluid out of the pores and through free surfaces during a volume change causes much more relaxation than does a shear deformation. Consider the relation between Poisson's ratio, shear modulus and bulk modulus for an elastic material.

$$\nu = \frac{3B-2G}{6B+2G}.$$

Considering a bulk compliance $\kappa = 1/B$, observing that for $\nu = 1/3$, $G\kappa = 0.462$, and using a Taylor expansion of the denominator to express the Poisson's ratio in terms of products,

$$\nu = \frac{\frac{1}{2}-\frac{1}{3}G\kappa}{1+\frac{1}{3}G\kappa} \approx \{\frac{1}{2} - \frac{1}{3}G\kappa\}\{1 - \frac{1}{3}G\kappa\} = \frac{1}{2} - \frac{1}{6}G\kappa + \frac{1}{9}G^2\kappa^2.$$

Applying the correspondence principle with $G$ assumed constant,

$$s\nu(s) \approx \frac{1}{2} - \frac{1}{6}Gs\kappa(s) + \frac{1}{9}G^2s^2\kappa^2(s).$$

Dividing by $s$ and transforming back,

$$\nu(t) \approx \frac{1}{2} - \frac{1}{6}G\kappa(t) + \frac{1}{9}G^2 \int \kappa(t-\tau)\frac{d\kappa(\tau)}{d\tau}d\tau.$$

The third term is small compared to the second unless the bulk creep is very fast. Since $\kappa(t)$ is an increasing function, the second term causes the Poisson's ratio to decrease with time, so the stretched rod in this case gets fatter.

## Example 8.6

Suppose a piezoelectric plate is provided with an electrically resistive path so that stress-generated charge can flow with time. Relate the coupling coefficient $K$ with the relaxation strength $\Delta$. For a Debye peak, what is the relationship between the loss tangent and the coupling coefficient?

**Solution**

By definition, in terms of the energies $W$ in Figure 8.32, $K^2 = \frac{W_1}{W_1+W_2}$. If stress is applied to the plate slowly, its stiffness is $E_s = 1/J^{\mathcal{E}}$ referring to Figure 8.32 since for sufficiently long time, current flow through the resistance forces the electric field to zero, a constant. Recall that $J^{\mathcal{E}}$ is compliance at constant electric field. Similarly for fast loading, $E_f = 1/J^{\mathcal{D}}$. Express the energies in terms of the stiffness values and the strains as shown in Figure 8.32:

$$W_2 = \frac{1}{2}E_f\epsilon_3^2,$$

$$W_1 + W_2 = \frac{1}{2}E_s\epsilon_2^2, \text{ but } E_s = \sigma_0/\epsilon_2 \text{ and } E_f = \sigma_0/\epsilon_3, \text{ so,}$$

$$K^2 = \frac{\frac{1}{E_s}-\frac{1}{E_f}}{\frac{1}{E_s}} = 1 - \frac{E_s}{E_f}.$$

Consider the definition of *relaxation strength* as the change in stiffness during relaxation divided by the stiffness at long time, referring to Equation 2.40 and Equation 2.49, $\Delta = \frac{E_1}{E_2}$. This may also be written in terms of the stiffness values for fast and slow loading, $E_f = E_1 + E_2$ and $E_s = E_2$. Then, $\Delta = \frac{E_f-E_s}{E_s}$.

Combining, $\Delta = \frac{K^2}{1-K^2}$. Observe that the range for the coupling coefficient $K$ is from zero to one, while the range for the relaxation strength $\Delta$ is from zero to infinity.

If we have a Debye peak, referring to Equation 3.31,

$$\tan\delta_{max} = \frac{1}{2}\frac{\Delta}{\sqrt{1+\Delta}}, \text{ then}$$

$$\tan\delta_{max} = \frac{1}{2}\frac{K^2}{\sqrt{1-K^2}}.$$

Here, the limiting aspects of fast and slow loading have been considered, not the time scale of relaxation that is governed by the electrical time constant $\tau_{el} = RC$ with $R$ as the resistance and $C$ as the capacitance. As for a comparison with

Equation 8.127, we observe that an external resistor was not considered in the development of that equation, which takes into account phase angles in the piezoelectric and dielectric coefficients. Resistance within the piezoelectric element itself is subsumed in the imaginary part of the dielectric coefficient.

**Example 8.7**

How is negative damping (acoustic amplification) achieved in materials?

**Answer**

An energy source must be provided. In semiconductors, an electric current is applied. The current can couple with phonons which are quanta of acoustic energy. This coupling gives rise to the amplification. Active materials and systems are essential to the study of viscoelasticity since in order to apply a force or deformation to a passive material, an active material or system is needed.

**Example 8.8**

Does a stretched rod made of metal held at constant extension get fatter or thinner with time? Suppose that the thermoelastic relaxation dominates.

**Answer**

The thermoelastic relaxation, as with the fluid-flow relaxation, is a volumetric effect. Relaxation due to this effect occurs in the bulk modulus but not in the shear modulus. Therefore, as in Example 8.5, Poisson's ratio decreases in magnitude so the rod becomes fatter. The change will be rather small since the thermoelastic relaxation gives rise to a much smaller relaxation strength than fluid flow effects or the glass–rubber transition in a polymer. Polymers in the transition region, by contrast, exhibit a Poisson's ratio that increases in magnitude, from near 0.3 in the glassy region to near 0.5 in the rubbery region.

**Example 8.9**

How large is the thermoelastic damping peak for rubber at small strain?

**Solution**

The case of a glassy polymer was considered in Example 8.1. Physical properties of rubber at small strain are similar with the exception of the modulus which is about a thousand times smaller than that of a glassy polymer. The maximum $\tan\delta$ due to thermoelastic coupling is therefore on the order $10^{-6}$ due to the large compliance of rubber. Effects will be larger for stretched rubber due to the nonlinearity in thermal expansion with strain. Indeed, following [69], the slope of the stress temperature curve is about 36 times larger in magnitude at 370 percent extension than at 3 percent extension. So the thermal expansion is also 36 times larger in magnitude, so the thermoelastic damping is magnified by $36^2$ or more than 1,000 but that ($\tan\delta \approx 10^{-3}$) is still considerably smaller than the overall damping of rubber.

**Example 8.10**

Show the details in the derivation of Equation 8.5.

**Solution**

Put $\xi(t) = \xi_0\sigma_0(e^{-t/\tau})$ in $\epsilon = J\sigma + \kappa\xi$. The stress is $\sigma_0$ after time zero, because we have assumed creep. Substituting, $\epsilon(t) = J(\infty)\sigma + \kappa\xi_0\sigma_0(e^{-t/\tau})$. For long time, $t \to \infty$, $\epsilon(t) = J(\infty)\sigma_0$. For $t = 0$, $\epsilon = \sigma_0(J(\infty) + \kappa\xi_0)$. Recall the definition of the

relaxation strength $\Delta = [J(\infty) - J(0)]/J(0)$. So $\xi_0 = -\Delta J(0)/\kappa$, so, recognizing $J_\Xi = J(0)$ and $J = J(\infty)$,
$\epsilon(t) = \sigma_0 J[(1 - \frac{J(0)}{J(\infty)}\Delta e^{-t/\tau})]$ as desired.

## 8.13 Problems and Questions

8.1. Consider fluid flow based relaxation as a coupled-field process. Ordinarily such relaxation is observed in compression or tension, since a volume change is required to induce fluid motion. Can you design a material that exhibits relaxation due to fluid flow in shear?

8.2. The Zener relaxation in brass occurs at a relatively high temperature. Under what circumstances would the Zener relaxation occur at room temperature? Can you find an example? What might such a material be used for?

8.3. How may an understanding of viscoelastic mechanisms be used in the selection of materials to resist creep?

8.4. Demonstrate the piezocaloric effect as follows: grip a thick rubber band at both ends. Subjectively evaluate the temperature of the band by briefly touching it to a free area of skin such as the bottom of your nose. Suddenly stretch the band and again evaluate the temperature. Does it get warmer or cooler? Keep it stretched for about a minute so it comes into equilibrium with the environment. Evaluate its temperature, suddenly release the tension, and repeat. This time, does it get warmer or cooler? Discuss.

8.5. What is the effect of a phase angle in thermal expansion [156] upon thermoelastic damping of a material? Consider the thermal expansion coefficient to be complex: $\alpha^* = \alpha'(1 + i \tan\delta_\alpha)$.

8.6. There are currently no standard materials for calibration of viscoelastic instrumentation, because the physical processes that give rise to viscoelastic behavior tend to depend on specimen preparation methods as well as on environmental variables such as temperature and humidity. Moreover, for many materials there are multiple relaxation mechanisms simultaneously active in a given frequency range. Develop a candidate standard material based on known relaxation mechanisms discussed in this chapter.

8.7. Describe several ways in which the damping of rubber can be increased.

### BIBLIOGRAPHY

[1] Zener, C., *Elasticity and Anelasticity of Metals*, Chicago: University of Chicago Press, 1948.

[2] Nowick, A. S., and Berry, B. S., *Anelastic Relaxation in Crystalline Solids*, New York: Academic, 1972.

[3] Wert, C. A., Internal friction in solids, *J Appl Phys*, 1888–1895, 1986.

[4] Nye, J. F., *Physical Properties of Crystals*, Oxford: Oxford University Press, 1976.

[5] Zener, C., Internal friction in solids I – Theory of internal friction in reeds, *Phys Rev*, 52, 230–235, 1937.

[6] Zener, C., Internal Friction in Solids II. General Theory of Thermoelastic Internal Friction, *Phys Rev*, 53, 90–99, 1938.

[7] Sokolnikoff, I. S., *Mathematical Theory of Elasticity*, Malabar, FL: Krieger, 1983.

[8] Zener, C., Otis, W., and Nuckolls, R., Internal Friction in Solids. III. Experimental Demonstration of Thermoelastic Internal Friction, *Phys Rev*, 53, 100–101, 1938.

[9] Milligan, K. B., and Kinra, V. K., On the Thermoelastic Damping of a One-Dimensional Inclusion in a Uniaxial Bar, *Mech Res Comm*, 20, 137–142, 1993.

[10] Alblas, J. B., On the General Theory of Thermoelastic Damping, *Appl Sci Res*, 10, 349–362, 1961.

[11] Alblas, J. B., A Note on the theory of Thermoelastic Damping, *J Thermal Stresses*, 4, 333–335, 1981.

[12] Kinra, V. K., and Bishop, J. E., Elastothermodynamic Analysis of a Griffith Crack, *J Mech Phys So*, 44, 1305–1336, 1996.

[13] Bishop, J. E., and Kinra, V. K., Thermoelastic Damping of a Laminated Beam in Flexure and Extension, *J Reinforced Plastics and Composites*, 12, 210–226, 1993.

[14] Bishop, J. E., and Kinra, V. K., Elastothermodynamic Damping in Composite Materials, *Mech Compos Mat Struct*, 1, 75–93, 1994.

[15] Bishop, J. E., and Kinra, V. K., Analysis of Elastothermodynamic Damping in Particle-Reinforced Metal-Matrix Composites, *Metall and Mat Trans*, 26A, 2773–2783, 1995.

[16] Kinra, V. K., and Milligan, K. B., A Second Law Analysis of Thermoelastic Damping, *J Appl Mech*, 61, 71–76, 1994.

[17] Biot, M. A., Thermoelasticity and Irreversible Thermodynamics, *J Appl Phys*, 27, 240–253, 1956.

[18] Bishop, J. E., and Kinra, V. K., Equivalence of the Mechanical and Entropic Descriptions of Elastothermodynamic Damping in Composite Materials, *Mech Compos Mat Struct*, 3, 83–95, 1996.

[19] Bennewitz, K., and Rötger, H., On the Internal Friction of Solid Bodies: Absorption Frequencies of Metals in the Acoustic Region, *Phys Z*, 37, 578, 1936.

[20] Randall, R. H., Rose, F. C., and Zener, C., Intercrystalline Thermal Currents as a Source of Internal Friction, *Phys Rev*, 53, 343–348, 1939.

[21] Terzhagi, K., *Theoretical Soil Mechanics*, New York: J. Wiley, 1943.

[22] Wang, H., *Theory of Linear Poroelasticity*, Princeton NJ: Princeton University Press, 2000.

[23] Biot, M. A., General Theory of Three-Dimensional Consolidation, *J Appl Phys*, 12, 155–164, 1941.

[24] Biot, M. A., and Willis, D. G., The Elastic Coefficients of the Theory of Consolidation, *J Appl Mech*, 594–601, 1957.

[25] Biot, M. A., Theory of Propagation of Elastic Waves in a Fluid Saturated Porous Solid. I. Low Frequency Range, *J Acoust Soc Am*, 28, 168–178, 1956.

[26] Biot, M. A., Theory of Propagation of Elastic Waves in a Fluid Saturated Porous Solid. II. Higher Frequency Range, *J Acoust Soc Am*, 28, 179–191, 1956.

[27] Biot, M. A., Generalized Theory of Acoustic Propagation in Porous Dissipative Media, *J Acoust Soc Am*, 34, 1254–1264, 1962.

[28] Biot, M. A., Mechanics of Deformation and Acoustic Propagation in Porous Media, *J Appl Phys*, 33, 1482–1498, 1962.

[29] Kenyon, D. E., Consolidation in Transversely Isotropic Solids, *J Appl Mech*, 46, 65–70, 1979.

[30] Plona, T. J., Observation of a Second Bulk Compressional Wave in a Porous Medium at Ultrasonic Frequencies, *Appl Phys Let*, 36, 259–261, 1980.

[31] Berryman, J. G., Confirmation of Biot's Theory, *Appl Phys Let*, 37, 382–384, 1980.

[32] Lakes, R. S., Yoon, H. S., and Katz, J. L., Slow Compressional Wave Propagation in Wet Human and Bovine Cortical Bone, *Science*, 220, 513–515, 1983.

[33] Zeevaert, L., *Foundation Engineering for Difficult Subsoil Conditions*, New York: Van Nostrand, 1972.

[34] Gent, A. N., and Rusch, K. C., Viscoelastic Behavior of Open Cell Foams, *Rubber Chem and Technology*, 39, 388–396, 1966.

[35] Hilyard, N. C., *Mechanics of Cellular Plastics*, New York: Macmillan, 1982.

[36] AliMed®, Inc., Dedham, MA.

[37] Snoek, J. L., Effect of Small Quantities of Carbon and Nitrogen on the Elastic and Plastic Properties of Iron, *Physica*, 8, 711–733, 1941.

[38] Woodruff, E., A study of the Effects of Temperature upon a Tuning Fork, *Phys Rev*, 16, 325–355, 1903.

[39] Gibala, G., and Wert, C. A., The Clustering of Oxygen in Solid Solution in Niobium. I. Experimental, *Acta Metall*, 14, 1095–1105, 1966.

[40] Weller, M., The Snoek Relaxation in *bcc* Metals – from Steel Wire to Meteorites, *Mat Sci Eng*, A442, 21–30 2006.

[41] Cannelli, G., Cantelli, R., and Cordero, F., New anelastic Relaxation Effect in Y-Ba-Cu-O at Low Temperature: A Snoek-Type Peak Due to Oxygen Diffusion, *Phys Rev B*, 38, 7200–7202, 1988.

[42] Woirgard, J., Sarrazin, Y., and Chaumet, H., Apparatus for the Measurement of Internal Friction as a Function of Frequency between $10^{-5}$ and 10 Hz, *Rev Sci Instr*, 48, 1322–1325, 1977.

[43] Zener, C., Internal Friction of an Alpha Brass Crystal, *Trans Amer Inst of Mining and Metallurgical Eng* (AIME), 152, 122–126, 1943.

[44] Seraphim, D. P., and Nowick, A. S., Magnitude of the Zener Relaxation Effect. III. Anisotropy of the Relaxation Strength in Ag-Zn and Li-Mg solid solutions, *Acta Metall*, 9, 85 1961.

[45] Nowick, A. S., and Seraphim, D. P., Magnitude of the Zener Relaxation Effect. I. Survey of Alloy Systems. *Acta Metall*, 9, 40–48 1961.

[46] Mazot, P., Saissi, I., Halbwachs, M., and Woirgard, J., Etude des effets de frottement intérieur liés à l'apparition d'un ordre directionnel auto-induit dans un alliage Au 30 % Ni (Investigation of Internal Friction Effects Related to Orientation Ordering Induced in a Au-Ni 30% alloy), *J de physique*, Colloque C9, 44, 271–278, 1983.

[47] Li, C. Y., and Nowick, A. S., Magnitude of the Zener Effect. II. Temperature Dependence of the Relaxation Strength in α Ag-Zn, *Acta Metall*, 9, 49–58, 1961.

[48] Gorsky, W. S., Theory of Elastic Aftereffect in Unordered Mixed Crystals, *Z Physik*, SU8, 457–471, 1935.

[49] Schaumann, G., Völkl, J., and Alefeld, G., Relaxation Process Due to Long Range Diffusion of Hydrogen and Deuterium in Niobium, *Phys Rev Let*, 21, 891–893, 1968.

[50] Sinning, H. R., Mechanical Damping by Intercrystalline Diffusion of Hydrogen in Metallic Polycrystals, *Phys Rev Let*, 85, 3201–3204, 2000.

[51] Ferry, J. D., *Viscoelastic Properties of Polymers*, 2nd ed., New York: J. Wiley, 1970.

[52] Ferry, J. D., Landel, R. F., and Williams, M. L., Extensions of the Rouse Theory of Viscoelastic Properties to Undiluted Linear Polymers, *J Appl Phys*, 26, 359–362, 1955.

[53] Bagley, R. L., The Thermorheologically Complex Material, *Int J Eng Sci*, 29, 797–806, 1991.

[54] Bendler, J. T., and Schlesinger, M. F., Generalized Vogel Law for Glass Forming Liquids, *J Statistical Phys*, 53, 531–541, 1988.

[55] Baer, E., Hiltner, A., and Keith, H. D., Hierarchical structure in polymeric materials, *Science*, 235, 1015–1022, 1987.

[56] Blonski, S., Brostow, W., and Kubát, J., Molecular Dynamics Simulations of Stress Relaxation in Metals and Polymers, *Phys Rev B*, 49, 6494–6500, 1994.

[57] Jonscher, A. K., A New understanding of the Dielectric Relaxation of Solids, *J Mat Sci*, 16, 2037–2060, 1981.

[58] Jonscher, A. K., The 'universal' Dielectric Response, *Nature*, 267, 693–679, 1977.

[59] Dissado, L. A., and Hill, R. M., Non-Exponential Decay in Dielectrics and Dynamics of Correlated Systems, *Nature*, 279, 685–689, 1979.

[60] Palmer, R. G., Stein, D. L., Abrahams, E., and Anderson, P. W., Models of Hierarchically Constrained Dynamics for Glassy Relaxation, *Phys Rev Let*, 53, 958–961, 1984.

[61] Heijboer, J., Secondary Loss Peaks in Glassy Amorphous Polymers, *Int J Polym Mat*, 6, 11–37, 1977.

[62] Schwarzl, F., and Staverman, A. J., Time–Temperature Dependence of Linear Viscoelastic Behavior, *J Appl Phys*, 23, 838–843, 1952.

[63] Ward, I. M., and Hadley, D. W., *Mechanical Properties of Solid Polymers*, New York: J. Wiley, 1993.

[64] Plazek, D. J., Oh, Thermorheological Simplicity, Wherefore Art Thou? *J Rheology*, 40, 987–1014, 1996.

[65] Sperling, L. H., Sound and Vibration Damping with Polymers, in *Sound and Vibration Damping with Polymers*, R. D. Corsaro and L. H. Sperling, eds., Washington, DC: American Chemical Society, 1990.

[66] Wood-Adams, P., and Costeux, S., Thermorheological Behavior of Polyethylene: Effects of Microstructure and Long Chain Branching, *Macromolecules*, 34, 6281–6290, 2001.

[67] Plazek D. J., Anomalous Viscoelastic Properties of Polymers: Experiments and Explanations, *J Non-Crystalline Solids*, 353, 3783–3787, 2007.

[68] Treloar, L. R. G., *The Physics of Rubber Elasticity*, Oxford: Oxford University Press, 1975.

[69] Anthony, R. L., Caston, R. H., and Guth, E., Equations of State for Natural and Synthetic Rubber-Like Materials. I. Unaccelerated Natural Soft Rubber, *J Physical Chem*, 46, 826–840, 1942.

[70] Kelvin, Lord, (Thompson, W.) Collected Works, Vol. I, Cambridge, UK: Cambridge University Press, 1882, p. 291.

[71] Wood, L. A., and Roth, F. L., Stress-Temperature Relation in a Pure Gum Vulcanizate of Natural Rubber, *J Appl Phys*, 15, 781–780, 1944.

[72] Weiner, J. H., Entropic versus Kinetic Viewpoints in Rubber Elasticity, *Am J Phys*, 55, 746–749, 1987.

[73] Mooney, M., Wolstenholme, W. E., and Villars, D. S., Drift and Relaxation of Rubber, *J Appl Phys*, 15, 324–337, 1944.

[74] Kê, T. S., Experimental Evidence of the Viscous Behavior of Grain Boundaries in Metals, *Phys Rev*, 71, 533–546, 1947.

[75] Zener, C., Theory of the Elasticity of Polycrystals with Viscous Grain Boundaries, *Phys Rev*, 60, 906–908, 1941.

[76] Crossman, F. W., and Ashby, M. F., The Non-Uniform Flow of Polycrystals by Grain-Boundary Sliding Accommodated by Power Law Creep, *Acta Metall*, 23, 425–440, 1975.

[77] Bell, R. L., and Langdon, T. G., An Investigation of Grain Boundary Sliding During Creep, *J Mat Sci*, 2, 313–323, 1967.

[78] Zelin, M. G., and Mukherjee, A. K., Deformation Strengthening of Grain-Boundary Sliding in A Pb-62wt%Sn Alloy, *Phil Mag Let*, 68, 201–206, 1993.

[79] Zelin, M. G., Yang, H, Valiev, R. Z., and Mukherjee, A. K., Interaction of High-Temperature Deformation Mechanisms in a Magnesium Alloy with Mixed Fine and Coarse Grains, *Met Trans*, 23A, 3135–3140, 1992.

[80] Zelin, M. G., and Mukherjee, A. K., Common Features of Intragranular Dislocation Slip and Intergranular Sliding, *Phil Mag Let*, 68, 207–214, 1993.

[81] Gittus, J., ed. *Cavities and Cracks in Creep and Fatigue*, London: Applied Science Publishers, 1981.

[82] Nelson, D. J., and Hancock, J. W., Interfacial Slip and Damping in Fibre Reinforced Composites, *J Mat Sci*, 13, 2429–2440, 1978.

[83] Lakes, R. S., and Saha, S., Cement Line Motion in Bone, *Science*, 204, 501–503, 1979.

[84] Park, H. C., and Lakes, R. S., Cosserat Micromechanics of Human Bone: Strain Redistribution by a Hydration-Sensitive Constituent, *J Biomechanics*, 19, 385–397, 1986.

[85] Cremer, L., and Heckl, M., *Structure-Borne Sound*, 2nd ed., E. E. Ungar, trans., Berlin: Springer Verlag, 1988.

[86] Snoek, J., Mechanical After-Effect and Chemical Constitution, *Physica*, 6, 591–592, 1939.

[87] Zener, C., Stress-Induced Preferential Ordering of Pairs of Solute Atoms in Metallic Solid Solution, *Phys Rev*, 71, 34–38, 1947.

[88] Childs, B. G., and Le Claire, A. D., Relaxation Effects in Solid Solutions Arising from Changes in Local Order. I. Experimental, *Acta Metall*, 2, 718–726, 1954.

[89] Spears, C. J., and Feltham, P., On the Amplitude-Independent Internal Friction in Crystalline Solids, *J Mat Sci*, 7, 969–971, 1972.

[90] Feltham, P., Internal Friction in Concentrated Solid Solutions, *Phil Mag A*, 50, L35–L38, 1984.

[91] Granato, A., and Lücke, K., Theory of Mechanical Damping due to Dislocations, *J Appl Phys*, 27, 583–593, 1956.

[92] Lücke, K., and Granato, A. V., Simplified Theory of Dislocation Damping Including Point Defect Drag. I. Theory of Drag by Equidistant Point Defects, *Phys Rev B*, 24, 6991–7006, 1981.

[93] Robinson, W. H., Amplitude-Independent Mechanical Damping in Alkali Halides, *J Mat Sci*, 7, 115–123, 1972.

[94] Bordoni, P. G., Nuovo, M., and Verdini, L., Relaxation of Dislocations in Copper, *Nuovo Cimento*, 14, 273–314, 1959.

[95] Read, T. A., The Internal Friction of Single Metal Crystals, *Phys Rev*, 58, 371–380, 1940.

[96] Swift, I. H., and Richardson, J. E., Internal Friction of Zinc Single Crystals, *J Appl Phys*, 18, 417–425, 1947.

[97] Alers, G. A., and Thompson, D. O., Dislocation Contributions to the Modulus and Damping of Copper at Megacycle Frequencies, *J Appl Phys*, 32, 283–293, 1961.

[98] Lücke, K., and Granato, A. Simplified Theory of Dislocation Damping including Point Defect Drag. I. Theory of Drag by Equidistant Point Defects, *Phys Rev B*, 24, 6991–7006, 1981.

[99] Bordoni, P. G., Nuovo, M., and Verdini, L., Relaxation of Dislocations in Copper, Il, *Nuovo Cimento*, 14, 273–314, 1959.

[100] Seeger, A., On the Theory of the Low Temperature Internal Friction Peak Observed in Metals, *Phil Mag*, 8(1), 651–662, 1956.

[101] Marchesoni, F., Internal Friction by Pinned Dislocations: Theory of the Bordoni Peak, *Phys Rev Let*, 74, 2973–2976, 1995.

[102] Hasiguti, R., The Structure of Defects in Solids, *Ann Rev Mat Sci*, 2, 69–92, 1972.

[103] Hasiguti, R., Igata, M., and Kamoshita, G., Internal Friction Peaks in Cold-Worked Metals, *Acta Metall*, 10, 442–447, 1962.

[104] Salje, E. K. H., *Phase Transformations in Ferroelastic and Co-elastic Crystals*, Cambridge, UK: Cambridge University Press, 1990.

[105] Landau, L. D., On the Theory of Phase Transitions, in *Collected papers of L. D. Landau*, D. Ter Taar, ed., London: Gordon and Breach and New York: Pergamon, 1965.

[106] Arlt, G., Twinning in Ferroelectric and Ferroelastic Ceramics: Stress Relief, *J Mat Sci*, 25, 2655–2666, 1990.

[107] Falk, F. Model Free energy, Mechanics, and Thermodynamics of Shape Memory Alloys. *Acta Metall*, 28, 1773–1780, 1980.

[108] Jaglinski, T., Stone, D. S., Kochmann, D, and Lakes, R. S., Materials with Viscoelastic Stiffness Greater Than Diamond, *Science*, 315, 620–622, 2007.

[109] Benoit, W. Thermodynamics of Phase Transformations, in *Mechanical Spectroscopy $Q^{-1}$*, R. Schaller, G. Fantozzi, and G. Gremaud, eds., Switzerland: Trans Tech Publications, pp. 341–360, 2001.

[110] Zhang, J. X., Fung, P. C. W., and Zeng, W. G., Dissipation Function of the First-Order Phase Transformation in Solids via Internal Friction Measurements, *Phys Rev B*, 52, 268–277, 1995.

[111] Berlincourt, D. A., Curran, D. R., and Jaffe, H., Piezoelectric and Piezomagnetic Materials and Their Function in Transducers, in *Physical Acoustics*, E. P. Mason, ed., Vol. 1A, New York: Academic, 1964, pp. 169–270.

[112] Ikeda, T., The Internal Friction of Barium Titanate Ceramics, *J Physical Soc Japan*, 13, 809–818, 1958.

[113] Cheng, B. L., Gabbay, M., Maglione, M., Jorand, Y., Fantozzi, G., Domain Walls Motions in Barium Titanate Ceramics, *J de Physique IV*, Colloque C8, 6, 647–650, 1996.

[114] Postnikov, V. S., Pavlov, V. S., Gridnev, S. A., and Turkov, S. K., Interaction between the 90° Domain Walls and Point Defects of the Crystal Lattice in Ferroelectric Ceramics, *Sov Phys Sol State*, 10, 1267–1270, 1968.

[115] Zhang, J. X., Yang, Z. H., and Fung, P. C. W., Dissipation Function of the First-Order Phase Transformation in $VO_2$ Ceramics Solids by Internal Friction Measurements, *Phys Rev B*, 52, 278–284, 1995.

[116] Bidaux, J. E., Schaller, R., and Benoit, W., Study of the H.C.P–F.C.C. Phase Transition in Cobalt by Acoustic Measurements, *Acta Metall*, 37, 803–811, 1989.

[117] Harrison, R. J., Simon A. T., and Redfern, S. A. T., The Influence of Transformation Twins on the Seismic-Frequency Elastic and Anelastic Properties of Perovskite: Dynamical Mechanical Analysis of Single Crystal $LaAlO_3$, *Physics of the Earth and Planetary Interiors*, 134, 253272, 2002.

[118] Golovin, I. S., and Rivière, A., Zener Relaxation in Ordered and Disordered Fe-(22–28 %) Al Alloys, *Mat Sci Eng*, A, 442, 86–91, 2006.

[119] McCrum, N. G., Read, B. E., and Williams, G., *Anelastic and Dielectric Effects in Polymeric Solids*, NY: Dover, 1991, p. 46.

[120] Lulay, J., and Wert, C., Internal Friction in Alloys of Mg and Cd, *Acta Metall*, 4, 627–631, 1956.

[121] Cook, L. S., and Lakes, R. S., Viscoelastic Spectra of $Cd_{0.67}Mg_{0.33}$ In Torsion and Bending, *Metall Trans*, 26A, 2037–2039, 1995.

[122] Litov, E., and Uehling, E. A., Polarization Relaxation in the Ferroelectric Transition Region of $KD_2PO_4$, *Phys Rev Let*, 21, 809–812, 1968.

[123] Brody, E. M., and Cummins, H. Z., Brillouin-Scattering Study of the Ferroelectric Transition in $KH_2PO_4$, *Phys Rev Let*, 21, 1263–1266, 1968.

[124] Fantozzi, G. Ferroelectricity, in *Mechanical Spectroscopy $Q^{-1}$*, R. Schaller, G. Fantozzi, and G. Gremaud, eds. Switzerland: Trans Tech Publications, 2001, pp. 361–381.

[125] Pérez-Sáez, R. B., Recarte, V., Nó, M. L., and San Juan, J., Anelastic Contributions and Transformed Volume Fraction during Thermoelastic Martensitic Transformations, *Phys Rev B*, 57, 5684–5692, 1997.

[126] Siefert, A. V., and Worrell, F. T., The Role of Tetragonal Twins in the Internal Friction of Copper Manganese Alloys, *J Appl Phys*, 22, 1257–1259, 1951.

[127] Van Humbeeck, J. Stoiber, J., Delaey, L., and Gotthardt, R., The High Damping Capacity of Shape Memory Alloys, *Z Metallk*, 86, 176–183, 1995.

[128] Fung, P. C., Zhang, J. X., Lin, Y., Liang, K. F., Lin, Z. C., Analysis of Dissipation of a Burst-Type Martensite Transformation in a Fe-Mn Alloy by Internal Friction Measurements, *Phys Rev B*, 54, 7074–7083, 1996.

[129] Jaglinski, T., Frascone, P., Moore, B., Stone, D., and Lakes, R. S., Internal Friction due to Negative Stiffness in the Indium-Thallium Martensitic Phase Transformation, *Phil Mag*, 86 (27-21), 4285–4303, 2006.

[130] Schoeck, G., Bisogni, E., and Shyne, J., The Activation Energy of High Temperature Internal Friction, *Acta Metall*, 12, 1466–1468, 1964.

[131] Cagnoli, G., Gammaitoni, L., Marchesoni, F., and Segoloni, D., On Dislocation Damping at Low Frequency, *Philosophical Magazine A*, 68, 865–870, 1993.

[132] Brodt, M., Cook, L. S., and Lakes, R. S., Apparatus for Determining the Properties of Materials over Ten Decades of Frequency and Time: Refinements, *Rev Sci Instr*, 66, 5292–5297, 1995.

[133] Lakes, R. S., and Quackenbush, J., Viscoelastic Behaviour in Indium Tin Alloys over a Wide Range of Frequency and Time, *Phil Mag Let*, 74, 227–232, 1996.

[134] Hirscher, M., Schweitzer, E., Weller, M., and Kronmüller, H., Internal Friction in NiAl Single Crystals, *Phil Mag Let*, 74, 189–194, 1996.

[135] Marchesoni, F., and Patriarca, M., Self-Organized Criticality in Dislocation Networks, *Phys Rev Let*, 72, 4101–4104, 1994.

[136] Braginsky, V. B., Mitrofanov, V. P., and Panov, V. I., *Systems with Small Dissipation*, Chicago: University of Chicago Press, 1985.

[137] Papadakis, E. P., Ultrasonic Attenuation Caused by Scattering in Polycrystalline Media, in *Physical Acoustics*, IVB, 269–328, 1968.

[138] Zener, C., Internal friction in Solids: V. General theory of Macroscopic Eddy Currents, *Phys Rev*, 53, 1010–1013, 1938.

[139] Coronel, V. F., and Beshers, D. N., Magnetomechanical Damping in Iron, *J Appl Phy*, 64, 2006–2015, 1988.

[140] Mason, W. P., Rotational Relaxation in Nickel at High Frequencies, *Rev Mod Phys*, 25, 136–140, 1953.

[141] Ganganna, H. V., Fiore, N. F., and Cullity, B. D., Microeddy Current Magnetic Damping in Nickel, *J Appl Phy*, 42, 5792–5797, 1971.

[142] Berry, B. S., and Pritchet, W. C. $\Delta$E-Effect and Macroeddy-Current Damping in Nickel, *J Appl Phys*, 49, 1983–1985, 1978.

[143] Zener, C., Internal Friction in Solids. V. General Theory of Macroscopic Eddy Currents, *Phys Rev*, 53, 1010–1013, 1938.

[144] Bodriakov, V. Yu., Kikitin, S. A., and Tishin, A. M., Magnetoelastic Properties of Gadolinium, *J Appl Phys*, 72, 4247–4249, 1992.

[145] Roberts, J. T. A., and Barrand, P., Magnetomechanical Damping Behaviour in Pure Nickel and a 20 wt. % copper-nickel alloy, *Acta Metall*, 15, 1685–1697, 1967.

[146] Lakes, R. S., Shape-Dependent Damping in Piezoelectric Solids, *IEEE Trans Sonics, Ultrasonics*, SU27, 208–213, 1980.

[147] Martin, G. E., Dielectric, Piezoelectric, and Elastic Losses in Longitudinally Polarized Segmented ceramic Tubes, *U.S. Navy J. Underwater Acoustics*, 15, 329–332, April 1965.

[148] Furukawa, T., and Fukada, E. Piezoelectric Relaxation in Composite Epoxy-PZT System Due to ionic Conduction, *Japan J Appl Phys*, 16, 453–458, March 1977.

[149] Bur, A. J., Measurements of the Dynamic Piezoelectric Properties of Bone as a Function of Temperature and Humidity, *J Biomechanics*, 9, 495–507, 1976.

[150] Holland, R. Representation of Dielectric, Elastic, and Piezoelectric Losses by Complex Coefficients, *IEEE Trans Sonics Ultrason*, SU-14, 18–20, January 1967.

[151] Jaffe, H., and Berlincourt, D. A., Piezoelectric Transducer Materials, *Proc IEEE*, 53, 1372–1386, 1965.

[152] Cady, W. G., *Piezoelectricity*. New York: Dover, 1964.

[153] Joffe, A. F., *The Physics of Crystals*. New York: McGraw-Hill, 1928.

[154] IRE Standards on Piezoelectric Crystals: Measurements of Piezoelectric Ceramics, *Proc IRE*, 49, 1161–1169, July 1961.

[155] IEEE Standard on Piezoelectricity, IEEE 176-1978; New York: Inst. Electrical, Electronics Engineers, 1978.

[156] Lakes, R. S., Thermoelastic Damping in Materials with a Complex Coefficient of Thermal Expansion, *J Mech Behav Mts*, 8, 201–216, 1997.

[157] Palmer, R. G., Stein, D. L., Abrahams, E., and Anderson, P. W., Models of Hierarchically Constrained Dynamics For Glassy Relaxation, *Phy Rev Let*, 53, 958–961, 1984.

[158] Klafter, J., and Schlesinger, M. F., On the Relationship among Three Theories of Relaxation in Disordered Systems, *Proc Natl Acad Sci USA*, 83, 848–851, 1986.

[159] Jonscher, A. K., The 'Universal' Dielectric Response, *Nature*, 267, 693–679, 1977.

[160] Ngai, K. L., Universality of Low-Frequency Fluctuation, Dissipation, and Relaxation Properties of Condensed Matter. I, Comments, *Solid State Physics*, 9, 127–140, 1979.

[161] Scher, H., Shlesinger, M. F., and Bendler, J. T., Time Scale Invariance in Transport and Relaxation, *Physics Today*, 44, 26–34, 1991.

[162] Quinn, J. J., Speake, C. C., and Brown, L. M., Materials Problems in the Construction of Long-Period Pendulums, *Phil Mag A*, 65, 261–276, 1992.

[163] Cottrell, A. H., Andrade Creep, *Phil Mag Let*, 73, 35–37, 1996.

[164] Cottrell, A. H., Logarithmic and Andrade Creep, *Phil Mag Let*, 75, 301–307, 1997.

[165] Jonscher, A. K., *Dielectric Relaxation in Solids*, London: Chelsea Dielectrics Press, 1982, p. 294.

[166] Schiener, B., Böhmer, R., Loidl, A., and Chamberlin, R. V., Nonresonant Spectral Hole-burning in the Slow Dielectric Response of Supercooled Liquids, *Science*, 274, 752–754, 1996.

[167] Shi, X., and McKenna, G. B., Spectral Hole Burning Spectroscopy: Evidence for Heterogeneous Dynamics in Polymer Systems, *Phys Rev Let*, 94, 157801, 2005.

[168] Ashby, M. F., A First Report on Deformation Mechanism Maps, *Acta Metall,* 20, 887–897, 1972.

[169] Frost, H. J., and Ashby, M. F., *Deformation Mechanism Maps*, Oxford: Pergamon, 1982.

[170] Hynes, A., and Doremus, R., Theories of Creep in Ceramics, *Critical Rev Solid State and Materials Science,* 21, 129–187, 1996.

[171] Karato, S., and Wu, P., Rheology of the Upper mantle: A synthesis, *Science,* 260, 771–778, 1993.

[172] Martin, J. L., Creep in Pure Metals, in *Creep Behaviour of Crystalline Solids,* B. Wiltshire and R. W. Evans, eds., Pineridge Press (U.K.), 1985, pp. 1–33; adduced diffusion results of Sherby and Miller, 1979.

[173] Oikawa, H., and Langdon, T. G., The Creep Characteristics of Pure Metals and Metallic Solid Solution Alloys, in *Creep Behaviour of Crystalline Solids,* B. Wiltshire and R. W. Evans, eds., UK: Pineridge Press, 1985, Swansea, pp. 33–82.

[174] Ritchie, I. G., Pan, Z. L., and Goodwin, F. E., Characterization of the Damping Properties of Die-Cast Zinc-Aluminum Alloys, *Met Trans,* 22A, 617–622, 1991.

[175] Ritchie, I. G., and Pan, Z. L., High Damping Metals and Alloys, *Met Trans,* 22A, 607–616, 1991.

[176] Laddha, S., and Van Aken, D. C., On the Application of Magnetomechanical Models to Explain Damping in an Antiferromagnetic Copper-Manganese Alloy, *Met Trans,* 26A, 957–964, 1995.

[177] Nabarro, F. R. N., and de Villiers, H. L. *The Physics of Creep*, London: Taylor and Francis, 1995.

[178] Dyson, F., Time without End – Physics and Biology in an Open Universe, *Rev Mod Phys,* 51, 447–460, 1979.

[179] Ashby, M. F., On the Engineering Properties of Materials, *Acta Metall,* 37, 1273–1293 (1989).

[180] *Modern Plastics Encyclopedia*, Vol. 51 No. 10A, New York: McGraw Hill, 1974–1975.

# 9

# Viscoelastic Composite Materials

## 9.1 Introduction

Composite materials are those that contain two or more distinct constituent materials or phases, on a microscopic or macroscopic size scale larger than the atomic scale, and in which physical properties are significantly altered in comparison with those of a homogeneous material. In this vein, fiberglass and other fibrous materials are viewed as composites, but alloys, such as brass, are not. Semicrystalline polymers, such as polyethylene, have a heterogeneous structure, which can be treated via composite theory. Biological materials also have a heterogeneous structure and are known as natural composites. Composites may contain solid, liquid, and gas phases. For example, composites of gas and liquid include mist and foam; composites of solid and gas include foam and smoke. Composites with a structural role have several solid phases or a connected solid phase with gas or liquid in the interstices (structural foam and honeycomb).

## 9.2 Composite Structures and Properties

### 9.2.1 Ideal Structures

The properties of composites are greatly dependent upon *microstructure*. Composites differ from homogeneous materials in that considerable control can be exerted over the larger scale structure; and hence over the desired properties. In particular, the properties of a composite depend upon the *shape* of the heterogeneities, upon the *volume fraction* occupied by them, and upon the *interface* between the constituents. Volume fraction refers to the ratio of the volume of a constituent to the total volume of a composite specimen. The shape of the heterogeneities in a composite is classified as follows. The principal inclusion shape categories (Figure 9.1) are the particle, with no long dimension; the fiber, with one long dimension; and the platelet (flake, lamina), with two long dimensions. The inclusions may vary in size and shape within a category. For example, particulate inclusions may be spherical, ellipsoidal, polyhedral, or irregular. *Cellular solids* are those in which the "inclusions"

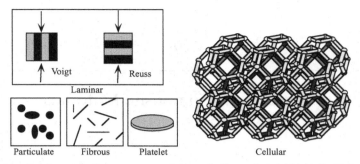

Figure 9.1. Idealized composite structures. Arrows indicate force. Top left: Voigt and Reuss structures consist of laminations of two solid phases; black and gray. Bottom left: inclusions (dark) in a matrix (white) can have different shapes.

are voids or cells, filled with air or liquid. Cellular solids include honeycombs, in which the structure is largely two-dimensional, and foams, in which the structure is fully three-dimensional. Foams can be *open-cell*, in which the foam has a structure of "ribs," with no barrier between adjacent cells; or *closed cell*, in which plate or membrane elements separate adjacent cells. Open-cell foams are of particular interest in the context of viscoelasticity, since viscoelastic damping can arise as a result of the viscosity of fluids (such as water or air) moving through the pore structure as discussed in §8.3. Hierarchical composites have structural elements within structural elements. The *coated spheres* morphology has multiple length scales. The entire volume is filled with particles of one phase coated with a layer of a second phase. This morphology is used in theoretical analyses of extremal behavior. A detailed treatment of structure property relations is presented in the excellent book by Milton [1].

### 9.2.2 Anisotropy due to Structure

Composites may be isotropic or anisotropic. Properties of anisotropic materials depend on direction; isotropic materials have no preferred direction. Composites containing spherical particulate inclusions distributed randomly in an isotropic matrix are macroscopically isotropic. Unidirectional fibrous composites and laminates are highly anisotropic if the phases differ greatly in modulus. Laminates containing fibrous layers can be prepared with a controlled degree of anisotropy. Anisotropy can be dealt with using a tensorial stress–strain relation (see also §2.8). For elastic materials, Hooke's law of linear elasticity is given by

$$\sigma_{ij} = C_{ijkl}\epsilon_{kl}, \tag{9.1}$$

with $C_{ijkl}$ as the elastic modulus tensor, and the usual Einstein summation convention assumed in which repeated indices are summed over Sokolnikoff [2], Timoshenko and Goodier [3], and Fung [4]. In the compliance formulation, $\epsilon_{ij} = J_{ijkl}\sigma_{kl}$ with $J_{ijkl}$ as the compliance tensor.

There are 81 components of $C_{ijkl}$ (or of the compliance $J_{ijkl}$), but taking into account the symmetry of the stress and strain tensors, only 36 of them are independent.

The constant $C_{1111}$ is a modulus representing the ratio of stress to strain in the one or $x$ direction, with all other strain components zero. It is revealed by a confined compression test or an ultrasonic test at high frequency for which the Poisson strain is zero. $C_{1111}$ is not equal to a Young's modulus unless Poisson's ratio is zero. The constant $J_{1111} = 1/E_1$ is the inverse of the Young's modulus $E_1$ in the one direction because all other stress components, except $\sigma_{11}$, are zero. The constant $C_{2323}$ represents a shear modulus because it is the ratio of shear stress to a corresponding shear strain. The constant $J_{1212} = -v_{12}/E_1$ is related to Poisson's ratio $v$. The modulus and compliance tensors are at times written [5] in a six by six matrix form by using the correspondence $11 \rightarrow 1, 22 \rightarrow 2, 33 \rightarrow 3, 23 \rightarrow 4, 13 \rightarrow 5, 12 \rightarrow 6$. In this reduced notation, $C_{2323} \rightarrow C_{44}$. The modulus does not transform (under coordinate changes) as such a matrix; it still transforms as a fourth rank tensor.

If the elastic solid is describable by a strain energy function, the number of independent elastic constants is reduced to 21 by virtue of the resulting symmetry $C_{ijkl} = C_{klij}$. An elastic modulus tensor with 21 independent constants describes an anisotropic material with the most general type of anisotropy, triclinic symmetry. Such materials, including triclinic crystals, exhibit nonzero values of elastic constants such as $J_{1123}$ which give rise to shear deformation in response to axial stress. Bend-twist coupling can occur in such an asymmetric material. Materials with orthotropic symmetry are invariant to reflections in two orthogonal planes and are describable by nine elastic constants. Materials with axisymmetry, also called transverse isotropy or hexagonal symmetry, are invariant to 60° rotations about an axis and are describable by five independent elastic constants. For example, a unidirectional fibrous material may have orthotropic symmetry if the fibers are arranged in a rectangular packing, or hexagonal symmetry if the fibers are packed hexagonally. Wood and bovine plexiform bone have orthotropic symmetry while human compact bone has hexagonal symmetry. Materials with cubic symmetry are invariant to 90-degree rotations about each of three orthogonal axes. Cross-ply fibrous composites can exhibit cubic symmetry provided the concentration of fibers in each orthogonal direction is the same. Cubic materials are describable by three elastic constants. Isotropic materials, with properties independent of direction are describable by two independent elastic constants.

The elastic properties of composites have been studied extensively. The upper and lower bounds of *stiffness* of two phase and multi- phase composites have been obtained by Hashin, and by Hashin and Shtrikman in terms of volume fraction of constituents [6, 7]. Much work on composites is reviewed by Hashin [8].

In linearly viscoelastic materials, the Boltzmann superposition integral has the following form in the tensorial modulus formulation:

$$\sigma_{ij}(t) = \int_0^t C_{ijkl}(t-\tau)\frac{d\epsilon_{kl}}{d\tau}d\tau. \qquad (9.2)$$

Each component of the relaxation modulus tensor $C_{ijkl}(t)$ can have not only a different value but also a different time dependence. Similarly each component of the complex modulus tensor $C_{ijkl}^*$ can have a different frequency dependence.

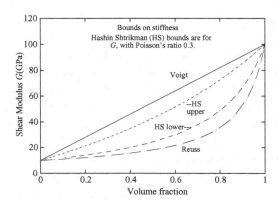

Figure 9.2. Modulus volume fraction of a constituent for several elastic composites. HS refers to the Hashin–Shtrikman bounds for isotropic composites.

For viscoelastic materials, we may have an asymmetric [9, 10] modulus tensor $C_{ijkl}(t) \neq C_{klij}(t)$ since there is no strain energy function for viscoelastic materials. However, $C_{ijkl}(0) = C_{klij}(0)$, and $C_{ijkl}(\infty) = C_{klij}(\infty)$. No effect of asymmetry in the modulus or compliance tensor is seen in isotropic materials, however in some anisotropic materials, certain equalities of Poisson's ratio terms seen in elastic materials do not obtain in viscoelastic materials.

As for viscoelastic composites, if the phases are linearly viscoelastic the effective relaxation and creep functions, or the complex dynamic modulus or compliance, of the composite can be calculated from constituent properties by the correspondence principle of the theory of linear viscoelasticity [11, 12]. In some cases explicit results in terms of linear viscoelastic matrix properties were given, permitting direct use of experimental information [13]. If the inclusions have an effect on the dislocation density in a crystalline or polycrystalline matrix, then the matrix itself is changed by the presence of inclusions, and behavior will differ from the correspondence solution.

## 9.3 Prediction of Elastic and Viscoelastic Properties

### 9.3.1 Basic Structures: Correspondence Solutions

Composite structures such as those of the Voigt and Reuss composites give rise to exact analytical solutions for the elastic moduli; these solutions can be subjected to the correspondence principle to obtain solutions for viscoelastic composites. Similarly, properties of composites containing dilute concentrations spherical or platelet inclusions are also known in terms of constituent properties. Moreover, for the simplest case of an *elastic* two-phase composite, the stiffness of Voigt and Reuss composites (Figure 9.1) described below represent rigorous upper and lower bounds on the Young's modulus for a given volume fraction of one phase (Figure 9.2). The Hashin–Shtrikman equations presented below represent upper and lower bounds on the elastic stiffness of *isotropic* composites. Behavior of viscoelastic composites is predicted by the correspondence principle [14–16, 18], by which the relationship between constituent and composite elastic properties can be converted to a steady state harmonic viscoelastic relation by replacing moduli such as the Young's moduli $E$ by $E^*(i\omega)$ or $E^*$, in which $\omega$ is the angular frequency of the harmonic loading.

In this section, we explore the viscoelastic properties of several two phase (with volume fractions $V_1 + V_2 = 1$) composites of well defined structure, in terms of the assumed properties of the constituents [19]. As for the Voigt and Reuss structures, we will find that curves representing the viscoelastic properties enclose a region in a stiffness-loss map. However, these curves are not proven to be bounds on the viscoelastic behavior.

### 9.3.2 Voigt Composite

The Voigt composite can contain laminations as shown in Figure 9.1; the strain in each phase is the same. In a simple, one-dimensional view, one considers the shear modulus $G$ or neglects any complications due to mismatch in the Poisson effect to evaluate Young's modulus $E$. For elastic materials with no slip between the phases,

$$G_c = G_1 V_1 + G_2 V_2, \tag{9.3}$$

in which $G_c$, $G_1$ and $G_2$ refer to the shear modulus of the composite, phase 1 and phase 2, and $V_1$ and $V_2$ refer to the volume fraction of phase 1 and phase 2 with $V_1 + V_2 = 1$. The dependence of stiffness on volume fraction is shown in Figure 9.2. Use of the correspondence principle gives

$$G_c^* = G_1^* V_1 + G_2^* V_2, \tag{9.4}$$

with $G^* = G' + iG''$ and loss tangent $\tan\delta = G''/G'$. Taking the ratio of real and imaginary parts, the loss tangent of the composite $\tan\delta_c = G_c''/G_c'$ is given by

$$\tan\delta_c = \frac{V_1 \tan\delta_1 + V_2 \frac{G_2'}{G_1'} \tan\delta_2}{V_1 + \frac{G_2'}{G_1'} V_2}. \tag{9.5}$$

The relation between stiffness and damping is shown in the stiffness-loss map in Figure 9.3, in which phase 1 is stiff; phase 2 is high loss. In the time domain, the correspondence principle gives, in the Laplace plane, $sG_c(s) = sG_1(s)V_1 + sG_2(s)V_2$, so,

$$G_c(t) = G_1(t)V_1 + G_2(t)V_2. \tag{9.6}$$

### 9.3.3 Reuss Composite

The geometry of the Reuss model structure is shown in Figure 9.1; each phase experiences the same stress but different strain. For elastic materials,

$$\frac{1}{G_c} = \frac{V_1}{G_1} + \frac{V_2}{G_2}. \tag{9.7}$$

The dependence of stiffness on volume fraction is shown in Figure 9.2. Since the constituents are aligned, this composite is *structurally* anisotropic. Because the Reuss laminate is identical to the Voigt laminate except for orientation with respect to the stress, the laminate is *mechanically* anisotropic. Again, using the correspondence principle, the viscoelastic relation is obtained as

$$\frac{1}{G_c^*} = \frac{V_1}{G_1^*} + \frac{V_2}{G_2^*}. \tag{9.8}$$

Again, separating the real and imaginary parts of $G_c^*$, the loss tangent of the composite $\tan\delta_c$ is obtained:

$$\tan\delta_c = \frac{(\tan\delta_1 + \tan\delta_2)[V_1 + V_2\frac{G_1'}{G_2'}] - (1 - \tan\delta_1 \tan\delta_2)[V_1 \tan\delta_2 + V_2 \tan\delta_1 \frac{G_1'}{G_2'}]}{(1 - \tan\delta_1 \tan\delta_2)[V_1 + V_2\frac{G_1'}{G_2'}] + (\tan\delta_1 + \tan\delta_2)[V_1 \tan\delta_2 + V_2 \tan\delta_1 \frac{G_1'}{G_2'}]}. \tag{9.9}$$

In the compliance formulation, the Reuss relation can be written more simply in terms of the compliances $J^* = 1/G^*$,

$$J_c^* = J_1^* V_1 + J_2^* V_2. \tag{9.10}$$

The corresponding analysis of the loss tangent is also simpler.

$$\tan\delta_c = \frac{V_1 \tan\delta_1 + V_2\frac{J_2'}{J_1'} \tan\delta_2}{V_1 + \frac{J_2'}{J_1'} V_2}. \tag{9.11}$$

The relation between stiffness and damping is shown, in comparison with the Voigt composite, in the stiffness-loss map in Figure 9.3. In the time domain, a correspondence solution for the compliance is simple as in the above case for the Voigt modulus.

### 9.3.4 Hashin–Shtrikman Composite

Allowing for "arbitrary" phase geometry, the upper and lower bounds on the elastic moduli of an *isotropic* composite as a function of composition were developed using variational principles. The lower bound for the elastic shear modulus $G_L$ of the composite was given as [7]

$$G_L = G_2 + \frac{V_1}{\frac{1}{G_1 - G_2} + \frac{6(K_2 + 2G_2)V_2}{5(3K_2 + 4G_2)G_2}}, \tag{9.12}$$

in which $K_1$, $G_1$, $V_1$; $K_2$, $G_2$ and $V_2$ are the bulk modulus, shear modulus and volume fraction of phases 1 and 2, respectively. Here, $G_1 > G_2$, so that $G_L$ represents the lower bound on the shear modulus. Interchanging the numbers 1 and 2 in

Figure 9.3. Stiffness-loss map. Calculated behavior of several composites.
Each point corresponds to a different volume fraction. Reuss: Solid triangles, Reuss. Solid diamonds: Voigt.
Open triangles: Hashin–Shtrikman, different values of Poisson's ratio of compliant phase. Hashin–Shtrikman lower
(top), $G_L$, Equation 9.13, $\nu = 0.3$; and
$\nu = 0.45$. $\triangle$ Open squares $\nu = 0.3(1 +$
$0.1i)$. Diagonal squares: H–S upper, $G_U$,
Equation 9.14, $\nu = 0.3$, x; Plus $+$: $\nu =$
0.45.

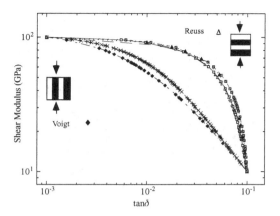

Equation 9.12 results in the upper bound $G_U$ for the shear modulus. Bounds on
the stiffness are shown in Figure 9.2.

As for viscoelastic materials, the correspondence principle is again applied [19].
The complex viscoelastic shear moduli of the composite $G_L^*$ and $G_U^*$ are obtained as

$$G_L^* = G_2^* + \frac{V_1}{\frac{1}{G_1^*-G_2^*} + \frac{6(K_2^*+2G_2^*)V_2}{5(3K_2^*+4G_2^*)G_2^*}}, \tag{9.13}$$

and

$$G_U^* = G_1^* + \frac{V_2}{\frac{1}{G_2^*-G_1^*} + \frac{6(K_1^*+2G_1^*)V_1}{5(3K_1^*+4G_1^*)G_1^*}}. \tag{9.14}$$

In these cases, the loss tangent is more complicated to write explicitly, so it is
more expedient to graphically display computed numerical values in the stiffness
loss map in Figure 9.3.

The lower bound for bulk modulus of an elastic material is as follows [8]:

$$K_L = K_2 + \frac{V_1(K_1 - K_2)(3K_2 + 4G_2)}{(3K_2 + 4G_2) + 3V_2(K_1 - K_2)}. \tag{9.15}$$

Here, $K_1$ and $K_2$ are the bulk moduli of the two phases, $G_1$ and $G_2$ are the shear
moduli, and $V_1$ is the volume fraction of the first phase. In the viscoelastic case, these
moduli become complex: $K_1 \rightarrow K_1^*$.

Because the Hashin–Shtrikman formulae represent exact solutions for hierarchical composites, the corresponding viscoelastic properties are also attainable.
These formulae are not necessarily bounds for viscoelastic properties.

### 9.3.5 Spherical Particulate Inclusions

For a small volume fraction, $V_1 = 1 - V_2$, of spherical elastic inclusions (particles) in a continuous phase of another elastic material, the shear modulus of the

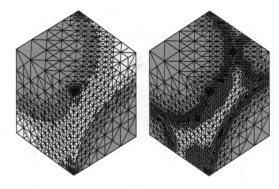

Figure 9.4. Concentrated spherical inclusions in a matrix: finite-element mesh (adapted from [17]). Left: single particle size. Right: two particle sizes.

composite $G_c$ was given as [16]

$$\frac{G_c}{G_1} = 1 - \frac{15(1 - \nu_1)(1 - \frac{G_2}{G_1})V_2}{7 - 5\nu_1 + 2(4 - 5\nu_1)\frac{G_2}{G_1}}. \tag{9.16}$$

in which $\nu_1$ is the Poisson's ratio of phase 1, and phase 1 and phase 2 represent the matrix material and the inclusion material respectively. The stiffness of such a composite, as a function of volume fraction of inclusions, is close to the Hashin–Shtrikman lower bound for isotropic materials. For larger volume fractions of inclusions, analysis is more complicated as a result of the interaction of stress fields around nearby inclusions. Stiff spherical inclusions are less efficient, per volume, in achieving a stiff composite than fiber or platelet inclusions. Conversely, soft spherical inclusions have the least effect in reducing the stiffness, in comparison to other inclusion shapes. The case of a large concentration of spherical inclusions is not amenable to simple analytical solution, so finite-element analyses are done [17], Figure 9.4. The matrix and the particulate reinforcement phase may be analyzed first as linearly elastic solids. Once the effective elastic properties of the composite are obtained by finite element analysis, a curve fit is generated and the correspondence principle is used to obtain effective viscoelastic properties of the composite.

Using the correspondence principle again [19], Equation 9.16 becomes

$$G_c^* = G_1^* - \frac{15(1 - \nu_1^*)(G_1^* - G_2^*)V_2}{7 - 5\nu_1^* + 2(4 - 5\nu_1^*)\frac{G_2^*}{G_1^*}}, \tag{9.17}$$

for the complex shear modulus of the composite. The loss tangent again is complicated to write explicitly, but can readily be computed numerically assuming there is no relaxation in Poisson's ratio, as presented in Figure 9.5. Behavior of a composite with soft particles approximates the Voigt relation in the stiffness-loss map; by contrast a composite with a dilute concentration of stiff spherical inclusions behaves more like a Reuss composite. In the time domain, relaxation in Poisson's ratio complicates the solution as discussed in Chapter 5.

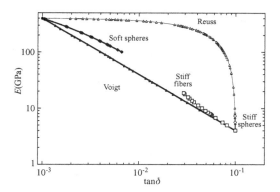

Figure 9.5. Stiffness-loss map. Calculated behavior of several composites with assumed phase properties different from those in Figure 9.3. Each point corresponds to a different volume fraction. Effect of phase geometry for Voigt (Solid triangles), Reuss (open triangles), stiff spheres (open diamonds), soft spheres (solid diamonds), and randomly oriented stiff fibers (squares).

### 9.3.6 Fiber Inclusions

For a three dimensional dilute concentration of randomly oriented fiber elastic inclusions of phase 2 in a matrix of phase 1, the Young's modulus $E_c$ of the composite is as follows [15], assuming a Poisson's ratio of both fiber and matrix to be 1/4:

$$E_c = \frac{1}{6} E_2 V_2 + E_1 \frac{1 + \frac{1}{4} V_2 + \frac{1}{6} V_2^2}{1 - V_2}, \qquad (9.18)$$

using the correspondence principle,

$$E_c^* = \frac{1}{6} E_2^* V_2 + E_1^* \frac{1 + \frac{1}{4} V_2 + \frac{1}{6} V_2^2}{1 - V_2}. \qquad (9.19)$$

At the low-volume fraction $V_2$ of randomly oriented fibers, the stiffening effect of the fibers is one-sixth of the value obtained in the Voigt composite in which the fibers are all aligned. However for viscoelastic composites, the curve for random fibers in the stiffness-loss map in Figure 9.5 is close to the curve for the Voigt solid. Corresponding volume fractions are not, however, identical.

### 9.3.7 Platelet Inclusions

For a dilute concentration of randomly oriented platelet elastic inclusions (flakes) of phase 2 in a matrix of phase 1, the shear modulus $G_c$ of the composite [16] was given as

$$G_c = G_1 + \frac{V_2(G_2 - G_1)}{15} \left[ \frac{9K_2 + 4(G_1 + 2G_2)}{K_2 + \frac{4}{3} G_2} + 6 \frac{G_1}{G_2} \right]. \qquad (9.20)$$

Stiff platelets are more efficient, per volume, in achieving a stiff composite than fiber or particle inclusions. At low volume fraction $V_2$ of randomly oriented platelets, the stiffening effect of the fibers is one-half of the value obtained in the Voigt composite: $E_c \approx E_1 V_1 + (\frac{1}{2}) E_2 V_2$. The predicted stiffness of a random platelet reinforced composite is identical to the Hashin–Shtrikman upper bound for isotropic materials [20], however this limiting stiffness will only be achieved if the platelets are infinitely thin [21].

Again, using the correspondence principle, Equation 9.20 becomes

$$G_c^* = G_1^* + \frac{V_2(G_2^* - G_1^*)}{15}[\frac{9K_2^* + 4(G_1^* + 2G_2^*)}{K_2^* + \frac{4}{3}G_2^*} + 6\frac{G_1^*}{G_2^*}], \qquad (9.21)$$

for the complex shear modulus of the composites. In viscoelastic composites, the curve for platelet inclusions in the stiffness-loss map is close to the Voigt and Hashin–Shtrikman upper-bound curves, and corresponds to minimal damping for given composite stiffness.

We may consider "unintentional" platelet structure in materials, such as gray cast iron, as described by Millett et al. [22]. Gray cast iron exhibits a loss tangent of about 0.011 over a range of frequencies, and it contains platelet shaped inclusions of graphite with a loss tangent of 0.015. The graphite gives rise to most of the loss; the cause is thought to be dislocation-loop motion.

### 9.3.8 Stiffness-Loss Maps

Predicted properties of viscoelastic composites are plotted as stiffness-loss maps (plots of $|E^*|$ vs. $\tan\delta$) as shown in the following figures. Since the dynamic properties depend on frequency, each map corresponds to a particular frequency.

In the stiffness-loss map in Figure 9.3, the Reuss curve which had been on the bottom in the plot of stiffness versus volume fraction in Figure 9.2 now is on top; the Voigt curve becomes the lower curve in the stiffness loss map. Therefore, while the Voigt geometry is most favorable in terms of attaining a stiff composite from the least amount of stiff inclusions, it gives the least damping for given stiffness. Moreover, the lower and upper two-phase Hashin-Shtrikman composites behave similarly to the Voigt and Reuss composites, respectively, in the stiffness-loss map even though they differ greatly in a plot of stiffness vs volume fraction. As for the physical attainment of the Voigt and Reuss composites, simple laminates can be made as in Figure 9.1, but these are anisotropic. In the Reuss structure each phase carries the full stress, so that a composite of this type will be weak if, as is usual, the soft phase is weak.

The effect of inclusion shape is shown in the stiffness loss map in Figure 9.5. The composite containing soft spherical inclusions is also found to behave similarly to the Voigt composite in that a small volume fraction of soft, viscoelastic material has a comparatively small effect on the loss tangent [19]. A small amount of stiff spherical inclusions confers a Reuss-like effect of increasing the stiffness while not changing the damping much. Stiff fibers are more Voigt-like in reducing the damping as they increase the stiffness. In a composite containing soft platelet inclusions, it is found that the results are similar to those of the Reuss structure. A small volume fraction of platelet inclusions as phase 2 results in a large increase in loss tangent without significant stiffness reduction. However, soft platelets resemble

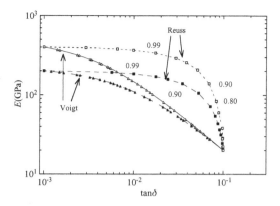

Figure 9.6. Stiffness-loss map showing the effect of the stiffness of the stiff phase [37]. Each point corresponds to a different volume fraction, indicated by numbers.

penny-shaped cracks in the matrix, so that such a composite would be weaker than the matrix, particularly if the matrix were brittle.

The effect of increasing the stiffness of the stiff phase, from 200 GPa, corresponding to steel, to 400 GPa, corresponding to tungsten, is shown in Figure 9.6. For both composites, $\tan\delta$ of the stiff inclusions is assumed to be 0.001, a value representative of results reported in the literature. Observe that if the stiff phase is made stiffer, the resulting composite can be made both stiff and lossy (with a large value of $\tan\delta$).

Figure 9.7 shows the effect of changing the properties of the lossy phase. As would be expected, a lossy phase of the highest possible damping gives rise to a composite with both high stiffness and high loss [37]. Moreover, both the Reuss and Voigt curves are convex to the right when the lossy phase is stiff.

As for high concentrations of spherical inclusions, Figure 9.8 shows the results of finite-element analysis [17] of properties of a composite with stiff, low damping spherical inclusions to approximate the Hashin–Shtrikman lower curve (upper in a stiffness loss map) if the concentration is small and to be intermediate between the Hashin–Shtrikman curves if the concentration is large.

Plotted viscoelastic behavior of the Voigt and Reuss composites encloses a region on the stiffness-loss map, and the properties of the other composites considered

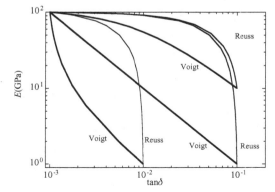

Figure 9.7. Stiffness-loss map showing the effect of the properties of the soft phase [37].

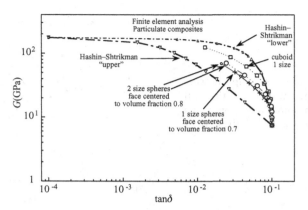

Figure 9.8. Stiffness loss map for concentrated spherical inclusions in a matrix (adapted from Kim, Swan, and Lakes [17]). Volume fractions are in increments of 0.1 up to 0.9 unless otherwise stated. Upright triangles denote Hashin–Shtrikman upper formulae; inverted triangles, Hashin–Shtrikman lower formula; smaller triangles, volume fraction 0.92, 0.94, 0.96, 0.98; squares denote cubical particles; large open circles denote single-sized spherical particles; and small open circles denote multiple-sized spherical particles.

lie within that region. However, no proof has been given that these curves constitute bounds upon the viscoelastic behavior. That issue is considered in §9.4.

In a related vein, the effect of inclusion shape upon viscoelastic response in the time domain has been considered, with the aim of minimizing creep in composites subjected to constant stress [23]. Composites containing stiff fibers are predicted to exhibit less creep than composites with stiff particles. Many composites currently in use have a rich and complicated hierarchical structure in which fibrous layers are organized into laminae. Analysis of such materials, even in the elastic case, can be quite complicated as a result of such structure. The transition to the viscoelastic case is facilitated by use of the dynamic correspondence principle in which elastic constants are replaced by complex functions of frequency [24]. Viscoelastic composites have also been analyzed via the finite-element method [25] and the dynamic correspondence principle, with application to fibrous materials. Complex phase geometries that are not easily amenable to analytical solution may be treated by this approach. As for the time domain, Poisson's ratio, if time dependent, complicates the analysis considerably as discussed in §5.6.2, 5.7. Results obtained via the correspondence principle do not account for damping due to unintentional cracks, pores, or chemical compounds formed at interfaces.

Results of this section are summarized as follows:

(1) In a stiffness-loss map, the two-phase Hashin composites corresponding to upper and lower elastic bounds behave similarly to the Voigt and Reuss composites, respectively.

(2) Reuss laminates and materials filled with stiff, low-loss platelets in a compliant high loss phase exhibit high stiffness combined with high loss tangent. However, in the Reuss structure each phase carries the full stress, so that a composite of this type will not be strong if, as is usual, the compliant phase is weak.

Figure 9.9. Coated sphere composite structure.

(3) A composite containing soft lossy spherical inclusions in a stiff matrix behaves similarly to the Voigt composite: low loss and a reduction in stiffness.
(4) If one desires a composite with maximal stiffness for given volume fraction, fibrous or platelet geometries are the best. Such geometries also tend to minimize damping and creep. However, if one desires a composite with high stiffness combined with high viscoelastic damping, a Reuss geometry or a concentrated suspension of stiff particles will give better results.

## 9.4 Bounds on the Viscoelastic Properties

As for bounds on elastic properties, the bounds on bulk modulus can be attained by a coated sphere composite morphology (Figure 9.9). Exact attainment of the Hashin–Shtrikman shear modulus formula is possible via a hierarchical laminate morphology [27] (Figure 9.10). The coated sphere is a neutral inclusion: it does not disturb the assumed hydrostatic stress state in the surrounding medium, assumed to have a bulk modulus equal to that of the coated inclusion. Milton and Serkov [26] show that since the stress and strain fields are undisturbed in the process, the loads and displacements at the boundary are undisturbed, hence the effective bulk modulus of the medium remains unchanged by the progressive addition of inclusions.

As for bounds on the viscoelastic properties, the curves for the Voigt and Reuss composites enclose a region in the stiffness-loss map, as do the curves for the upper and lower Hashin–Shtrikman composites. They represent extremes of composites

Figure 9.10. Hierarchical laminate (adapted from Milton [27]).

that can be fabricated via known structures for which analytical solutions are known, however the issue of bounds on the behavior is distinct. Roscoe [28] has mathematically established bounds for the real and imaginary parts $E'$ and $E''$ of the complex modulus of composites and has shown them to be equivalent to the Voigt and Reuss relations. Therefore, the stiffness $|E^*|$ of the composite is bounded from above by the Voigt limit and cannot exceed the stiffness of the stiff phase. This is not quite the same as establishing bounds for a stiffness-loss map because it is not obvious whether a maximum in $\tan\delta = E''/E'$ could be obtained simultaneously with a maximum in $E'$. In particular, we can construct $\tan\delta_c = E''_{\text{Voigt}}/E'_{\text{Reuss}} > E''_{\text{Reuss}}/E'_{\text{Reuss}}$ and be within the bounds of Roscoe.

Bounds on the viscoelastic behavior in bulk deformation have been developed for fixed volume fractions of constituents [45] and for arbitrary volume fractions [46], incorporating the three-dimensional aspects of deformation. Construction of curves for the viscoelastic bounds is more complicated than for the elastic bounds. The lower bound for bulk modulus of an elastic material is given in Equation 9.15. In the viscoelastic case, these moduli become complex: $K_1 \rightarrow K_1^*$. Since the shear moduli are also involved, the bounds on the complex bulk moduli depend on the phase Poisson's ratios. Increasing the Poisson's ratio of the soft phase increases both the upper and lower bounds on the composite damping. If the Poisson's ratio is complex, which is certainly possible, the composite damping can increase. The loss tangent $\tan\delta_B$ for bulk deformation of the composite can exceed the bulk loss tangent of either phase. However the bulk loss tangent of the composite is bounded from above by the largest of $\tan\delta_B$ and $\tan\delta_G$ of the phases and bounded from below by the smallest of these. In particular cases, the viscoelastic bounds coincide or are close to the curves for the Hashin–Shtrikman coated-sphere composite morphology. Bounds on the viscoelastic behavior in shear deformation [47, 48] are also known. The bounds for arbitrary phase volume, hence a stiffness-loss map, are close to or in some cases identical to the curves for the Hashin–Shtrikman composites.

The bounds for viscoelastic composites, and in particular the microstructures identified as extremal, can be used to guide the design and fabrication of composites with specified damping, which may be high or low, as required for specific applications.

## 9.5 Extremal Composites

One can design composites that attain bounds such as those of Hashin–Shtrikman via hierarchical structures as discussed in §9.4. The hierarchical structures may be challenging to fabricate, so a distribution of sizes of stiff particles may be embedded in a high damping matrix to approximate the coated sphere morphology.

Early experiments on high damping composites disclosed elevated damping but modest stiffness as follows. Composites made by casting indium [35] into the interstices of copper foams of different structure exhibited high damping from the indium but relatively low stiffness. Indium-aluminum materials [36] with embedded particles of indium exhibited elevated damping near the melting point of indium but not at ambient temperature. Figure 9.5 shows that soft high damping inclusions in

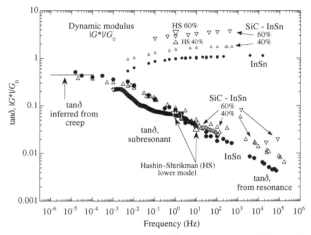

Figure 9.11. Torsional damping tanδ and normalized shear modulus $|G^*|/G_0$ versus frequency of 40 percent by volume (triangles) and 60 percent by volume (inverted triangles) SiC in InSn compared to InSn (solid circles). $G_0 = 7.5$ GPa. Properties were measured directly at small strain with no appeal to time temperature superposition (adapted from Ludwigson, Swan, and Lakes [38]).

particulate form have the least effect upon the composite damping, but stiff inclusions in a high damping matrix are most favorable to maximize $E$ tanδ. Experiment confirms the notion of achieving high modulus and high damping simultaneously in a composite by appropriate choice of stiff laminar [37] or particulate [38] inclusions and a high damping metal matrix stiffer than a polymer. The inclusion of stiff particles increases the modulus but has little effect on the damping over much of the frequency range as shown in Figure 9.11. The elevated damping at high frequency is attributed to processes, for example, damping from dislocations due to differential thermal shrinkage, or thermoelastic damping, not anticipated in the composite analysis. The product $E$ tanδ (a figure of merit for damping layers) can be made a factor of 10 to 30 greater than the maximum attainable in polymer damping layers (Figure 7.8). Metal matrix composites, based on powders of CuAlNi shape memory alloys (SMAs) embedded in an indium matrix, were designed to exhibit high mechanical damping [39]. The material exhibits, in some ranges of temperature and frequency, internal friction as high as 0.54.

It is possible to exceed such bound predictions by relaxing one or more of the assumptions used to derive them. The bounds tacitly assume both phases to be of positive modulus. If one phase has negative elastic modulus, then the composite is predicted have a modulus greater than that of either constituent, and damping, which can diverge [40]. Experiments [41] on constrained lumped unit cells containing buckled tubes indeed show damping which diverges as the negative stiffness phase is tuned to match the phase of positive stiffness. Further experiments on composites with a dilute concentration of particulate inclusions, which can undergo phase transformation, showed large damping greater than that of either constituent [42]; even a Young's modulus substantially greater than that of diamond [43]. Materials with negative values of bulk modulus or shear modulus are unstable if

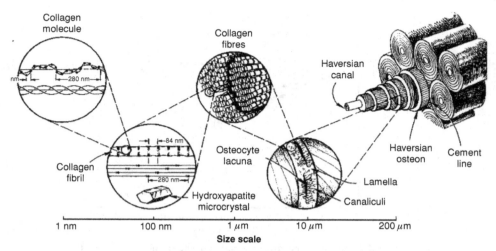

Figure 9.12. Hierarchical structure of human bone [79].

unconstrained. Rigid constraint of the surfaces can stabilize a material with a negative bulk modulus. A stiff, but not perfectly rigid constraint, such as provided by a composite matrix, can also provide stability to a particle of negative bulk modulus [44]. Materials based on these concepts require tuning to achieve extreme properties. Phase transforming materials can be tuned to achieve negative moduli by varying the temperature. This negative modulus arises from stored elastic energy in inclusions. This is in contrast to composites which exhibit negative properties over a narrow frequency range via inertial resonant effects.

## 9.6 Biological Composite Materials

Most composites of biological origin exhibit a rich hierarchical structure [79]. Hierarchical solids contain structural elements which themselves have structure. Human compact bone is a natural composite which exhibits a complex hierarchical structure [49, 50] as shown in Figure 9.12. In bone, the presence of proteinaceous or polysaccharide phases can give rise to significant viscoelasticity, as discussed in §7.5. Observe in Figure 7.22 that the loss tangent of compact bone attains a broad *minimum* over the frequency range associated with most bodily activities. The mineral phase of bone is crystalline hydroxyapatite ($Ca_{10}(PO_4)_6(OH)_2$), which is virtually elastic; it provides the stiffness of bone [51]. On the microstructural level are the osteons [52], which are large ($\sim 200\,\mu$m diameter) hollow fibers composed of concentric lamellae and of pores. The lamellae are built of fibers, and the fibers contain fibrils (smaller fibers). At the ultrastructural level (nanoscale) the fibers are a composite of the mineral hydroxyapatite and the protein collagen, which has a triple helix structure. Specific structural features have been associated with properties, such as stiffness via the mineral crystallites [49], creep via the cement lines between osteons [53] (§8.5), and toughness via osteon pull-out at the cement lines [54]. Lacunae are ellipsoidal pores with dimensions on the order $10\,\mu$m that provide spaces for the osteocytes (bone cells) that maintain the bone and allow it to adapt to changing conditions of stress by mediating growth or resorption of bone in response to stress. Haversian canals contain blood vessels that nourish the tissue, and nerves for sensation. Flow of fluid within the pore space in bone is important in the nutrition of bone cells. Stress generated fluid flow can give rise to mechanical damping in bending or in tension/compression via the Biot mechanism. It appears that this occurs at ultrasonic frequency well above the frequency range for normal bodily activity. Two level hierarchical analytical models involving the osteon as a large fiber has been used to understand anisotropic elasticity [55] and viscoelasticity [56] of bone.

Plant tissues such as wood [57] and bamboo [58] are cellular solids with complex hierarchical structures. The cell walls themselves contain small fibers. Viscoelastic properties of wood, discussed in §7.5, have been associated with those of the lignin within the wood, which exhibits a glass transition temperature near 100°C. Bamboo contains fiber-like structural features known as bundle sheaths [58] as well as oriented porosity along the stem axis. Bamboo, moreover, has functional gradient properties in which there is a distribution of Young's modulus across the culm (stem) cross-section. Plant fibrous materials including jute, bamboo, sisal, and bamboo contain lignin and cellulose, in which cellulose microfibrils are embedded in a matrix of lignin and hemicellulose. In dry bamboo [59] $\tan\delta$, was about 0.01 in bending and 0.02 to 0.03 in torsion, with little dependence on frequency in the audio range. Wet bamboo exhibited somewhat greater $\tan\delta$: 0.012 to 0.015 in bending and 0.03 to 0.04 in torsion. These figures are comparable to those for wood.

Soft tissues, presented in §7.5, contain vascular space which contains tissue fluid. Viscoelasticity in soft tissues results in part from the intrinsic viscoelasticity of biological molecules, and in part from stress-induced fluid flow, analyzed in §8.3.

## 9.7 Poisson's Ratio of Viscoelastic Composites

Poisson's ratio increases in polymers because the shear modulus relaxes much more than the bulk modulus. In composites, each phase may relax at a different rate. Complex Poisson's ratios can occur in viscoelastic composites [46]. The sign of the imaginary part can be positive or negative. In the time domain, the viscoelastic Poisson's ratio can increase or decrease [60, 61]. Since the Poisson's ratio is a cross property with no energy density associated with it, such effects can occur in passive materials.

Control of the time dependence of Poisson's ratio in composites is illustrated by envisaging a negative Poisson's ratio honeycombs or foams that have concave or "re-entrant" cells [62], or laminates [63] with a negative Poisson's ratio. The foams can be isotropic and have Poisson's ratio as small as −0.8. The laminates can be isotropic if the lamina thickness and angles are chosen appropriately. In the laminates Poisson's ratio can approach the lower limit −1 if the material represented by the shaded regions in Figure 9.13 is made very soft. The foam interstices may be filled with a honeycomb or with a foam with a positive Poisson's ratio. By making one cellular structure elastic and the other viscoelastic, one can achieve time dependent Poisson's ratios which increase or decrease with time. Similarly in the laminate

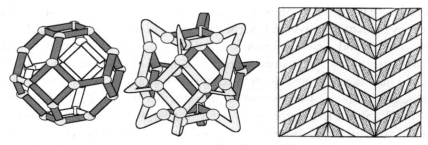

Figure 9.13. Left: foam cell, idealized normal. Center: a negative Poisson's ratio [62]. Right: hierarchical laminate of Milton [63] with a negative Poisson's ratio.

each phase can have a different viscoelastic time or frequency dependence. Complex Poisson's ratios have been determined for several matrix materials and particulate composites [64].

Viscoelastic composite Poisson's ratio may occur in composites even if each phase has a real (elastic) Poisson's ratio. Since Poisson's ratio governs the curvature of plates in bending, the viscoelastic Poisson's ratio of composite laminates can influence the performance of plate structures.

## 9.8 Particulate and Fibrous Composite Materials

### 9.8.1 Structure

Composites used in engineering applications may have a fibrous [65, 66], cellular, or particulate structure. Practical fibrous composites commonly have a low order of hierarchical structure [79] in which fibers are embedded in a matrix to form an anisotropic sheet or lamina; such laminae are bonded to form a laminate (Figure 9.14). In the analysis of fibrous composites, the fibers and matrix are regarded as continuous media in the analysis of the lamina; the laminae are then regarded as continuous in the analysis of the laminate. The stacking sequence of laminae and the orientation of fibers within them governs the composite anisotropy. Practical particulate composites commonly have a distribution of particle sizes, either by design or unintentionally (Figure 9.15).

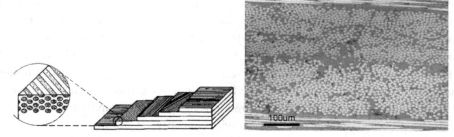

Figure 9.14. Left: hierarchical structure of a practical fibrous laminate [79]. Right: angle ply graphite epoxy fibrous composite, cross-section image (after L. Dong, University of Wisconsin, with permission).

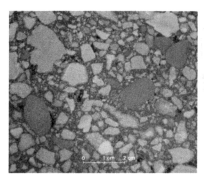 

Figure 9.15. Structure of particulate composites; note difference in scale. Left: asphalt (after R. Delgadillo, U. Wisconsin, with permission). Right: dental composite (adapted from Park and Lakes [71], after Y. Papadogiannis, with permission).

### 9.8.2 Particulate Polymer Matrix Composites

In many particulate composites [67], a polymer matrix is reinforced by stiff inclusions. The stiffness of a composite with spherical inclusions, as a function of volume fraction of inclusions, is experimentally found to be close to the Hashin–Shtrikman lower bound for isotropic materials. Most particulate inclusions used ordinarily exhibit little viscoelastic response. Dental composite filling resins, for example, contain particles of stiff mineral such as silica, in a polymer matrix [68, 69]. The silica inclusions confer stiffness and abrasion resistance comparable to that of the tooth. Even though these materials contain a high concentration of filler, significant long term creep, corresponding to a change of a factor of four in stiffness with time, can occur [70].

### *Filled Polymers and Nanofillers*

Fillers consisting of comparatively large particulate inclusions in a more compliant polymer matrix tend to increase the modulus and have little effect on the damping in agreement with composite theory. If the inclusions are sufficiently small, phenomena occur on the molecular scale, not anticipated in composite theory. For example, rubber filled with carbon black inclusions of a diameter on the order 30 nm exhibits viscoelastic properties quite different from those of unfilled rubber [72]. The filler has a minimal effect in the glassy regime but it shifts the $\alpha$ transition to a lower frequency by about one decade. The equilibrium modulus (at long time) is markedly increased by the filler. This is understood by comparing the inclusion separation 10 nm in a heavily filled rubber, with the similar distance between cross-links in the rubber in question. The filler particle acts as a further cross-link between molecules. Filled rubber tends to be highly nonlinear, in which high-amplitude dynamic strain reduces the stiffening effect of the filler [73].

Fine grain inclusions may behave differently from larger scale inclusions due to scale-dependent interactions and molecular scale phenomena. Such inclusions have been used in practical filled rubbers for many years. Research studies in nanoscale

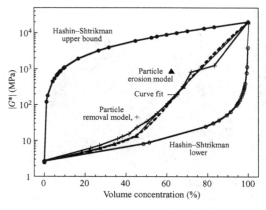

Figure 9.16. Shear modulus versus inclusion concentration based on finite-element analysis of asphalt structure, and comparison with bounds (adapted from Lakes, Kose, and Bahia [80]).

inclusions include the following. Polymer matrix nanocomposites may exhibit an increase or reduction in glass-transition temperature ($T_g$) in comparison to the pure polymer [74, 75]. Above $T_g$ the polymer is comparatively compliant and extensible; below, it is much stiffer. In silica – polystyrene nanocomposites the $T_g$ decreases by 11 K with increasing filler content. In thin polymer layers, a gradient in $T_g$ extends tens of nanometers into the film [76]. In PMMA with aluminum oxide inclusions 39 nm in size, $T_g$ decreases by about 25°C for filler concentration greater than 0.5 percent [77]. The underlying cause of the $T_g$ suppression in constrained geometries remains a subject of inquiry. Interfaces and surfaces also slow the rate of physical aging, also known as structural relaxation, in polymethyl methacrylate (PMMA) [78]. The reduction of aging rate is a factor of 2 near a free surface and a factor of 15 at an interface with silica.

## Asphalt

Asphalt is a particulate, porous composite (Figure 9.15). The particles are sand and gravel; the matrix, known as binder, is a petroleum derivative. Air-filled pores are unintentionally introduced during the mixing. The viscoelastic behavior is dominated by that of the matrix because asphalt is a composite with stiff particles. Because the particles are irregular, even angular, finite element analysis is helpful [80]. Moreover, the particles have a large concentration so dilute approximations are inadequate for the full range. In this analysis, binder properties at low frequency or high temperature were assumed. For a dilute concentration of particles, the modulus approximates the lower bound as shown in Figure 9.16. For a dilute concentration of inclusions, the modulus and damping approximate the lower Hashin–Shtrikman formula as shown in Figure 9.17, but, for higher concentrations, there is a deviation. As with analytical solutions, the correspondence principle may be applied to the results of elastic finite element analysis to obtain a prediction of viscoelastic response. As with all stiffness-loss maps, the map changes with frequency since the constituent properties are frequency dependent. Viscoelastic properties of asphalt, presented in §7.2.6, indicate large softening at low frequency, which is suggestive of a viscoelastic liquid.

Figure 9.17. Stiffness-loss map for asphalt based on finite-element analysis; comparison with Hashin–Shtrikman formulae (adapted from Lakes, Kose, and Bahia [80]).

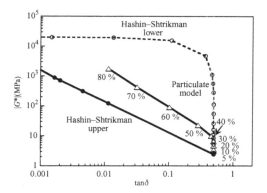

### 9.8.3 Fibrous Polymer Matrix Composites

In fibrous polymer matrix composites, fibers made of a stiff, strong material are embedded in a polymer matrix. Such composites are considerably lighter in weight than structural metals of comparable stiffness and strength. For example [5], a graphite–epoxy composite with 63 percent by volume of aligned fibers has a longitudinal Young's modulus $E$ of about 160 GPa compared with 200 GPa for structural steel, a tensile strength $\sigma_{ult}$ of 1.7 GPa compared with 0.5 GPa for structural steel, and a density of 1.6 g/cm$^3$, compared with 7.8 g/cm$^3$ for structural steel. The transverse properties of such a unidirectional composite are less: $E = 11$ GPa, $\sigma_{ult} = 42$ MPa. Therefore, if stiffness and strength are needed in directions other than the longitudinal, cross-ply or angle-ply laminated microstructures are fabricated. Each lamina may contain fibers in a different direction.

Most materials used as fibers, such as boron, graphite, and glass, exhibit much less creep or damping than the matrix polymers [81]. Polymer matrix materials, such as epoxy, are highly viscoelastic materials. Viscoelastic behavior of the composite as a whole depends on the stress experienced by the matrix [81]. Consequently polymer matrix unidirectional fiber composites creep little if they are loaded along the direction of the fibers, and creep considerably if loaded transversely or in shear (Figure 9.18). As for laminates of fibrous layers, fiber-dominated graphite-epoxy lay-ups such as $[0]_{48}$ and $[0/45/0/-45]_{6s}$ exhibit little viscoelastic response and little redistribution of strain due to creep [82]. The subscripts in the lamination code represent the number of plies, and $s$ represents a laminate symmetric about its midplane; the quantities in the brackets represent the angle of each ply with respect to a given direction. By contrast, matrix-dominated lay-ups, such as $[90]_{48}$ and $[90/-45/90/-45]_{6s}$, exhibited significant creep and strain redistribution.

The $[\pm45]_s$ stacking sequence of graphite–epoxy laminates is considered matrix dominated since much of the applied load is transmitted through the matrix. Master curves for these composites disclose relaxation approaching zero stiffness at sufficiently long time or sufficiently high temperature. The shift factor depends on moisture content as well as temperature [83]. In graphite–epoxy composites, fiber spacings of 2 to 7 $\mu m$ occur. This length scale is comparable to the size of

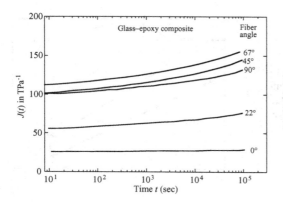

Figure 9.18. Experimental creep curves for various angles of stress with respect to fiber orientation for glass–epoxy composite (adapted from Pomeroy [81]).

structural features such as shear bands, crazes or spherulites in polymer matrix materials, therefore the behavior of the pure matrix (neat resin) may differ from the same matrix material as a constituent of a composite [85]. In this situation use of the correspondence principle with data for pure matrix can be expected to generate errors.

In some applications, such as vibration absorption, the designer wishes to maximize the viscoelastic response of structural members or added layers. One way to do this is to fabricate composites with a controlled amount of slip between the fibers and the matrix [84]. For steel fibers and a rubber matrix, fibers can be strongly bound by adding a priming agent or weakly bound by omitting the primer. By varying the surface roughness of the fibers one can control the degree of slip. Large damping and hysteresis is so obtained [84]. Lamination theory applied to graphite epoxy laminates agrees reasonably with experimental measurement of modulus and damping [29].

### 9.8.4 Metal–Matrix Composites

Metal–matrix composites [86] include aluminum stiffened with inclusions of alumina ($Al_2O_3$) particles, and aluminum reinforced with graphite, boron, or silicon carbide. They are of particular interest in applications involving high temperature, for which a polymer matrix composite would be inappropriate. Such metal–matrix composites exhibit increased stiffness, strength and wear resistance. Particulate inclusions do not have as much stiffening effect as fibers, as discussed in §9.3, but they are less expensive than fibers, so they are more often used. Materials used for inclusions tend to be stiff and of low damping. Silicon carbide (SiC), for example, exhibits $E = 450$ GPa and $\tan\delta \approx 10^{-4}$ at 20°C [87] compared with 70 GPa for aluminum. Aluminum itself exhibits low damping, typically below $10^{-3}$ for aluminum and below $10^{-5}$ for some aluminum alloys.

As for damping, magnesium–matrix composites may exhibit relatively high damping (to $\tan\delta \approx 0.01$) principally as a result of the high damping exhibited by magnesium [88]. For metal–matrix composites with particulate inclusions of ceramic at room temperature, $\tan\delta \approx 0.005$ for boron–aluminum, $\tan\delta \approx 0.007$ for alumina–aluminum, and $\tan\delta \approx 0.004$ for SiC-aluminum [90]. As for creep at high stress and high temperature (up to 350°C), a high-volume fraction of fine-particle reinforcement can significantly reduce creep [91]. Damping in laminated composites of this type depends on the ply angles as it does in polymer matrix composites.

Damping in metal–matrix composites can arise due to damping in one or both of the phases. Use of a high-loss phase such as magnesium can give rise to a high-loss composite. In some particulate composites, an excess of damping is attributed to high concentrations of dislocations near the metal–ceramic interface in SiC-Al [92] and in SiC-AlLi [93]. Damping due to thermoelastic coupling (§8.2) may be important in some metal matrix composites. Thermoelastic damping in composite materials arises due to the heterogeneity of the thermal and mechanical properties of such materials, leading to heat flow between constituents: hence, mechanical energy dissipation. Graphite-aluminum composites can attain $\tan\delta$ up to $4 \times 10^{-3}$ due to thermoelastic damping [89]. Such damping is *not* accounted for in treatments (§9.3) via the correspondence principle because the overall actual damping contains a contribution from a coupled-field interaction between the constituents.

Laminated composites containing two metal phases have been studied in an effort to simultaneously achieve strength, toughness, and a reasonably high damping [94]. A loss tangent of about 0.001 was observed at 30 Hz, and it increased linearly with frequency.

Including flake graphite in up to 10 percent volume fraction into 6061-T6 aluminum alloy by spray-deposition gives rise to $E = 44$ GPa (lower than that of aluminum alloy, 69 GPa) and $\tan\delta \approx 0.01$ at low audio frequency [95]. The damping is attributed to sliding of the planar structure of graphite. Including alumina ($Al_2O_3$) fibers increases the stiffness of aluminum but reduces the damping at 80 kHz and almost completely suppresses the amplitude-dependent response [96]. In pure aluminum, $\tan\delta$ is about $2.5 \times 10^{-4}$ for strains up to $10^{-5}$; the damping increases at higher strains due to dislocation motion. Addition of tungsten fibers to aluminum increases the damping.

Fullerene inclusions in a magnesium alloy have a modest effect at 1 percent concentration in increasing the damping [97], from 1.5 to $4 \times 10^{-3}$ at 950 Hz, and a strain amplitude of $10^{-4}$.

## 9.9 Cellular Solids

Cellular solids [98] are composites in which one phase is empty space or a fluid, such as water or air. Cellular solids include honeycombs, foams and other porous materials. A representative stress–strain curve for an open-cell elastomer foam (Figure 9.19) is shown in Figure 9.20. At small strains the foam deforms by bending of the ribs or struts in the foam structure. At a compressive strain above about 5 percent, the ribs undergo buckling, which gives rise to a plateau region in which the foam deforms progressively at near constant stress. At a sufficiently high compressive stress, the cell ribs come in contact and the foam densifies, which gives rise to a rapid increase in apparent stiffness. Viscoelasticity in the foam manifests itself in

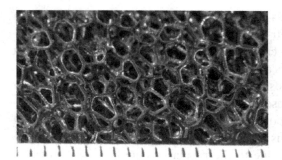

Figure 9.19. Open cell polymer foam (scale marks 1 mm).

creep under transient tests, and energy dissipation in dynamic studies. Foams exhibit linear viscoelasticity, hence elliptic stress–strain diagrams, under small strain amplitude dynamic loading. Under large deformation, foams behave nonlinearly, and the stress–strain diagram becomes distorted, as shown in Figure 9.20.

In low-density, closed-cell foams, the pressure of air or other gas in the pores contributes to the overall stiffness. In open cell foams, air in the pores is free to escape as the material is stressed. Under quasistatic conditions or at low frequency, air flow has little effect. Under these conditions the correspondence principle may then be applied to the foam as a composite with one mechanically active phase. So, if an elastic open cell foam of Young's modulus $E$ and the density $\rho$ is governed by [98]

$$\frac{E}{E_s} = [\frac{\rho}{\rho_s}]^2,$$  (9.22)

(with $\rho_s$ is the density and $E_s$ is the modulus of the solid material of which the foam is made), then, for a corresponding viscoelastic foam,

$$\frac{E^*}{E_s^*} = [\frac{\rho}{\rho_s}]^2.$$  (9.23)

Similarly, for honeycomb, Figure 9.21, $\frac{E}{E_s} = [\frac{\rho}{\rho_s}]^3$ in-plane, and $\frac{E}{E_s} = [\frac{\rho}{\rho_s}]$, out-of-plane, for tension or compression perpendicular to the honeycomb. Under such circumstances the loss tangent of the foam or honeycomb is the same as that of

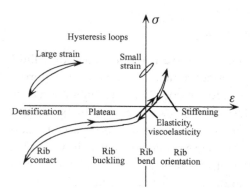

Figure 9.20. Stress–strain curve for an open-cell elastomer foam. Hysteresis loops for dynamic loading at small and large strain amplitudes are shown offset for clarity.

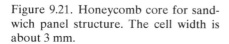

Figure 9.21. Honeycomb core for sandwich panel structure. The cell width is about 3 mm.

the solid phase, assuming no chemical changes occur during the foaming process. Similarly, the creep of the cellular solids is the same as that of the solid phase. In reality, it is possible that the foaming process might alter the chemical and physical properties of the solid phase. Moreover, the correspondence principle only applies to linear behavior.

Foams easily undergo large deformations which cause nonlinear behavior [98–100]. Flexible foams, when compressed sufficiently, exhibit buckling of the cell ribs, followed by contact. Frictional contact between ribs provides an additional damping mechanism for heavy load. Elastomeric polymer foams, such as those used in seat cushions, exhibit incremental properties, which depend on the degree of superposed static precompression [100]. For a representative seat cushion foam, the damping was $\tan\delta \approx 0.2$ and the dynamic Young's modulus was $E \approx 20$ kPa for preompressions 20 percent to 50 percent at frequencies of 1 to 10 Hz. For small precompression, the foam was stiffer and had less damping. Similar damping was observed in Scott Industrial Foam [101]; hysteresis loops changed from nearly elliptical at 2 percent strain amplitude to banana shaped at 15 percent strain amplitude. Damping exhibited a maximum for about 15 percent static precompression.

Fluid such as air or water in interconnected pores can move under stress. At sufficiently high frequency, significant viscoelastic loss can occur in these materials as a result of the viscosity of the air or water in the pores, as we have examined in §8.3. For flexible open cell polymer foams, the loss tangent (Figure 8.6) exhibits a peak, which can have a magnitude greater than 0.5, depending on the characteristics of the foam and fluid [102]. One can extract the material constants for the theoretical model from relatively simple experiments [102, 103]. Sandstone is a natural porous material and the presence of fluid in the pores increases the loss [104].

At ultrasonic frequencies, one can generate in porous media a second type of compressional wave which is slower than the normal compressional wave. This acoustic slow wave arises from a dynamic interaction between the solid and fluid phases. Experimental results have been obtained for fused glass bead media and for bone [104–109]. The attenuations for the fast and slow waves need not be equal.

For polymer foams, at sufficiently high frequency it is possible to set the ribs or walls of the foam into vibration, resulting in additional loss and dispersion [110]. For flexible polymer foams, this occurs at about 1 kHz. Moreover, acoustic absorption in negative Poisson's ratio foams is higher than that in foams of conventional structure [111]. Under quasistatic circumstances, below such frequencies, the loss tangent of open-cell cellular solids is that of the solid material of which it is made, following the correspondence principle. In closed-cell foams, the air contained within the cells can support some load. If a steady load is applied in a creep test, air is lost from the cells due to diffusion through the cell walls. This gives rise to creep [112].

## 9.10  Piezoelectric Composites

Piezoelectric composites find use in electromechanical transducers and in transducers intended for "smart" materials. Piezoelectric composites are available with particles of lead titanate piezoelectric ceramic embedded in a polychloroprene rubber [113]. The overall damping properties of such composites are, as anticipated in §9.3, dominated by the damping of the polymer matrix which exhibits a large peak due to the glass transition. There is also a peak in the piezoelectric coupling and in the dielectric loss at the glass transition temperature. Piezoelectric composites have also been considered with the aim of achieving high damping in stiff structural composites [114]. It is considered possible to achieve a peak damping of 0.12 and a longitudinal modulus of 43 GPa with 30 percent by volume of resistively shunted piezoelectric fibers in a hybrid glass–epoxy composite. Experiments with ceramic–epoxy composites containing one ring-shaped inclusion of resistively shunted piezoelectric ceramic disclosed damping exceeding 0.10 over a narrow frequency range [115].

## 9.11  Dispersion of Waves in Composites

In composites, there are several physical processes which give rise to dispersion (frequency dependence of the wave speeds or moduli inferred from them). Viscoelastic damping, as we have seen in §3.3, is linked to dispersion via the Kramers–Kronig relations. Dispersion also can arise due to microscopic resonance or standing wave effects in the structural elements of the composite. This is referred to as geometric dispersion [116]. Dispersion due to viscoelasticity always entails increase of wave speed with frequency because the damping in passive materials is positive. Geometric dispersion can give rise to a decrease of wave speed with frequency [117]. In typical fibrous composites, such phenomena occur at frequencies in the MHz range and in foams in the kHz range or below. If the wavelength becomes comparable to the size of structural elements in the material, extreme dispersion combined with large wave attenuation can occur [118]. Composites with periodic structure can exhibit cut-off frequencies. If particulate inclusions are distributed randomly, there are no cut-off frequencies [119]. In strongly scattering media, the group velocity [120] associated with pulses of waves can differ substantially from the phase velocity associated with individual peaks associated with waves.

In a composite structure, which is not entirely regular, wave propagation is impeded by the irregularities, so that there is an exponential decay of amplitude, even if there is no dissipation [30]. Experimental results support the concept.

### Localized Resonance

Resonance of microstructure in composites has been shown to result in extreme dielectric properties, even if the inclusions have a small concentration [31]. Similar resonances can be envisaged in viscoelastic composites. Localized resonance can lead to cloaking: invisibility of objects within wave fields of a particular frequency [32]. Composites that exhibit localized resonance have been called *metamaterials*, and can exhibit negative dielectric or magnetic properties over a narrow band of frequency [33]. The consequences of negative dielectric properties in refraction of lenses was originally analyzed by Veselago [34]. Negative properties can give rise to lenses that perform better than the usual theoretical limit.

## 9.12 Summary

The viscoelastic properties of a composite depend on the properties of its phases. For cases of simple geometry, an exact calculation of these properties can be obtained by use of the correspondence principle.

## 9.13 Examples

**Example 9.1**

Consider a composite containing stiff elastic particles embedded in a viscoelastic matrix. We want to reduce creep. Should we use stiffer particles? What are some applications of such composites?

**Solution**

At the simplest level one may approximate the behavior as that of a Reuss composite according to Equation 9.10. If $V_1 = 0.5$,

$$\tan\delta_c \approx \frac{\tan\delta_1 + \frac{J_2'}{J_1'}\tan\delta_2}{1 + \frac{J_2'}{J_1'}}.$$

Suppose phase 1 is soft and high loss, then $J_1 \gg J_2$, so $\tan\delta_2 = 0$ (elastic inclusions), $\tan\delta_c \approx \tan\delta_1[1 - \frac{J_2'}{J_1'}]$, so stiffer inclusions slightly reduce the $\tan\delta$, hence a slight reduction in the creep because the slope of the creep curve on a log–log scale is proportional to $\tan\delta$. One may pursue a more sophisticated approximation based on the Hashin–Shtrikman formulae or on a spherical inclusion model. Silica particles in a polymer matrix are used in composite dental fillings. Creep can be a problem, but wear is usually more of a problem. The hard mineral particles in a high concentration reduce the wear. A very high concentration of stiff elastic particles would

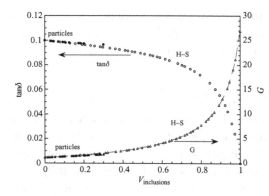

Figure 9.22. Effect of spherical silica inclusions in an epoxy matrix.

be needed to significantly reduce the creep. It may be impractical to achieve such a high concentration. It may be more promising to seek a polymer which exhibits less creep, and to allow time for aging of the polymer after it is polymerized.

### Example 9.2

Derive $E_c = E_1 V_1 + E_2 V_2$, for the elastic Voigt composite microstructure, in one dimension, neglecting the Poisson effects.

### Solution

Since it is elastic, Hooke's law for phases 1 and 2 is

$\sigma_1 = E_1 \epsilon_1$, and $\sigma_2 = E_2 \epsilon_2$

The loads $F$ in terms of the cross-sectional areas $A$ are

$F_1 = A_1 \sigma_1 = E_1 \epsilon_1 A_1$, and $F_2 = A_2 \sigma_2 = E_2 \epsilon_2 A_2$.

In view of the geometry, the load $F_c$ on the composite block must be the sum of the loads on the constituents,

$F_c = F_1 + F_2$.

$F_c = \sigma_c A_c = \sigma_1 A_1 + \sigma_2 A_2$, so, dividing by $A_c$, and recognizing the volume fraction for this geometry as the ratio of cross-sectional areas,

$\sigma_c = \sigma_1 \frac{A_1}{A_c} + \sigma_2 \frac{A_2}{A_c} = \sigma_1 V_1 + \sigma_2 V_2$.

Finally, divide by the strain, which is the same in both constituents provided they are perfectly bonded,

$E_c = E_1 V_1 + E_2 V_2$,

as desired.

The viscoelastic version of this may be obtained without appeal to the correspondence principle by following the same steps assuming the moduli to be complex quantities. The volume fractions are always real quantities based on their definition.

### Example 9.3

Consider an isotropic particulate composite containing an epoxy matrix with $E = 3$ GPa and $\tan\delta = 0.1$ and silica inclusions with $E = 70$ GPa and $\tan\delta = 10^{-5}$. Determine and discuss the stiffness and damping of such a composite. For the purpose of calculation assume a Poisson's ratio of 0.3. What are the prospects of achieving minimal viscoelasticity in such a composite?

**Solution**

Stiffness and damping based on the Hashin–Shtrikman lower bound on stiffness, Equation 9.13 and Equation 9.17 for a small volume fraction of spherical inclusions are plotted in Figure 9.22. Observe that stiff spherical inclusions are inefficient in creating a stiff composite: the stiffness is close to the Hashin–Shtrikman lower bound on the stiffness. Moreover, for readily attainable volume fractions of inclusions, the damping of the composite is close to that of the matrix itself. If one wishes to make a stiff composite with low damping, the particulate morphology is problematical. A fibrous or platelet morphology is more promising for such a goal.

**Example 9.4**

Calculate the fluid flow relaxation strength in a cellular solid. Assume a foam with a polymer skeleton with $G = 100$ kPa. The solid volume fraction is 0.03. Let the fluid in the interstices be air, and consider the surfaces of the foam to freely allow fluid flow. Consider relaxation in shear, bulk deformation, and uniaxial compression.

**Solution**

Consider separately the compressibility of the air and of the polymer foam rather than via the Biot theory because (1) we have the phase properties but not the Biot coefficients, (2) the cellular solid is of low density, (3) only the relaxation strength is requested, not the time constant.

Air has a pressure of $P = 14.7$ psi or 100 kPa under normal sea level conditions. Air as a fluid has a shear modulus of zero, and a bulk modulus governed by the ideal gas equation,

$PV = nRT$. So,

$\frac{dP}{dV} = -\frac{nRT}{V^2}$, so,

$\frac{dP}{\frac{dV}{V}} = P$.

So, the bulk modulus is $B = \frac{\sigma}{\epsilon} = \frac{dP}{\frac{1}{3}\frac{dV}{V}} = 3P$.

For air at the given pressure, $B = 300$ kPa. The factor of three arises from the fact that the volumetric deformation depends on the product of deformations in three directions.

The bulk modulus in terms of the shear modulus $G$ and Poisson's ratio $v$ is

$B = 2G\frac{1+v}{3(1-2v)}$.

The foam has a Poisson's ratio of about 0.3, so $B = 2.167G = 217$ kPa.

For an elementary treatment of this problem, neglect the deformation of the solid phase as pressure is increased in the fluid which permeates the pores. Then the bulk stiffness at zero time is the sum of the stiffness of the foam skeleton and that of the air, which at zero time has no time to escape. For an exact solution, the full Biot analysis of §8.3 is called for:

$B(0) = 217\,\text{kPa} + 300\,\text{kPa}.$

The relaxation strength for bulk deformation is
$$\Delta_B = \frac{B(0)-B(\infty)}{B(\infty)} = \frac{300+217-217}{217} = 1.32.$$
The peak damping for bulk deformation, assuming a Debye peak is
$$\tan\delta = \frac{1}{2}\frac{\Delta}{\sqrt{1+\Delta}} = 0.45.$$
This is a substantial damping.

As for shear, air has zero stiffness in shear. Moreover, the viscosity of air is low so that damping due to global shear is negligible, except at very high frequency. Therefore, the corresponding relaxation strength and damping of the foam in shear due to the air in the pores, are zero.

As for damping in axial deformation observe that Young's modulus is given by
$$E = 3B(1-2v) = \frac{9GB}{3B+G}.$$
So, with $G(0) = G(\infty) = 100\,\text{kPa}$ and $B(0) = 517\,\text{kPa}$ and $B(\infty) = 217\,\text{kPa}$, the relaxation strength for axial deformation is
$$\Delta_E = 0.084.$$
This is considerably less than in the case of bulk deformation, because uniaxial tension deformation is considerably influenced by the shear properties, which, in this example, do not relax.

## 9.14 Problems

9.1. Show how complex Poisson's ratios $v^*$ can be achieved in composites in which each phase has a real Poisson's ratio.

9.2. Why does a unidirectional graphite epoxy fibrous composite have low damping and creep in the fiber direction?

9.3. Derive the composite stiffness in terms of constituent stiffness values for the Reuss composite microstructure. Neglect Poisson effects.

9.4. Calculate the fluid flow relaxation strength assuming a polymer foam skeleton with $G = 100$ kPa. Let the fluid be water.

9.5. Consider the Voigt composite in three dimensions as a laminate. What is its symmetry? How many elastic constants are there if each phase is elastic? How many have been calculated in this chapter? How many independent functions are there, if each phase is viscoelastic? Can one really neglect Poisson's ratio? What might be done to allow for a nonzero Poisson's ratio in the calculation?

BIBLIOGRAPHY

[1] Milton, G., *The Theory of Composites*, Cambridge, UK: Cambridge University Press, 2002.
[2] Sokolnikoff, I. S., *Theory of Elasticity*, Malabar, FL: Krieger, 1983.
[3] Timoshenko, S. P., and Goodier, J. N., *Theory of Elasticity*, New York: McGraw Hill, 1982.
[4] Fung, Y. C., *Principles of Solid Mechanics*, Englewood Cliffs, NJ: Prentice Hall, 1968.
[5] Agarwal, B. D., and Broutman, L. J., *Analysis and Performance of Fiber Composites*, 2nd ed., New York: J. Wiley, 1990.

[6] Hashin, Z., The Elastic Moduli of Heterogeneous Materials, *J Appl Mech, Trans ASME*, 84E, 143–150, 1962.

[7] Hashin Z., and Shtrikman, S., A Variational Approach to the Theory of the Elastic Behavior of Multiphase Materials, *J Mech Phys Solids*, 11, 127–140, 1963.

[8] Hashin, Z., Analysis of Composite Materials – A Survey, *J Appl Mech*, 50, 481–505, 1983.

[9] Day, W. A., Restrictions on Relaxation Functions in Linear Viscoelasticity, *Quart J Mech Appl Math*, 24, 487–497, 1971.

[10] Rogers, T. G., and Pipkin, A. G., Asymmetric Relaxation and Compliance Matrices in Linear Viscoelasticity, *Zeits für Angewandte Math. & Physik (ZAMP)* 14, 334–343, 1963.

[11] Hashin, Z., Viscoelastic Behavior of Heterogeneous Media, *J Appl Mech Trans ASME*, 32E, 630–636, 1965.

[12] Schapery, R. A., Stress Analysis of Viscoelastic Composite Materials, *J Composite Mat*, 1, 228–267, 1967.

[13] Hashin, Z., Viscoelastic Fiber Reinforced Materials, *AIAA J*, 4, 1411–1417, 1966.

[14] Hashin, Z., Complex Moduli of Viscoelastic Composites: I. General Theory and Application to Participate Composites, *Int J Solids, Structures*, 6, 539–552, 1970.

[15] Christensen, R. M., *Mechanics of Composite Materials*, New York: John Wiley & Sons, 1979.

[16] Christensen, R. M., Viscoelastic Properties of Heterogeneous Media, *J Mech Phys Solids*, 17, 23–41, 1969.

[17] Kim, H. J., Swan, C. C., and Lakes, R. S., Computational Studies on High Stiffness, High Damping SiC-InSn Particulate Reinforced Composites, *Int J Solids, Structures*, 39, 5799–5812, 2002.

[18] Christensen, R. M., *Theory of Viscoelasticity*, Academic, NY, 1982.

[19] Chen, C. P., and Lakes, R. S., Analysis of High Loss Viscoelastic Composites, *J Mat Sci*, 28, 4299–4304, 1993.

[20] Christensen, R. M., Isotropic Properties of Platelet Reinforced Media, *J Eng Mat Tech*, 101, 299–303, 1979.

[21] Norris, A. N., The Mechanical Properties of Platelet Reinforced Composites, *Int J Solids, Structures*, 26, 663–674, 1990.

[22] Millett, P., Schaller, R., and Benoit, W., Internal Friction Spectrum and Damping Capacity of Grey Cast Iron, in *Deformation of Multi-Phase and Particle Containing Materials, Proceedings of the 4th Riso International Symposium on Metallurgy and Materials Science*, J. B. Bilde-Sorensen, N. Hansen, A. Horsewell, T. Leffers, and H. Linholt, eds., Roskilde Denmark: Riso National Laboratory, 1983.

[23] Wang, Y. M., and Weng, G. J., The Influence of Inclusion Shape on the Overall Viscoelastic Behavior of Composites, *J Appl Mech*, 59, 510–518, 1992.

[24] Zinoviev, P. A., and Ermakov, Y. N., *Energy Dissipation in Composite Materials*, Technomic, Lancaster, PA: 1994.

[25] Brinson, L. C., and Knauss, W. G., Finite Element Analysis of Multiphase Viscoelastic Solids, *J Appl Mech*, 59, 730–737, 1992.

[26] Milton, G. W., and Serkov, S. K., Neutral Coated Inclusions in Conductivity and Anti–Plane Elasticity. *Proceed Roy Soc*, 457, 1973–1997, 2001.

[27] Milton, G. W., Modeling the Properties of Composites by Laminates, in J. L. Erickson, D. Kinderlehrer, R. Kohn, J. L. Lions, eds., *Homogenization and Effective Moduli of Materials and Media*, Berlin: Springer Verlag, 1986, pp. 150–175.

[28] Roscoe, R., Bounds for the Real and Imaginary Parts of the Dynamic Moduli of Composite Viscoelastic Systems, *J Mech Phys Sol*, 17, 17–22, 1969.

[29] Adams, R. D., Mechanisms of Damping in Composite Materials, *J de Physique*, Colloque C9, 44, 29–37, 1983.

[30] Hedges, C. H., and Woodhous J., Vibration Isolation from Irregularity in a Nearly Periodic Structure: Theory and Measurements, *J Acoust Sec Am*, 74(3), 894–905, 1983.

[31] Nicorovici, N. A. Mcphedran, R. C., and Milton, G. W. Optical and Dielectric Properties of Partially Resonant Systems, *Phys Rev B*, 49, 8479–8482, 1994.

[32] Nicorovici, N. A., and Milton, G. W., On the Cloaking Effects Associated with Anomalous Localized Resonance, *Proceed Roy Soc London, A*, 462, 3027–3059, 2006.

[33] Linden, S., Enkrich, C., Wegener, M., Zhou, J., Koschny, T., and Soukoulis, C. M., Magnetic Response of Metamaterials at 100 Terahertz, *Science*, 306, 1351–1353, 2004.

[34] Veselago, V. G., The Electrodynamics of Substances with Simultaneouly Negative Values of $\epsilon$ and $\mu$, *Sov Phys Uspekhi*, 10, 509–514, 1968.

[35] Chen, C. P., and Lakes, R. S., Viscoelastic Behaviour of Composite Materials with Conventional or Negative Poisson's Ratio Foam as One Phase, *J Mat Sci*, 28, 4288–4298, 1993.

[36] Malhotra, A. K., and Van Aken, D. C., Experimental and Theoretical Aspects of the Internal Friction Associated with the Melting of Embedded Particles, *Acta Metall Mater*, 41, 1337–1346, 1993.

[37] Brodt, M., and Lakes, R. S., Composite Materials which Exhibit High Stiffness and High Viscoelastic Damping, *J Composite Mat*, 29, 1823–1833, 1995.

[38] Ludwigson, M. Swan, C. C., and Lakes, R. S., Damping and Stiffness of Particulate SiC–InSn Composite, *J Compos Mate*, 36, 2245–2254, 2002.

[39] San Juan, J., and Nó, M. L., Internal Friction in a New Kind of Metal Matrix Composites, *Mat Sci Eng A*, 442, 429–432, 2006.

[40] Lakes, R. S., Extreme Damping in Composite Materials with a Negative Stiffness Phase, *Phys Rev Let*, 86, 2897–2900, 2001.

[41] Lakes, R. S., Extreme Damping in Compliant Composites with a Negative Stiffness Phase, *Philos Mag Let*, 81, 95–100, 2001.

[42] Lakes, R. S., Lee, T., Bersie, A., and Wang, Y. C., Extreme Damping in Composite Materials with Negative Stiffness Inclusions, *Nature*, 410, 565–567, 2001.

[43] Jaglinski, T., Stone, D. S., Kochmann, D., and Lakes, R. S., Materials with Viscoelastic Stiffness Greater than Diamond, *Science*, 315, 620–622, 2007.

[44] Drugan, W. J., Elastic Composite Materials Having a Negative Stiffness Can Be Stable, *Phys Rev Let*, 98, 055502, 2007.

[45] Gibiansky, L. V., and Milton, G. W., On the Effective Viscoelastic Moduli of Two Phase Media: I. Rigorous Bounds on the Complex Bulk Modulus, *Proc Roy Soc London*, 440, 163–188, 1993.

[46] Gibiansky, L. V., and Lakes, R. S., Bounds on the Complex Bulk Modulus of a Two Phase Viscoelastic Composite with Arbitrary Volume Fractions of the Components, *Mech Mat*, 16, 317–331, 1993.

[47] Gibiansky, L. V., Milton, G. W., and Berryman, J. G., On the Effective Viscoelastic Moduli of Two-Phase Media: III. Rigorous Bounds on the Complex Shear Modulus in Two Dimensions, *Proc Roy Soc London*, A, 455, 2117–2149, 1999.

[48] Gibiansky, L. V., and Lakes, R. S., Bounds on the Complex Bulk and Shear Moduli of a Two-Dimensional Two-Phase Viscoelastic Composite, *Mech of Mat*, 25, 79–95, 1997.

[49] Hancox, N. M., *Biology of Bone*, Cambridge, UK: Cambridge University Press, 1972.

[50] Currey, J., *The Mechanical Adaptations of Bones*, Princeton, NJ: Princeton University Press, 1984.

[51] Katz, J. L., Hard Tissue as a Composite Material – I. Bounds on the Elastic Behavior, *J Biomechanics*, 4, 455–473, 1971.

[52] Frasca, P., Harper, R. A., and Katz, J. L., Isolation of Single Osteons and Osteon Lamellae, *Acta Anatomica*, 95, 122–129, 1976.

[53] Lakes, R. S., and Saha, S., Cement Line Motion in Bone, *Science*, 204, 501–503, 1979.

[54] Piekarski, K., Fracture of Bone, *J Appl Phys*, 41, 215–223, 1970.

[55] Katz, J. L. Anisotropy of Young's Modulus of Bone, *Nature*, 283, 106–107, 1980.

[56] Gottesman, T., and Hashin, Z., Analysis of Viscoelastic Behaviour of Bones on the Basis of Microstructure, *J Biomechanics*, 13, 89–96, 1980.

[57] Thomas, R. J., Wood: Formation and Morphology in *Wood Structure and Composition*, M. Lewin and I. S. Goldstein, eds., New York: Marcel Dekker, 1991.

[58] Amada, S., Hierarchical Functionally Gradient Structures of Bamboo, Barley, and Corn, *MRS Bulletin*, 20, 35–36, 1995.

[59] Amada, S., and Lakes, R. S., Viscoelastic Properties of Bamboo, *J Mat Sci*, 32, 2693–2697, 1997.

[60] Lakes, R. S., The Time Dependent Poisson's Ratio of Viscoelastic Cellular Materials Can Increase or Decrease, *Cellular Polymers*, 10, 466–469, 1992.

[61] Lakes, R. S., and Wineman, A., On Poisson's Ratio in Linearly Viscoelastic Solids, *J Elasticity*, 85, 45–63, 2006.

[62] Lakes, R. S., Foam Structures with a Negative Poisson's Ratio, *Science*, 235 1038–1040, 1987.

[63] Milton, G., Composite Materials with Poisson's Ratios Close to $-1$, *J Mech Phys Solids*, 40, 1105–1137, 1992.

[64] Agbossou, A., Bergeret, A., Benzarti, K., and Alberola, N., Modelling of the Viscoelastic Behaviour of Amorphous Thermoplastic–Glass Beads Composites Based on the Evaluation of the Complex Poisson's Ratio of the Polymer Matrix, *J Mat Sci*, 28, 1963–1972, 1993.

[65] Hashin Z., and B. W., Rosen, The Elastic Moduli of Fiber-Reinforced Materials, *J Appl Mech, Trans ASME*, 31, 223–232, 1964.

[66] Hashin, Z., On Elastic Behaviour of Fibre Reinforced Materials of Arbitrary Transverse Phase Geometry, *J Mech Phys Solids*, 13, 119–134, 1965.

[67] Ahmed, S., and Jones, F. R., A Review of Particulate Reinforcement Theories for Polymer Composites, *J Mat Sci*, 25, 4933–4942, 1990.

[68] Cannon, M. L., Composite Resins, in *Encyclopedia of Medical Devices and Instrumentation*, J. G. Webster, ed., New York: J. Wiley, 1988.

[69] Craig, R., Chemistry, Composition, and Properties of Composite Resins, in H. Horn, eds., *Dental Clinics of North America*, Philadelphia, PA: Saunders, 1981.

[70] Papadogianis, Y., Boyer, D. B., and Lakes, R. S., Creep of Conventional and Microfilled Dental Composites, *J Biomed Mat Res*, 18, 15–24, 1984.

[71] Park, J. B., and Lakes, R. S. *Biomaterials*, 3rd ed., New York: Springer, 2007.

[72] Payne, A. R., in *Rheology of Elastomers, Temperature-frequency relationships of dielectric and mechanical properties of rubber*, P. Mason, and N. Wookey, eds., London: Pergamon, 1958, pp. 86–112.

[73] Payne, A. R., A Note on the Existence of a Yield Point in the Dynamic Modulus of Loaded Vulcanizates, *J Appl Polymer Sci*, 3, 127, 1960.

[74] Bansal, A., Yang, H., Li, C., Cho, K., Benicewicz, B. C., Kumar, S. K., and Schadler, L. S., Quantitative Equivalence between Polymer Nanocomposites and Thin Polymer Films, *Nature Mat*, 4, 693–698, 2005.

[75] Mayes, A. M., Nanocomposites: Softer at the Boundary, *Nature Mat*, 4, 651–652, 2005.

[76] Ellison, C. J., and Torkelson, J. M., The Distribution of Glass-Transition Temperatures in Nanoscopically Confined Glass Formers, *Nature Mat*, 2, 695–700, 2003.

[77] Ash, B. J., Schadler, L. S., and Siegel, R. W., Glass Transition Behavior of Alumina–Olymethylmethacrylate Nanocomposites, *Mat Let*, 55, 83–87, 2002.

[78] Priestley, R. D., Ellison, C. J., Broadbelt, L. J., and Torkelson, J. M., Structural Relaxation of Polymer Glasses at Surfaces, Interfaces and in between, *Science*, 309, 456–459, 2005.

[79] Lakes, R. S., Materials with Structural Hierarchy, *Nature*, 361, 511–515, 1993.

[80] Lakes, R. S., Kose, S., and Bahia, H., Analysis of High Volume Fraction Irregular Particulate Damping Composites, *ASME J Eng Mat Tech*, 124, 174–178, 2002.

[81] Sturgeon, J. B., Creep of Fibre Reinforced Thermosetting Resins, in *Creep of Engineering Materials*, C. D. Pomeroy, ed., London: Mechanical Engineering Publications, Ltd., 1978.

[82] Tuttle, M. E., and Graesser, D. L., Compression Creep of Graphite–Epoxy Laminates Monitored using Moiré Interferometry, *Optics and Lasers in Engineering*, 12, 151–171, 1990.

[83] Flaggs, D. L., and Crossman, F. W., Analysis of the Viscoelastic Response of Composite Laminates during Hygrothermal Exposure, *J Compos Mat*, 15, 21–40, 1981.

[84] Nelson, D. J., and Hancock, J. W., Interfacial Slip and Damping in Fibre Reinforced Composites, *J Mat Sci*, 13, 2429–2440, 1978.

[85] Sternstein, S. S., Srinavasan, K., Liu, S., and Yurgartis, S., Viscoelastic Characterization of Neat Resins and Composites, *Polymer Preprints*, 25, 201–202, 1984.

[86] Taya, M., and Arsenault, R. J., *Metal Matrix Composites*, Oxford: Pergamon, 1989.

[87] Wolfenden, A., Rynn, P. J., and Singh, M., Measurement of Elastic and Anelastic Properties of Reaction-Formed Silicon Carbide Materials, *J Mat Sci*, 30, 5502–5507, 1995.

[88] Wren, G. G., and Kinra, V. K., Axial Damping in Metal-Matrix Composites. II: A Theoretical Model and Experimental Verification, *Experimental Mechanics*, 32, 172–178, 1992.

[89] Kinra, V. K., Wren, G. G., Rawal, S., and Misra, M., On the Influence of Ply-Angle on Damping and Modulus of Elasticity of a Metal-Matrix Composite, *Metall Trans*, 22A, 641–651, 1991.

[90] Wong, C. R., and Holcomb, S., Damping Studies of Ceramic Reinforced Aluminum, in ASTM STP 1169, Mechanics and Mechanisms of Material Damping, DRCRC-SME91/15, David Taylor Research Center, Bethesda, MD, 1991.

[91] Lloyd, D. J., Particle Reinforced Aluminium and Magnesium Matrix Composites, *Int Mat Rev*, 39, 1–23, 1994.

[92] Ledbetter, H. M., and Datta, S. K., Young's Modulus and the Internal Friction of an SiC Particle Reinforced Aluminum Composite, *Mat Sci Eng*, 67, 25–30, 1984.

[93] Gutiérrez-Urrutia, I., Nó, M. L., San Juan, J., Internal Friction Behavior in Sic Particle Reinforced 8090 Al-Li Metal Matrix Composite, *Mat Sci Eng*, A 370, 555–559, 2004.

[94] Bonner, B. P., Lesuer, D. R., Syn, C. K., Brown, A. E., and Sherby, O. D., Damping Measurements for Ultra-High Carbon Steel/Brass Laminates, *Damping of Multiphase Inorganic Materials*, ASM Materials Week, Chicago, November 2–5, 1992.

[95] Zhang, J., Perez, R. S., Gungor, M. N., and Lavernia, E. S., Damping Characterization of Graphite Particulate Reinforced Aluminum Composite, in *Developments in Ceramic and Metal-Matrix Composites*, K. Upadhya, ed., TMS 1992 meeting, San Diego; Minerals, Metals, and Materials Society, 1992.

[96] Wolfenden, A., and Wolla, J. M., Mechanical Damping and Dynamic Modulus Measurements in Alumina and Tungsten Fibre–Reinforced Aluminium Composites, *J Mat Sci*, 24, 3205–3212, 1989.

[97] Watanabe, H., Sugioka, M., Fukusumi, M., Ishikawa, K., Suzuki, M., and Shimizu, T., Mechanical and Damping Properties of Fullerene Dispersed AZ91 Magnesium Alloy Composites Processed by a Powder Metallurgy Route, *Mat Trans*, 47(4), 999–1007, 2006.

[98] Gibson, L. J., and Ashby, M. F., *Cellular Solids*, Oxford: Pergamon, 1988; Cambridge, 2nd ed., UK: Cambridge University Press, 1997.

[99] Hilyard, N. C., Hysteresis and Energy Loss in Flexible Polyurethane Foams, in N. C. Hilyard, and A. Cunningham, eds., *Low Density Cellular Plastics*, London: Chapman and Hall, 1994.

[100] Cunningham A., Huygens E., and Leenslag J. W., MDI Comfort Cushioning for Automotive Applications, *Cellular Polymers*, 13, 461–472, 1994.

[101] Martz, E. O., Lakes, R. S., and Park, J. B. Hysteresis Behaviour and Specific Damping Capacity of Negative Poisson's Ratio Foams, *Cellular Polymers*, 15, 349–364, 1996.

[102] Gent, A. N., and Rusch, K. C., Viscoelastic Behavior of Open Cell Foams, *Rubber Chemistry and Technology*, 39, 388–396, 1966.

[103] Kim, Y. K., and Kingsbury, H. B., Dynamic Characterization of Poroelastic Materials, *Exp Mech*, 17, 252–258, 1979.

[104] Fatt, I., The Biot–Willis Elastic Coefficients for a Sandstone, *J Appl Mech*, 26, 296–297, 1959.

[105] Plona, T. J., Observation of a Second Bulk Compressional Wave in a Porous Medium at Ultrasonic Frequencies, *Appl Phys Let*, 36, 259–261, 1980.

[106] Berryman, J. G., Confirmation of Biot's Theory, *Appl Phy Let*, 37, 382–384, 1980.

[107] Johnson, D. L., Plona, T. J., Scala, C., Pasierb, F., and Kojima, H., Tortuosity and Acoustic Slow Waves, *Phys Rev Let*, 49, 1840–1844, 1982.

[108] Johnson, D. L., and Plona, T. J., Acoustic Slow Waves and the Consolidation Transition, *J Acoust Soc Am*, 72, 556–565, 1982.

[109] Lakes, R. S., Yoon, H. S., and Katz, J. L., Slow Compressional Wave Propagation in Wet Human and Bovine Cortical Bone, *Science*, 220, 513–515, 1983.

[110] Chen, C. P., and Lakes, R. S., Dynamic Wave Dispersion and Loss Properties of Conventional and Negative Poisson's Ratio Polymeric Cellular Materials, *Cellular Polymers*, 8, 343–359, 1989.

[111] Howell, B., Prendergast, P., and Hansen, L., Acoustic Behavior of Negative Poisson's Ratio Materials, DTRC-SME-91/01, David Taylor Research Center, March 1991.

[112] Mills, N. J., Time Dependence of the Compressive Response of Polypropylene Bead Foam, *Cellular Polymers*, 16, 194–215, 1997.

[113] Rittenmyer, K., Temperature Dependence of the Electromechanical Properties of 0-3 $PbTiO_3$ Polymer Piezoelectric Composite Materials, *J Acoust Soc Am*, 96, 307–318, 1994.

[114] Lesieutre, G. A., Yarlagadda, S., Yoshikawa, S., Kurtz, S. K., and Xu, Q. C., Passively Damped Structural Composite Materials using Resistively Shunted Piezoceramic Fibers, *J Mat Eng and Perform*, 2, 887–892, 1993.

[115] Law, H. H., Rossiter, P. L., Koss, L. L., and Simon, G. P., Mechanisms in Damping of Mechanical Vibration by Piezoelectric Ceramic-Polymer Composite Materials, *J Mat Sci*, 30, 2648–2655, 1993.

[116] Sutherland, H. J., On the Separation of Geometric and Viscoelastic Dispersion in Composite Materials, *Int J Sol, Structures*, 11, 233–246, 1975.

[117] Sutherland, H. J., Dispersion of Acoustic Waves by Fiber-Reinforced Viscoelastic Materials, *J Acoust Soc Am*, 57, 870–875, 1975.

[118] Kinra, V. K., and Anand, A., Wave Propagation in a Random Particulate Composite at Long and Short Wavelengths, *Int J Solids, Structures*, 18, 367–380, 1982.

[119] Kinra, V. K., Ultrasonic Wave Propagation in a Random Particulate Composite, *Int J Solids, Structures*, 16, 301–312, 1980.

[120] Page, J. H., Sheng, P., Schriemer, H. P., Jones, I., Jing, X., and Weitz, D. A., Group Velocity in Strongly Scattering Media, *Science*, 271, 634–637, 1996.

# 10

## Applications and Case Studies

### 10.1 Introduction

Viscoelasticity can enter in the application of materials in many ways. In some applications one must deal with *natural* materials, such as stone, earth, or wood in the case of building construction, or bone and soft tissue in the case of biomedical engineering. In these cases, the viscoelastic behavior of the natural materials should be known. *Artificial* materials used in engineering applications may exhibit viscoelastic behavior as an unintentional side effect. Finally, one may *deliberately* use the viscoelasticity of certain materials in the design process, to achieve a particular goal.

### 10.2 A Viscoelastic Earplug: Use of Recovery

A foam earplug [1, 2] was designed to be easily fitted into the ear by making use of the controlled viscoelastic behavior of the polymer from which it is made. The earplug serves to attenuate sound entering the ear to protect the ear from damage from excessive noise, and also to alleviate suffering and reduce human fatigue.

To insert the earplug, the user rolls it into a narrow cylindrical shape, then inserts it into the ear canal. Insertion is easier if the outer ear is pulled outward and upward, since that straightens the ear canal. The earplug then gradually expands as a result of viscoelastic recovery to fill and contact the ear canal, and it then effectively blocks noise.

The earplug is [2] cylindrical, somewhat larger than the ear canal, and made of a foamed polymeric material with a recovery from 60 percent compression to 40 percent compression occurring in 1 to 60 seconds and an equilibrium stiffness at 40 percent compression from 0.2 to 1.3 p.s.i. (1.4 kPa to 9 kPa). In the preferred embodiment of the invention, the recovery occurs in 2 to 20 seconds. Detailed specifications for size were a diameter range of 3/8 in to 3/4 in (9.5 mm to 19 mm) and optimally 9/16 in to 11/16 in (14 mm to 17 mm) and a length from 1/2 in to 1 in (13 mm to 25 mm).

As for materials, any flexible polymeric material which can be foamed and which has the desired viscoelastic characteristics may be used [2]. The inventor considers the use of polymers of ethylene, propylene, vinyl chloride, vinyl acetate, diisocyanate, cellulose acetate, or isobutylene. A composition of vinyl chloride homopolymers and copolymers comprising at least 85 percent by weight of vinyl chloride and up to 15 percent of other monomers is favored. The viscoelastic behavior is governed by the content of organic plasticizer. The relatively slow recovery behavior confers to the user the ability to initially compress the earplug and provide sufficient time for insertion into the ear canal. Slow recovery allows the earplug to gradually expand and attempt to regain its former shape and thus conform to the ear canal.

Commercially available foam earplugs are about 13 mm in diameter and 18 to 20 mm long. The material has a Poisson's ratio approaching zero at large compressive strain. Representative earplugs exhibit power law behavior in creep from about 10 seconds to $10^6$ seconds at small stress, typically $J(t) \propto At^{1/8}$, with $J^{-1}(10 \text{ sec}) \approx$ 100 kPa [3]. Different earplugs differ in their mechanical properties. The observed creep behavior does not correspond to a single recovery time described in the patent. Evidently the inventor considered that only a limited aspect of the full viscoelastic behavior was relevant to the application. At larger stress, earplug creep compliance depends on stress level. Nevertheless, for strain as high as 24 percent, recovery follows creep. Recovery does not follow creep at 70 percent strain. Recovery proceeds to completion given enough time. As for Poisson's ratio, we initially considered the zero Poisson's ratio to be advantageous in that it implies a zero shearing stress upon the ear canal during recovery. Any shear stress on the ear canal might be perceived as a tickle, especially if slip occurred. However there is no mention of this in the patent, so the near-zero Poisson's ratio is most likely a side effect of the nonlinear behavior of the foam used.

## 10.3 Creep and Relaxation of Materials and Structures

### 10.3.1 Concrete

Concrete used in structural applications can exhibit significant creep (§7.4.2). Excessive deformation of concrete structures, which may in part result from creep, can lead to the following [4]. Visually objectionable sag may occur; the sag may result in ponding of water on roofs which can result in accelerated deterioration of the building. Sag may also interfere with the operation of sliding doors or other machinery. Partition walls may crack as a result of structural deformation; doors may stick or jam. Excess deformation can cause cracks at joints, permitting the entrance of water into the structure. Therefore, design of concrete structures for adequate rigidity must take into account the creep of concrete. Concrete, moreover, exhibits aging, and this must be taken into account in the analysis and design of concrete structures [4].

### 10.3.2 Wood

Wood used in construction can exhibit significant creep, as shown in Figure 10.1. Over sufficiently long time intervals, particularly if stresses are high, this creep can

Figure 10.1. Sag under the prolonged action of gravity. Left: sag of a wood building structure. Right: sag of lead pipes (after Ashby and Jones [16], with permission).

lead to visible sag in wooden boats and building structures [5]. Structural failure or collapse can occur as a result of excess creep [7]. Since structures made of wood or other materials may be in service for many years, long-term creep behavior is determined. Wood does not obey time–temperature superposition, therefore tests are done over long periods of time. Representative results over 12 years are shown in Figure 10.2. In these studies the wood was loaded with dead weights with a rise time of about one minute, exposed to natural variations in environmental humidity, and its deformation measured with a digital gauge. Periods of more rapid creep were associated with higher relative humidity. Creep fracture of wood occurs if sufficiently high stress is maintained for sufficiently long periods of time.

### 10.3.3 Power Lines

Electric power lines become warm from the current in them. Because new power lines may be met with public opposition, electric utilities attempt to obtain as much capacity as is practical from existing infrastructure [8]. One way to do this is to pass more current through the wire than was specified in the original design; this causes additional heating in the wire. Creep [9] occurs in the wire. Creep due to elevated

Figure 10.2. Creep in a wood beam bent by intended 10-year design load corresponding to a stress of 8.8 MPa (1190 psi) (adapted from Gerhards [7]).

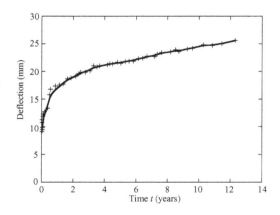

Figure 10.3. Rutting due to progressive deformation in an asphalt road from vehicle traffic (courtesy of R. Delgadillo, University of Wisconsin, with permission).

temperature in the wire should be evaluated if its temperature exceeds 75°C. Excess creep results in reduced clearance between the power line and nearby objects. Copper conductors and conductors reinforced by steel are less susceptible to creep at elevated temperature than aluminum.

### 10.3.4 Glass Sag: Flowing Window Panes

Glass is considered to be a viscoelastic liquid. When heated sufficiently it softens and flows at a time scale accessible to the observer. Since heating will accelerate relaxation processes, it is presumed that glass at ambient temperature also flows, but more slowly than hot glass. The stained glass windows of twelfth century cathedrals appear to be thicker at the bottom than at the top. This has been widely attributed to slow flow of the glass over the centuries at ambient temperature. Calculations, however, suggest that the time scale for flow exceeds $10^7$ years [10] or even $10^{32}$ years [11], far beyond historical time scale or any time scale directly observable by humans. The calculation is complicated by the fact the structure of most types of glass varies with temperature, so the usual Arrhenius expression for the effect of temperature does not apply. Instead, viscosity $\eta$ as a function of temperature for glass is described by an empirical Vogel–Fulcher Tamman formulation $\log\eta = A + B/(T - T_0)$, with $A, B, T_0$ as empirical parameters which depend on glass composition. The thickness gradient in glass from old cathedrals is attributed to the method of manufacture. Specifically, the molten glass was spun by hand at the end of a metal bar, forming a disc of glass of nonuniform thickness.

### 10.3.5 Indentation: Road Rutting

Asphalt [12] roads can become indented by the passage of vehicle traffic during warm weather; this is called rutting [13] as shown in Figure 10.3. The indented regions can hold rain water and can interfere with a driver's control, particularly on highways. A particular segment of road experiences a series of stress pulses from tires rolling over it. This dynamic stress history has a static component corresponding to the average force, because the force from the tire is always downward, never upward. Therefore, the creep behavior of the asphalt is important in determining the degree of the rutting problem. Viscoelasticity of asphalt is highly temperature dependent as discussed in §7.2.6, consequently the rutting is more severe in warm weather. Because the compliance increases with time or with the inverse of frequency, rutting is more severe when vehicles move slowly or are parked. Asphalt road material is a composite of gravel smaller than 25 mm and fine filler as small as 0.075 mm, called aggregate, and a petroleum-based binder. If the material is adequately compacted, creep is reduced. Asphalt also exhibits aging which has been attributed to oxidation and the evaporation of volatile components or progressive collapse of free volume on the molecular scale. The creep compliance decreases with the time of aging. Slow shrinkage occurs at low temperature; such shrinkage is implicated in the cracking of road surfaces during winter.

### 10.3.6 Leather

Creep can cause stress to be redistributed because the highly stressed parts or regions creep the most. The fact that old shoes are more comfortable than new ones is attributed to this phenomenon [5].

### 10.3.7 Creep-Resistant Alloys and Turbine Blades

Creep-resistant alloys are used in high-temperature applications including heat exchangers, furnace linings, boiler baffles, bolts at high temperature, power plants, exhaust systems, and components for gas turbines [14]. Turbine blades in jet engines are particularly demanding of material creep resistance in that they are subject to large tensile stress of centrifugal origin, and high temperature. The designer of the engine has an incentive to raise the operating temperature to improve the engine's thermodynamic efficiency as well as the power-to-weight ratio. If the temperature is too high, metals can creep. Significant creep in this application is unacceptable since the end of each blade must closely fit the stationary engine housing, else compressed gas in the engine will escape, lowering engine efficiency. Therefore, materials that resist creep at high temperature are desirable.

Development of creep-resistant alloys [14–16] has been facilitated by an understanding of viscoelastic mechanisms (§8.9.3). At high temperature, creep in metals results from several mechanisms including dislocation motion leading to power-law creep, and viscous slip at the grain boundaries. Nickel-based superalloys contain significant amounts of other elements, as shown for a representative detailed composition given in §7.3. These elements impede the motion of dislocations by the formation of a solid solution as well as by the formation of hard precipitates [16], for example, $Ni_3Al$, $Ni_3Ti$, $MoC$, on the scales of $\mu m$. Small inclusions of refractory oxides are also known to improve the strength and creep-resistance of metals [14]. Use of such inclusions is known as *dispersion strengthening*. Dislocation motion is impeded by such inclusions since the dislocation cannot penetrate the particle readily.

Grain boundary slip can also be reduced by improvements in alloy chemistry. Specifically [15] a high volume fraction of a second phase (such as nickel aluminide precipitate) improves grain strength and tailored grain boundary structure containing carbides and borides reduces grain boundary slip. Current super-alloys for turbine blades and disks contain nickel, cobalt, chromium, aluminum, titanium, and often tungsten and molybdenum; trade names for these alloys include Inconel and Mar-M. Despite advances in alloys, creep associated with grain boundaries can be excessive. One solution to the problem is *directional solidification* in which columnar grains are created so that they are aligned with the centrifugal stress in the blade and therefore experience no shear or tensile stress across them. Directional solidification is achieved by solidifying an investment casting of a blade in the presence of a thermal gradient. In current engines, creep of the blades is reduced significantly by forming each blade from a single crystal of metal, eliminating the grain boundaries entirely. The blades are grown by a modified directional solidification method, in which grain reduction and selection procedures are used in the early stages of solidification. The consequence of this control of viscoelastic behavior is that people can travel by air at reduced cost and with increased safety.

### 10.3.8 Loosening of Bolts and Screws

Bolts and screws can become loose with time, due to viscoelasticity in either the bolt or in the device it holds together. If the bolted joint fails to perform properly, people may be injured or lose access to services on which they depend. Examples include engines, pressure vessels, and reactors. Polymer bolts and screws exhibit substantial viscoelastic behavior at ambient temperature. Metal bolts in hot structures, such as engines and power plants, may be at a sufficient fraction of their melting point in absolute temperature that substantial creep occurs. Although structural metals, such as steel and aluminum alloys, are approximately elastic at room temperature, they exhibit significant viscoelastic behavior at elevated temperature. At sufficiently high temperature, even steel will creep. Therefore, after the bolt is tightened, it can become loose with time in the hot structure. In a small engine, for example, if a cylinder head bolt becomes loose, compression is lost, and the engine begins to perform poorly or not at all. It is impractical to use creep-resistant superalloys here since they are too expensive. There is, however, a choice of affordable alloys, and they differ in their creep resistance [6]. The manufacturer can extend the useful life of the device by choice of alloy. Because available engineering data usually include modulus and strength but not creep, testing is required to obtain the creep properties. The bolt and the surrounding metal may be comparable in their structural stiffness, so the condition is neither creep nor relaxation; it is intermediate. Therefore, analysis of the joint is called for as well to predict its performance. If the materials are linear (as is likely in polymers), then the analysis of Example 6.1 is applicable to the bolted joint as well, since the load is shared. For the nonlinear behavior of metals, a simplified analysis assuming secondary creep is given below. Secondary creep in metals or ceramics is always nonlinear, but a full nonlinear analysis of the bolted joint problem has not yet been done. Even so, alloys which resist creep also resist relaxation so one can make a choice of material.

As an example, if temperature and the bolt stress is sufficiently high, secondary creep proceeds by a power law in stress, with $B_0$ as a constant, which depends on the material,

$$\frac{d\epsilon}{dt} = B_0 \sigma^n, \tag{10.1}$$

the following calculation [16] can be performed to determine how often one must re-tighten the bolt. The total strain $\epsilon$ is regarded as the sum of an elastic part $\epsilon_{el} = \sigma/E$, and a creep part $\epsilon_{creep}$.

$$\epsilon = \epsilon_{el} + \epsilon_{creep}. \tag{10.2}$$

In the bolt, the total strain is constant. Differentiating the above and substituting,

$$\frac{1}{E}\frac{d\sigma}{dt} = -B_0 \sigma^n. \tag{10.3}$$

Integrating from an initial stress $\sigma_0$ at time $t = 0$ to a final stress $\sigma_f$ at time $t$,

$$\frac{1}{\sigma_f^{n-1}} - \frac{1}{\sigma_0^{n-1}} = B_0 E(n-1)t. \tag{10.4}$$

The final stress is

$$\sigma_f = \{B_0 E(n-1)t + \frac{1}{\sigma_0^{n-1}}\}^{-1/(n-1)}. \tag{10.5}$$

Now the bolt must carry a minimum load to hold the machine together. For the sake of doing an example, assume that half the initial stress is sufficient: $\sigma_f = \frac{1}{2}\sigma_0$. Then, the time required for this amount of relaxation is

$$t = \frac{2^{n-1} - 1}{E\sigma_0^{n-1}(n-1)B_0}. \tag{10.6}$$

This would be the minimum time interval between retightening of the bolts by a maintenance worker.

The above analysis seems simple in view of the fact that the interrelation between creep and relaxation for linear materials (which are simpler than nonlinear ones) involves a convolution integral as studied in §2.4. The simplifying assumption in the above analysis is that the same relation between stress and creep strain rate is valid both under conditions of constant stress (specifically, secondary creep) and constant strain (relaxation). The material may not in fact behave this way. Complex behavior can occur in nonlinear viscoelastic materials. Moreover, the above results for final stress and for the time give rise to indeterminate forms for $n = 1$, which corresponds to a linearly viscoelastic material. Nevertheless, such simple analyses can be of practical use. Interrelation between creep and relaxation for weakly nonlinear systems is discussed in §2.12.

### 10.3.9 Computer Disk Drive: Case Study of Relaxation

A computer disk drive was introduced some time ago, and after about six months of use, the drives experienced tracking errors and were returned to the factory [17]. The vendor examined some of the failed drives and found reduced screw-torque holding the interrupter flag to the stepper motor shaft within the drive. At first the problem was thought to be a thermal expansion mismatch between the steel of the shaft and the aluminum thought to be used in the flag, but calculation revealed a deformation of this origin would be too small to be a problem. It was discovered that the flag was in fact made of zinc. Zinc was chosen because, it could be die cast cheaply and easily. However, zinc exhibits significant creep and relaxation at room temperature. This was in fact the source of the problem. The following diagnostic test was performed. The interrupter flag set screws were torqued to the design specification value of 3.0 in-pounds. The disk drives were placed in an environmental chamber and the temperature raised to the maximum allowable storage temperature of 60°C at less than the specified maximum ramp rate of 10°C per hour. The temperature was then lowered to 40°C, the maximum allowable operating temperature. None of the drives failed after 10 days under these conditions, however that was not unexpected because the drives that failed in service did so after a longer period of time. However, measurement of the set screw torque disclosed that it had relaxed by an average of 54 percent, so that the clamping force was reduced to less than the design value. It was concluded that the material was inappropriate for the application. The following caveat is from the Zinc Institute:

> There are, however two factors peculiar to zinc which must be borne in mind in the engineering design of threaded connections. First, the relationship between the torque applied to a fastener and the retention or clamping load is erratic, apparently due to the tendency of zinc alloy threads to gall under thread contact pressures. A prescribed tightening torque will develop retention or clamping forces that may vary widely from one joint to the next. Second, zinc alloys tend to cold flow, even at room temperatures, so that the stresses in the threads relax and the clamping force is reduced. The rate of relaxation, which is a function of the elasticity of the joint, diminishes rapidly as the load decreases, tending to level off at a low stress level. At high stresses, such as those developed at 50 percent to 75 percent of the thread stripping load, noticeable relaxation occurs within a few days. The long term retention strength of a zinc alloy is therefore lower than other metals, even though the published value for tensile and shear strength of the zinc alloy may be greater. It is essential to provide for relaxation, especially when converting components to zinc alloys from aluminum, copper, or ferrous alloys.

A calculation of the required clamping force disclosed that after relaxation, the force was less than half the necessary amount. Consequently the attempt to save money by choosing zinc for this purpose was unwise.

Figure 10.4. Tilt of a retaining wall due to creep in the soil.

## 10.3.10 Earth, Rock, and Ice

### *Coupled Fields in Geology: Stress-Induced Fluid Flow*

Creep of soils can give rise to the tilt of retaining walls (Figure 10.4), settlement of foundations, movement of buildings [18, 25], or even tilt of large buildings, as in the Leaning Tower of Pisa. Prediction of such creep requires consideration of the fact that the creep depends on the hydration of the soil and its temperature and, hence, on weather and climatic conditions. Effective Young's moduli for initial deformation of soils is from about 1–3 MPa for soft clay to about 70 MPa for dense sand [18]. Time scales for time-dependent settlement can range from months to years. Soil is actually a fluid–solid composite. Creep in soils may be understood in part by Biot-type poroelastic analysis as discussed in §8.3. The time scale of the creep or relaxation, expressed as the time constant $\tau$, increases as the square of the size of the region subject to consolidation. Therefore the coupled-field equations are used directly in such problems. The Leaning Tower of Pisa (Figure 10.5) has inclination that has been increasing progressively over the years [19] due to the time-dependent deformation of the supporting soil. The Tower is is founded on weak, highly compressible soils subject to consolidation. Pressure of water in the pores was suggested to play a role in the lean [20], as was soil plasticity and a ratchet effect from changes in water table level [19]. Concern that the tower might reach leaning instability led to efforts to stop the progressive lean. Cement grout was injected in 1932 but that failed to prevent increasing lean. Following analysis, earth was excavated from the north side, stabilizing the lean.

A region of land near Galveston, Texas sank and became submerged in the sea after large amounts of oil were extracted from the pores in the underlying rock [21]. Levels of fresh drinking water in wells are known to vary with ocean tides and even with the passage of trains nearby.

Figure 10.5. The Leaning Tower of Pisa has undergone progressive lean.

Earthquakes and subsidence can result from extraction of oil, gas, or water from underground reservoirs. For example, hundreds of shallow, small to moderate earthquakes [22] have occurred near the Lacq deep gas field in southwestern France since 1969. The relationship between average pressure drop in the reservoir and subsidence (sinking of the ground) is remarkably linear, lending support to the linear poroelastic model. Fluid injected into underground strata can also cause earthquakes. For example, waste fluid injected into a well near Denver, Colorado generated small earthquakes. The fluid supports part of the normal stress on the rock and reduces the frictional resistance to sliding. This hypothesis was tested by injecting water under control into an oil field in Rangely, Colorado [23].

As for rock, creep in rock can give rise to redistribution of stress around mine tunnels and is therefore of some interest to mining engineers. Creep at very slow strain rates occurs in continental drift. Movement of the Earth's plates is associated with steady state creep at strain rates of $10^{-20}$ to $10^{-29}$ per second [24].

As for ice, its viscoelasticity (§7.2) is of interest in connection with glaciers and with structures built where there is moving ice. Topographic features associated with glaciers are a result of the viscoelasticity of the ice. Flow of liquid water gives rise to different topography; if the ice were an elastic solid, it would not flow.

### 10.3.11 Solder

Solders are alloys of low melting point (120°C to 320°C) used to join metal parts in plumbing (Figure 10.1) or to achieve electrical connections [26–29]. Elements commonly used for electrical solders are tin, lead, silver, bismuth, indium, antimony,

Table 10.1. *Solder properties*

| Material alloy | Solidus T($°$C) | Liquidus T($°$C) | Creep resistance rank | Tensile strength (MPa) |
|---|---|---|---|---|
| 63Sn/37Pb | 183 | 183 | moderate | 35 |
| 60Sn/40Pb | 183 | 190 | low | 28 |
| 95Sn/5Sb | 235 | 240 | higher | 56 |
| 52In/48Sn | 118 | 131 | n/a | n/a |
| 60In/40Sn | 122 | 113 | low | 7.6 |
| 96.5Sn/3.5Ag | 221 | 221 | high | 58 |

and cadmium. Lead-tin alloys are commonly used. In view of the toxicity of lead, other alloys, such as tin-antimony solders, have been developed. Because solders are used at high homologous temperatures, several relaxation mechanisms are active. Dislocation movement and grain boundary slip are considered to be the most important mechanisms [28]. Most solders exhibit considerable creep, and this creep can be of concern when it is desired that electrical connections be maintained over long periods of time. Development of alloys that exhibit reduced creep, yet have a low melting point, is challenging. One possibility is dispersion hardening by magnetically distributed iron particles [29]. Moreover, thermal cycling occurs in electrical equipment which is turned on and off, and this cycling can affect the performance of the connections. Failure of solder joints can occur via creep, fatigue, or by their interaction. Some solder alloys and selected properties are as follows [26], in Table 10.1.

### 10.3.12 Filaments in Light Bulbs and Other Devices

Filaments in light bulbs are made of tungsten, a metal with a high melting point (>3,300°C) [16]. The filament can therefore be electrically heated to a temperature high enough for light emission. The filament is made as hot as possible ($\approx$2,000°C), to improve efficiency. Although tungsten at room temperature exhibits little creep, it does creep at elevated temperature. This creep is a principal cause of the failure of the light bulb: the filament sags, and its coils touch each other [16] leading to a localized short circuit. The creep resistance of tungsten can be improved by alloying. For example, "nonsag" tungsten contains microscopic bubbles of elemental potassium [30]. In an illuminated light bulb, the potassium becomes a pressurized gas. The gas bubbles prevent recrystallization of the tungsten, which is responsible for much of its high-temperature creep. Light bulbs can also be made to last longer if they are operated at a lower temperature. The user can apply a lower average voltage by placing a diode in series with the bulb. However, the light is then redder and dimmer, and conversion of electricity into light is less efficient.

Electrodes in ion lasers such as the argon-ion laser get hot since the power density in the laser is high. The electrodes therefore exhibit creep. Sag of the electrodes under their own weight is a major failure mechanism in these lasers.

### 10.3.13 Tires: Flat-Spotting and Swelling

Tires in a car parked in cold weather are comparatively stiff, and exhibit more noticeable creep as a result of shifting of the time scale of retardation processes due to
the low temperature. Consequently, the flat portions of the tires in contact with the
road may persist for a while after the car is driven, so the driver perceives repetitive
thumps from the tires. This is known as "tire flatspotting," and it is related to the
creep behavior of the tire materials, particularly nylon fibers [31–33]. Nylon fibers
also creep considerably during service, so that the tire grows gradually larger [33].
Deformation due to this creep is large enough that size adjustments are made in the
retreading of tires for commercial vehicles.

### 10.3.14 Cushions for Seats and Wheelchairs

Polymer foams used in seat cushions are viscoelastic. Consequences include: (1) progressive conformation of the cushion to the body shape, (2) damping of vibration,
which may be transmitted to people via seats in automobiles and other vehicles,
and (3) progressive densification of the foam due to long term creep. The perceived
comfort of a foam cushion depends on the compliance of the foam and particu-
lars of its stress–strain behavior, discussed in §9.9. Foams which are too compliant
may bottom out (undergo densification). In this nonlinear regime, the incremental
stiffness is much higher than the initial stiffness near zero strain [34]. It has been
suggested that foams with a bimodal distribution of cell sizes are more comfortable
than foams with a single cell size, but the nonlinear stress-strain curves are similar
[35]. Moreover, the damping of the cushion governs the transmission of vibration to
the person, and this perception of vibration is important to users of seats in automobiles, trucks, and heavy equipment.

   The properties of cushions are crucially important in reducing illness and suffering in people who are confined to wheelchairs or hospital beds for long periods.
Prolonged pressure on any part of the body can obstruct circulation in the capillaries sufficient to cause a sore or ulcer known as a *pressure sore*, also called a *bed sore*.
In its most severe manifestation, a pressure sore can form a deep crater-like ulcer
in which underlying muscle or bone is exposed [36, 37]. To reduce the incidence
of pressure sores, the maximum pressure on any part of the body surface should
not exceed 32 mmHg (4.3 kPa) for long periods. This pressure is comparable to the
pressure of blood in capillaries. Too much external pressure prevents blood flow and
damages the tissue. Moreover, pressure should be uniformly distributed, adequate
air flow should be provided to the skin, and frictional forces should be minimized.
Several cushion materials have been tried to minimize the incidence and severity of
pressure sores [38, 39]. Viscoelastic foam allows the cushion to progressively conform to the body shape. However, foam densification due to creep results in a stiffer
cushion. Too stiff a cushion increases the risk of pressure sores, therefore current
foam cushions must be replaced after six months use. Efforts to develop better materials continue.

### 10.3.15 Artificial Joints

In severe cases of arthritis, pain in joints can prevent a person from moving about. Knees and hips are replaced with synthetic materials in such cases. For example, the hip joint has a ball and socket structure that allows considerable mobility of the leg. To alleviate severe pain and disability it has become commonplace to surgically replace the joint with a total hip prosthesis consisting of a ball joint fixed in the upper end of the femur (thigh bone), and a socket of ultrahigh molecular weight polyethylene cemented into the pelvis. Over years of use, the ball migrates slowly into the socket, due to both wear and creep. Most of the migration appears to be due to creep [40]. Loosening of the ball joint portion in the thigh bone is usually more of a problem clinically. The knee serves as a hinge and has in addition a degree of translational mobility. Artificial knees are generally successful but can become loose after years of use.

### 10.3.16 Dental Fillings

Dental composite tooth fillings, which consist of inorganic mineral particles in a polymeric matrix, offer the cosmetic benefit of resembling tooth structure, and do not release metallic ions into the oral environment. The polymer phase is viscoelastic, and some studies of viscoelastic behavior of dental composites have been reported [41–44]. The study of viscoelasticity in these materials is relevant in elucidating polymer properties such as crosslink density, segmental motion, and degree of polymerization, and composite properties such as interfacial bonding. Viscoelastic behavior of dental materials governs some aspects of their performance in that the forces of mastication have a static component, which over time can lead to an indentation of the restoration as a result of its creep. Wear processes also cause indentation of the filling. The creep is greatest soon after the composites are prepared, and as polymerization proceeds, the creep becomes less. Silver amalgam fillings [44] also exhibit creep as well as aging due to continued chemical reactions in the filling after it is prepared. For that reason it is best to avoid chewing stiff foods for a while after receiving a filling. Extrusion of amalgam dental fillings at the margin with the tooth has been observed [45] both in clinical settings and in laboratory tests. Such extrusion is deleterious because it can encourage adhesion of bacterial plaque as well as provoke fracture of the margin of the restoration, which loses support.

### 10.3.17 Food Products

Transient mechanical loads are applied to food products during harvesting, transportation, and storage as well as during chewing. Viscoelastic behavior of various fruits and vegetables have been studied with the aim of achieving better quality of food products in the marketplace. To that end, viscoelastic properties have been measured [46, 47] and some results are discussed in §7.5. Such data can aid in the design of food-handling equipment and procedures.

Perceived freshness of some foods depends on the glass transition temperature $T_g$ of constituents. For example, a fresh cookie [48] had a $T_g$ of 60°C. It would be consumed near room temperature, in the glassy state, perceived as crunchy. Exposure to the atmosphere resulted in a weight gain of 5 percent and a reduction of the $T_g$ to 10°C, most likely due to the plasticizing effect of the adsorbed moisture. The texture would then be perceived as leathery or rubbery.

Viscoelasticity is the basis of informal tests of the ripeness of melons. Many people tap melons before buying them, to judge from the sound whether they are ripe. This is an example of free-decay of vibration following an impulse.

### 10.3.18 Seals and Gaskets

*Rubbery* materials are commonly used in various kinds of seals and gaskets which prevent the flow of pressurized gas or liquid through the gap between two adjacent solid objects. The gap may change with time, and the seal must accommodate this change. If the gap changes too rapidly for the seal to follow, the pressurized fluid will not be contained. An example is the catastrophic explosion of the space shuttle Challenger in 1986. O-rings were used to prevent hot exhaust gas from escaping through joints in the solid rocket boosters. The shuttle was launched on a cold day. The low temperature caused O-ring material to acquire a more leathery rather than rubbery consistency, with more marked viscoelastic response (See §7.2). The material therefore could not rapidly adapt to changes in the spacing of the joint it was intended to seal. Hot gas escaped through the side of the booster causing the spacecraft's fuel tank to explode, resulting in death of the astronauts. The leathery viscoelastic response was described as a loss of resiliency. Relaxation and recovery following a transient are slow under these conditions. The change in behavior with temperature was graphically demonstrated by physicist Richard Feynman who, in a press conference, immersed a segment of O-ring rubber in ice water, squeezed it with a clamp, and observed slow recovery upon release of the clamp [71]. Feynman demonstrated not only viscoelastic behavior but also physical insight, in general. Feynman earned his Nobel Prize in particle physics and was not a specialist in viscoelastic materials. Engineers were aware of the role of thermoviscoelasticity in the O-rings and had warned against launching the spacecraft. The disaster was a consequence of management decisions [71].

### 10.3.19 Relaxation in Musical Instrument Strings

Players of stringed instruments are aware that nylon strings rapidly go out of tune [49]; in comparison, metal strings retain their tuning. The pitch of the note generated by the string is calculated as follows. The speed of sound $c$ in a string of mass $m$ and length $L$, under tensile force $F$ is $c = \sqrt{\frac{F}{m/L}}$. This may be expressed in terms of the stress $\sigma$ and density $\rho$, $c = \sqrt{\frac{\sigma}{\rho}}$. The frequency $f$ of the lowest (fundamental) natural frequency is $f = \frac{c}{\lambda} = \frac{1}{\lambda}\sqrt{\frac{\sigma}{\rho}}$.

This frequency governs the pitch of the tone. The length of the string corresponds to half a wavelength λ, because the string is fixed at each end. The musician tunes the instrument by adjusting the extensional displacement of the string by means of a screw. After tuning, the string is under essentially constant axial strain, and it undergoes relaxation of stress, which causes the pitch of tones derived from the string to decrease with time. Therefore, the musician finds it necessary to again tune strings made of a polymer such as nylon. Strings made of catgut contain natural polymers, also undergo relaxation. Wood in the instrument is also viscoelastic, but provided the instrument is much stiffer than the strings, displacements due to creep in the wood are small. Strings also can go out of tune as a result of temperature changes. As for the free decay of vibration in a plucked string, energy is radiated as sound, and this radiation can be the predominant cause of the free decay of the vibration in a string in the instrument.

### 10.3.20 Winding of Tape

Materials stored in the form of rolls include paper, plastic sheets, and magnetic tape for computers and for audio and video recording. As for tapes, the requirement for high-quality reproduction of stored information implies that the dimensions of the tape be essentially unchanged during and between read and write operations [50]. When wound, the sheet or tape is under tension and is applied one layer at a time over a core. The materials involved are viscoelastic. Therefore, the final stress state in the layers will depend on the speed of the winding, time of storage, and any starts or stops in the process. A fairly complicated analysis of this problem has been presented [50].

## 10.4 Creep and Recovery in Human Tissue

All tissue is viscoelastic as discussed in §7.5. Some examples of application of tissue viscoelasticity are as follows. Detection of viscoelasticity by impulse response for the purpose of medical diagnosis is discussed in §10.10.6.

### 10.4.1 Spinal Discs: Height Change

The discs in the human spinal column are viscoelastic and exhibit creep [51]. One result of the creep under normal body weight is that most people are about 1 cm taller in the morning than in the evening. For young people, the difference in height can be as much as 2 percent, which corresponds to a 1.4 inch (3.7 cm) variation in the height of a 6-foot (183 cm) person [52]. The increase in height after bed rest is due to creep recovery in the discs. Astronauts under microgravity conditions gained 5 cm in height, then suffered backaches. One can measure stature to within 0.4 mm. With refined measurements of this kind, classic creep and recovery curves can be obtained from a living person under gravitational load, following a change in sitting position or following placement of a weight on the shoulders. In one study, five

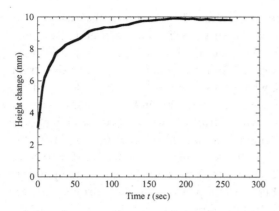

Figure 10.6. Height change in a living human subject due to tissue creep (adapted from Magnusson et al. [54]).

minutes of sitting gave rise to an average of 0.2 percent reduction in body height for young adults, and 0.4 percent for people aged 60 to 65 [53]. Height change occurs in living human subjects due to tissue creep [54] measured by transducers on the skin, as shown in Figure 10.6. Spinal creep measured by magnetic resonance imaging of vertebrae [55] discloses 7 to 9 percent creep of the combined height of discs in the lower back over 20 minutes. The disc height of the lower back is about one-third of the total height. Creep appears greater in an extended posture than in a flexed posture. Such experiments are done to better understand the effects of posture and exercise on the spine. Such results also may be of use in minimizing back pain in ergonomic design, such as design or choice of chairs. Dynamic mechanical damping also occurs in spinal discs [56] resulting in absorption of energy in cyclic loading. Specifically, the viscoelasticity of spinal discs is also thought to play a role in the damping of impulses [57] from walking and running.

### 10.4.2 The Nose

The noses of older people may appear to droop (Figure 10.7). This change in appearance is not necessarily due to viscoelasticity. The nose actually continues to grow during adulthood [58]. Moreover, other soft and hard tissues in the face undergo growth during adulthood. Knowledge of change in these tissues is useful to orthodontists and reconstructive surgeons.

### 10.4.3 Skin

Skin is viscoelastic, therefore it has a delayed recovery in response to transient deformation. The skin's ability to recover its original shape is important clinically [59]

Figure 10.7. *Droop* in biological and synthetic materials. Apparent droop of the human nose is due to growth. Structural metals do not droop at ambient temperature and over humanly accessible time scales because crystal imperfections are largely immobile in metals with a high melting point.

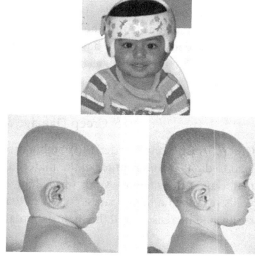

Figure 10.8. Baby heads before (left) and after (right) treatment with helmet (top) (adapted from Cranial Technologies [67], with permission).

in the context of plastic surgery. Moreover, many skin diseases result in changes in the mechanical properties of skin, and differential diagnosis may be possible based on mechanical tests [60]. Quantitative measurements of skin recovery can be made in the living subject. An informal test of recovery may be done by pinching the skin on the back of the hand for several seconds. In a healthy young person, the skin snaps back in less than one second but in older people and in cigarette smokers, recovery following this deformation may require many seconds to several minutes.

### 10.4.4 The Head

Babies have deformable heads. The head may be misshapen following the birth process, but the shape returns to normal after about six weeks. Abnormal head shape can result from prematurity, because the cranium stiffness increases substantially in the last weeks of pregnancy, from the pressure of multiple fetuses in utero, as well as from prolonged sleeping of the baby on the back. Extended use of car seats and carriers can also be problematical [65]. If the baby sleeps in a single position, the long-term load component can cause an abnormal head shape [61] due to creep deformation. Such deformation was observed after it was recommended that infants be positioned to sleep on their backs or sides to reduce risk of sudden infant death syndrome [62]. Following that publication, many children were observed with abnormal head shapes. Abnormal head shape can be prevented if the parents frequently reposition the infant's head. Abnormal head shape in infants can be corrected by a helmet-like device worn on the head [63, 64]. The device provides a gentle pressure to gradually restore the normal shape of the head. An illustration of shape change achieved by helmet treatment is shown in Figure 10.8. Curiously, some medical practitioners have done elaborate surgery on such babies [66] in a four-hour operation, which, at best, causes bleeding, pain, distress to the family and economic hardship, and leaves a scar from one ear to the other. Other surgeons contend there is indeed an epidemic of misdiagnosis. Helmet treatment [67], which takes advantage of the natural viscoelasticity of the tissue, is clearly more humane. Heads of infants have been deliberately deformed by ancient Egyptians as early as 2000 BC as well as by aboriginal peoples in Africa and North and South America.

## 10.5 Creep Damage and Creep Rupture

Creep, which occurs under the combination of a large stress and a long time, can lead to damage and rupture of the material. During tertiary creep in polycrystalline materials such as metals, voids form along the grain boundaries [16]. Growth of the voids results in a higher stress for given load, hence a higher creep strain rate. The strain rate increases until the material breaks.

### 10.5.1 Vajont Slide

Perhaps the most dramatic case history of creep rupture is the Vajont slide [68] in which a large mass of rock on a steep slope suddenly collapsed, and slid down the slope into a reservoir. This created a water wave that, obliterated the town of Longeroné at the cost of 2,500 lives. The tragedy occurred in 1963, however slow creep of the rock had been observed for three years prior to the creep rupture. Creep rupture is also a matter of concern in metals exposed to high stress and high temperature for extended periods. Optimal design of structural elements in this regime is complicated by the intrinsically nonlinear nature of secondary and tertiary creep [69].

### 10.5.2 Collapse of a Tunnel Segment

A roof panel in a tunnel in Boston (the Big Dig project) broke free, crushing a passing car, killing one person and injuring another. The ceiling panels were installed in a grid suspended by hangers drilled into the top of the reinforced concrete tunnel and grouted with epoxy adhesive [70]. Efforts at cost-cutting led to a design with heavy panels, fewer supports than in the original design, and no redundancy. Creep in the polymer led to the failure.

## 10.6 Vibration Control and Waves

### 10.6.1 Analysis of Vibration Transmission

An example of the use of compliant viscoelastic materials in vibration isolation is presented. Vibration isolators are used to protect people or machinery from the harmful effects of vibration from engines, road travel, or other sources. Suppose that a substrate (Figure 10.9) undergoes vibration given by $x_1(t) = ae^{i\omega t}$, (in which

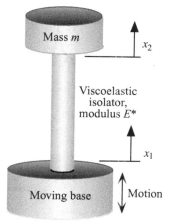

Figure 10.9. Sketch of a mass $m$ isolated from a moving support by a viscoelastic bar of stiffness $E^*$ and length $l$.

$a$ is the amplitude of the displacement) and upon the substrate is a block of modulus $E^*$ of length $l$ and cross-sectional area $A$ supporting a mass $m$. In the following, a simple one-dimensional analysis is conducted, with effects due to Poisson's ratio neglected. If such effects were included, an additional geometrical factor would appear multiplying the modulus. The analysis is similar to that in §3.5 pursued for a lumped system in which the dynamic compliance of the system was calculated; here the issue of interest is the transmission of vibratory motion from the substrate to the supported mass. This problem is amenable to the correspondence principle, however since the solution is simple, it is treated directly in the frequency domain. The motion of mass $m$ is described by $x_2(t) = b^* e^{i\omega t}$, in which the response amplitude $b^*$ may have a phase and so be describable as a complex number. The system is considered to have one degree of freedom. In applications involving vibration isolation, the system *transmissibility* $T$ is of interest. The transmissibility is defined as follows [72]:

$$T = |\frac{b^*}{a}|. \tag{10.7}$$

The transmissibility represents the dimensionless ratio of the displacement amplitude of the driven mass to the displacement amplitude of the moving substrate. It is usually desirable to minimize the transmissibility to protect the mass from too much vibration.

For this system, Newton's second law of motion becomes as follows [72]:

$$m\frac{d^2 x_2(t)}{dt^2} = \frac{A}{l} E^*(x_1 - x_2). \tag{10.8}$$

Substituting the harmonic time dependence,

$$-\omega^2 m b^* = \frac{A}{l} E^*(a - b^*). \tag{10.9}$$

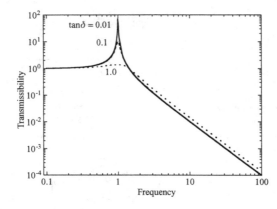

Figure 10.10. Transmissibility of lumped mass-isolator system, of single degree of freedom, for various $\tan\delta$, versus normalized angular frequency $(\omega/\omega_0)$.

So, the complex ratio of response displacement to driving displacement is, with $E^* = E'(1 + i\tan\delta)$,

$$\frac{b^*}{a} = \frac{\frac{A}{l}E^*}{\frac{A}{l}E^* - \omega^2 m} \tag{10.10}$$

$$\frac{b^*}{a} = \frac{1 + i\tan\delta}{1 + i\tan\delta - \frac{\omega^2}{\omega_0^2}}, \tag{10.11}$$

with

$$\omega_0^2 = \frac{\frac{A}{l}E'}{m} \tag{10.12}$$

as the natural angular frequency. Recall that the angular frequency is given by $\omega = 2\pi f$, with $f$ as frequency. Moreover, both $E'$ and $\tan\delta$ depend on frequency.

Separating this into magnitude and phase $\theta$,

$$\mathcal{T} = \left|\frac{b^*}{a}\right| = \frac{\sqrt{1 + \tan^2\delta}}{\sqrt{(1 - \frac{\omega^2}{\omega_0^2})^2 + \tan^2\delta}} \tag{10.13}$$

$$\theta = \tan^{-1}\left\{\frac{-\frac{\omega^2}{\omega_0^2}\tan\delta}{1 - \frac{\omega^2}{\omega_0^2} + \tan^2\delta}\right\}. \tag{10.14}$$

The transmissibility $\mathcal{T}$ tends to 1 at frequencies far enough below the natural frequency, and it decreases with frequency as $(\omega_0/\omega)^2$ for frequencies well above the natural angular frequency $\omega_0$, (Figure 10.10). Therefore, if the transmissibility is to be minimized at the higher frequencies, the natural frequency should also be minimized. To make a detailed plot of $\mathcal{T}$ versus $\omega$, the frequency dependence of $E'$ as it influences $\omega_0$ must be taken into account. Although some writers [72] present plots based on different assumptions about $E'(\omega)$, recall that the Kramers–Kronig relations developed in §3.3 prescribe a unique relationship between the real and imaginary parts of a complex mechanical property function. Further examples and analyses are given in Snowdon [72] and Frolov and Furman [73].

In practical systems, the isolator, represented by the block of stiffness $E^*$, is chosen so that the natural frequency is as far as possible below the frequencies of excitation of the substrate, so that the transmissibility is minimized in the frequency range of interest. The natural frequency is reduced by making the isolator more compliant (Equation 10.12). If the isolator is too compliant, any force upon the mass itself will cause excessive motion or even bottoming. In some cases it may be unavoidable that excitation motion appears at or near the natural frequency. For such cases, the isolator must be viscoelastic, to reduce the amplitude of the response at resonance. Because vibration can affect the human body (Figure 10.11), isolation systems including shock absorbers and seat cushions are used to protect the body as described below in §10.6.11.

## 10.6.2 Resonant (Tuned) Damping

If damping is required over a limited range of frequency, resonant (or tuned) damping may be used [74–76]. An example of a situation involving narrow-band excitation is an electrical transformer. Transformers produce acoustic noise at harmonics of the power line frequency, as a result of magnetostriction of the iron alloys in them. The axial vibration configuration in Figure 10.9 may be used to achieve a viscous or resistive mechanical input impedance at the base. Alternatively one may use a configuration based on a shearing resonance. Here, the objective is to maximize the dissipation of power from the vibrating "base" rather than to minimize the acceleration of the mass as considered above. Therefore the material requirements differ. The viscoelastic element should have the smallest possible loss tangent if the dissipation is to be maximized at the resonance frequency. Under such conditions, the supported mass will undergo high acceleration, but that is immaterial because it is a dead weight rather than a fragile object to be protected.

## 10.6.3 Rotating Equipment Vibration

Computer disk drives make use of viscoelastic materials to protect the disk from shock and impacts which occur when the device is transported, to reduce acoustic noise that disturbs the user, and to increase data density by improving tracking. For example, constrained layer damping is used for the disk housing [77, 78] to reduce noise and vibration, and compliant viscoelastic buffers are used to reduce the effect of impact on the disk as well as vibration. Even so, some people have been bothered sufficiently by computer noise that they retrofit their computers [80] with additional materials and components to reduce noise.

Audio turntables have been isolated from outside vibration by suspending them from a relatively complex system of springs. An alternative design for an isolator uses a viscoelastic platform for support rather than a costly and complex spring based system [81]. The viscoelastic material, referred to as "a dry goo" by a manufacturer's representative, is sandwiched between layers of fiberboard.

Viscoelastic elastomer is available to be used as rubber "feet" [1] or base material to support scientific apparatus or other equipment that may be perturbed by vibration transmitted through the laboratory floor. Gaskets made of such materials can function not only as seals but also can reduce vibration in structures.

Viscoelastic elastomers are also used in the form of inertial dampers to reduce settling time during rapid angular accelerations of head positioning motors in computer disk drives [1], and to reduce resonant vibration in high speed robots. *Nutation* motion of spinning rotors mounted upon gimbals (pivot joints) occurs when a perturbing torque is suddenly applied. Nutation is an oscillatory motion of the spin axis. In some gyroscopic, aerospace and satellite applications it is desirable to damp this nutation motion. Such damping may be achieved by attaching masses to the rotor by a compliant, viscoelastic stalk or annulus [82, 83]. The nutation motion causes an oscillatory strain in the viscoelastic material, which dissipates energy and damps the motion.

### 10.6.4 Large Structure Vibration: Bridges and Buildings

Bridge and building structures are dynamically loaded by the wind and by earthquakes, and the resulting vibration can be problematical [84]. Sway of buildings may annoy or nauseate the occupants, and oscillation of bridges can cause structural damage. Damping in a building arises from viscosity of the air surrounding the building; viscoelasticity of the building materials; friction damping at joints [85] and contact points; hysteresis if the loads in the building or ground exceed the elastic range; and damping due to radiation of elastic waves through the supporting ground [86].

Oscillations can be induced in bridges by traffic motion, which is usually random. The input may become synchronized with the bridge fundamental natural frequency if the traffic consists of people walking. The Millennium Bridge in London began to sway unexpectedly when it was opened to foot traffic; the people did not consciously attempt to drive any resonance. The sway was attributed to the fact that a biological system, in this case human gait, is modified by the environment. Specifically the bridge motion is understood to modify the gait of each person [87]. The analysis gives rise to a system of coupled equations which may be of use determining the damping needed in the design of such bridges.

Viscoelastic dampers are used in some tall buildings to damp out sway oscillations of the building caused by wind. Viscoelastic dampers were used in the World Trade Center in New York [86], and viscoelastic dampers consisting of steel plates coated with a polymer compound are used in the 76-story Columbia Center building in Seattle. The viscoelastic dampers are connected to some of the diagonal bracing members [88]. Vibration damping may be achieved by means other than viscoelastic solids. For example, the tuned mass damper is used in other buildings such as the City Corp Center in New York and the John Hancock Tower in Boston. The tuned mass damper consists of a large concrete block weighing about 2 percent of the entire building. It is near the top of the building, upon a smooth concrete surface, and linked by a spring and macroscopic shock absorber to the building structure [86]. The mass and spring constant are tuned so that the resonant frequency of the system corresponds to the principal sway resonance frequency of the building. If building accelerations reach a threshold (such as 0.003 of the acceleration of gravity) due to a wind storm, pumps are activated that float the block on a layer of oil so that it can oscillate in synchrony with the vibration, and damp it out. This is considered an active damping method.

### 10.6.5 Damping Layers for Plate and Beam Vibration

Flexural vibration in thin plate elements of structures can result in noise which is objectionable or harmful to people. Such vibration can be reduced by applying a damping layer [74, 89] of viscoelastic polymer of loss modulus $E_d''$ and thickness $t_d$, to the metal plate, which has storage modulus $E_p'$ and thickness $t_p$ [89].

For *axial* vibration of a plate with a damping layer, the effective loss $\tan\delta_{eff}$ is [76]

$$\tan\delta_{eff} = \frac{E_d'\tan\delta_d t_d}{E_p' t_p + E_d' t_d}. \tag{10.15}$$

If the damping layer is much more compliant than the plate, and it is not too thick,

$$\tan\delta_{eff} \approx \frac{E_d'\tan\delta_d}{E_p'}\frac{t_d}{t_p}. \tag{10.16}$$

For *bending* vibration of a plate with a damping layer the effective loss $\tan\delta_{eff}$ is

$$\tan\delta_{eff} \approx E_d'\tan\delta_d \frac{t_p\frac{1}{4}(t_d + t_p)^2}{\frac{1}{12}E_p't_p^3 + E_d't_d\frac{1}{4}(t_d + t_p)^2}. \tag{10.17}$$

Again if the damping layer is not too thick,

$$\tan\delta_{eff} \approx 3\frac{E_d'\tan\delta_d}{E_p'}\frac{t_d}{t_p}. \tag{10.18}$$

in which $t_p$ is thickness of the plate. The effective damping for a plate with two damping layers was derived via the correspondence principle in Example 5.1.

Consequently the loss modulus $E'' = E_d'\tan\delta_d$ of the unconstrained damping layer should be maximized for both axial and bending vibration. Axial vibration is the most difficult to damp; in bending there is the advantage that the softer damping layer is some distance from the neutral axis, hence more effective from a strain energy perspective.

If a thin layer of metal foil is cemented to the polymer layer, it is called a *constrained layer* [75]. Substantial shear strains can be induced in the polymer material during the bending of the plate, increasing the damping of the structure. Analysis of constrained layer damping is more complicated than that for free layers [76]. Damping depends on the wavelength of bending waves. Damping in the constrained layer method is strongly frequency dependent as a result of dependence on the wavelength. Even if the damping $\tan\delta_d$ of the polymer layer is constant, the overall damping $\tan\delta_{eff}$ of the composite plate exhibits a peak, less than two decades-wide, in frequency. As temperature changes, the shear modulus of the layer changes substantially, and so the frequency of the peak shifts too. Design of a constrained layer damping system is more involved than for free layers. The peak in the damping versus frequency can be significantly broadened on the low-frequency end by segmenting the constraining layer [74]. Other variants of the damping-layer method include multiple layers with different properties, or use of a filler in the polymer to provide an internal constraint. Enhanced overall damping may also be achieved by mounting the damping layer upon stand-off spacers [90]. The rationale is to move the damping layer as far from the neutral axis as possible to maximize strain levels in the layer.

Tapes containing both a polymer and a metal layer are available for damping applications. Damping with the constrained layer method increases with the damping, $\tan\delta_d$, of the damping layer and the thickness of the constraining layer [75]. If a minimum weight design is desired, the product $E \tan\delta$ enters the design process indirectly. Methods involving polymer layers are well known [91–93]. These layers offer acceptable performance in suppressing bending vibration of thin plates. They have the disadvantages of temperature sensitivity, flammability, a narrow effective range of temperature and frequency, and a comparative inability to control vibration in deformation modes other than bending. For materials of the highest damping ($\tan\delta > 1$), the full width at half of the maximum of the damping peak at constant frequency may be only about 18°C [94] although the peak may be broadened [95, 96] somewhat by nonstoichiometric compositions. This breadth is adequate for machinery that operates near room temperature, or at another nearly constant temperature, but is insufficient for aircraft for which the skin temperature may vary over a considerable range.

In some applications it is impractical to use polymer laminates. Several high-damping metals such as CuMn are available, as described in §7.3, and some of their applications are described in §10.6.6.

## 10.6.6 Structural Damping Materials

### *Rationale*

High-loss, flexible polymer layers may be used, as described in §10.6.5, to reduce flexural vibration of plates. It is more difficult to reduce axial vibration, since the attached layer would require both high stiffness and high damping. Moreover, some environments are hostile to polymers and in some designs, the extra space occupied by a polymer layer would be problematical. Figure 7.2 shows that high stiffness combined with high damping does not occur among common materials. There are some alloys, as discussed in §7.3.3, which offer high stiffness with moderate damping.

These materials, such as CuMn alloy, have the disadvantage of being nonlinear: the damping is much less at small strain than at large strain. A linear material with high stiffness and high loss would be useful in many applications involving vibration reduction, particularly in the aerospace field [100].

### High-Loss Metals

High-loss metals are used in situations in which polymer layers are inappropriate. A classic example is the use of copper-manganese alloy to reduce vibration and radiated noise from naval ship propellers [101]. This alloy has been used to replace a low damping metal for valves, which had previously failed by fatigue, in a pump [102]. It has also been used to reduce noise in a pneumatic rock crusher. The damping capacity of cast iron makes it preferable to steel for large machine tool frames. Zinc-aluminum alloys have been used to reduce vibration in hand-held pneumatic hammers, engine mounts, and camshaft drive pulleys in engines [103].

### High-Loss Composites

Damping layers of improved performance have been made of composite materials. Stiff inclusions [104–106] embedded in a high damping polymer matrix can give rise to enhanced performance of the damping layer. Such composites are in harmony with the principles articulated in §9.3. Compliant, high-loss layers have been embedded in composites as constrained layers subject to shear [107].

### New Materials

The viscoelastic mechanisms discussed in Chapter 8 have been explored primarily for the purpose of understanding existing materials; however they may also be used to guide the synthesis of new materials. For example, in one modality, composite materials may be designed to achieve a measure of thermoelastic damping [108], by proper choice of constituents. Damping achieved in this way is advantageous in that it is hardly dependent on temperature. However, it is difficult to achieve much thermoelastic damping ($\tan\delta$ greater than 0.01) using common materials as is evident from Table 8.1. Piezoelectric materials are more promising as a second modality in that the coupling between the stress and electric field can be much stronger than the coupling between the stress and temperature field for available materials as discussed in §8.7.2. Vibration can be actively controlled by detecting it with sensors, amplifying and conditioning the signal, and driving actuators to neutralize the vibration [109–111] (§10.7). If there are multiple modes of vibration, multiple electronic feedback channels must be provided. A resistive circuit element can be attached to a piezoelectric inclusion to achieve substantial damping [112]. A current challenge is how to make a fine grained composite with resistively damped piezoelectric elements. A third modality is the development of composite materials with optimal architecture for high damping, as discussed in §9.3 and §9.4. The phase which provides the high damping need not be stiff or strong provided that the composite geometry permits the other phase to contribute stiffness and strength. Composites with $E = 161$ GPa (more than twice as stiff as aluminum) and $\tan\delta = 0.1$ (comparable to a glassy polymer) have been made [113].

### Damping by Granular Materials

It has long been known that granular materials, such as sand or lead shot, can damp vibration [76]. Sand placed in the cavities of concrete blocks is effective in damping noise and vibration in building structures. Effective loss factors can exceed 0.1. Indeed, budget conscious workers in holography have supported sensitive optical components in sand rather than expensive honeycomb optical isolation tables [97]. Sand has also been used within control levers to reduce vibration transmitted to the operators of heavy trucks and other equipment [98]. The weight penalty of sand can be avoided by using low-density granular materials such as perlite (a volcanic silica glass) or polyethylene beads [99]. The low speed of sound (100 m/s or less) in such materials is considered advantageous in coupling with low-frequency bending modes of structures.

### 10.6.7 Piezoelectric Transducers

Piezoelectric materials exhibit coupling between the electric polarization and mechanical deformation. They are used in vibration as frequency standards, as spark generators in stoves, and in transducers. For materials used in ultrasonic transducers, a high damping is often desirable for the following reasons. In applications, such as ultrasonic flaw detection and in medical diagnostic ultrasound, a short pulse is transmitted which contains only a few oscillations at the transducer's resonant frequency. The reason is that a short pulse permits greater resolution of defects. The Fourier transform of a short pulse in the time domain contains a wide distribution of frequency. Recall from §3.5.1 that $\Delta\omega/\omega_0 \approx [\sqrt{3}]\tan\delta$. Therefore, a large loss tangent is desirable to obtain response over a relatively wide band of frequency. Quartz is low loss and is unsuitable for this type of application. Some piezoelectric ceramics, such as lead metaniobate, exhibit a high loss and are used in ultrasonic transducers. Damping can also be introduced by means of a backing layer applied to the transducer; this layer may contain lead particles in a polymer matrix. For high power piezoelectric emitters, the damping should be low otherwise the device can overheat. In addition to transducers [114, 115], there are many other applications of piezoelectric materials [116].

### 10.6.8 Aircraft Noise and Vibration

Engines and associated machinery used in aircraft can produce considerable noise and vibration. Several examples follow.

### Case Study: Vibration Fatigue of Jet Engine Inlet

In the TF-30-P100 jet engine used in military aircraft, the entering air first encounters stationary inlet guide vanes that guide air into the first stage of the engine's turbine [75, 117]. Cracks were observed in the inlet guide vanes. Flight test measurements disclosed high vibratory stresses in the vanes, leading to high cycle fatigue. Vibration measurements disclosed high stresses corresponding to resonant torsion and bending modes from 3 to 4 kHz. A viscoelastic damper was designed to reduce the vibration. The design was driven by the fact that the inlet air varies in temperature from 0° to 125°F under most conditions, and that on occasion deicing procedures generated transient temperatures exceeding 400°F. A constrained-layer damping approach was chosen. Laboratory tests compared the effectiveness of several compositions of layer material. The final damping wrap consisted of a constraining layer of aluminum 0.005″ (0.13 mm) thick and two viscoelastic layers, 0.002″ (0.05 mm) thick, of ISD-830 material (3M Corp.) for the concave surface and ISD-112 for the convex surface. An additional aluminum layer, 0.002″ (0.05 mm) thick under the viscoelastic layer was incorporated to avoid air entrapment between layers; this was bonded to the titanium surface of the vane with structural epoxy, type AF 126, nominally 0.005″ (0.13 mm) thick. The damping wraps were installed on engines, and no further cracks were observed in the vanes.

### Case Study: Helicopter Noise

A rescue helicopter, designated HH-53C and manufactured by the Sikorsky Aircraft Company, was noisy: sound levels approaching 120 dB (with respect to 0.00002 $\mu$ bar) were measured in the cabin. The noise came from the turbine and from gears, but was exacerbated by resonance of the fuselage skin. Noise can be reduced by placing standard acoustic blankets consisting of a layer of fiberglass sandwiched between layers of vinyl cloth inside the fuselage. This approach was not used because it would interfere with maintenance, inspection, and repair of the helicopter. Therefore, a damping layer approach was explored [75]. The vibration and temperature environment was first characterized via accelerometers and thermocouples applied to the fuselage. A constrained layer approach was chosen. Although only a small portion of the structure was covered with damping treatments, fuselage skin accelerations were reduced by up to 12 dB and high-frequency noise was reduced by 5 to 11 dB in the cabin.

### Case Study: Helicopter Exhaust Stack

Some helicopters (type CH-54) have experienced problems of vibration-induced fatigue of exhaust stacks [75, 118]. Attempts were made to reduce vibration amplitudes by increasing structural stiffness but these failed since the noise spectrum from the engines was broadband. The exhaust stacks exhibited many lightly damped

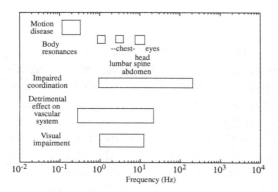

Figure 10.11. Frequency regions for effects of vibration on the human body (adapted from Frolov and Furman [73]).

($\delta = 0.001$ to $0.1$) resonances over the audio range. Polymer damping layers were not practical in view of the high temperatures involved, up to 800°F. The final damping material chosen was a vitreous enamel (Corning Glass No. 8363) which survives high temperature and exhibits a damping peak of about 0.5 at 100 Hz, near 800°F. Testing included trials at room temperature with polymer dampers, tests of enamel on a steel cantilever at high temperature, and finally tests on the actual helicopter exhaust stack. A free layer design 0.25 mm thick was chosen. The free layer approach was facilitated by the relatively high dynamic Young's modulus of the glass, about 17 GPa, at the damping peak. The composite peak loss factor was about 0.15 at 100 Hz and 800°F. This was sufficient to prevent the metal fatigue of the pipe.

### 10.6.9 Solid Fuel Rocket Vibration

Energy released during the burning of solid rocket propellant has the side-effect of amplifying acoustic waves in the rocket chamber [119]. This amplification can cause oscillations of sufficient amplitude to crack or break up the solid propellant. The solid propellant has, by virtue of its binder, viscoelastic properties [120, 121]; therefore it attenuates stress waves. Higher values of tan$\delta$ have the effect of stabilizing combustion in the rocket and preventing fracture of the propellant. One propellant exhibited $E = 1.1$ GPa and a large tan$\delta$ of 0.18 at acoustic frequencies [120].

### 10.6.10 Sports Equipment Vibration

Damping layers have been used to reduce vibration in sports equipment such as golf clubs [122, 123]. Piezoelectric damping elements have been used commercially in skis [124] to damp vibration, can cause the skier to lose control at high speeds.

### 10.6.11 Seat Cushions and Automobiles: Protection of People

#### *Effects of Vibration on the Body*

People in moving vehicles can suffer a variety of detrimental effects from dynamic accelerations as shown in Figure 10.11. Whole body vibration can adversely influence the ability of people to perform various tasks [125]. For example, vision can be degraded by vibration of the observed object, of the eyes, or both. The eye can track the motion of objects at low speeds or low frequencies, but a rapidly vibrating object will appear blurred. People who perceive vibration may be annoyed or distracted. The body perceives acceleration most easily near 7 Hz at which the perception threshold is less than 0.01 m/s$^2$, but at lower or higher frequency it is more difficult to perceive acceleration. The threshold approaches 0.1 m/s$^2$ at 0.1 Hz and at 100 Hz. People can experience motion sickness and may vomit if oscillated at low frequency. The range 0.1 Hz to 0.5 Hz is most provocative of motion sickness. It is desirable to reduce vibration and noise within vehicles to reduce suffering in those people who must travel. To that end, viscoelastic materials can be of use.

The low-frequency oscillations that give rise to motion sickness are at too low a frequency to be attenuated by seat cushions or suspension systems. Motions sickness may be prevented or reduced by location of ship passengers or crew where translational movement is minimum and by design of the person's visual environment.

### Reduction of Body Exposure to Vibration

Seat cushions can play a role in reducing the body's exposure to vibration. Seats in vehicles have compliant cushions for several reasons [125]. Cushions distribute pressure around the bony prominences (ischial tuberosities) in the pelvis; they are soft to drop into; they usually have a low heat capacity so extreme temperatures are not so objectionable to the user; they provide friction to prevent sliding; and they attenuate vibration. These benefits are not necessarily achieved with the same physical characteristics. Moreover, the optimum dynamic behavior of a seat depends on the spectrum of vibration frequencies and on the desired criterion, for example, comfort or preservation of health. To analyze the isolation of the human body from vibration, one cannot simply use a single degree of freedom model, such as the one presented above, because the body itself is not a rigid mass. The body cannot be regarded as rigid above about 2 Hz. Transmissibility coefficients have been determined for pairs of points on the human body, such as the seat and the head of a seated person. The body is heavily damped, so that the transmissibility attains a peak of no more than 1.5 at about 8 Hz. As one might expect, the curves of transmissibility versus frequency depend on posture. Vibrations in vehicles contain a spectrum of frequencies; in general there will be some stimulus at the resonance frequency (typically near 4 Hz) associated with the compliance of the seat cushion and the mass of the body. Unless adequate damping is present, the cushion will actually make matters worse by amplifying the amplitude of vibration.

As for low frequency isolation of the automobile itself, the suspension system of automobiles contains springs and "shock absorbers" that are viscous dampers or dashpots. The resonance of such a system is at a relatively low frequency and is heavily damped, as can be demonstrated by abruptly pressing down on a fender of a parked car, and observing the free decay of vibration. The natural frequency of the car's suspension is well below the principal frequency content contained in excitation by bumps. Viscoelastic damping in the tires also contributes to the isolation of the auto body from bumps.

Reduction in automobile interior noise has been achieved by laminates consisting of layers of polymer resin and steel for automobile structural parts [126, 127]. Interior noise reduction of 5 dB has been achieved by replacing a conventional dash panel with such laminates. The resin layer consists of a thermosetting polyester resin blended with spherical nickel particles as a filler. The filler permits the laminate to be spot welded by forming electrically conductive paths between the steel layers.

### 10.6.12 Vibration in Scientific Instruments

Many scientific instruments are vulnerable to vibration [128]. For example, scanning probe microscopes, such as the scanning tunneling microscope and atomic force microscope, are capable of resolving individual atoms on a size scale below 0.2 nm. An atomically sharp tip supported by a microscopic cantilever is scanned across the specimen. Tip deflection is measured via a reflected laser beam and is held constant by a feedback circuit driving a piezoelectric actuator. In view of the high resolution desired, vibration must be minimized. Structural beam elements within the instrument may be made of magnesium, which has a low density $\rho$ and comparatively high damping. Stresses are low in such an instrument, so the limited strength of magnesium is not a problem. Moreover, magnesium has a high value of $E/\rho^2$, which is favorable for bending stiffness per unit weight. As another example consider the optical tables used to support laser equipment used in holography, interferometry, and other optical procedures which are sensitive to vibration. The table is supported upon pneumatic legs in which pressurized air supports the table via force upon a piston. The legs provide compliant support so that the table-leg system has a low natural frequency $f_0$ near 1 Hz. Following Equation 10.13, the transmissibility for vibration goes as $(f_0/f)^2$ for frequencies $f$ well above the natural frequency. Therefore, little vibration above 1 Hz gets through to the table. The table is made using a honeycomb sandwich construction which confers a high rigidity per unit mass, hence a high fundamental resonance frequency. That resonance is damped by a tuned damper [129] consisting of a mass supported by flexure springs and immersed in a viscous fluid.

### 10.6.13 Waves

#### *Waves in the Earth*

In geophysics, one can use seismic waves to infer cross-sectional maps of properties of matter in the earth: seismic tomography [130]. In the upper mantle of the earth, $Q \approx 100$ ($\tan\delta \approx 0.01$), and in the lower mantle, $Q \approx 300$ ($\tan\delta \approx 0.003$) for shear waves. Temperature increases with depth, therefore the derivative of wave speed with temperature is important in interpretation. By virtue of the Kramers–Kronig relations developed in §3.3, damping gives rise to frequency dependence of the wave speed; moreover since the damping processes are thermally activated, changes in temperature alter damping as well as velocity. In another geophysical example, creep of rock structures in the earth is measured [131] and related to changes in stress associated with earthquakes. Ultimately, such measurements may be of use in earthquake prediction. In yet another example, viscoelasticity of moon rocks, inferred from seismic studies, was used to better understand the nature of those rocks as discussed in §7.4.

### Ultrasonic Testing

Ultrasonic waves are used for materials characterization as discussed in §6.9, as a method of nondestructive evaluation to detect flaws in machine parts, and in medical diagnosis. Flaws or diseased tissues are detected via the reflections caused by mismatch in acoustic impedance. Waves at higher frequencies offer superior resolution since the corresponding short wavelength allows small features to be visualized. However, wave attenuation due to tan$\delta$ tends to increase with frequency, and this attenuation, combined with the depth of the flaw, limits the highest frequency that can be practically used [132]. For that reason higher frequencies are used in the diagnosis of shallow organs, such as the eye, and lower frequencies are used for larger organs, such as the liver. In some tests, such as evaluation of fatigue-crack damage in machine parts [133, 134], damping or attenuation may be associated with the structural feature under study.

## 10.7 "Smart" Materials and Structures

### 10.7.1 "Smart" Materials

#### Overview

"Smart" materials include materials that allow coupled fields (§8.1.3), in which one field variable is controlled by the user. Coupling involving stress, strain and temperature is particularly useful. For example, an ordinary thermostat for a heater or refrigerator typically contains a coiled strip of metals of different thermal expansion. Temperature changes cause the strip to deform, and activate an electrical switch. The thermostat itself and the building which contains it, is actually a smart structure. Smart *structures* contain sensors and actuators, and appropriate control electronics. Analytical tools from electrical engineering are used to design such structures [135]. Smart *materials* may be used to make the sensors and actuators in a smart structure or by itself to achieve a desired function, including vibration damping.

Materials may be designed in imitation of, or inspired by, biological composites with structural hierarchy [136]; such materials have also been named smart. Even so, no such material has approached the adaptability seen in the plant kingdom. Hierarchical composites allow the designer to achieve extremal values of material properties, such as strength to weight ratio, damping layer performance, or

Poisson's ratio. Hierarchical composites also permit multiple functions to be carried out by the same material. In bone, for example, the skeleton provides the structural framework for the body and also provides a reservoir of calcium for ionic balance, as well as a source of new blood cells. As for vibration damping, a polymer layer provides damping but minimal stiffness or strength, but a high-damping metal can provide damping and stiffness and strength. If a piezoelectric material is used as a damping element, it can provide the additional function of providing an electrical signal with which the user can monitor vibration and other aspects of structural performance.

### 10.7.2 Shape Memory Materials

Shape memory materials can provide a larger actuation force than bi-material strips. Shape memory alloys such as Ni-Ti have been used in temperature controlled valves [137]. Also, tube coupling devices made of Ni-Ti alloy are expanded in the low-temperature state, then the tube is inserted, followed by heating, which causes the coupling metal to shrink, securing the joint. Deformation of shape memory alloy wires can be controlled by passing a (large) electric current to heat them. This approach was tried in a robotic hand, but response was slow due to the time required for cooling. The heated wire approach was successfully used in a controllable endoscope used to visualize the interior of the human colon for medical purposes. Overall, thermally driven actuators using shape memory alloys are capable of large strains, but they require considerable power to drive them, and their response is slow.

Shape memory ceramics [138] include piezoelectric, ferroelectric ceramics such as barium titatate and lead titanate zirconate. Many applications such as ultrasonic transducers, microphones, speakers, and micromotors depend on the piezoelectric effect rather than the shape memory effect. These ceramics are also useful based on shape memory effect in clamps and latch relays which maintain their shape due to hysteresis even after the driving electric signal has been turned off. Ceramics respond more rapidly than alloys and require less power to operate.

Shape-memory polymers [139] are based on the glass-rubber transition. For example an amorphous polymer, such as PMMA, can be deformed near the glass transition temperature $T_g$ (no more than 10°C over); if heated above $T_g$, it returns to its original shape. A shape memory effect in polyethylene, a crystalline polymer is widely used for heat-shrinkable tubes. The polymer is heated above about 80°C at which the micro-crystalline parts melt. The polymer is stretched but flow is resisted by cross-links. The stretched shape is fixed by cooling. When the polymer is again heated, it returns to its original shape. Such tubes are used for sealing electrical connections.

Shape memory materials are considered as high damping materials due to the motion of twin boundaries. As with alloys such as CuMn, behavior is nonlinear, so that damping increases markedly with strain amplitude. Therefore, they are most effective when used to reduce vibration of high amplitude.

### 10.7.3 Self-Healing Materials

Cracks form in materials subject to overload or to repeated deformation in fatigue. These cracks degrade the physical properties and can lead to fracture. Self-repair of cracks in polymers has been achieved by embedding micro-capsules of healing agent in the polymer [140]. Intrusion of a crack breaks one or more capsules, releasing the viscous monomer healing agent that polymerizes on contact with embedded catalyst. This bonds the crack faces and restores the properties of the material.

### 10.7.4 Piezoelectric Solid Damping

Vibration in skis has been damped by piezoelectric inclusions with attached electric circuits as discussed in §10.6.10. Mechanical damping is essential in ultrasonic transducers intended for non-destructive testing, since a well defined acoustic pulse is called for as discussed in §10.6.7. In addition to the damping built into the transducer, additional damping can be tuned by adjusting an external electrical resistor in the transmitter or receiver electronics.

### 10.7.5 Active Vibration Control: "Smart" Structures

Active methods have been used in some buildings to reduce sway due to wind or earthquakes. For example, as discussed in §10.6.4, a tuned damper [86] in a building is activated by pumps triggered by accelerometers which sense a threshold acceleration. Hybrid systems for buildings [141] incorporate both tuned mass dampers and weights, driven by computer-controlled actuators, in response to measured accelerations. Active methods have also been used in the design of truss structures to be used in space stations and other spacecraft. Piezoelectric and magnetostrictive materials have been used as actuators.

In bridges, vibration of cables or other structural components can be a problem, particularly in longer bridges which are more vulnerable to instability on windy days. Such vibration can be damped passively [142] or actively [143, 144]. The active control system consists of a sensor for deformation, velocity, or load, a feedback amplifier, and an actuator to generate an opposing force. Analysis of active damping is simplified if one uses collocated actuator and sensor pairs [145, 146]. Since the actuators in an active damping system are driven via an external energy source, it is possible that such a system can become unstable and go into oscillation. Therefore the development of control modalities, which guarantee absorption of vibration energy, is attractive and aided by use of collocated actuator and sensor pairs.

### 10.8 Rolling Friction

Rolling friction arises principally from viscoelasticity in the materials which are in rolling contact [147]. This is physically reasonable since rolling results in periodic variation in stress; in a viscoelastic material, dissipation of energy occurs, and that dissipation is manifested as rolling friction. The physical origin of rolling friction was not immediately obvious to early scientists [148]. Coulomb [149] believed that rolling friction was due to bumps or asperities in one of the surfaces. Experiments later showed that, contrary to such a notion, rolling friction depends on load, speed, and other variables. Microscopic slip was then considered as a possible cause, but experiments by Bowden and Tabor [150] demonstrated that such slip has a minimal effect on rolling friction. Specifically, they drilled a hole in a roller that was painted to leave an imprint on a rubber substrate. The imprint was observed to be elliptical, indicating deformation of the rubber, and sharp, indicating minimal slip.

### 10.8.1 Rolling Analysis

The problem of determining the coefficient of rolling friction from viscoelastic properties is not amenable to the correspondence principle, and so is difficult to solve in exact form. Even so, some scaling arguments are sufficient to elucidate several aspects of rolling resistance. The coefficient of rolling friction $\lambda_{\mathrm{roll}}$ is defined as the ratio of the component of tangential force, which decelerates the object, to normal force, which supports its weight. Since $\lambda_{\mathrm{roll}}$ is dimensionless, and since it must be zero in the absence of dissipative processes, one might naively suppose it to be proportional to $\tan\delta$. However, $\lambda_{\mathrm{roll}}$ cannot be arbitrarily large, for the following reason. The net force vector must arise at some point within the contact region of radius $r_c$ assuming nonsticky contact. Therefore, for a ball of radius $R$,

$$\lambda_{\mathrm{roll}} \leq \frac{r_c}{R}. \tag{10.19}$$

If the size of the contact region is known from observation, an upper bound on the rolling resistance may be calculated [151]. The size of the contact region may also be determined via elasticity theory. For a rigid ball in contact with a flat deformable surface of Young's modulus $E$ and Poisson's ratio $\nu$, by force $F$, the contact region radius $a$ is given by elasticity theory [152] as

$$r_c = \{\frac{3}{4}F\frac{1-\nu^2}{E}R\}^{1/3}. \tag{10.20}$$

This is a special case of contact of two deformable spheres. If the contact region in the viscoelastic case has a similar size, the coefficient of rolling friction $\lambda_{\mathrm{roll}}$ is bounded by

$$\lambda_{\mathrm{roll}} \leq \{\frac{3}{4}(1-\nu^2)\}^{1/3}F^{1/3}\frac{1}{E^{1/3}}\frac{1}{R^{2/3}}. \tag{10.21}$$

Here, the stiffness $E$ is interpreted as a dynamic modulus at the frequency of rolling. Flom and Bueche [153] presented a solution for the rolling resistance of a sphere on a viscoelastic substrate. It was developed assuming a material describable by the Maxwell model, and is of a cumbersome form. The rolling resistance arises due to an asymmetric pressure distribution, which develops in viscoelastic substrate materials. Some particular cases are of interest. Here, $\varphi_c$ is defined as the ratio of contact region radius to ball radius. $G$ is the shear modulus,

$$\lambda_{\mathrm{roll}} \approx \tan\delta \tag{10.22}$$

for $\tan\delta \ll \varphi_c = \frac{r_c}{R}$,

$$\lambda_{\mathrm{roll}} = 0.434\varphi_c = 0.243\,F^{1/3}\frac{1}{G^{1/3}}\frac{1}{R^{2/3}}, \tag{10.23}$$

for $\tan\delta = \varphi_c = \frac{r_c}{R}$,

$$\lambda_{\mathrm{roll}} = 0.510\varphi_c = 0.248\,F^{1/3}\frac{1}{G^{1/3}}\frac{1}{R^{2/3}}, \tag{10.24}$$

for $\tan\delta = 2\varphi_c = \frac{2r_c}{R}$.

For the particular case of large $\tan\delta$, $\lambda_{\mathrm{roll}} \approx 0.590\varphi_c$ so the inequality in Equation 10.19 is satisfied in all cases. Moreover, for large $\tan\delta$, the dependence of rolling resistance on ball radius, vertical load, and substrate stiffness is the same as anticipated above in Equation 10.21. Analysis of rolling of rigid cylinders on a viscoelastic substrate [154] disclosed results similar to the above. Rolling resistance attains a peak value as a function of rolling speed. Experiments on balls of various kinds of rubber [155] confirmed the general features of the viscoelastic theory of rolling resistance.

### 10.8.2 Rolling of Tires

Rolling friction is important in tires, because such friction influences fuel consumption. Representative values [148] for the coefficient of rolling friction are 0.004 for bicycle tires, 0.008 for truck tires, 0.01 for radial automobile tires, and 0.015 for bias ply automobile tires. Sliding friction in polymers is also linked to viscoelastic properties [31] in that the making and breaking of contacts is governed by the same kinds of molecular processes as are responsible for viscoelasticity. In tires, it is desirable to minimize rolling friction to reduce fuel consumption. For that purpose a low-loss rubber is best [156]. It is also desirable to maximize sliding friction to prevent skidding (and to prevent suffering due to possible injury to the driver or to others) during braking, turning, and acceleration. For that purpose a high-loss rubber is best. Tires made of high-loss rubber are also relatively quiet and smooth riding because vibrations are heavily damped. Because these requirements are contradictory, different rubber compositions are used for racing and for normal driving. A degree of compromise can be achieved by using a low-loss rubber in highly strained regions of the tire, and a higher loss rubber for the portion in contact with the road. Rolling friction also depends upon the air pressure within the tire because pressure alters the cyclic deformation during rolling. Specifically high pressure reduces the size of the contact patch during rolling, hence the rolling friction. High air pressure in the tire also increases the structural stiffness of the tire and contributes to a bumpier ride. Conversely, an increase in the load upon the tire increases the contact patch size, hence increases the rolling friction.

Tires undergo two principal types of deformation [157]. The first consists of the flattening of the bottom of the tire tread against the road. During highway driving at 100 km/hr (62 mph) the tire rotates about 20 times per second, so the fundamental frequency of oscillatory strain in the tire is about 20 Hz. The second deformation consists of small-scale indentations of the tire by irregularities in the road surface. Tires grip the road with the aid of these microscopic conformations. Frequencies associated with this deformation are high, up to 1 MHz in a skidding tire. Therefore, rolling resistance could be reduced without compromising sliding friction by reducing $\tan\delta$ at low frequency but maintaining $\tan\delta$ at high frequency. Use of silica ($SiO_2$) rather than carbon as particulate inclusions in the rubber, has such an effect since, unlike carbon, silica forms chemical bonds with the polymer chains in the rubber, reducing the damping at the lower frequencies.

## 10.9  Uses of Low-Loss Materials

### 10.9.1  Timepieces

Watches and clocks use vibrating objects to generate periodic signals, which may be divided into intervals for timing purposes. In earlier mechanical designs, torsional oscillation is set up in a pivoted wheel held by a torsional spring. The ticking sound from such timepieces comes from the escapement mechanism used to convert oscillatory motion into the uniform circular motion used to display the time. In quartz timepieces, vibration at ultrasonic frequency is set up in a crystal of quartz ($SiO_2$). Quartz is a piezoelectric material: it deforms in response to an electric signal, and when stressed it generates an electric polarization. The piezoelectric property makes quartz useful as a tuning element to control the frequency of an electronic circuit. There are other piezoelectric materials, notably the ceramics, which are less costly than quartz and offer a stronger piezoelectric effect. Quartz is used for timepieces and other frequency standards because its loss tangent is extremely small in comparison with that of other available piezoelectrics (see §7.1.2, §7.4.5). The peak in the resonance curve of compliance versus frequency is therefore sharp, and the resonance frequency is well-defined. Moreover, little power is required to maintain vibration at the resonant frequency, which, for watches, is typically 32.768 kHz [158]. As a result of the small loss tangent of quartz, timepieces can be made which are precise and accurate and require little power. However, the quality factor $Q$ (Recall that $\tan\delta \approx Q^{-1}$) for practical resonators is 50,000 to 120,000 minimum [159], lower than expected ($10^6 - 10^7$) from the intrinsic damping of the quartz. The difference is attributed to the resonator mounting that allows vibratory energy to leak away in the supporting structure. That mount must be more robust in a watch, which is subject to jolts, than in an experimental apparatus. In addition, the stability of quartz in relation to temperature, time, and environmental changes is essential in this application [158].

## 10.9.2 Frequency Stabilization and Control

Vibrating piezoelectric crystals are also used to control the frequency of radio transmitters and receivers. As in timepieces, it is advantageous to use a material with as small a loss tangent as possible [159]. Other desirable characteristics of materials include minimal aging (change of properties with time) and minimal change of properties with temperature.

Tuning forks are used as frequency standards in music, in testing of the ear, and in physics [161]. To a first approximation the fork can be considered as two cantilever bars joined with an offset at the base. Tuning forks are commonly made of aluminum or aluminum alloy. The low loss tangent of aluminum allows the fork to continue vibrating for a long period after it is struck. Some energy is lost through radiation of sound, in addition to the loss due to viscoelasticity in the material.

## 10.9.3 Gravitational Measurements

The detection of gravitational waves is a problem of great interest in modern physics. Gravitational waves are undulatory deformations in the structure of space-time. They are caused by cosmic events such as the collapse of massive stars. One method that has been used in attempts to detect gravitational waves is to compare the vibration of resonant bar detectors placed in different geographical locations. If both detectors register a signal, it must be from a remote source, possibly cosmological, not a local one, such as passing trucks or earthquakes. The sensitivity of such detectors depends inversely on the loss tangent of the resonant bar [159], therefore low loss materials have been used to make them. Recent scientific apparatus for the detection of gravitational waves makes use of materials with high $Q$ to support mirrors in interferometers. Wires and fibers of several metals of high melting point as well as sapphire and fused silica were compared [160]. Each wire (0.1 to 0.5 mm diameter) was hung from a clamp; normal modes were excited by a capacitor plate. Results from 1 Hz to 1 kHz at ambient temperature, 300 K, showed damping dominated by thermoelastic peaks from 100 Hz to 1 kHz for many of the materials. The tan$\delta$ of most of the metals was from $10^{-4}$ to $10^{-3}$; fused silica had the lowest damping at $10^{-6}$ at the higher frequencies.

Measurement of the Newtonian gravitational constant **G** are performed with a Cavendish balance in which a dumbbell shaped pair of masses is suspended from a slender wire or fiber. Fixed masses are then placed at different locations near the dumbbell. The period of oscillation is measured for two positions of the attracting masses, and **G** is inferred. A key assumption in the interpretation of results is that the spring constant of the torsion fiber is independent of frequency. Individual measurements have estimated errors of about 0.1 percent but different measurements yield values differing by several parts per thousand. Systematic errors in this measurement have recently been attributed to viscoelasticity at low frequency in the suspension wire of the Cavendish torsion balance [162–164]. This viscoelasticity, even in supposedly elastic materials such as fused silica, tungsten, or

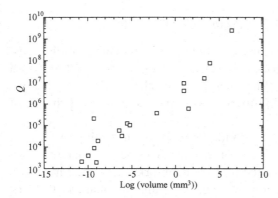

Figure 10.12. Maximum quality factor $Q$ in resonators as a function of volume based on results from a variety of authors (adapted from Ekinci and Roukes [165]).

copper-beryllium, gives rise to a spring stiffness which depends on frequency contrary to the usual assumptions. The frequency dependence is small but can nevertheless be large enough to interfere with a high precision measurement.

### 10.9.4 Nanoscale Resonators

The microscale is usually considered to be the size scale accessible via the light microscope, 1 mm to 1 $\mu$m; the nanoscale, also called the ultrastructural scale, corresponds to dimensions below 100 nm; some authors consider the scale to be below 1 $\mu$m. Nanoscale properties are of interest since surface effects that are predominant at small size allow new degrees of freedom for design of materials and systems.

Vibrating systems at the nanoscale [165] by virtue of their small size, can exhibit natural frequencies of 10 MHz to 1 GHz, of potential interest as frequency standards for radio frequency oscillators and signal processing devices. It is desirable to have a high $Q$ (a low damping) in such applications; mechanical resonators are of interest in this context since they can exhibit higher $Q$ than electrical resonators. The difficulty is that the maximum attainable $Q$ tends to scale with the linear dimension, so that $Q$ exceeding $10^9$ can be achieved in macroscopic resonators, but it is difficult to attain $Q$ greater than $10^3$ or $10^4$ at the nanoscale, even with high purity single crystals, as shown in Figure 10.12. Effects at the surfaces contribute to the energy dissipation, and the ratio of surface to volume increases as the size decreases. Thermoelastic damping is also pertinent [166], because as the resonator becomes smaller, not only do its natural frequencies increase, but the thermal relaxation characteristic frequency increases as well.

## 10.10 Impulses, Rebound, and Impact Absorption

### 10.10.1 Rationale

The response of an object to impulsive force is different in viscoelastic materials in comparison with elastic ones. The effect of viscoelasticity can be used to reduce the force of impact, to reduce ringing or vibration after an impact, and in the design of bumpers, buffers, pads, shoe insoles and helmets. Impact response also affects the rebound of a ball and is therefore pertinent to various sports.

### 10.10.2 Analysis

Viscoelastic materials can be used to reduce the force of impact [167]. In the analysis of the impact, a compliant elastic layer is first placed between the impactor and the surface to be protected, and its effect is analyzed. Then, the layer is assumed to be viscoelastic, and its optimal viscoelastic properties are determined, with the aim of minimizing the impact force.

Consider first an elastic impact buffer considered as a one-dimensional mass-spring system. The massless linearly elastic buffer in one dimension is mechanically equivalent to a spring. The displacement $u$ in free vibration of the mass $m$ is

$$u(t) = A \sin(\omega_0 t + \theta_b), \tag{10.25}$$

in which the spring is a massless axial spring, the natural angular frequency is $\omega_0 = \sqrt{Ebc/mh}$, $E$ is the axial modulus of the elastic block, $b$ and $c$ are its cross-section dimensions, $h$ is its axial length, and $A$ and $\theta_b$ are amplitude and phase constants respectively depending on the initial displacement and velocity of $m$. In the case of impact of a moving mass $m$ of velocity $V$ on a spring, Equation 10.25 is used to describe the motion of $m$. The contact of $m$ and the spring occurs at $t = 0$. The displacement, $u$ of $m$ is, therefore, $u(t) = \frac{V}{\omega_0} \sin(\omega_0 t)$.

The impact force $F$ is obtained as $F(t) = m\omega_0 V \sin(\omega_0 t)$.

The impact force is proportional to the spring deflection, or the displacement of $m$. The maximum impact force is $F_{max} = V\sqrt{mEbc/h}$ with respect to the maximum spring deflection $u_{max} = V\sqrt{mh/Ebc}$ at $t = \frac{\pi}{2}\sqrt{mh/Ebc}$. $F_{max}$ is related to $u_{max}$ by $F_{max} = Ebcu_{max}/h$, or $F_{max} = Ebc\epsilon$ in which $\epsilon$ is the maximum compressive engineering strain of the spring. Decreasing the value of $E$ results in a decrease of the maximum impact force. However, if the buffer is too soft, it will bottom out, corresponding to $\epsilon = 1$. A value of

$$E = \frac{mV^2}{bch\epsilon^2} \tag{10.26}$$

is the optimal stiffness to minimize $F_{max}$ if the buffer layer is elastic [167].

Consider [167] a one-dimensional linearly viscoelastic buffer upon a substrate impacted by a moving mass. To investigate the impact behavior in terms of the viscoelastic properties of the buffer directly, we approximate the impact as one half cycle of free decay oscillation. The governing equation for the one dimensional forced oscillation of a massive viscoelastic block with an attached mass $m$ is similar to that given by Christensen [168] for the torsional case, and is as follows:

$$F_{ext}(t) - u(t)bc\rho\omega^2 h\frac{ctn(\Omega^*)}{\Omega^*} = m\frac{\partial^2 u(t)}{\partial t^2}, \tag{10.27}$$

in which $F_{ext}(t)$ is the external force, $F_{ext}(t) = F_0 \sin(\omega t)$, $u(t)$ is the displacement of $m$, $b$ and $c$ are the cross-section dimensions of the viscoelastic buffer, $h$ its axial length, $\rho$ its mass density, $\omega$ the angular frequency of harmonic oscillation, $E^*(i\omega)$ or $E'(i\omega)(1 + i\tan\delta)$ the uniaxial complex modulus, $E'(i\omega)$ is the storage modulus, $\tan\delta$ the loss tangent, and $\Omega^* = \sqrt{\frac{\rho\omega^2 h^2}{E^*(i\omega)}}$.

The lateral restrictions at the ends of the buffer are neglected. This is considered appropriate since the analysis is on the basis of a one-dimensional problem. Now suppose the buffer mass is much less than that of the impactor as a result of a small density $\rho$ or axial length $h$. Then, we can approximate $\Omega^* \to 0$, and, therefore $\text{ctn}(\Omega^*) \to 1/\Omega^*$. The buffer is therefore a massless viscoelastic "spring" and the induced deflection and stress are uniform in the axial direction. Then,

$$F_{ext}(t) - u(t)E^*(i\omega)\frac{bc}{h} = m\frac{\partial^2 u(t)}{\partial t^2}. \tag{10.28}$$

The external force $F(t)$ vanishes in the situation when the mass has struck the buffer and is vibrating freely on it. The impact, from contact to peak force, is a one-quarter cycle. The solution is then,

$$u(t) = Ae^{\omega_0(-\alpha_b + i\beta)t + \theta_b i}, \tag{10.29}$$

in which $\omega_0 = \sqrt{\frac{E'(i\omega)bc}{mh}}$; $\alpha_b = (1 + \tan^2\delta)^{1/4}\sin\frac{\delta}{2}$; $\beta = (1 + \tan^2\delta)^{1/4}\cos\frac{\delta}{2}$; and $A$ and $\theta_b$ are constants depending on the initial displacement and velocity of $m$. This solution for free vibration of the mass-viscoelastic buffer system is identical to that for one dimensional impact of mass $m$ on a viscoelastic buffer. Based on the initial condition the displacement of $m$ can be written as

$$u(t) = \frac{V}{\beta\omega_0}\sin(\beta\omega_0 t)e^{-\alpha_b\omega_0 t}, \tag{10.30}$$

in which $V$ is the initial velocity of moving mass $m$, and the contact of $m$ and the buffer occurs at $t = 0$ again. Accordingly, the force induced on $m$ is $F(t) = m\frac{d^2 u(t)}{dt^2}$,

$$F(t) = \frac{mV}{\beta}\omega_0\{\beta^2\sin(\beta\omega_0 t) + 2\alpha_b\beta\cos(\beta\omega_0 t) - \alpha_b^2\sin(\beta\omega_0 t)\}e^{-\alpha_b\omega_0 t}. \tag{10.31}$$

By equating $du(t)/dt$ and $dF(t)/dt$ to zeros respectively, the maximum displacement $u_{max}$ and the maximum impact force $F_{max}$ are derived in terms of the design parameters as

$$u_{max} = V\sqrt{\frac{mh}{E'(i\omega)bc}}\frac{1}{(1 + \tan^2\delta)^{1/4}}e^{-\tan(\delta/2)(\pi/2 - \delta/2)}, \tag{10.32}$$

at $t_0 = \frac{1}{\omega_0}\frac{\pi - \delta}{2(1 + \tan^2\delta)^{1/4}\cos(\frac{\delta}{2})}$,
and

$$F_{max} = V\sqrt{\frac{mE'(i\omega)bc}{h}}(1 + \tan^2\delta)^{1/4}e^{-\tan(\delta/2)(\pi/2 - 3\delta/2)}, \tag{10.33}$$

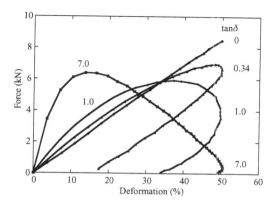

Figure 10.13. Effect of $\tan\delta$ on impact force, including the transient term, with the storage modulus held constant (adapted from Chen and Lakes [167]).

at time $t = \frac{1}{\omega_0} \frac{\pi - 3\delta}{2(1 + \tan^2\delta)^{1/4}\cos(\frac{\delta}{2})}$, for $\tan\delta \le 1.73$, or

$$F_{max} = 2V\sqrt{\frac{mE'(i\omega)bc}{h}}(1 + \tan^2\delta)^{1/4}\sin(\frac{\delta}{2}), \qquad (10.34)$$

at $t = 0$, for $\tan\delta \ge 1.73$. From Equation 10.32 the optimal stiffness is

$$E'(i\omega) = \frac{mV^2h}{bc}(\frac{1}{u_{max}})^2 \frac{1}{(1 + \tan^2\delta)^{1/2}}e^{-\tan(\delta/2)(\pi - \delta)}, \qquad (10.35)$$

if there is a limit $u_{max}$ on the maximum deflection. Substituting, $F_{max}$ becomes

$$F_{max} = \frac{mV^2}{u_{max}}e^{-\tan(\delta/2)(\pi - 2\delta)}, \qquad (10.36)$$

for $\tan\delta \le 1.73$, or

$$F_{max} = 2\frac{mV^2}{u_{max}}\sin(\frac{\delta}{2})e^{-\tan(\delta/2)(\pi/2 - \delta/2)}, \qquad (10.37)$$

for $\tan\delta \ge 1.73$.

Curves of force versus deformation for different values of $\tan\delta$ are shown in Figure 10.13.

Differentiating with respect to $\delta$, one finds the optimal loss tangent to minimize the impact force to be
$\tan\delta = 0.4$
if the stiffness is held constant, and
$\tan\delta = 1.1$
if the stiffness is treated as a design variable. For $\tan\delta = 1.1$ the optimal value of $E'(i\omega)$ is 24 percent of the value for the optimized buffer, based on Equation 10.35. The resulting impact force is 52 percent of the value found in the optimized elastic buffer, based on Equation 10.26. So, the effect of buffer viscoelasticity is to substantially reduce the peak impact force in comparison with an elastic buffer of the same thickness.

More sophisticated analysis [167] including the transient term does not change the numerical results much, and it does not change the conclusion. Specifically, the angular frequency at which one refers to tan$\delta$ is $\omega_0$, the natural angular frequency of the mass-buffer system. The impact actually contains a distribution of frequencies, so a more complete analysis must include assumptions or input data regarding the frequency dependence of the damping. Such a requirement gives rise to a considerable complication, requiring a numerical solution. If the indenter is spherical rather than one-dimensional, the optimum loss tangent is greater than in the above case, because the force-deformation characteristic for this case is a nonlinear function, concave up. Again, the peak impact force can be reduced by almost a factor of two by using a viscoelastic material rather than an elastic material for the buffer.

If foam is used as a buffer, its nonlinearity (§9.9) deserves consideration. The nonlinearity is usually beneficial for buffers since the plateau region of foam deformation accommodates more energy input for given deformation than a linearly elastic spring.

The conclusion is that force associated with impact from a moving object can be minimized by an elastic buffer of appropriate stiffness. The force can be further reduced by an additional factor of about two if the buffer is made viscoelastic. The optimal loss tangent for this application is large; it can be achieved in viscoelastic elastomers.

### 10.10.3 Bumpers and Pads

As for specific practical applications of viscoelastic materials in impact absorption, viscoelastic elastomers are used in automobile bumpers [169] and as bumpers and crash stops to isolate computer disk drives from mechanical shock during power failures and during shipment, as well as to protect equipment which could be damaged by being dropped. Car bumpers may contain lumped element damping units [170]. Moving parts in computer disk drives are also damped [171] to improve tracking and response speed. Viscoelastic materials are also used as bushings in printing presses to minimize shock forces. Viscoelastic foams are used for seats for airplanes, spacecraft, bicycles, and trucks to reduce the user's discomfort. Viscoelastic foams are also used in padding for athletic equipment, for example, helmets, ski boots, wrestling mats, and weight-training equipment.

### *Helmets*

Helmets may contain foam or rubbery pads in contact with the head, and an outer shell, which can absorb energy by crushing under a heavy impact. An essentially viscous alternative has recently been developed for helmets. The absorber should be as thick as possible to minimize the peak force, however the material used takes up some space, which limits the available distance for deceleration. If one uses a collapsible thin-wall polymer chamber [172] with a hole to allow viscous escape of air, the space occupied by the absorber when it is fully compressed is minimal. So the effective thickness for the absorber is used most effectively. The behavior of such a system is dominated by a viscous dashpot effect, so that deceleration force is larger for a more rapid impact. This is beneficial since a rapid impact could otherwise cause the absorber to bottom out. Helmets using this approach contain as many as 18 shock absorbers.

### 10.10.4 Shoe Insoles, Athletic Tracks, and Glove Liners

Viscoelastic elastomers are used in shoe insoles to reduce impacts transmitted to a person's skeleton. Some measurements have been reported of the effect of shoe mechanical properties on forces on or vibration in the human body. Accelerations of the tibia and skull during walking show peaks [173] of about 5 $g$ and 0.5 $g$, respectively, when hard heels are worn ($g$ is the acceleration due to gravity). Peak accelerations are reduced by about a factor of two when compliant heels are worn. Viscoelastic inserts in shoes reduced leg accelerations by up to a factor of 1.8 [174]. Some investigators report a reduction in back and joint pain and other problems in athletes who use shoes with viscoelastic isolators; moreover, people suffering from pain due to arthritis [175] or occupational exposure to prolonged standing on hard surfaces [176] observed significant reduction in pain after using the viscoelastic inserts. People with joint diseases appear to have insufficient shock absorbing capacity in the tissues [177]. A variety of the available viscoelastic elastomer materials (with trade names Sorbothane, Implus, Noene) have been compared [178] for damping and stiffness.

Compliant polymers have also been used for the surfaces of running tracks. There is considered to be an optimal compliance, which results in the fastest speed of runners in racing events [179]. The improvement in speed for distance events upon the "tuned" track was about 3 percent compared with a hard surface. Compliant track surfaces also have been reported to be more comfortable for runners and to result in fewer injuries to competitive athletes. A running surface which is too compliant, as one might expect, slows down the runners. Analyses of tracks thus far have incorporated surface compliance but not damping.

A related application of viscoelastic elastomers is in liners of work gloves used by operators of tools which generate severe vibration, such as jackhammers and chipping tools. Vibration can cause circulatory disturbances in the hands and arms. Long exposure to severe vibration can result in numbness and blanching of the fingers [180–182] a condition known as Raynaud's syndrome, or "vibration white finger." The vibration suppresses blood circulation to the extent that fingers turn white and numb. Use of high-damping metals in the tool itself has also been tried to reduce suffering due to this problem.

### 10.10.5 Toughness of Materials

Rubber, though it is compliant, is tougher than most materials. The high energy required to propagate a crack in rubber may be a million times greater than the surface energy of rubber. Therefore, the classic Griffith theory for toughness is not directly applicable to rubbery materials. The strong enhancement of the toughness

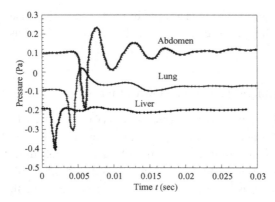

Figure 10.14. Waveforms associated with response to diagnostic percussion impulses upon different parts of the human body (adapted from Murray and Neilson [187]).

of rubber is attributed to viscoelastic energy dissipation in front of the crack tip. This dissipation causes heating near the crack tip [183], enhancing the viscoelastic dissipation process. Moreover, [184] fracture in rubber takes place intermittently. The crack grows at high speed and then stops. A possible mechanism of crack stopping is by splitting at the crack tip. Also, strength of adhesion depends upon the rheology of an adhesive as well as upon its interaction with a substrate. Peel strength [185] of polymers exhibits maxima corresponding to the change from liquid-like consistency to rubbery consistency, and to the change from rubbery to glassy behavior.

Viscoelastic behavior also enhances toughness in other materials, particularly those which do not exhibit much plasticity [186]. Energy is dissipated by viscoelastic damping as a crack extends, slowing or stopping the crack growth. Peaks in both damping and in toughness have been observed in ceramics at elevated temperature. Viscoelastic toughening is beneficial provided it is due to a restricted motion of microstructural elements; if the viscoelasticity arises from flow processes, then the material will creep excessively.

### 10.10.6 Tissue Viscoelasticity in Medical Diagnosis

Physicians are able to obtain important diagnostic information by pressing the patient's body with one finger and tapping (percussing) with another finger. The sounds produced and the tactile sensations felt reveal to the examiner the condition of the underlying tissue [187, 188]. Percussion as a diagnostic tool dates from a publication in 1761 by Auenbrugger [189].

Oscilloscope tracings of percussion sounds [187] disclose damped oscillations similar to the curves for free-decay of vibration described in §3.6. Different tissues, and comparison of diseased and healthy states of a single tissue, reveal waveforms of considerable difference in the rate of decay of vibration. The percussion sound [187] described by a physician as "dull" corresponded to a single pressure impulse of less than 3 ms; *resonant* sound consisted of two to three oscillations with a duration of 15 ms; while *tympany* consisted of a damped sinusoid with center frequency 200 to 600 Hz, and lasting more than 40 ms, as shown in Figure 10.14. Lung lesions up to 50 mm deep and 20 to 30 mm in diameter can be detected by diagnostic percussion [190]. Percussion has also been used to evaluate the degree of bone fracture healing [191].

Diagnostic percussion is an informal test of structural resonance in the human body. As we have seen in §3.6, the free decay of vibration in a structure depends on the loss tangent of the material. The waveforms described in the paragraph above indicate that structures in the body are heavily damped.

## 10.11 Rebound of a Ball

### 10.11.1 Analysis

A ball dropped on a viscoelastic substrate rebounds to a height lower than the height from which it was dropped. In this section, the height of rebound is related to the loss tangent of the material [167, 192].

Tan$\delta$ of viscoelastic materials can be determined from the degree of rebound in a one-dimensional rebound test via the solutions [167] of $U(t)$ and $F(t)$ given above. A mass $m$ dropped from height $H_0$ on a massive block of material rebounds back with velocity $V_1$ following an impact time $t_{imp}$ and back to height $H_1$. The ratio of velocities is defined as the *coefficient of restitution*, $e$. Specifically,

$$e = \frac{velocity_{\text{separation}}}{velocity_{\text{approach}}}. \tag{10.38}$$

Since the potential energy is $mgH$ (with $g$ as the acceleration due to gravity) and the kinetic energy is $\frac{1}{2}mv^2$, the height ratio is

$$\frac{H_1}{H_0} = e^2 \equiv \mathbf{f}. \tag{10.39}$$

The coefficient of restitution depends on the properties of both the ball and the material upon which it bounces. If one material is much stiffer than the other, strain energy in the stiff material is negligible, and the damping properties of the more compliant material dominate the behavior. The impact time $t_{imp}$ is obtained by equating $F(t)$ to zero [167] in the above analysis, Equation 10.31, so that

$$t_{imp} = \frac{1}{\omega_0} \frac{\pi - \delta}{(1 + \tan^2\delta)^{1/4}\cos(\frac{\delta}{2})}, \tag{10.40}$$

with $\omega_0 = \sqrt{\frac{Ebc}{mh}}$.

$V_1$ is obtained by substituting $t_{imp}$ in the time derivative of $u(t)$. Tan$\delta$ is therefore related to the rebound height ratio $H_1/H_0$ by

$$\frac{H_1}{H_0} = (\cos\delta + \tan\frac{\delta}{2}\sin\delta)e^{-2\tan(\delta/2)(\pi-\delta)}. \tag{10.41}$$

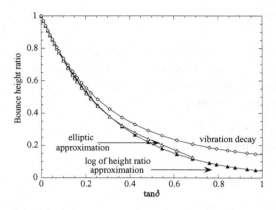

Figure 10.15. Diagram of several approximations for the relationship between height ratio and loss tangent for ball rebound. The elliptic approximation, which neglects the initial transient, is not shown over the range in which the approximation radically breaks down.

Again, more sophisticated analysis is possible [167] but the results are similar (Figure 10.15). If tan$\delta$ is small, the relation above can be approximated as follows:

$$\tan\delta \cong \frac{1}{\pi}\ln(\frac{H_0}{H_1}).$$
(10.42)

This is the form obtained from a study of the approximate solution for free-decay of vibration as done in Example 3.3. To conclude, a rigid ball dropped on a viscoelastic substrate bounces high if the loss tangent is small, and does not bounce much if the loss tangent is large. Similar results can be expected for a viscoelastic ball on a rigid substrate.

### 10.11.2 Applications in Sports

In many sports, the rebound of a ball following an impact is highly relevant to the character of the sport. For example in baseball [193, 194], the batter likes to hit the ball as far as possible. The baseball itself consists of a winding of wool and cotton yarn around a core of cork, and covered by cowhide [193]. The ball is rather inelastic: it exhibits a height ratio of 0.3 in a drop test. This corresponds to a coefficient of restitution of only about 0.55 [194]. The coefficient of restitution of a major league baseball is required to be $e = 0.546 \pm 0.032$ [195] based on a standard impact with a block of ash wood. Even such a small allowable variation can result in as much as a 15 foot (4.6 m) difference in the range of a well-hit baseball. It is not really meaningful to speak of a loss tangent for a baseball, since it is actually a heterogeneous structure. Since the ball's constituents are natural polymers, the viscoelasticity depends on temperature and humidity. Adair [193] adduces several apocryphal stories of efforts by home teams to gain advantage by cooling some or all of the baseballs prior to the game. The rules have been changed so that all balls must be provided to the umpires two hours before game time. The rebound velocity of the ball depends on the properties of both the bat and ball. Since the bat is stiffer than the ball, its viscoelastic properties are less important. Even so, since aluminum bats dissipate less mechanical energy than wood ones, it is possible to hit the ball further with an aluminum bat. The mechanical damping of the bat also manifests itself in the sound of the impact: a "ping" sound in the case of an aluminum bat which has low damping.

Baseball games are played in regions which differ widely in both temperature and humidity. Some people have considered a low humidity environment to be advantageous. Measurements of the coefficient of restitution of baseballs [196] showed a change from 0.55 at zero humidity to about 0.5 at 100 percent humidity. This full humidity range corresponds to a change in the distance of a batted ball by as much as 30 feet.

Some writers have suggested that the baseballs themselves have become more lively in recent years, a possible cause for the increase in home runs [197]. Recent balls have a higher proportion of synthetic fibers than older ones. Their cores seem to have more rebound, but such a test is not definitive since natural and synthetic polymers undergo physical aging.

In a hollow ball, the rebound depends on the pressure of the gas within the ball. The reason is that the compressed gas acts as an elastic spring element. Rebound resilience of a hollow ball depends on the internal pressure by virtue of the contact region size which depends on pressure. Tennis balls, for example, are pressurized to increase rebound resilience. The rebound of a tennis ball decreases over a period of weeks due to diffusion of gas through the rubber. The balls are stored in pressurized cans to prevent any loss of internal pressure prior to use. Rebound of tennis balls used in tournament play is restricted: the height ratio must be from 0.53 to 0.58 [198, 199]. Footballs and basketballs are large enough to accommodate a valve so that they can be inflated with air to control rebound characteristics.

Golf balls [199] were typically made of rubber thread wrapped under tension around a core; more recent golf balls contain concentric layers of rubbery material. The coefficient of restitution $e$ for a drive in golf is about 0.7. Since the golf club has been observed to exhibit no appreciable deformation during impact, the value of $e$ depends almost entirely on the ball. Over a temperature range of 0°C to 27°C, the value of $e$ for the ball changes from 0.64 to 0.75. For this reason a golf shot does not carry as far on a cold day as on a hot day. Some golfers keep golf balls warm in a pocket, or even in a battery-powered electric heater, on cold days. Rubber has such a low thermal conductivity that the thermal time constant for a golf ball is about an hour. Therefore, it is insufficient to prewarm the ball briefly. As in the case of baseball and tennis, there is a restriction on the coefficient of restitution of golf balls considered legal for sporting competition.

High rebound balls include Ping-pong balls, handballs, and superballs. A Ping-pong ball dropped on a concrete floor exhibited a coefficient of restitution $e$ of 0.85, and for a handball, 0.82 [200]. A 'super ball' is even more responsive; it has $e = 0.94$ corresponding to a low-loss tangent (From Equation 10.39 and Equation 10.42, $\tan\delta \approx 0.04$). It is an interesting toy. The super ball can bounce in unexpected directions owing to coupling between translational and rotational degrees of freedom [201, 202].

The game of squash, by contrast, requires use of a viscoelastic ball with a low degree of rebound. The squash ball is composed of a proprietary blend of rubber containing as many as fifteen ingredients [203]. A "slow" (yellow dot) squash ball, in a drop test [204] from 1 m, exhibited a height ratio of 0.2, corresponding to a high effective loss tangent. By Equation 10.41, $\tan\delta \approx 0.7$. The actual $\tan\delta$ of the rubber in the squash ball may actually be higher than this since the ball is hollow, and the air inside, though not pressurized, offers some elastic response. The World Squash Federation specifies a rebound height ratio (given as a percentage) of 12 percent minimum at 23°C, and 26 percent to 33 percent at 45°C assuming a drop height of 2.54 m (100"). A specification is given at 45°C since such a temperature has been measured in balls during vigorous play [203]. Most of the energy of impact between the ball and racket or ball and playing surface is converted into heat as a result of viscoelasticity in the ball.

## 10.12 Applications of Soft Materials

### 10.12.1 Viscoelastic Gels in Surgery

Viscoelastic pastes and gels are used to facilitate various kinds of surgery. Methylcellulose has a long history of use in ophthalmic procedures. Sodium hyaluronate is a large polysaccharide molecule which forms a viscoelastic gel with water. It occurs in the connective tissues of vertebrates [205] and in the vitreous humor of the eye. Viscoelastic gels based on methylcellulose or purified sodium hyaluronate has been used in surgery, for cell protection, maintenance of tissue spaces, tissue lubrication, and tissue manipulation. Use of such viscoelastic pastes has been of particular use in ophthalmic surgery [206], to maintain tissue spaces, replace the vitreous humor [207], and to manipulate and protect the delicate tissues of the eye. Manipulation of tissue in a space filled with viscoelastic paste is facilitated by the damping of disturbances. Moreover bleeding is minimized by the viscous resistance of the paste. Purified sodium hyaluronate has been used in tendon repair [208], and a gel of carboxymethylcellulose and polyethylene oxide for spine surgery and neurosurgery [209]. The gel is used to prevent scar tissue formation after surgery.

### 10.12.2 Hand Strength Exerciser

Viscoelastic putty [210], a silicone polymer, is used for exercising the gripping strength of the hand, either for rehabilitation after injury, or for strength development for sports, such as rock climbing. The putty exhibits substantial creep and at long times is a viscoelastic liquid. It conforms well to the shape of the hand when squeezed slowly, but is stiff when squeezed rapidly. The response of the material also encourages the user to squeeze it.

### 10.12.3 Viscoelastic Toys

Silly Putty® is a highly viscoelastic silicone polymer [211] sold as a toy. It exhibits extreme rate dependence. When dropped on a hard surface it bounces back to a

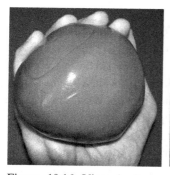 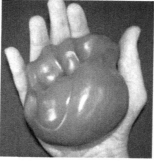 

Figure 10.16. Viscoelastic hand strength exercise putty before (left) and after (center) squeezing. Left on the table for a few hours, the material sags under gravity (right).

substantial fraction of the original height. When stretched slowly, it draws out like taffy. The material flows like a liquid when left alone for some time.

Gelatinous materials have also been used as toys. For example, a transparent gelatinous material has been fabricated as a melt blend admixture of poly (styrene-ethylene-butylene-styrene) triblock copolymer with high levels of a plasticizing oil [212, 213]. This material has been marketed under the name GlueSlug®; it exhibits a large maximum elongation combined with low stiffness. Other applications, such as flexible lenses, light pipes, hand exercise grips, and acoustical isolators have been suggested in the patents.

### 10.12.4 No-Slip Flooring, Mats, and Shoe Soles

Cork is at times used for floor surfaces which are claimed not to become slippery even if wet or polished [214]. Viscoelastic rubbery materials are used as mats in bathtubs to prevent falls on wet and soapy surfaces. The friction between a shoe or foot and a floor surface arises from two sources: adhesion due to atomic bonds between the surfaces, and resistance due to irreversible loss of mechanical energy due to viscoelastic deformation in one or both of the surfaces. Friction due to adhesion is abolished by polishing or by lubrication by water or oil, therefore, polished or lubricated hard surfaces are slippery. If one surface is compliant and viscoelastic, the energy lost in deforming it will provide sliding friction, even if the surfaces are polished or lubricated. Use of viscoelastic materials in these applications can reduce suffering due to injury from falls.

## 10.13 Applications Involving Thermoviscoelasticity

When one cuts or machines stiff polymers, the cutting process causes a rise in temperature, which reduces the stiffness and increases the loss tangent at frequencies associated with the cutting process. Continued deformation of the polymer gives rise to a further temperature rise as mechanical energy is converted into thermal energy by viscoelastic loss. If the heat cannot escape rapidly enough, a thermal runaway process can occur in which the polymer part to be machined overheats and melts. The machinist soon learns to cut slowly enough to prevent this from occurring.

In the making of tempered glass, a slab of red-hot glass is extruded between rollers. Jets of air are directed at the free surfaces to cool the glass. The effect of thermal contraction, heat conduction, and thermoviscoelasticity is that the cooled glass acquires a distribution of residual stress in which the free surfaces are under compression. Such a distribution is beneficial in a brittle material such as glass since fracture initiates at cracks or defects under tension. Since most of the defects are at the surface as a result of scratches or other damage, a superposed compressive stress improves the overall strength of the glass plate when it is bent.

Determination of viscoelastic properties can be used as a probe into various microphysical phenomena in materials. For example, the true and effective diffusion coefficients can differ as a result of differences in the thermodynamic activity of the diffusing substance [215]; calculations based on measurement of concentration versus distance give the effective diffusion coefficient. True diffusion coefficients in alloys have been determined indirectly by evaluation of relaxation times via measurement of damping over a wide range of temperatures.

## 10.14 Satellite Dynamics and Stability

Early satellites were given a spinning motion to stabilize them. The spin was about an axis of minimum inertia, similar to the spin of a rifle bullet. For example, the Explorer 1 satellite was given a spin of this type. The satellite had a stable orientation for a few minutes after which it gradually tumbled out of alignment. The cause is the viscoelasticity of the materials in the satellite. The satellite angular momentum $\mathbf{L} = I\omega$ with $I$ as mass moment of inertia and $\omega$ as spin angular velocity is constant because, in orbit, there are no external torques. The kinetic energy of rotation about a principal axis is $T_{\mathcal{KE}} = I\omega^2/2$. The energy for spin about an axis for large inertia is smaller then the energy for spin about an axis for small inertia at constant angular momentum as can be seen by substitution. Therefore, if the satellite can dissipate energy it will in time spin about the axis for large inertia.

The stability of a spinning object such as a satellite is examined in detail in view of the Euler angles $\theta$, $\phi$, and $\psi$, [216] based on Thomson and Reiter [217] analysis.

In a freely spinning body of revolution with principal moments of inertia $I_A$, $I_A$, and $I_C$, precession occurs if the spin axis is not aligned with a principal axis of inertia. The spin angular velocity is defined as $\frac{d\phi}{dt}$, the precession angular velocity as $\frac{d\psi}{dt}$, and $\theta$ as the angle between the angular momentum $\mathbf{L}$ and the 3 axis,

$$\frac{d\psi}{dt} = \frac{I_C}{(I_A - I_C)\cos\theta}\frac{d\phi}{dt}. \tag{10.43}$$

The angular velocity about the $z$ (or 3) axis is, from the definition of the Euler angles,

$$\omega_3 = \frac{d\phi}{dt} + \frac{d\psi}{dt}\cos\theta. \tag{10.44}$$

The angular momentum $\mathbf{L}$ is constant because it is assumed that there are no external torques.

Consider first the question of stability. For small misalignment,

$$\mathbf{L} = I_C \omega_0 \tag{10.45}$$

$$\omega_3 = \omega_0 \cos\theta,$$

$$\frac{d\psi}{dt} = \frac{I_C}{I_A}\omega_0. \tag{10.46}$$

The kinetic energy $T_{K\mathcal{E}}$ of rotation is

$$T_{K\mathcal{E}} = \frac{1}{2}I_A(\omega_1^2 + \omega_2^2) + \frac{1}{2}I_C\omega_3^2, \tag{10.47}$$

but for small misalignment $\theta$,

$$T_{K\mathcal{E}} = \frac{1}{2}I_C\omega_0^2[1 + (\frac{I_C}{I_A} - 1)\sin^2\theta]. \tag{10.48}$$

If the kinetic energy $T_{K\mathcal{E}}$ is dissipated due to viscoelasticity in the satellite, there must be a change in $\theta$. Differentiating, the rate of change is

$$\frac{dT_{K\mathcal{E}}}{dt} = I_C\omega_0^2(\frac{I_C}{I_A} - 1)\sin\theta\cos\theta\frac{d\theta}{dt}. \tag{10.49}$$

Since $T_{K\mathcal{E}}$ decreases with time in a dissipative material,

$$\frac{d\theta}{dt} < 0, \quad if \quad \frac{I_C}{I_A} > 1, \quad but \quad \frac{d\theta}{dt} > 0, \quad for \quad \frac{I_C}{I_A} < 1. \tag{10.50}$$

So spin about the principal axis of minimum moment of inertia ($I_C/I_A < 1$) is unstable in the presence of dissipation.

Now consider the rate at which the angle $\theta$ changes as a result of damping in the satellite. The precession generates a time varying stress in the object with a frequency $2\pi\{\frac{d\phi}{dt}\}^{-1}$. To calculate that stress, consider a simple specific model of a spinning satellite, consisting of two identical parallel rigid disks of mass $m$ connected at their centers by a viscoelastic tube of length $L$, dynamic Young's modulus $E$, and loss tangent $\tan\delta$. The mass moment of inertia of each disk is $I_{C1}$ about its polar axis and $I_{A1}$ about its diametric axis. The gyroscopic moment about each disk is

$$M_G = I_{C1}[\frac{d\phi}{dt} + \frac{d\psi}{dt}\cos\theta]\frac{d\psi}{dt}\sin\theta - I_{A1}(\frac{d\psi}{dt})^2\sin\theta\cos\theta, \tag{10.51}$$

in which the moments of inertia about the center of mass are
$I_C = 2I_{C1}$,
$I_A \approx 2(I_{A1} + mL^2)$, so,

$$M_G = \frac{1}{2}\{I_C(\frac{d\phi}{dt} + \frac{d\psi}{dt}\cos\theta)\frac{d\psi}{dt}\sin\theta - I_A(\frac{d\psi}{dt})\sin\theta\cos\theta\} + mL^2(\frac{d\psi}{dt})^2\sin\theta\cos\theta, \tag{10.52}$$

but the term in the { } braces is the moment about the center of mass; that moment is zero for a freely spinning object. So, the gyroscopic moment is

$$M_G = mL^2(\frac{d\psi}{dt})^2\sin\theta\cos\theta = FL\cos\theta, \tag{10.53}$$

with $F$ as the centripetal force of the precessing disk. The bending moment on the connecting tube is

$$M_z = M_G\frac{z}{L}, \tag{10.54}$$

and the maximum stress is

$$\sigma = \frac{M_z y}{I} = mL^2(\frac{d\psi}{dt})^2\sin\theta\cos\theta\frac{z}{L}\frac{y}{I}. \tag{10.55}$$

Since,

$$\frac{d\psi}{dt} = \frac{I_C}{I_A}\omega_0, \tag{10.56}$$

$\sigma = \frac{1}{2}mL^2(\frac{I_C}{I_A})^2\omega_0^2\frac{z}{L}\frac{y}{I}\sin\theta\cos\theta$. The energy dissipation per unit volume per cycle is

$$\Gamma_d = \frac{1}{2}\tan\delta\frac{\sigma^2}{E}, \tag{10.57}$$

so the total energy dissipation for the volume $V$ is

$$\int \Gamma_d dV = \frac{\tan\delta}{48\pi E}(\frac{mL^2 y}{I})^2 V(\frac{I_C}{I_A})^4(\frac{I_C}{I_A} - 1)\omega_0^5 \sin^2\theta \cos^3\theta, \tag{10.58}$$

with $V$ as the volume of material. Since $d\theta/dt$ is proportional to this $\Gamma_d$ via Equation 10.49, tumbling is initiated by a small initial misalignment, and increases to a peak value, then decreases as $\theta$ approaches 90 degrees [217, 218].

An instability of this type was observed in the early Explorer I satellite, which was given a spin about its longitudinal axis of minimum moment of inertia. In about 90 minutes, $\theta$ went from near 0 degrees to about 60 degrees. Later satellites were stabilized by spinning about an axis of maximum moment of inertia: hence, they tend to be oblate in shape rather than prolate.

## 10.15 Summary

We have considered the effect of viscoelastic behavior in the performance of materials in particular applications, as well as the implications in design. The influence of viscoelasticity can be beneficial, as in cases when it is used explicitly in design to achieve design goals. Cases have been presented in which high or low mechanical damping were used to achieve specific objectives. If viscoelasticity is ignored in the design process, the results of design calculations may not correspond to physical reality, leading to an unsuccessful design. Proper use of viscoelastic materials can aid in the design of devices that serve to reduce suffering.

## 10.16 Examples

### Example 10.1

Consider the use of piezoelectric materials as stiff damping elements. A piezoelectric ceramic is assumed to be connected to an external electric circuit. Determine the maximum value of $\tan\delta$ attainable with available piezoelectric materials connected to a resistive circuit and discuss. A one dimensional treatment will suffice for this example.

### Solution

In Example 8.6 the energy concepts used in the definition of coupling coefficient $\mathcal{K}$ were related to the limiting stiffnesses considered in the definition of relaxation strength to obtain for a Debye peak,

$$\tan\delta_{max} = \frac{1}{2}\frac{\mathcal{K}^2}{\sqrt{1 - \mathcal{K}^2}}. \tag{10.59}$$

Piezoelectric elements have been studied and used as damping elements. A form identical to Equation 10.59 was obtained by a different approach for a piezoelectric element loaded with a resistor [112].

If the piezoelectric element is loaded by a short circuit, it is at constant (zero) electric field; if it is electrically free, it is at constant electric displacement. At low frequency, the piezoelectric element is considered equivalent to a capacitor of value $C$. If an external resistor is connected, any transient charge on that capacitor decays with a time constant, $\tau = RC$. The largest available $\mathcal{K}$ in common piezoelectric materials is about 0.75 in lead titanate zirconate ceramics, for load in the 3 direction and polarization in the 3 direction. This gives $\tan\delta_{max} = 0.43$ for a Debye peak. This is a large loss tangent in view of the relatively high stiffness $E$ (about 70 GPa) of piezoelectric ceramics. Overall damping would not be so high since only a small part of a structure would consist of piezoelectric damping elements. Moreover, in some applications the designer desires a high $\tan\delta$ over a wider range of frequency than a Debye peak. Even so, the concept is attractive for structural damping.

### Example 10.2

How fast does a golf ball leave the club during a drive shot? How much energy is dissipated? How much does the golf ball warm up as a result of the energy dissipation? What is the effective $\tan\delta$ of the golf ball? Assume that [199] the club head has a mass of 0.2 kg and is moving at 50 m/sec, that the ball is initially stationary and has a mass of 0.046 kg, that the heat capacity of rubber is 1,700 J/kg°C , and that the coefficient of restitution is $e = 0.7$.

### Solution

Consider [199] conservation of momentum: the total momentum of ball and club must be the same before and after the impact (denoted by underlined symbols).

$$m_{ball}v_{ball} + m_{club}v_{club} = m_{ball}\underline{v}_{ball} + m_{club}\underline{v}_{club}.$$

Writing the definition of coefficient of restitution in symbolic form:

$$e = \frac{\underline{v}_{ball} - \underline{v}_{club}}{v_{club} - v_{ball}}.$$

Solving for the velocities,

$$\underline{v}_{\text{ball}} = \frac{m_{\text{club}}\{v_{\text{club}}(1+e)-ev_{\text{ball}}\}+m_{\text{ball}}v_{\text{ball}}}{m_{\text{ball}}+m_{\text{club}}},$$

$$\underline{v}_{\text{club}} = \frac{m_{\text{ball}}\{v_{\text{ball}}(1+v_{\text{club}}e)-e\}+m_{\text{club}}v_{\text{club}}}{m_{\text{ball}}+m_{\text{club}}}.$$

Incorporating the assumption that the ball is initially stationary,

$$\underline{v}_{\text{ball}} = \frac{m_{\text{club}}v_{\text{club}}(1+e)}{m_{\text{ball}}+m_{\text{club}}},$$

$$\underline{v}_{\text{club}} = \frac{v_{\text{club}}(m_{\text{club}}-em_{\text{ball}})}{m_{\text{ball}}+m_{\text{club}}}.$$

Substituting the assumed numbers,

$\underline{v}_{\text{ball}} = 70\,\text{m/sec}$, $\underline{v}_{\text{club}} = 34$ m/sec.

The loss in energy is

$$\Delta U = \tfrac{1}{2}m_{\text{club}}v_{\text{club}}^2 - \tfrac{2}{2}m_{\text{club}}\underline{v}_{\text{club}}^2 - \tfrac{2}{2}m_{\text{ball}}\underline{v}_{\text{ball}}^2.$$

Substituting,

$$\Delta U = \frac{m_{\text{ball}}m_{\text{club}}v_{\text{club}}^2(1-e^2)}{2(m_{\text{ball}}+m_{\text{club}})} = 23 \text{ joules.}$$

Considering the mass of the ball and the given heat capacity, the temperature rise is $\Delta T = 0.3°\text{C}$, which is unimportant.

The effective $\tan\delta$ of the golf ball is given as a function of the height ratio for rebound of a dropped ball (Equation 10.42)

$\tan\delta \cong \frac{1}{\pi}\ln(\frac{H_0}{H_1})$,

but the height ratio is $e^2 = 0.49$, from the given value $e = 0.7$, so

$\tan\delta \cong 0.24$.

Actually, the coefficient of restitution was given for a golf club impact, which involves large deformation of the ball under which materials behave nonlinearly. Moreover, the ball is not a homogeneous material. Therefore, one speaks of an "effective" loss tangent.

**Example 10.3**

Traditionally, applicants for police training must be above a threshold height. If a person is one-half inch or one inch too short to qualify, is there any hope for acceptance?

**Solution**

Creep in spinal discs results in a considerable variation in a person's height during the day. People are taller in the morning than the evening since the spinal discs are substantially unloaded during sleep. The applicant may try to take the height test in the morning. If more height is needed, a few days of bed rest will allow further recovery of creep strains in the spine.

**Example 10.4**

How does three-dimensional deformation influence the use of viscoelastic rubber in such applications as shoe insoles to reduce impact force in running, or wrestling mats to reduce impact force in falls?

**Solution**

Refer to Example 5.9, in which deformation under transverse constraint is analyzed. Rubbery materials are much stiffer when compressed in a thin layer geometry than they are in shear or in simple tension; they are too stiff to perform the function of reducing impact. Compliant layers can be formed by corrugating the rubber to provide room for lateral expansion or by using an elastomeric foam, which typically has a Poisson's ratio near 0.3. Corrugated rubber is used in shoe insoles and in vibration isolators for machinery. Foam is used in shoes and in wrestling mats.

## 10.17 Problems

10.1. Calculate the rolling resistance of a tire. Specifically, infer the effective horizontal force of resistance by determining the power dissipated in the tire due to its viscoelasticity. The coefficient of rolling friction is defined as the ratio of the horizontal resistance force to the vertical force. Assume that $E^*$ of the rubber is known. This problem may be considerably simplified by treating the tire as a thin-walled cylindrical pressure vessel with a flat region corresponding to contact with the road. How does the rolling resistance depend on the air pressure in the tire? How does it depend on the speed of the vehicle?

10.2. Calculate the effective tension–compression damping of a steel plate with a single layer of viscoelastic polymer glued to it. Assume all thicknesses and material properties are known.

10.3. Calculate the effective flexural damping of a steel plate with a single layer of viscoelastic polymer glued to it. Assume all thicknesses and material properties are known.

10.4. Obtain an approximate value for the optimal stiffness for a knee pad intended to minimize impact force upon the knee in sports. Assume a reasonable value for the thickness of the pad. Is it more beneficial to make the pad of a viscoelastic material or to incorporate a plateau region in the stress–strain curve (such as by elastic collapse of a foam)?

10.5. Suppose a material were developed with $\tan\delta < 0$. What causal mechanisms might be used to achieve such a result? What might such a material be used for? Design a sport that could take advantage of a ball with such behavior.

10.6. If a squash ball is available, infer the $\tan\delta$ of the rubber in the ball via a rebound test. How does rebound depend on temperature? Are your results consistent with the expected behavior of a polymer in the rubbery and leathery regions. If you have access to test equipment, measure the $\tan\delta$ of the rubber. Compare with the results of a rebound test.

10.7. If you have a guitar or other stringed instrument that can use nylon or catgut strings, a microphone, and a method to determine frequency, then perform the following evaluation of stress relaxation. Install and tune a new string, and measure the frequency as a function of time after tuning [49]. Compare with a metal string. Interpret your results in terms of what you now know about viscoelastic behavior.

10.8. How does the rebound resilience of a hollow ball differ from that of a solid one?

## BIBLIOGRAPHY

[1] EAR Corporation, Div., Cabot Corp., 7911 Zionsville Rd, Indianapolis, IN 46268 USA.

[2] Gardner, R., Jr., U.S. Patents 3,811,437, 1974; RE 29,487 Earplugs, 1977.

[3] Calcagno, B., Lopez Garcia, M. D. C., Kuhns, M., and Lakes, R. S., On the Non-linear Creep and Recovery of Open Cell Earplug Foams, *Cellular Polymers*, 27, 165–178, 2008.

[4] Bazant, Z., *Mathematical Modeling of Creep and Shrinkage of Concrete*, New York: J. Wiley, 1988.

[5] Gordon, J. E., *Structures*, Harmondsworth, Middlesex, UK: Penguin, 1983.

[6] Jaglinski, T., and Lakes, R. S., Creep behavior of Al-Si die-cast alloys, *J Eng Mat and Technology*, 126, 378–382, 2004.

[7] Gerhards, C. C., Bending Creep and Load Duration of Douglas Fir 2 × 4s under Constant Load for up to 12 + years, *Wood and Fiber Science*, 32, 489–501, 2000.

[8] Reding, J. L., Draft guide for Determining the Effects of High Temperature Operation on Conductors, Connectors, and Accessories, IEEE P1285/D7.01, June 2004.

[9] Harvey J. R., and Larson, R. E., Use of Elevated Temperature Creep Data in Sag-Tension Calculations, *IEEE Trans*, PAS-89, (3), 380–386, 1970.

[10] Stokes, Y. M., Flowing Windowpanes: Fact or Fiction? *Proc Roy Soc Lond A*, 455, 2751–2756, 1999.

[11] Zanotto, E. D., Do Cathedral Glasses Flow? *Am J Phys*, 66, 392–395, 1998.

[12] Krishnan, J. M., and Rajagopal, K. R., Review of the Uses and Modeling of Bitumen from Ancient to Modern Times, *Appl Mech Rev*, 56, 149–214, 2003.

[13] Collop, A. C., Cebon, D., and Hardy, M. S. A, Viscoelastic Approach to Rutting in Flexible Pavements, *J Transp Eng*, 121, 82–93 1995.

[14] Nabarro, F. R. N., and de Villiers, H. L. *The Physics of Creep*, London: Taylor and Francis, 1995.

[15] Backman, D. G., and Williams, J. C., Advanced Materials for Aircraft Engine Applications, *Science*, 255, 1082–1087, 1992.

[16] Ashby, M. F., and Jones, D. R. H., *Engineering Materials*, Oxford: Pergamon, 1980.

[17] Willett, F., The Case of the Derailed Disk Drives, *Mech Eng*, 110, 42–44, Jan. 1988.

[18] Das, B. M., *Principles of Geotechnical Engineering*, 2nd ed., Boston: PWS-Kent, 1990.

[19] Burland, J. B., Jamiolkowski, M., and Viggiani, C., The Stabilisation of the Leaning Tower of Pisa, *Soils and Foundations*, 43(5), 63–80, 2003.

[20] Anderson, J. G. C., and Trigg, C. F., *Case Studies in Engineering Geology*, London: Elek Science, 1976.

[21] Wang, H. F., *Theory of Linear Poroelasticity*, Princeton, NJ: Princeton University Press, 2000.

[22] Segall, P., Grasso, J. R., and Mossop, A., Poroelastic Stressing and Induced Seismicity Near the Lacq Gas Field, Southwestern France, *J Geophys Res*, 99, (B8), 15423–15438, 1994.

[23] Raleigh, C. B., Healy, J. H., and Bredehoeft, J. D., An Experiment in Earthquake Control at Rangely, Colorado, *Science*, 191, (4233) 1230–1237, 1976.

[24] Bassett, R. H., Time-dependent Strains and Creep in Rock and Soil Structures, in *Creep of Engineering Materials*, C. D. Pomeroy, ed., London: Mechanical Engineering Publications, Ltd., 1978.

[25] Terzaghi, K., and Peck, R. B., *Soil Mechanics in Engineering Practice*, 2nd ed., New York: J. Wiley, 1967.

[26] Hwang, J. S., Soldering and Solder Paste Technology, Harper, C., ed., *Electronics Packaging and Interconnection Handbook*, New York: McGraw Hill, 1991.

[27] Hwang, J. S., and Vargas, R. M., Solder Joint Reliability–Can Solder Creep? *Soldering and Surface Mount Technology*, 5, 38–45, 1990.

[28] Morris, J. W., Jr., Goldstein, J., and Mei, Z., Microstructure and Mechanical Properties of Sn-In and Sn-Bi Solders, *JOM*, 45, 25–27, 1993.

[29] McCormack, M., and Jin, S., Progress in the Design of New Lead-Free Solder Alloys, *JOM*, 45, 36–40, 1993.

[30] Bartha, L., Lassner, E., Schubert, W. D., and Lux, B., eds., *The Chemistry of Non-Sag Tungsten*, Amsterdam: Elsevier, 1995.

[31] Ferry, J. D., *Viscoelastic Properties of Polymers*, 2nd ed., New York: J. Wiley, 1970.

[32] Howard, W. H., and Williams, M. L., Viscoelasticity and Flatspotting, *Rubber Chem and Tech*, 40, 1139–1146, 1967.

[33] Setright, L. J. K., *Automobile Tyres*, London: Chapman and Hall, 1972.

[34] Cunningham A., Huygens E., and Leenslag J. W., MDI Comfort Cushioning for Automotive Applications, *Cellular Polymers*, 13, 461–472, 1994.

[35] Dementjev, A. G., Deformation of Flexible Polyether Polyurethane Foams with Bimodal Cell Structure, *Cellular Polymers*, 15, 155–171, 1996.

[36] Dinsdale, S. M., Decubitus Ulcers: Role of Pressure and Friction in Causation, *Archives of Physical Medicine and Rehabilitation*, 55, 147–152, 1974.

[37] Yarkony, G. M., Pressure Ulcers: A review, *Archives of Physical Medicine and Rehabilitation*, 75, 908–917, 1994.

[38] Palmieri, V. R., Haelen, G. T., and Cochran, G. V., A Comparison of Sitting Pressures on Wheelchair Cushions as Measured by Air Cell Transducers and Miniature Electronic Transducers, *Bull Prosthetics Res*, 10–33, 5–8, 1980.

[39] Garber, S. L., Wheelchair Cushions: A Historical Review, *Am J Occup Therapy*, 39, 453–459, 1985.

[40] Rose, R. M., Nusbaum, H. J., Schneider, H., Ries, M., Paul, I., Crugnola, A., Simon, S. R., and Radin, E. L., On the True Wear Rate of Ultra High Molecular Weight Polyethylene in the Total Hip Prosthesis, *J Bone Joint Surg*, 62-A, 537–549, 1980.

[41] Von Finger, W., Elastizitat von Composite- fullungsmaterialein, *Dtsch Zahnaerztl Z*, 30, 345–349, 1975.

[42] Ruyter, I. E., and Oysaed, H. Compressive Creep of Light Cured Resin-Based Restorative materials, *Acta Odontol Scand*. 40, 319–324, 1982.

[43] Papadogianis, Y., Boyer, D. B., and Lakes, R. S., Creep of Conventional and Microfilled Dental Composites, *J Biomed Mate Res*, 18, 15–24, 1984.

[44] Papadogianis, Y., Boyer, D. B. and Lakes, R. S., Creep of Amalgam at Low Stress, *J Dental Res*, 66, 1569–1575, 1987.

[45] Mahler, D. B., and Van Eysden, J., Dynamic Creep of Dental Amalgam, *J Dental Res*, 48, 501–508, 1969.

[46] Lu, R., and Puri, V. M., Characterization of Nonlinear Creep Behavior of Two Food Products, *J. Rheology*, 35, 1209–1233, 1991.

[47] Rao, M., and Steffe, J. F., *Viscoelastic Properties of Foods*, London and New York, Elsevier Applied Sciences, 1992.

[48] Roos, Y. H., Karel, M., and Kokini, J. L., Glass Transitions in Low Moisture and Frozen Foods: Effects on Shelf Life and Quality, *Food Technology*, 50, 95–105, 1996.

[49] Vilela, P. M., Moscoso, R. A., and Thompson, D., What Every Musician Knows about Viscoelastic Behavior, *Am J Phys* 65, 1000–1003, 1997.

[50] Lin, J. Y., and Westmann, R. A., Viscoelastic Winding Mechanics, *J Appl Mech*, 56, 821–827, 1989.

[51] Eagle, R., A Pain in the Back, *New Scientist*, 84, 170–172, October 18, 1979.

[52] Althoff, I., Brinckmann, P., Frobin, W., Sandover, J., and Burton, K., An Improved Method of Stature Measurement for Quantitative Determination of Spinal Loading, *Spine*, 17, 682–693, 1992.

[53] Magnusson, M., Hult, E., Lindström, I., Lindell, V., Pope, M., and Hansson, T., Measurement of Time-Dependent Height Loss During Sitting, *Clin Biomech*, 5, 137–142, 1990.

[54] Magnusson M., Aleksiev A., Spratt K., Lakes R. S., and Pope M. H., Hyperextension and Spine Height Changes, *Spine*, 21(22), 2670–2675, 1996.

[55] Hedman, T. P., and Fernie, G. R., In vivo Measurement of Lumbar Spinal Creep in Two Seated Postures Using Magnetic Resonance Imaging, *Spine*, 20, 178–183, 1995.

[56] Kasra, M., Shirazi-adl, A., and Drouin, G., Dynamics of Human Lumbar Intervertebral Joints: Experimental and Finite Element Investigations, *Spine*, 17, 93–102, 1992.

[57] Galante, J. O., Tensile Properties of the Human Lumbar Annulus Fibrosus, *Acta Orthopaedica Scandinavica*, Supplementum 100, 191, 1967.

[58] Formby, W. A., Nanda, R., and Currier, G. F., Longitudinal Changes in the Adult Facial Profile, *Am J Orthodontics and Dentofacial Orth*, 105, 464–476, 1994.

[59] Gunner, C. W., Hutton, W. C., and Burlin, T. E., An Apparatus for Measuring the Recoil Characteristics of Human Skin *in vivo*, *Med and Biol Eng and Computing*, 17, 142–144, 1979.

[60] Payne, P. A., Measurements of Properties and Function of Skin, *Clin Phys Physiol Meas*, 12, 105–129, 1991.

[61] Largo, R. H., and Duc, G., Head Growth and Changes in Head Configuration in Healthy Preterm and Term Infants During the First Six Months of Life, *Hel Paediatr*, 32, 431–442, 1977.

[62] AAP Task Force on Infant Positioning and SIDS: Positioning and SIDS, *Pediatrics* 89, 1120–1126, 1992.

[63] Clarren, S., Smith, D., Hanson, J., Helmet Treatment for Plagiocephaly and Congenital Muscular Torticollis, *J Pediatr*, 94, 443, 1979.

[64] Pomatto, J. K., Littlefield, T. R., Manwaring, K., and Beals, S. P., Etiology of Positional Plagiocephaly in Triplets and Treatment using a Dynamic Orthotic Cranioplasty device, *Neurosurg Focus*, 2, 1–4, 1997.

[65] Littlefield, T. R., Kelly, K. M., Reiff, J. L., Pomatto, J. K., Car Seats, Infant Carriers, and Swings: Their Role in Deformational Plagiocephaly, *J Prothestics and Orthotics*, 15, 102–106, 2003.

[66] Ortega, B. Unkind Cut: Some Physicians do Unnecessary Surgery on Heads of Infants, *Wall Street Journal*, February 23, 1996.

[67] Cranial Technologies, Phoenix, AZ, www.cranialtech.com.

[68] Bassett, R. H., Time-Dependent Strains and Creep in Rock and Soil Structures, in *Creep of Engineering Materials*, C. D. Pomeroy, ed., London: Mechanical Engineering Publications, Ltd., 1978.

[69] Zyczkowski, M., Optimal Structural Design under Creep Conditions, *Appl Mech Rev*, 49, 433–446, 1996.

[70] Snyder, G., *Michigan Construction News*, Dragline, www.michiganconstructionnews.com, accessed November 3, 2006.

[71] Feynman, R. P., *What Do You Care What Other People Think?* New York: W. W. Norton, 1988.

[72] Snowdon, J. C., *Vibration and Shock in Damped Mechanical Systems*, New York: J. Wiley, 1968.

[73] Frolov, K. V., and Furman, F. A., *Applied Theory of Vibration Isolation Systems*, New York: Hemisphere/Taylor and Francis, 1990.

[74] Kerwin, E. M., Jr. and Ungar, E. E., Requirements Imposed on Polymeric Materials in Structural Damping Applications, in *Sound and Vibration Damping with Polymers*, R. D. Corsaro and L. H. Sperling, eds., Washington, DC: American Chemical Society, 1990.

[75] Nashif, A. D., Jones, D. I. G., and Henderson, J. P., *Vibration Damping*, New York: John Wiley, 1985.

[76] Cremer, L., Heckl, M. A., and Ungar, E. E., *Structure Borne Sound*, 2nd ed., Berlin: Springer, Verlag, 1988.

[77] U.S. Patent 5,757,580.

[78] U.S. Patent 5,666,239.

[79] U.S. Patent 6,809,898.

[80] http://quietpc.com/. See also *New York Times*, Circuits, October 11, 2007.

[81] Fantel, H., Turntables Rise to the Challenge of the CD Era, Regarding Ariston Equipment, *New York Times*, May 3, 1987.

[82] Miles, J. W., On the Annular Damper for a Freely Precessing Gyroscope, II, *J Appl Mech*, 30, 189–192, 1963.

[83] Chang, C. O., and Chen, M. P., Elastomer Damper for a Freely Precessing Dual Spin Seeker, *J Guidance*, 16, 221–224, 1992.

[84] Branson, D., *Deformation of Concrete Structures*, New York: McGraw Hill, 1977.

[85] Gaul, L., and Nitsche, K., The Role of Friction in Mechanical Joints, *Appl Mech Rev*, 54, 93–105, 2001.

[86] Taranath, B. S., *Structural Analysis and Design of Tall Buildings*, New York: McGraw Hill, 1988.

[87] Strogatz, S. H., Abrams, D. M., McRobie, A., Eckhardt, B., and Ott, E., Crowd Synchrony on the Millennium Bridge, *Nature*, 438, 43–44, 2005.

[88] Schueller, W., *The Vertical Building Structure*, New York: Van Nostrand, 1990.

[89] Wetton, R. E., Design of Elastomers for Damping Applications, in *Elastomers: Criteria for Engineering Design*, C. Hepburn, R. J. W., Reynolds, eds., London: Applied Science Publishers, Ltd., 1979.

[90] Rogers, L., and Parin, M., Experimental Results for Stand Off Passive Vibration Damping Systems, in *Smart Structures and Materials 1995: Passive Damping*, Proc. SPIE Volume 2445, C. D. Johnson, ed., Bellingham, WA: Society of Photo-Optical Instrumentation Engineers, 1995, pp. 374–383.

[91] Oberst, H., Über die Dämpfung der Biegeschwingungen dünner bleche durch fest haftende Beläge, *Acustica 2*, Beih 4, AB 181–194, 1952.

[92] Pipkin, A. C., *Lectures on Viscoelasticity Theory*, Heidelberg: Springer Verlag; London: George Allen and Unwin, 1972.

[93] Oberst, H., Über die Dämpfung der Biegeschwingungen dünner bleche durch fest haftende Beläge II, *Acustica 4*, Beih 1, AB 433, 1954.

[94] Capps, R. N., and Beumel, L. L. Dynamic Mechanical Testing, Application of Polymer Development to Constrained Layer Damping, in *Sound and Vibration Damping with Polymers*, R. D. Corsaro and L. H. Sperling, eds., Washington DC: American Chemical Society, 1990.

[95] Hostettler, F., Energy Attenuating Polyurethanes, U.S. Patent 4,722,946.

[96] Tiao, W.Y., and Tiao, C. S., Methods for the Manufacture of Energy Attenuating Polyurethanes, U.S. Patent 4,980,386.

[97] Saxby, G., *Practical Holography*, Englewood Cliffs, NJ: Prentice Hall, 1988.

[98] U.S. Patent 3,990,535.

[99] Fricke, J. R., Lodengraf Damping – an Advanced Vibration Damping Technology, *Sound and Vibration*, 34, 22–27, 2000.

[100] Fishman, S. G. Damping in Metal Matrix Composites, Overview and Research Needs, in *Role of Interfaces on Material Damping*, Proceedings of an International Symposium Held in Conjunction with ASM's Materials Week and TMS/AIME Fall Meeting, October 13–17 1985, Toronto, B. B. Rath and M. S. Misra, eds., ASM, 1985, 33–41.

[101] Ritchie, I. G., and Pan, Z. L., High Damping Metals and Alloys, *Met Trans*, 22A, 607–616, 1991.

[102] James, D. J., High Damping Metals for Engineering Applications, *Mat Sci Eng*, 4, 1–8, 1969.

[103] Ritchie, I. G., Pan, Z. L., and Goodwin, F. E., Characterization of the Damping Properties of Die-Cast Zinc-Aluminum Alloys, *Met Trans* 22A, 617–622, 1991.

[104] Alberts, T. E., and Chen, Y., U.S. Patent 5,256,223, 1993.

[105] Mifune, N., U.S. Patent 5,300,355, 1994.

[106] Miller, D., U.S. Patent 3,894,169, 1975.

[107] Hutin, P., U.S. Patent 5,316,298, 1993.

[108] Bishop, J. E., and Kinra, V. K., Elastothermodynamic Damping in Composite Materials, *Mechanics of Composite Materials and Structures*, 1, 75–93, 1994.

[109] Forward, R. L., Electronic Damping of Vibrations in Optical Structures, *J Appl Opt* 18, 690–697, 1979.

[110] Forward, R. L., and Swigert, C. J., Electronic Damping of Orthogonal Bending Modes in a Cylindrical Mast-Theory, *J Spacecraft and Rockets*, 18, 5–10, 1981.

[111] Forward, R. L., Electronic Damping of Orthogonal Bending Modes in a Cylindrical Mast–Experiment, *J Spacecraft and Rockets*, 18, 11–17, 1981.

[112] Hagood, N. W., and von Flotow, A., Damping of Structural Vibrations with Piezoelectric Materials and Passive Electrical Networks, *J Sound and Vibration*, 146, 243–268, 1991.

[113] Brodt, M., and Lakes, R. S., Composite Materials which Exhibit High Stiffness and High Mechanical Damping, *J Composite Mat*, 29, 1823–1833, 1995.

[114] Fukumoto, A., The Application of Piezoelectric Ceramics in Diagnostic Ultrasound Transducers, *Ferroelectrics*, 40, 217–230, 1982.

[115] Berlincourt, D. A., Curran, D. R., and Jaffe, H., Piezoelectric and Piezomagnetic Materials and Their Function in Transducers, in *Physical Acoustics*, E. P. Mason, ed. Vol. 1A, New York: Academic, 1964, pp. 169–270.

[116] Rosen, C. Z., Hiremath, B., and Newnham, R., eds.. *Piezoelectricity*, New York: American Institute of Physics, 1982.

[117] Henderson, J. P., Damping Applications in Aero–Propulsion Systems, in *Damping Applications for Vibration Control*, P. J. Torvik, ed., New York: ASME Publication AMD-38, 1980.

[118] DeFelice, J. J., and Nashif, A. D., Damping of an Engine Exhaust Stack, *Shock Vib Bull*, 48, 75–84, 1978.

[119] McClure, F. T., Hart, R. W., and Bird, J. F., Acoustic Resonance in Solid Propellant Rockets, *J Appl Phys*, 31, 884–896, 1960.

[120] Landel, R. F., and Smith, T. L., Viscoelastic Properties of Rubberlike Composite Propellants and Filled Elastomers, *ARS J*, 31, 599–608, 1961.

[121] Nall, B. N., Acoustic Attenuation of a Solid Propellant, *AIAA J*, 1, 76–79, 1963.

[122] Sattinger, S. S., U.S. Patent 5,108,802, 1992.

[123] Artus, J. P., U.S. Patent 5,277,423, 1994.

[124] Ashley, S., Smart Skis, *Technology Review*, 99, 16, 1996.

[125] Griffin, M. J., *Handbook of Human Vibration*, New York: Academic, 1990.

[126] Staff, New and Improved Steel Products, *Advanced Materials and Processes*, 141, 52–56, 1992.

[127] Suzukawa, Y., Ikeda, K., Morita, J., and Katoh, A., Application of Vibration Damping Steel Sheet for Autobody Structural Parts, SAE Paper 920249, 1992.

[128] Cebon, D., and Ashby, M. F., Materials Selection for Precision Instruments, *Meas Sci Tech*, 5, 296–306, 1994.

[129] Staff, Tuned Damping – the Key to Effective Structural Damping, Newport Corporation, Irvine, CA, USA.

[130] Karato, S., Importance of anelasticity in the Interpretation of Seismic Tomography, *Geophysical Res Let*, 20, 1623–1626, 1993.

[131] Lienkaemper, J. J., Galehouse, J. S., and Simpson, R. W., Creep Response of the Hayward Fault to Stress Changes Caused by the Loma Prieta Earthquake, *Science*, 276, 2014–2016, 1997.

[132] Webster, J. G. *Medical Instrumentation*, Houghton-Mifflin, 1978.

[133] DiBenetto, A. T., Gauchel, J. V., Thomas, R. L., and Barlow, J. W., Nondestructive Determination of Fatigue Crack Damage Using Vibration Tests, *J Mat*, 7, 211–215, 1972.

[134] Schultz, A. B., and Warwick, D. N., Vibration Response: A Nondestructive Test for Fatigue Crack Damage in Filament Reinforced Composites, *J Composite Mat*, 5, 394–404, 1971.

[135] Banks, H. T., Smith, R. C., and Wang, Y., Smart Material Structures: Modeling, Estimation and Control, New York: J. Wiley; Paris: Masson, 1996.

[136] Lakes, R. S., Materials with Structural Hierarchy, *Nature*, 361, 511–515, 1993.

[137] Otsuka, K., and Kakeshita, T., Science and Technology of Shape Memory Alloys: New Developments, *MRS Bulletin*, 27, 91–98, February 2002.

[138] Uchino, K., Shape Memory Ceramics, in *Shape Memory Materials*, K. Otsuka and C. M. Wayman, ed., Cambridge UK: Cambridge University Press, 1998, Ch. 8, pp. 184–202.

[139] Irie, M., Shape Memory Polymers, in *Shape Memory Materials*, K. Otsuka and C. M. Wayman, eds., Cambridge UK: Cambridge University Press, 1998, Ch. 9, pp. 203–219.

[140] White, S. R., Sottos, N. R., Moore, J., Geubelle, P., Kessler, M., Brown, E., Suresh, S., and Viswanathan, S., Autonomic Healing of Polymer Composites, *Nature*, 409, 794–797, 2001.

[141] Culshaw, B., *Smart Structures and Materials*, Boston: Artech, 1996.

[142] Pacheco, B. M., Fujino, Y., and Sulekh, A., Estimation Curve for Modal Damping in Stay Cables with Viscous Damper, *J Struct Eng ASCE*, 119, 1961–1979, 1993.

[143] Yang, J. N., and Giannopoulos, F., Active Control and Stability of Cable Stayed Bridge, *J Eng Mech Div ASCE*, 105, 677–694, 1979.

[144] Achkire, Y., and Preumont, A., Active Tendon Control of Cable–Stayed Bridges, *Earthquake Eng and Structural Dynamics*, 25, 585–597, 1996.

[145] Cannon, R. H., and Rosenthal, D. E., Experiment in Control of Flexible Structures with Noncollocated Sensors and Actuators, *AIAA J Guidance*, 7, 546–553, 1984.

[146] Benhabib, R. J., Iwens, R. P., and Jackson, R. L., Stability of Large Space Structures Control Systems Using Positivity Concepts, *AIAA J Guidance*, 4, 487–494, 1981.

[147] Tabor, D., The Mechanism of Rolling Friction, *Phil Mag*, 43, 1055–1059, 1952.

[148] Schuring, D. J., The Rolling Loss of Pneumatic Tires, *Rubber Chem and Tech*, 53, 600–727, 1980.

[149] Coulomb, C. A., Théorie des Machines Simples, in Académie Royale des Sciences, *Mémoires de Mathématique et de Physique*, 10, 1785.

[150] Bowden, F. P., and Tabor, D., *Friction and Lubrication*, New York: J. Wiley, 1956.

[151] Witters, J., and Duymelinck, D., Rolling and Sliding Resistive Forces on Balls Moving on a Flat Surface, *Am J Phys* 54, 80–83, 1986.

[152] Timoshenko, S. P., and Goodier, J. N., *Theory of Elasticity*, New York: McGraw Hill, 1982.

[153] Flom, D. G., and Bueche, A. M., Theory of Rolling Friction for Spheres, *J Appl Phys*, 30, 1725–1730, 1959.

[154] Hunter, S. C., The Rolling Contact of a Rigid Cylinder with a Viscoelastic Half Space, *J Appl Mech*, 28, 611–617, 1961.

[155] Flom, D. G., Rolling Friction of Polymeric Materials. I. Elastomers, *J Appl Phy*, 31, 306–314, 1960.
[156] Luchini, J. R., ed., *Rolling Resistance of Highway Truck Tires*, Publication SP-546, Warrendale, PA: Society of Automotive Engineers, 1983.
[157] Mullins, J., Easy Rollers, *New Scientist*, 146, 31–33, 1995.
[158] Momosaki, E., and Kogure, S., The Application of Piezoelectricity to Watches, *Ferroelectrics*, 40, 203–216, 1982.
[159] Braginskii, V. B., Mitrofanov, V. P., and Panov, V. I., *Systems with Small Dissipation*, Chicago: University of Chicago Press, 1985.
[160] Cagnoli, G., Giammatoni, L., Kovalik, J. Marchesoni, F., and Punturo, M., Low Frequency Internal Friction in Clamped Free Thin Wires, *Phys Let A*, 255, 230–235, 1999.
[161] Rossing, T. D., Russell, D. A., and Brown, D. E., On the Acoustics of Tuning Forks, *Am J Phys*, 60, 620–626, 1992.
[162] Kuroda, K., Does the Time of Swing Method Give a Correct Value of the Newtonian Gravitational Constant? *Phys Rev Let*, 75, 2796–2798, 1995.
[163] Quinn, J. J., Speake, C. C., and Brown, L. M., Materials Problems in the Construction of Long-Period Pendulums, *Phil Mag A*, 65, 261–276, 1992.
[164] Maddox, J., Systematic Errors in 'Big G,' *Nature*, 377, 573, 1995.
[165] Ekinci, K. L., and Roukes, M. L., Nanoelectromechanical Systems, *Rev Sci Instrum*, 76, 061101, 2005.
[166] Lifshitz, R., and Roukes, M. L. Thermoelastic Damping in Micro- and Nanomechanical Systems, *Phys Rev B*, 61, 5600–5609, 2000.
[167] Chen, C. P., and Lakes, R. S., Design of viscoelastic Impact Absorbers: Optimal Material Properties, *Int J Solids, Structures*, 26, 1313–1328, 1990.
[168] Christensen, R. M. *Theory of Viscoelasticity*, New York: Academic, 1982.
[169] Bauer, W., U.S. Patent 4,427,225, 1984.
[170] Tuggle, R. E., U.S. Patent 3,715,139, 1973.
[171] Erpelding, A., U.S. Patent 5,606,447, 1997.
[172] U.S. Patent 20,070,190,293, Ferrara, V. R.; see also Schwartz, A., Helmet Absorbs Shock in a New Way, *New York Times*, A1, A10, October 27, 2007.
[173] Light, L. H., McLellan, G. E., and Klenerman, L., Skeletal Transients on Heel Strike in Normal Walking with Different Footwear, *J Biomech*, 13, 477–480, 1980.
[174] Voloshin, A., and Wosk, J., Influence of Artificial Shock Absorbers on Human Gait, *Clinical Orthopaedics*, 160, 52–56, 1981.
[175] Voloshin, A., and Wosk, J., An *in vivo* Study of Low Back Pain and Shock Absorption in the Human Locomotor System, *J Biomech*, 15, 21–27, 1982.
[176] Basford, J. R., and Smith, M. A., Shoe Insoles in the Workplace, *Orthopedics*, 11, 285–288, 1988.
[177] Voloshin, A., and Wosk, J., Shock Absorption of Meniscectomized and Painful Knees: A Comparative Study, *J Biomed Eng*, 5, 157–160, 1983.
[178] Garcia, A. C., Durá, J. V., Ramiro, J., Hoyos, J. V., and Vera, P., Dynamic Study of Insole Materials Simulating Real Loads, *Foot and Ankle International*, 15, 311–323, 1994.
[179] McMahon, T. A., and Greene, P. R., Fast Running Tracks, *Scientific American*, 239, 148–163, 1978.
[180] Loriga, G., Pneumatic Tools: Occupation and Health, in *Encyclopedia of Hygiene, Pathology, and Social Welfare*, Vol. 2, International Labour Office, Switzerland: Geneva, 1934.
[181] Wasserman, D., Reynolds, D., Behrens, V., and Samueloff, S., *Vibration White Finger Disease in U.S. Workers Using Pneumatic Chipping and Grinding Tools: II. Engineering Testing*, National Institute for Occupational Safety and Health, DHEW/NIOSH Publication 82-101, Cincinnati, 1981, pp. 1–89.

[182] Griffin, M. J., *Handbook of Human Vibration*, New York: Academic, 1990.

[183] Carbone, G., and Persson, B. N. J., Hot Cracks in Rubber: Origin of the Giant Toughness of Rubberlike Materials, *Phys Rev Let*, 96, 114301, 2005.

[184] Gent, A. N., Adhesion and Strength of Viscoelastic Solids. Is there a Relationship between Adhesion and Bulk Properties? *Langmuir*, 12 (19), 4492–4496, 1996.

[185] Gent, A. N., and Petrich, R. P., Adhesion of Viscoelastic Materials to Rigid Substrates, *Proceed Ro Soc London, A, Mathematical and Physical Sciences*, 310, (1502) 433–448, 1969.

[186] Schaller, R., and Fantozzi, G., Damping and Toughness, in *Mechanical Spectroscopy Q⁻¹*, R. Schaller, G. Fantozzi, and G. Gremaud, eds., Switzerland: Trans Tech Publications, 2001, pp. 615–620.

[187] Murray, A., and Neilson, J. M. M., Diagnostic Percussion Sounds: 1. A Qualitative Analysis, *Med Biological Engineering*, 13, 19–28, 1975.

[188] Murray, A., and Neilson, J. M. M., Diagnostic Percussion Sounds: 2. Computer-Automated Parameter Measurement for Quantitative Analysis, *Medical and Biological Eng*, 13, 29–38, 1975.

[189] Auenbrugger, L., *Inventum novum ex percussione thoracis humani ut signo abstrusos interni pectoris morbos detegendi*; reprinted in 1966, Dawsons, London; translated by Forbes, 1824.

[190] Delp, M. H., and Manning, R. T., in R. H. Major, *Major's Physical Diagnosis*, 7th ed., Philadelphia, PA: W. B. Saunder's, 1968, p. 102.

[191] Lippmann, R. K., The Use of Auscultatory Percussion for the Examination of Fractures, *J Bone Joint Surgery*, 14, 118–126, 1932.

[192] Calvit, H. H. Numerical Solution of the Problem of Impact of a Rigid Sphere onto a Linear Viscoelastic Half-Space and Comparison with Experiment. *Int J Solids, Structures*, 3, 951, 1967.

[193] Adair, R. K., *The Physics of Baseball*, New York: Harper Collins, 1994.

[194] Adair, R. K., The Physics of Baseball, *Physics Today*, 48, 26–30, May 1995.

[195] Kagan, D. T., The Effects of Coefficient of Restitution Variations on Long Fly Balls, *Am J Phys*, 58, 151–154, 1990.

[196] Kagan, D., and Atkinson, D., The Coefficient of Restitution of Baseballs as a Function of Relative Humidity, *Physics Teacher*, 42, 330–333, 2004.

[197] Rist, C., The Physics of Baseballs, *Discover*, May 2001. See also http://discovermagazine.com/ 2001/may/featphysics.

[198] Brody, H., The Tennis-Ball Bounce Test, *Physics Teacher*, 28, 407–409, 1990.

[199] Daish, C. B., *The Physics of Ball Games*, London: English Universities Press, 1972.

[200] Griffing, D. F., *The Dynamics of Sports*, 3rd ed. Oxford, Ohio: Dalog, 1987.

[201] Garwin, R. L., Kinematics of an Ultraelastic Rough Ball, *Am J Phys*, 37, 88–92, 1969.

[202] Armenti, A., Jr., ed., *The Physics of Sports*, New York: American Institute of Physics, 1992.

[203] Dunlop, Ltd., Barnsley, South Yorkshire, UK, private communication.

[204] Lakes, R. S., technical report, unpublished, 1996.

[205] Comper, W. D., and Laurent, T. C., Physiologic Function of Connective Tissue Polysaccharides, *Physiol Rev*, 58, 255–315, 1978.

[206] Pape, L. G., and Balazs, E. A., The Use of Sodium Hyaluronate (Healon®) in Human Anterior Segment Surgery, *Ophthalmol*, 87, 699–705, 1980.

[207] Pruett, R. C., Stephens, C. L., and Swann, D. A., Hyaluronic Acid Vitreous Substitute. A Six Year Clinical Evaluation, *Arch Ophthalmol*, 97, 2325–2330, 1979.

[208] St Onge, R., Weiss, C., Denlinger, J. L., and Balazs, E. A., A Preliminary Clinical Assessment of Na Hyaluronate Injection for Primary Flexor Tendon Repair in No-Man's Land, *Clin Orthopaedics*, 146, 269–275, 1980.

[209] Peanell, P. E., Blackmore, J. M., and Allen, M., U.S. Patent 5,156,839, 1992.

[210] Smith and Nephew Rolyan, Inc., N93 W14475 Whittaker Way, Menomonee Falls, WI 53051.

[211] Binney and Smith Co., Inc., Easton, PA.

[212] Chen, J. Y., Thermoplastic Elastomer Gelatinous Compositions, U.S. Patent 4,369,284, 1983.

[213] Chen, J. Y., Gelatinous Elastomeric Optical Lens, Light Pipe, Comprising a Specific Block Copolymer and an Oil Plasticizer, U.S. Patent 4,618,213, 1986.

[214] Gibson, L. J., and Ashby, M. F., *Cellular Solids*, Oxford: Pergamon, 1988; 2nd ed., Cambridge, UK: Cambridge University Press, 1997.

[215] Krishtal M. A., Determination of the True Diffusion Coefficients and of the Thermodynamic Activity by the Internal Friction Method, in *Internal Friction in Metals and Alloys*, V. S. Postnikov, F. N. Tavadze, and L. K. Gordienko, eds., New York: Consultants Bureau, 1967.

[216] Symon, K. R., *Mechanics*, 2nd ed., Reading, MA: Addison Wesley, 1964.

[217] Thomson, W. T., and Reiter, G. S., Attitude Drift of Space Vehicles, *J Astronautical Sciences*, 7, 29–34, 1960.

[218] Meirovitch, L., Attitude Stability of an Elastic Body of Revolution in Space, *J Astronautical Sciences*, 8, 110–119, 1961.

# Appendix

## A.1 Mathematical Preliminaries

### A.1.1 Introduction

The relationship between two physical quantities is commonly thought of as being describable by a function $f(x)$, which assigns a number to each numerical value of some independent variable $x$. In some physical systems, a mathematical description of greater generality is useful [1]. For example, it is impossible to measure a physical quantity at an exact value of $x$ but only over some small interval of $x$. As another example, the response of a system to an impulsive input is of interest in many contexts. A *functional* is a more general mathematical description that is useful in such situations. A functional is a rule that assigns a number to each function in a set of *testing functions*. A *distribution* is a particular type of functional. Distributions are useful in that a concept, such as the Dirac delta, which describes an impulsive stimulus, is correctly described as a distribution but not as a function. *Integral transforms* are extremely useful in the solution of problems in viscoelasticity.

### A.1.2 Functionals and Distributions

A *functional* is defined as an assignment $F(\mathbf{x})$ of a number to each element $\mathbf{x}$ of a vector space [1]. Examples of functionals include the scalar product and the norm of a vector. A functional is *linear* if $F(a_1\mathbf{x}_1 + a_2\mathbf{x}_2) = F(a_1\mathbf{x}_1) + F(a_2\mathbf{x}_2)$. The set of functions $x(t)$ on some interval $a \leq t \leq b$ may be regarded as a "space" $L_p$. Bounded linear functionals $F(x(t))$ in this space can be constructed as Stieltjes integrals constructed as limits of sums of step functions [2]. One can write a functional as $F(x) = \int_a^b x(t)dg(t)$ for continuous functions $x(t)$. A *distribution* is defined as a linear, continuous functional [1].

### A.1.3 Heaviside Unit Step Function

The Heaviside unit step function $\mathcal{H}(x)$ is defined as follows:
$\mathcal{H}(x) = 0$, if $x < 0$
$\mathcal{H}(x) = 1/2$, if $x = 0$
$\mathcal{H}(x) = 1$, if $x > 0$
so that
$\mathcal{H}(x - a) = 0$, if $x < a$
$\mathcal{H}(x - a) = 1/2$, if $x = a$
$\mathcal{H}(x - a) = 1$, if $x > a$

Observe that the function $\mathcal{H}(x)$ is discontinuous. Therefore, $d\mathcal{H}(x)/dx$ does not exist at $x = 0$. Because that derivative appears in the Boltzmann integral for a creep or relaxation procedure, a method of dealing with such derivatives is of use. Distributions, such as the Dirac delta, are used for that purpose and other purposes.

### A.1.4 Dirac Delta

The Dirac delta $\delta(x)$ is defined by

$$\int_{-\infty}^{\infty} \delta(x)dx = 1 \tag{A.1}$$

$$\delta(x) = 0, \quad x \neq 0.$$

There is no such function with such properties. The Dirac delta, although sometimes called the *delta function* or *impulse function*, is actually a distribution. Observe that $\int_{-\xi}^{\xi} \delta(x)dx = 1$, with $\xi$ as an arbitrarily small but nonzero number, the Dirac delta exhibits the sifting property:

$$\int_{-\infty}^{\infty} \delta(x)f(x)dx = f(0). \tag{A.2}$$

The sifting property of $\delta(x - a)$ with $a$ as a real number is demonstrated [3] by a change of variable $x = X + a$.
$\int_{-\infty}^{\infty} f(x)\delta(x - a)dx = \int_{-\infty}^{\infty} f(X + a)\delta(X)dX = f(X + a)|_{X=0} = f(a)$.
To develop a more formal understanding of the delta in connection with familiar functions, consider a sequence of functions $\phi_n(x)$. The sequence is called a *delta sequence* if

$$\lim_{n \to \infty} \int_{-\infty}^{\infty} \phi_n(x)f(x)dx = f(0). \tag{A.3}$$

Although $\phi_n(x)$ does not converge to any function, we say, in the sense of distributions, that $\phi_n(x)$ converges to $\delta(x)$. Examples of delta sequences include [1–5]:

$$\phi_n(x) = 0 \quad x < 0 \tag{A.4}$$

$$\phi_n(x) = \frac{n}{x_0} \quad 0 < x < \frac{x_0}{n},$$

$$\phi_n(x) = 0 \quad x > \frac{x_0}{n}.$$

$$\phi_n(x) = \frac{n}{\sqrt{\pi}} e^{-n^2 x^2}, \tag{A.5}$$

$$\phi_n(x) = \frac{n}{2} e^{-n|x|}, \tag{A.6}$$

$$\phi_n(x) = \frac{\sin nx}{\pi x}, \tag{A.7}$$

$$\phi_n(x) = \frac{n}{\pi}(1 + x^2 n^2). \tag{A.8}$$

As $n$ becomes large, each function attains a higher peak but becomes narrower in width.

The sifting property of the delta may be demonstrated [4] using the sequence given in Equation A.4.

$$\int_{-\infty}^{\infty} \phi_n(x) f(x) dx = \frac{n}{x_0} \int_0^{x_0/n} f(x) dx. \tag{A.9}$$

By the mean value theorem,
$\int_0^{x_0/n} f(x) dx = \frac{x_0}{n} f(\zeta \frac{x_0}{n}), 0 \leq \zeta \leq 1$, so $\int_{-\infty}^{\infty} \phi_n(x) f(x) dx = f(\zeta \frac{x_0}{n}), 0 \leq \zeta \leq 1$. Let
$n$ become large, $\int_{-\infty}^{\infty} \delta(x) f(x) dx = f(0)$.
This is the sifting property, valid for all functions $f$, which are continuous in the neighborhood of the origin.

We may also consider the *Heaviside sequences* $\mathcal{H}_n(x - a)$, such that

$$\frac{d\mathcal{H}_n(x - a)}{dx} = \phi_n(x - a). \tag{A.10}$$

Because of the properties of the Dirac delta as shown below, $\mathcal{H}_n(x)$ must converge to the Heaviside step function $\mathcal{H}(x)$. In this sense, we can regard the derivative of the step function as the delta. Integrating both sides,
$\int_{-\infty}^{\infty} \frac{d\mathcal{H}_n(x-a)}{dx} dx = \int_{-\infty}^{\infty} \phi_n(x - a) dx$, so
$\mathcal{H}_n(\infty) - \mathcal{H}_n(-\infty) = \int_{-\infty}^{\infty} \phi_n(x - a) dx$.
In the limit, $1 = \int_{-\infty}^{\infty} \delta(x) dx$.
The integral of the delta sequence members $\phi_n(x) = \frac{n}{\pi}(1 + x^2 n^2)$ is as follows:

$$\mathcal{H} = \frac{1}{2} + \frac{1}{\pi} \tan^{-1} nx. \tag{A.11}$$

Members of this delta sequence and the corresponding Heaviside sequence are shown in Figure A.1.

### A.1.5 Doublet

As an example of the use of delta sequences, let us determine the derivative of the delta. If $\phi_n(x) f(x)$ tends to zero for $x \rightarrow \pm \infty$, we may integrate by parts to obtain
$\lim_{n \to \infty} \int_{-\infty}^{\infty} \frac{d\phi_n(x)}{dx} f(x) dx = -\lim_{n \to \infty} \int_{-\infty}^{\infty} \phi_n(x) \frac{df(x)}{dx} dx = -\frac{df(x)}{dx}|_{x=0} = -\frac{df(x)}{dx}|_{x=0} \equiv$
$-f'(0)$.
We may express this, with the derivative of the delta called the *doublet*, $\psi$,
$\psi = \frac{d}{dx}\delta(x)$,

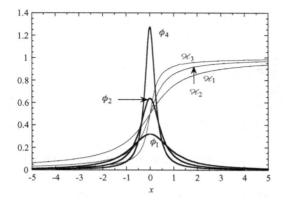

Figure A.1. Members of delta sequence $\phi_n$ and the corresponding Heaviside sequence $\mathcal{H}_n$.

as

$$\int_{-\infty}^{\infty} \psi(x) f(x) dx = -\frac{df(x)}{dx}\Big|_{x=0}. \tag{A.12}$$

So the doublet sifts the negative of the derivative of a function.

Members of the sequence

$\psi_n(x) = \frac{d}{dx}\phi_n(x)$, based on $\phi_n(x) = \frac{n}{\pi}(1 + x^2 n^2)$ are shown in Figure A.2.

The sifting properties of the doublet may also be illustrated via the definition of the derivative and the sifting properties of the delta [3].

$Z \equiv \int_{-\infty}^{\infty} \frac{\delta(x+h)-\delta(x)}{h} f(x) dx = \frac{1}{h}\{\int_{-\infty}^{\infty} \delta(x+h) f(x) dx - \int_{-\infty}^{\infty} \delta(x) f(x) dx\}$

$= \frac{1}{h}\{\int_{-\infty}^{\infty} \delta(X) f(X-h) dX - f(0)\} = \frac{1}{h}\{f(-h) - f(0)\}$

so $\lim_{h\to 0} Z = \int_{-\infty}^{\infty} \psi(x) f(x) dx = -\frac{df(x)}{dx}\Big|_{x=0}.$

Expressions involving the $n$th derivative $\delta^{(n)}(x)$ may be developed using similar arguments.

$\int_{-\infty}^{\infty} \delta^{(n)}(x) f(x) dx = (-1)^{(n)} \frac{d^n f(x)}{dx^n}\Big|_{x=0}.$

### Physical Interpretation of the Step, Delta, and Doublet

In the spatial domain, the step function can be used to represent distributed loads upon part of the length of a beam subjected to bending. In that context, the delta

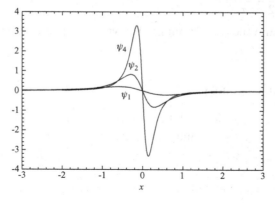

Figure A.2. Members of doublet sequence $\psi_n$.

represents a concentrated force, and the doublet represents a concentrated moment or torque.

In the time domain, the step function represents the stress history used in a creep test upon a viscoelastic material, or the strain history used in a relaxation test. The delta can be used to represent an impact force.

### A.1.6 Gamma Function

The *gamma function* $\Gamma(x)$ is given [10], for $x > 0$, by

$$\Gamma(x) = \int_0^\infty u^{x-1} e^{-u} du. \tag{A.13}$$

If $x$ is an integer $n$, then $\Gamma(n+1) = n!$. Some useful identities involving the gamma function are (with $p$ and $k$ as real numbers and $\mathcal{L}$ as a Laplace transform):

$$\Gamma(x)\Gamma(1-x) = \pi/\sin\pi x \qquad \Gamma(x+1) = x\Gamma(x) \tag{A.14}$$

$$\mathcal{L}[t^n] = \frac{n!}{s^{n+1}}, n = 0, 1, 2 \ldots; \tag{A.15}$$

$$\mathcal{L}[t^k] = \frac{\Gamma(k+1)}{s^{k+1}} \tag{A.16}$$

$$\int_0^\infty \frac{\sin x}{x^p} dx = \frac{\pi}{2\Gamma(p)\sin(p\pi/2)}, \tag{A.17}$$

for $0 < p < 1$.

$$\int_0^\infty \frac{\cos x}{x^p} dx = \frac{\pi}{2\Gamma(p)\cos(p\pi/2)}, \tag{A.18}$$

$0 < p < 1$.

Integrals, which may be useful, include
$\int_0^\infty e^{-ax} \sin bx\, dx = \frac{b}{a^2+b^2}$, $\int_0^\infty e^{-ax} \cos bx\, dx = \frac{a}{a^2+b^2}$ .

### A.1.7 Liebnitz Rule

Liebnitz Rule [6]:
$\frac{d}{dx} \int_a^b F(x,t)dt = \int_a^b \frac{\partial F(x,t)}{\partial x} dt + F(x,b)\frac{db}{dx} - F(x,a)\frac{da}{dx}$.

## A.2 Transforms

Transforms operate upon a function to give another function. Transforms are useful in solving problems in various branches of engineering and science including viscoelasticity theory, because they can be used to convert differential equations or integral equations into algebraic equations which are more easily solved. The most commonly used transforms are linear transforms [1, 7, 8]. Linear transforms $T[f(x)]$ obey the superposition rule

$$T[f(x) + g(x)] = T[f(x)] + T[g(x)]. \tag{A.19}$$

### A.2.1 Laplace Transform

The Laplace transform of a function $f(t)$ is defined as

$$\mathcal{L}[f(t)] \equiv F(s) \equiv \int_0^\infty f(t)e^{-st}dt, \tag{A.20}$$

with $s$ as the transform variable. Laplace transforms of selected functions are presented in §A.3.

The *convolution* of two functions $f$ and $g$ is defined as

$$\int_0^t f(t - \xi)g(\xi)d\xi. \tag{A.21}$$

The convolution theorem for Laplace transforms is

$$\mathcal{L}[\int_0^t f(t - \xi)g(\xi)d\xi] = \mathcal{L}[f(t)]\mathcal{L}[g(t)]. \tag{A.22}$$

The derivative theorem for Laplace transforms is

$$\mathcal{L}[\frac{df(t)}{dt}] = s\mathcal{L}[f(t)] - f(0). \tag{A.23}$$

The proof of these theorems is by direct integration using the definition.

### A.2.2 Fourier Transform

The Fourier transform is defined as follows [1] with the angular frequency $\omega = 2\pi\nu$, and $\nu$ as the frequency:

$$\mathcal{F}[f(t)] \equiv F(\omega) = \int_{-\infty}^\infty f(t)e^{-i\omega t}dt. \tag{A.24}$$

The inverse transform is given by

$$f(t) = \frac{1}{2\pi}\int_{-\infty}^\infty F(\omega)e^{i\omega t}d\omega. \tag{A.25}$$

There are different conventions [7] in the definition of Fourier transforms; some authors include $2\pi$ factors in the argument of the exponential to achieve symmetry; others define both the transform and the inverse transform with $1/\sqrt{2\pi}$ before the integral [10].

Consider several theorems in Fourier transforms:

*Shift theorem.* If $\mathcal{F}[f(t)] = F(\omega)$, then

$$\mathcal{F}[f(t - a)] = e^{-i\omega a}F(\omega). \tag{A.26}$$

To demonstrate this, use the definition

$$\int_{-\infty}^{\infty} f(t-a)e^{-i\omega t} dt = \int_{-\infty}^{\infty} f(t-a)e^{-i\omega(t-a)}e^{-i\omega a} d(t-a) = e^{-i\omega a}F(\omega). \quad (A.27)$$

*Convolution theorem.* If $\mathcal{F}[f(t)] = F(\omega)$ and $\mathcal{F}[g(t)] = G(\omega)$, then

$$\mathcal{F}[\int_{-\infty}^{\infty} f(t')g(t-t')dt'] = F(\omega)G(\omega). \quad (A.28)$$

To demonstrate this, use the definition and exchange orders of integration:

$$\int_{-\infty}^{\infty} \int_{-\infty}^{\infty} f(t')g(t-t')dt' e^{-i\omega t} dt = \int_{-\infty}^{\infty} f(t') \int_{-\infty}^{\infty} g(t-t')e^{-i\omega t} dt dt'. \quad (A.29)$$

Use the shift theorem, then use

$$\int_{-\infty}^{\infty} f(t')e^{-i\omega t'} G(\omega)dt' = F(\omega)G(\omega), \quad (A.30)$$

as desired.

*Derivative theorem.* If $\mathcal{F}[f(t)] = F(\omega)$, then

$$\mathcal{F}[\frac{df(t)}{dt}] = i\omega F(\omega). \quad (A.31)$$

The proof involves integration by parts, using the definition.

### A.2.3 Hartley Transform

The Hartley transform [7] resembles the Fourier transform; it differs in that the kernel is $cas(2\pi \nu t)$ rather than $\exp(-2\pi i \nu t)$. $H(s) = \int_{-\infty}^{\infty} f(x)cas(2\pi s x)dx$. The cas function is defined as $cas x = \cos x + \sin x$. It is a sinusoid of amplitude $\sqrt{2}$ shifted by one-eighth of a period. The Hartley transform is a real transform and the transform is identical to the inverse transform.

### A.2.4 Hilbert Transform

The Hilbert transform is defined as

$$F_{Hi}(x) = \frac{1}{\pi} \int_{-\infty}^{\infty} \frac{f(\xi)}{\xi - x}d\xi, \quad (A.32)$$

and the inverse transform is

$$f(x) = -\frac{1}{\pi} \int_{-\infty}^{\infty} \frac{F_{Hi}(\xi)}{\xi - x}d\xi. \quad (A.33)$$

The Hilbert transform is seen in the derivation of the Kramers–Kronig relations in §3.3. The Hilbert transform is also related to the concept of instantaneous phase via the following argument [7]. If $f(x) = \exp(-\pi x^2)\cos 4\pi x$, then $F_{Hi}(x) = -\exp(-\pi x^2)\sin 4\pi x$, so that the oscillations in the transform share the same envelope as those in the original function. However, the phase of an oscillatory function is explicitly revealed only where it crosses zero. The phase $\phi$ at intermediate points can be expressed as $\tan\phi = F_{Hi}(x)/f(x)$.

## A.3 Laplace Transform Properties

Some Laplace transform general properties are as follows [8–10]. Write the Laplace transform as

$$\mathcal{L}\{f(t)\} = F(s).$$

Linearity (superposition):

$$\mathcal{L}\{a_1 f_1(t) + a_2 f_2(t)\} = a_1 F_1(s) + a_2 F_2(s). \tag{A.34}$$

Derivative property:

$$\mathcal{L}\{\frac{df(t)}{dt}\} = sF(s) - f(0). \tag{A.35}$$

Integral property:

$$\mathcal{L}\{\int_0^t f(\tau)d\tau\} = \frac{F(s)}{s}. \tag{A.36}$$

Multiplication by time:

$$\mathcal{L}\{tf(t)\} = -\frac{d}{ds}F(s). \tag{A.37}$$

Division by time:

$$\mathcal{L}\{\frac{f(t)}{t}\} = \int_s^\infty F(u)du. \tag{A.38}$$

Multiplication by an exponential:

$$\mathcal{L}\{e^{-at} f(t)\} = F(s + a). \tag{A.39}$$

Time shift:

$$\mathcal{L}\{f(t - \mathbf{T})\mathcal{H}(t - \mathbf{T})\} = e^{-\mathbf{T}s} F(s). \tag{A.40}$$

Scale change:

$$\mathcal{L}\{f(at)\} = \frac{1}{a}F(s/a). \tag{A.41}$$

Convolution theorem:

$$\mathcal{L}\{\int_0^t f(\tau)g(t - \tau)d\tau\} = F(s) \cdot G(s). \tag{A.42}$$

Some Laplace transform pairs (Table A.1) are as follows:

Table A.1. *Laplace transform pairs [4, 10]*

| $f(t)$ | $F(s)$ |
|---|---|
| $\delta(t)$ | $1$ |
| $\delta(t-a)$ | $e^{-as}$ |
| $\mathcal{H}(t)$ | $1/s$ |
| $\mathcal{H}(t-a)$ | $\dfrac{1}{s}e^{-as}$ |
| $t\mathcal{H}(t)$ | $1/s^2$ |
| $e^{-at}$ | $\dfrac{1}{s+a}$ |
| $\dfrac{t^{n-1}e^{-at}}{(n-1)!}$ | $\dfrac{1}{(s+a)^n}$ |
| $\dfrac{1}{a}(1-e^{-at})$ | $\dfrac{1}{s(s+a)}$ |
| $\dfrac{[e^{bt}-e^{at}]}{(b-a)}$ | $\dfrac{1}{(s-a)(s-b)}$ for $a \neq b$ |
| $\dfrac{[be^{bt}-ae^{at}]}{(b-a)}$ | $\dfrac{s}{(s-a)(s-b)}$ for $a \neq b$ |
| $\cosh(at)$ | $\dfrac{s}{s^2-a^2}$ |
| $\sinh(at)$ | $\dfrac{a}{s^2-a^2}$ |
| $\cos(at)$ | $\dfrac{s}{s^2+a^2}$ |
| $\sin(at)$ | $\dfrac{a}{s^2+a^2}$ |
| $\dfrac{\sin at}{t}$ | $\tan^{-1}(a/s)$ |
| $t\cos(at)$ | $\dfrac{s^2-a^2}{(s^2+a^2)^2}$ |
| $t\sin(at)$ | $\dfrac{2as}{(s^2+a^2)^2}$ |
| $e^{-at}\sin\omega t$ | $\dfrac{\omega}{(s+a)^2+\omega^2}$ |
| $e^{-at}\cos\omega t$ | $\dfrac{s+a}{(s+a)^2+\omega^2}$ |
| $t^n$ | $\Gamma(n+1)s^{-n-1}$ |
| $t^n e^{-qt}$ | $\Gamma(n+1)(s+q)^{-n-1}$ |

## A.4 Convolutions

The *convolution* of two functions $f$ and $g$ is defined as

$$C(t) = f(t) * g(t) = \int_0^t f(\xi)g(t-\xi)d\xi. \qquad (A.43)$$

The range of integration for the causal variables of interest in viscoelasticity is from zero to $t$. Convolutions are also used to deal with spatially varying quantities for which causality is not an issue. In that case they are defined over a range of integration from $-\infty$ to $\infty$.

Convolutions have the *commutative property*, that is [11],

$$f(t) * g(t) = g(t) * f(t).$$

The proof is by a change of variable, $x = t - \xi$. Other properties of convolutions are associativity,

$$f(t) * [g(t) * h(t)] = [f(t) * g(t)] * h(t),$$

and distributivity,

$$f(t) * [g(t) + h(t)] = f(t) * g(t) + f(t) * h(t).$$

There is also the shift property:

if

$$C(t) = f(t) * g(t),$$

then

$$f(t) * g(t - \mathbf{T}) = f(t - \mathbf{T}) * g(t) = C(t - \mathbf{T}).$$

The integration can be carried out by graphical or by numerical means to aid in visualization or to perform calculations. Graphical convolution can be carried out as follows. Visualize the function $f(\xi)$, and keep it fixed. Obtain $g(-\xi)$ by flipping $g(\xi)$ about the vertical axis. Obtain $g(t_0 - \xi)$ for a particular time value by shifting $g(\xi)$ by an amount $t_0$ along the $t$ axis. Determine the area under the product $f(\xi)g(t_0 - \xi)$ to obtain the convolution $C(t_0)$ for the particular time value $t_0$. Repeat the procedure for different times to obtain $C(t)$ for all values of $t$.

The convolution integral, as with other integrals, can be defined as the limit of a sum.

$$C(t) = f(t) * g(t) = \lim_{dt \to 0} \sum_{mdt=0}^{t} f(mdt)g(t - mdt)dt. \tag{A.44}$$

Here, $dt$ is the time increment and $m$ is an integer specifying how many time increments are taken. If $t = kdt$, this can be written $C(t) = \lim_{dt \to 0}(dt) \sum_{m=0}^{k} f(m)g(k - m)$. The quantity $y(k) = \sum_{m=0}^{k} f(m)g(k - m)$ is called the convolution sum. The limits on the summation come from the causal nature of viscoelastic response. Since $f(k)$ is causal, $f(m) < 0$ for $m < 0$. Causality of $g(k)$ means that $g(k - m) = 0$ when $m > k$. So the product $f(m)g(k - m) = 0$ for $m < 0$ and for $m > k$, and it is nonzero only for $0 \le m \le k$.

Numerical convolution can be visualized by the *tape algorithm*. Numerical data for each function are written in sequence on tapes. The values of the convolution are extracted as follows from sums of products of the data on the tapes. The tapes are moved with respect to each other and the sum expanded. Continuing, one can obtain the numerical convolution of two functions over the full time domain. This procedure is identical to the graphical procedure described above.

The convolution sum can be written as a matrix equation $\mathbf{y} = \mathbf{g}\mathbf{f}$

$$\mathbf{g} = \begin{pmatrix} g(0) & 0 & 0 & \cdots \\ g(1) & g(0) & 0 & \cdots \\ \vdots & \vdots & \vdots & \ddots \end{pmatrix}.$$

This form suggests a numerical method for inverting the convolution process. The reverse of convolution is *deconvolution*. One can solve for $\mathbf{f}$ as $\mathbf{f} = \mathbf{g}^{-1}\mathbf{y}$, which requires an inversion of the matrix $\mathbf{g}$. The inversion can be carried out numerically. Such a procedure can be used to obtain the creep function from the relaxation function (or vice versa) using the convolution relation between these functions developed in Chapter 2. Deconvolution is also used in digital optics to remove blur from images.

## A.5 Interrelations in Elasticity Theory

For an isotropic elastic material, the elastic constants are related [12] as follows. Here, $E$ is Young's modulus, $G$ is the shear modulus, $B$ is the bulk modulus, $v$ is Poisson's ratio, $C_{1111}$ is an axial component of the elastic modulus tensor, and $\lambda$ and $\mu$ are the Lamé constants. See also Example 5.9.

$$G = \mu, \tag{A.45}$$

$$B = \lambda + \frac{2}{3}\mu, \, B = \frac{2G(1+v)}{3(1-2v)}, \tag{A.46}$$

$$B = \frac{E}{3(1-2v)}, \, B = \frac{GE}{3(3G-E)}, \tag{A.47}$$

$$v = \frac{E}{2G} - 1, \, v = \frac{3B-2G}{6B+2G}, \, v = \frac{1}{2} - \frac{E}{6B}, \tag{A.48}$$

$$v = \frac{\lambda}{2(\lambda+G)}. \tag{A.49}$$

$$E = 2G(1+v) \tag{A.50}$$

$$E = 3B(1-2v) = \frac{9GB}{3B+G}. \tag{A.51}$$

$$C_{1111} = \lambda + 2G = 2G[\frac{v}{1-2v} + 1]. \tag{A.52}$$

$$C_{1111} = B + \frac{4}{3}G. \tag{A.53}$$

$$C_{1111} = E\frac{1-v}{(1+v)(1-2v)}. \tag{A.54}$$

## A.6 Other Works on Viscoelasticity

The purpose of the present work is to provide an introduction to the theory of viscoelasticity with emphasis on the linear theory of both transient and dynamic behavior; solution of stress analysis problems; experimental methods; the properties of real materials; causal mechanisms; analysis of composites; and applications. The interested reader is referred to other works dealing with viscoelastic materials. Works emphasizing the mathematical aspects of the subject include the following. Christensen [13] presents the linear theory and includes the solution of advanced problems of research interest. Renardy, Hrusa, and Nohel [14] as well as Gurtin and Sternberg [15] present a postulational approach to the linear theory, emphasizing the proof of theorems. The works of Bland [16] and Flügge [17] are early introductions of the linear theory; Bland [16] emphasizes mechanical models with springs and dashpots. Golden and Graham [18] emphasize viscoelastic stress analysis, specifically methods for the solution of difficult boundary value problems not amenable to simple methods. Haddad [19] emphasizes the linear theory and provides examples of analysis of composites. Pipkin [20] succinctly presents a variety of pedagogic examples as well as a treatment of viscoelastic fluids. Gross [21] presents interrelations among the viscoelastic functions and a systematic conceptual organization of these functions. Findley, Lai, and Onaran [22] and Lockett [23] deal with nonlinear viscoelasticity and transient behavior. Shames and Cozzarelli [24] deal with viscoelasticity in the general context of inelastic phenomena; this work includes a discussion of nonlinear viscoelasticity and of plasticity. Tschoegl [5] presents an exhaustive treatment of the linear theory, with emphasis on constitutive equations. Gittus [25], Rabotnov [26], and Kraus [27] deal with creep, principally of metals to be used for structural applications. A review article by Bert [28] contains a review of work on dynamic properties and their determination. Wineman and Rajagopal's book [29] on the mechanical response of polymers provides an introduction to the linear theory with emphasis on transient response and applications.

Works dealing with the viscoelastic properties of particular classes of materials include books on polymers by Ferry [30]; McCrum, Read, and Williams [31]; Aklonis and MacKnight [32]; and Ward and Hadley [33]. McCrum, Read, and Williams [31] introduce structural aspects of polymers, molecular theories, and many experimental results for viscoelastic and dielectric relaxation in polymers. Ward and Hadley [33] present many aspects of polymer mechanics including linear and nonlinear viscoelasticity, as well as fracture. The classic works of Zener [34] and Nowick and Berry [35] deal with metals as crystalline materials. Nowick and Berry [35] deal with other crystalline materials as well. Creus [36] discusses viscoelasticity and material aging with applications to concrete structures and includes examples of viscoelastic finite-element analysis and other numerical techniques. Zinoviev and Ermakov [37] present analyses of viscoelasticity of fibrous and laminated composites. Lodge [38] presents viscoelastic liquids.

Works dealing with causal mechanisms include the works of Ferry [30] and McCrum, Read and Williams [31] on molecular theories for polymers; Zener [34] and Nowick and Berry [35] on crystalline materials, principally metals; and Nabarro and de Villiers [39] on the physics of creep including plasticity of crystalline materials. Works dealing largely with applications of viscoelastic materials include volumes on creep by Pomeroy [40] and Kraus [27]. Kraus [27] deals primarily with metals.

### BIBLIOGRAPHY

[1] Zemanian, A. H., *Distribution Theory and Transform Analysis*, New York: Dover, 1965.

[2] Schechter, M., *Principles of Functional Analysis*, New York: Academic, 1971.

[3] Hoskins, R. F., *Generalised Functions*, Chichester, England: Ellis Norwood, 1979.

[4] Sneddon, I., *The Use of Integral Transforms*, New York: McGraw Hill, 1972.

[5] Tschoegl, N. W., *The Phenomenological Theory of Linear Viscoelastic Behavior*, Berlin: Springer Verlag, 1989.

[6] Hildebrand, F. B., *Advanced Calculus for Applications*, 2nd ed., New York: Prentice Hall, 1976, p. 365.

[7] Bracewell, R. N., Numerical Transforms, *Science*, 248, 697–704, 1990.

[8] Sneddon, I. N., *Fourier Transforms*, New York: McGraw Hill, 1951.

[9] Swisher, G. M., *Introduction to Linear Systems Analysis*, Cleveland: Matrix, 1976.

[10] Spiegl, M. R., *Mathematical Handbook*, New York: McGraw Hill, 1968.

[11] Lathi, B. P., *Linear Systems and Signals*, Carmichael, CA: Berkeley–Cambridge Press, 1992.

[12] Sokolnikoff, I. S., *Mathematical Theory of Elasticity*, FL: Krieger, Malabar, 1983.

[13] Christensen, R. M. *Theory of Viscoelasticity*, New York: Academic, 1982.

[14] Renardy, M., Hrusa, W., and Nohel, W. J., *Mathematical Problems in Viscoelasticity*, Essex, UK: Longman; New York: John Wiley, 1987.

[15] Gurtin, M. E., and Sternberg, E., On the Linear Theory of Viscoelasticity, *Arch Rational Mech Anal*, 11, 291–356, 1962.

[16] Bland, D. R., *The Theory of Linear Viscoelasticity*, Oxford: Pergamon, 1960.

[17] Flügge, W., *Viscoelasticity*, Waltham, MA: Blaisdell, 1967.

[18] Golden, J. M., and Graham, G. A. C., *Boundary Value Problems in Linear Viscoelasticity*, Berlin: Springer-Verlag, 1988.

[19] Haddad, Y. M., *Viscoelasticity of Engineering Materials*, London: Chapman and Hall, 1995.

[20] Pipkin, A. C., *Lectures on Viscoelasticity Theory*, Heidelberg: Springer Verlag; London: George Allen and Unwin, 1972.

[21] Gross, B., *Mathematical Structure of the Theories of Viscoelasticity*, Paris: Hermann, 1968.

[22] Findley, W. N., Lai, J. S., and Onaran, K., *Creep and Relaxation of Nonlinear Viscoelastic Materials*, Amsterdam: North Holland, 1976.

[23] Lockett, F. J., *Nonlinear Viscoelastic Solids*, New York: Academic, 1972.

[24] Shames, I. H., and Cozzarelli, F. A., *Elastic and Inelastic Stress Analysis*, Englewood Cliffs, NJ: Prentice Hall, 1992.

[25] Gittus, J. Creep, *Viscoelasticity and Creep Fracture in Solids*, Halstead, NY: John Wiley; London: Applied Science, 1975.

[26] Rabotnov, Yu. N., *Creep Problems in Structural Members*, Amsterdam: North Holland, 1969.

[27] Kraus, H., *Creep Analysis*, New York: John Wiley, 1980.

[28] Bert, C. W., Material damping: An Introductory Review of Mathematical Models, Measures, and Experimental Techniques, *J Sound and Vibration*, 29, 129–153, 1973.

[29] Wineman, A., and Rajagopal, K. R., *Mechanical Response of Polymers*, Cambridge, UK: Cambridge University Press, 2000.

[30] Ferry, J. D., *Viscoelastic Properties of Polymers*, 2nd ed., New York: John Wiley, 1970.

[31] McCrum, N. G., Read, B. E., and Williams, G., *Anelastic and Dielectric Effects in Polymeric Solids*, New York: Dover, 1991.

[32] Aklonis, J. J., and MacKnight, W. J., *Introduction to Polymer Viscoelasticity*, New York: John Wiley, 1983.

[33] Ward, I. M., and Hadley, D. W., *Mechanical Properties of Solid Polymers*, New York: John Wiley, 1993.

[34] Zener, C., *Elasticity and Anelasticity of Metals*, Chicago: University of Chicago Press, 1948.

[35] Nowick, A. S., and Berry, B. S., *Anelastic Relaxation in Crystalline Solids*, New York: Academic, 1972.

[36] Creus, G. J., *Viscoelasticity–Basic Theory and Applications to Concrete Structures*, Berlin: Springer-Verlag, 1986.

[37] Zinoviev, P. A., and Ermakov, Y. N., *Energy Dissipation in Composite Materials*, Lancaster, PA: Technomic, 1994.

[38] Lodge, A. S., *Elastic Liquids*, New York: Academic, 1964.

[39] Nabarro, F. R. N., and de Villiers, H. L., *The Physics of Creep*, London: Taylor and Francis, 1995.

[40] Pomeroy, C. D., ed., *Creep of Engineering Materials*, London: Mechanical Engineering Publishers, Ltd., 1978.

# Symbols

## B.1 Principal Symbols

Principal symbols are identified below. Some symbols used within different communities have different meanings. For example, $\nu$ refers to frequency and Poisson's ratio. Where both meanings have to be used within a given section, alternate symbols (such as $f$ for frequency) are used, or a subscript is added to aid clarity.

Table B.1. *Greek letter symbols*

| Symbol | Meaning |
| --- | --- |
| $\alpha$ | thermal expansion |
| $\alpha$ | wave attenuation |
| $\delta$ | phase angle between stress and strain |
| $\delta(x)$ | Dirac delta function |
| $\delta_{ij}$ | Kronecker delta |
| $\Delta$ | relaxation strength |
| $\Delta t$ | increment of time |
| $\Delta \omega$ | width of resonance response curve |
| $\Gamma$ | normalized dynamic torsional compliance |
| $\Gamma(x)$ | gamma function |
| $\epsilon$ | strain |
| $\eta$ | viscosity |
| $\phi$ | phase angle |
| $\lambda$ | wavelength |
| $\Lambda$ | log decrement |
| $\nu$ | frequency |
| $\nu$ | Poisson's ratio |
| $\omega$ | angular frequency |
| $\varpi$ | angular frequency variable of integration |
| $\rho$ | density |
| $\sigma$ | stress |
| $\tau$ | time constant |
| $\tau_r$ | time constant in relaxation |
| $\tau_c$ | time constant in creep |
| $\tau$ | time variable of integration as in $d\tau$ |

Table B.2. *Script letter symbols*

| Symbol | Meaning |
|--------|---------|
| $\mathcal{D}$ | electric displacement |
| $\mathcal{E}$ | electric field |
| $\mathcal{G}$ | gain of amplifier |
| $\mathcal{H}(t)$ | Heaviside step function of time |
| $\mathcal{K}$ | Piezoelectric coupling coefficient |
| $\mathcal{L}(f(t))$ | Laplace transform of a function $f(t)$ |
| $\mathcal{F}(f(t))$ | Fourier transform of a function $f(t)$ |
| $\mathcal{R}$ | Energy coupling fraction in thermoelasticity |
| $\mathcal{T}$ | Transmissibility of vibration |
| $\mathcal{T}(f(x))$ | Transform of function $f(x)$ |

Table B.3. *Latin letter symbols*

| Symbol | Meaning |
|--------|---------|
| $a_T(T)$ | time temperature shift factor |
| $a$ | a constant |
| $A$ | cross-section area |
| $A$ | amplitude |
| $c$ | wave speed |
| $C_{ijkl}$ | elastic modulus tensor |
| $E$ | elastic Young's modulus |
| $E^*$ | complex dynamic Young's modulus |
| $E'$ | dynamic Young's storage modulus |
| $E''$ | dynamic Young's loss modulus |
| $E(t)$ | Young's relaxation modulus |
| $f$ | frequency |
| $\mathbf{f}$ | porosity |
| $G$ | elastic shear modulus |
| $G^*$ | complex dynamic shear modulus |
| $G'$ | dynamic shear storage modulus |
| $G''$ | dynamic shear loss modulus |
| $G(t)$ | shear relaxation modulus |
| $H(\tau)$ | relaxation spectrum |
| $J^*$ | complex dynamic compliance |
| $J'$ | dynamic storage compliance |
| $J''$ | dynamic loss compliance |
| $J(t)$ | creep compliance |
| $L(\tau)$ | retardation spectrum |
| $p$ | Laplace transform variable |
| $Q$ | quality factor |
| $R$ | gas constant |
| $R$ | a radius |
| $s$ | Laplace transform variable |
| $t$ | time |
| $T$ | temperature |
| $T$ | period of a wave |

# Index

Printed in the United States
By Bookmasters